NEW GLOBAL PERSPECTIVES ON ARCHAEOLOGICAL PROSPECTION

13TH INTERNATIONAL CONFERENCE ON ARCHAEOLOGICAL PROSPECTION
28 AUGUST - 1 SEPTEMBER 2019
SLIGO - IRELAND

Edited by James Bonsall

Archaeopress Publishing Ltd
Summertown Pavilion
18-24 Middle Way
Summertown
Oxford OX2 7LG

www.archaeopress.com

ISBN 978-1-78969-306-5
ISBN 978-1-78969-307-2 (e-Pdf)

Cover images:

Top Left:	Earth resistance data from a circular ditched monument SL014-209041-, Carrowmore Megalithic Cemetery, Co. Sligo.
Top right:	Electromagnetic induction survey of a fulacht fia, monument SL008-205----, Coney Island, Co. Sligo. Photograph: Ciarán Davis
Bottom left:	Megalithic Passage Tomb, monument SL014-209006-, Carrowmore Megalithic Cemetery, Co. Sligo
Bottom right:	Court Tomb, monument SL015-050----, Deerpark, Co. Sligo

This book is available direct from Archaeopress or from our website www.archaeopress.com

Contents

The 13th ICAP and the Importance of Archaeological Prospection in Ireland

James Bonsall
Centre for Environmental Research Innovation and Sustainability, Dept. of Environmental Science, Faculty of Science, Institute of Technology Sligo, Ireland

Introduction

The International Conference on Archaeological Prospection came to Ireland for the first time in 2019, when the 13th ICAP took place at the Institute of Technology Sligo (IT Sligo), in the west of Ireland. The conference featured a wide geographical spread of papers and individual presenters, reflecting the varied global approaches to archaeological geophysics and remote sensing in the 21st century. This volume presents 93 papers that were delivered to the conference as either a poster or an oral presentation. Each paper presented here has undergone a detailed peer review by three members of the ICAP Scientific Committee. On behalf of the ICAP 2019 Organising Committee and the Scientific Committee, I would like to thank the authors for their contributions to this volume and to all the delegates at the conference. I would also like to thank all of the members of the Scientific Committee for their diligent work during the peer review process.

Since the first ICAP Conference, which was held at the University of Bradford in 1995, the archaeological prospection community has grown exponentially with new research units and private sector companies emerging all over the world. Economic growth has fuelled construction projects globally, which in turn has emphasised the importance of development-led, rescue and preventive archaeology. Previously the realm of UK commercial archaeology, geophysical and remote sensing data are now gathered widely around the world to mitigate the impact of projects on previously known and unknown archaeological deposits. The conference, and thus the papers in this volume, feature case studies from 33 countries across Africa, Asia, Australasia, Europe and North America, reflecting current and global trends in archaeological prospection. Archaeological prospection techniques are relevant to all archaeologists no matter where they are located and no matter which time period they are studying; this volume includes papers of all time periods, from the Palaeolithic to the 20th century. No less than nine UNESCO World Heritage Sites are featured in research in this volume: two papers on the Monumental Earthworks of Poverty Point, USA; a paper each on the cities of Uruk-Warka and Ur (both parts of The Ahwar of Southern Iraq); two papers on Brú na Bóinne - Archaeological Ensemble of the Bend of the Boyne, Ireland; and papers on Dahshur (part of Memphis and its Necropolis – the Pyramid Fields from Giza to Dahshur), Egypt; Aksum, Ethiopia; Ephesus, Turkey; Prehistoric Pile Dwellings around the Alps; La Grand-Place, Brussels, Belgium; and Cerveteri (part of the Etruscan Necropolises of Cerveteri and Tarquinia), Italy.

Parts One to Five of this volume comprise research and case studies from different continents, illustrating the international reach of the volume and highlighting cultural similarities across certain regions. Research from a variety of ecosystems is presented, travelling far beyond the idealised 'flat pasture fields', with projects focused on deserts, forests, plains, grassland and uplands. Data has been collected from a variety of boats, vessels and specially developed floating sensor platforms in freshwater and marine environments, pushing researchers and their technologies in new and exciting directions. This volume also features research on natural disasters and their impact on archaeological sites, such as the deformation processes of earthquakes or inundation from flooding.

Part Six showcases technical and novel aspects of archaeological prospection, particularly in terms of data processing, convolutional neural networks, automated interpretive workflows and modelling as well as recent improvements in drone technology, remote sensing and visualisation. The science of geophysics and remote sensing is well established, but the means by which we acquire, use and re-use the data is constantly changing and improving with increased computer processing power, cloud-based solutions and cinematic visualisation methods. As we approach the end of the 2010s, we can clearly see the impact that multi-sensor cart and intregrated GPS surveys had on our discipline over the last decade. The advent of automated drone-acquired geophysical data is currently in its infancy, but case studies presented here represent a technological turning point for the way in which we prospect for archaeology, and suggest that the next decade will see increased technical improvements for automated data acquisition.

Part Seven contains a number of papers that review the legacy of prospection data, the curatorial outcomes of archaeological prospection research, both in terms of interpretation vs. ground observation and the promotion of new and collaborative networks to better understand the soils that we study. Included are papers from practitioners and curators who are developing innovative strategies to 'work smarter', encouraging new ways to assess success, promoting feedback between geophysicists and archaeologists and increasing the capability to harvest large amounts of commercially-acquired data to often short deadlines, via both technology and significant investment in the training of staff.

I believe this volume illuminates the varied ways in which practitioners have developed their methods of data collection and analysis across the globe, which differ from region to region in response to the local cultural needs and knowledge base. These different approaches have led to knowledge-sharing and collaboration on a global scale, as evidenced by the number of researchers and practitioners working in other countries or away from their host institutions. Some papers reflect the work of large and well-funded research units or multi-partner collaborations, whilst others reflect the work of individual researchers or community archaeology groups. Together, these papers illustrate the globalisation of archaeological prospection, demonstrating the familiarity and non-uniqueness of anomalies across the planet.

Archaeological Geophysics in Ireland

Archaeological geophysics on the island of Ireland has a long history, but one interrupted by significant periods of stagnation. Beginning with explorations at the early medieval Ráth na Ríogh on the Hill of Tara, Co. Meath, Professor Séan P. Ó Riordáin in 1952, used an early form of unrecorded electrical resistivity survey (Byrne 1995). Dr (later, Prof.) Martin Aitken carried out the first magnetic prospection surveys in 1959, at the medieval ecclesiastical sites of Downpatrick, Co. Down and Navan, Co. Armagh in Northern Ireland (Aitken 1959, Aitken *et al.* 1958). Aitken described the magnetism of the basalt geology of the Antrim plateau as "too violent to permit archaeological work" (1959: 206), which may have inadvertently discouraged further research in Ireland.

Dr Elizabeth K. Ralph of the University Museum's Applied Science Centre for Archaeology, University of Pennsylvania, carried out a small but successful magnetometer survey in 1968 at the prominent Iron Age royal site of Dún Ailinne, Co. Kildare (Wailes 1970), identifying occupational deposits, subsequently verified by excavation. There was, however, a lack of archaeological scientific research in Ireland during the 1960s and 1970s. Those seeking to improve the outlook in the 1980s were warned that there was no "continuous organised scientific support for strictly archaeological studies" (Mitchell 1978: 3). An absence of funds and little inclination on the part of the National Museum of Ireland and the Office of Public Works also contributed to the absence of archaeological science. The lack of geophysical research was further compounded by the absence of commercial archaeology in Ireland at this time; archaeological geophysics was restricted to a few interested researchers with limited funding and resources.

Ronnie Doggart (1983), a research student at Queens University Belfast (QUB), recognised that Ireland fell 10-15 years behind British and European archaeological prospection research. Doggart established that magnetic prospecting could work in Northern Ireland by exploiting advances in magnetometer and computing technology, focusing on a number of case studies to identify early medieval ringfort enclosure ditches as positive magnetic anomalies on sandstone geology. He also resolved some of the problems encountered by Aitken's (1959) experience of working on basalt geology, and successfully identified a ringfort enclosure ditch as a negative magnetic anomaly, rather than the expected positive magnetic response encountered on a sedimentary bedrock. Doggart also successfully identified Mesolithic and Neolithic occupation sites at Bay Farm and Lough na Trosk in Co. Antrim, and at Mount Sandel, Co. Derry – Ireland's earliest Mesolithic settlement. Importantly, Doggart concluded that researchers should use magnetometers extensively across Ireland to increase knowledge of sites and anomaly types. Through Doggart's work, published in 1983, Irish archaeological geophysicists began to catch up with the work of those European researchers that had been established since the mid-1960s.

A key driver for scientific research in Ireland was the rise of commercial archaeology, which saw an exponential increase in the number of excavations on the island from the 1970s and 1980s, leading to the wider use of geophysical surveys. The 1981-1982 Cork-Dublin gas pipeline (Cleary *et al.* 1987) was the largest archaeological project at the time and the first use of geophysics on an infrastructure project. Magnetometer and earth resistance surveys were used at selected areas beyond the pipeline corridor rather than for the prospection of unrecorded sites, such as the identification of an Early Bronze Age ring-ditch at Ballyveelish. Archaeologists were now alerted to the possibility of discovering ancient settlements beyond the limits of conventionally recognised monuments thanks to geophysical survey.

During the late 1980s, Martin Byrne, a postgraduate student at University College Cork, carried out earth resistance surveys across Co. Cork that located souterrains and lazy-bed activity (Monk 1989: 31). Elsewhere, geophysics continued to be used beyond areas of intrusive investigation such as Cooney's (1990) survey at Ballynee to identify the extent of a known souterrain. Byrne's MA thesis (Byrne 1995) was the first since Doggart's work to examine the state of archaeological geophysics in Ireland and focused on the use of earth resistance. Despite the developments of the 1980s, Irish geophysics was still in its infancy compared to work carried out in the UK.

The Irish prospection experience was fuelled by a combination of heritage-sensitive planning regulations that require scientific assessment of development-threatened archaeological sites and the so-called Celtic Tiger economic boom of the 1990s and early 2000s. Both homegrown Irish companies and those from the UK and Germany carried out a widerange of geophysical assessments across the country across a variety of soils and archaeological sites. By the end of the 2000s, geophysics had been embraced as a useful archaeological tool, helped in no small part by its use on national road schemes which gave the technique a wide exposure to archaeologists working on subsequent phases of investigation. The work of the Discovery Programme and others has also promoted the use of archaeological prospection at high-status Iron Age royal sites such as the Hill of Tara (Newman 1997), Dún Ailinne, (Corcoran 2007, Johnston *et al.* 2009) and Rathcroghan (Barton and Fenwick 2005, Waddell *et al.* 2009).

In the academic sector, Queen's University Belfast (QUB), the National University of Ireland Galway (NUIG) and the National University of Ireland Maynooth (NUIM) also developed archaeological geophysics research during the 2000s. Both QUB and NUIG carried out marine and terrestrial geophysics, and QUB became a research hub for the development of forensic geophysics (Ruffell *et al.* 2009). Research at the Department of Geography in NUIM was driven by the Environmental Geophysical Unit, under the leadership of Dr Paul Gibson, with contributions made to the study and mapping of high-status medieval monuments (Gibson 2007, Gibson and Breen 2005, Gibson and George 2006, O'Rourke and Gibson 2009).

The subsequent financial crisis period was one of reflection and provided the opportunity to study legacy data derived from the geophysical surveys and excavations of the 2000s to quantify the level of success offered by prospection surveys in Ireland (Bonsall *et al.* 2014a, Bonsall & Gaffney 2016). Formal guidance on the use of geophysics derived from that research (Bonsall *et al.* 2014b) has since been implemented by the state body Transport Infrastructure Ireland (the largest procurer of archaeological services in Ireland), and others. In the meantime, private and public sector practitioners, as well as third level institutes of education, embraced the new technological benefits that were developed during the recession, resulting in the use of UAV and multi-sensor platforms as a standard practice for archaeological prospection in Ireland. Today, a vibrant commercial sector is responsible for acquiring large datasets across thousands of hectares per year, with several independent archaeological geophysical companies trading in Ireland, and a number of archaeological consultancies with in-house specialists.

Archaeological Prospection at IT Sligo

IT Sligo has been at the forefront of archaeological prospection in Ireland for the last 16 years. In 2003 it became the first Institute of Technology in the country to offer degrees in archaeology and is still the only third level education body in Ireland where archaeology is taught as a science degree, rather than a humanities degree. The archaeology team at IT Sligo are embedded within the Department of Environmental Science in the Faculty of Science, alongside experts in forensics, geological and environmental science and nanotechnology. At the core of the Applied Archaeology programme are two dedicated modules on the theory and practice of archaeological geophysics, which has resulted in a strong geophysical education for an entire generation of upcoming archaeologists.

IT Sligo's Department of Environmental Science has built up an impressive suite of archaeological prospection equipment including UAVs, laboratory and field probes for magnetic susceptibility, ground-penetrating radar instruments, metal detectors, magnetometers and earth resistance meters. We are also part of the global earthquake study network thanks to a seismometer acquired in 2013. These instruments have enabled IT Sligo undergraduate students to conduct geophysical surveys of archaeological sites and landscapes and assess materials for conservation in unique ways (Dowd *et al.* forthcoming.). For example, Michael Gleeson manufactured his own iron gall ink and used magnetic susceptibility methods to explore the provenance, conservation and detectability of ancient writing technology. Deirdre Kelly used magnetometry, magnetic susceptibility and earth resistance along with a high-resolution UAV survey to study a prehistoric hengiform enclosure at Carrowmably, Co. Sligo. Sally Siggins and Ciarán Davis each used magnetometry and earth resistance to investigate post-medieval bastioned forts in Sligo Bay, providing a new understanding of the defence of Sligo harbour in the past.

As a result of our prospection knowledge and skillset, we were a key partner in the recovery and conservation of hundreds of conflict-related artefacts from the largest metal detection survey in Ireland, the international Longest Day Research Project, a study of the 1798 Battle of Vinegar Hill. The focus of recent geophysical research at IT Sligo has investigated the impacts of challenging environments on archaeological prospection, such as the use of electromagnetic induction (EMI) over a Late Bronze Age cooking site in the intertidal environment (Bonsall and Dowd 2017); UAV, gradiometry and earth resistance over an early medieval monastic site in a remote (and geologically challenging) upland area (Beglane *et al.* 2018); and the assessment of low- or no-contrast prehistoric and post-medieval middens across dunefields, using EMI, gradiometry, magnetic susceptibility, earth resistance and ground-penetrating radar (Napora *et al.* 2019).

In the summer of 2019, we collected a variety of geophysical, photogrammetry and terrestrial laserscan data in advance of excavations at Alice and Gwendoline Cave, Co. Clare, where a butchered bear patella dating to 10,860-10,736 cal. BC provided the first evidence for a human presence during the Upper Palaeolithic in Ireland (Dowd and Carden 2016). The data analysis is at a preliminary stage, but has demonstrated the benefits of using a magnetic susceptibility point sensor across the cave floor in the search for palaeohearths and the discrimination of undisturbed cave deposits using combined earth resistance, magnetometry and metal detection.

Carrowmore megalithic cemetery, Co. Sligo, has been the focus of geophysical surveys for cohorts of IT Sligo students for many years, and the work has produced a number of discoveries. Among these are features associated with an enigmatic prehistoric monument that was excavated in the summer of 2019 as part of our undergraduate training excavation.

The Importance of an International Conference on Archaeological Prospection for Ireland

Back in 1995, at the 1st ICAP at the University of Bradford, Martin Byrne, the only Irish delegate attending, reiterated a clear and long-term need to develop expertise and training initiatives in Irish universities. Now, 24 years later, Ireland has achieved this and more. Happily, the development of archaeological geophysics at IT Sligo has also propelled Irish research further and Ireland is now the beneficiary of a small but focused geophysical community. I am very pleased to note that the countries most frequently featured in the papers contained in this volume are tied, at eight papers each, to Ireland, the UK and Italy. Irish archaeological prospection has come a long way since 1952 and this gives me hope for those countries or practitioners that are just beginning to explore the use of geophysics and remote sensing techniques, that they will also be making important and impactful research in the coming years. To this end, it is vitally important that the ICAP community continues to meet every two years to discuss the latest trends, challenges, discoveries and methodologies across the world, sharing knowledge and experiences that are relevant to us all, and encouraging the next generation of researchers and practitioners to develop our discipline further.

References

Aitken, M.J. 1959. News and Notes - Magnetic Prospecting: An Interim Assessment. *Antiquity*. 33(131). 205-207.

Aitken, M.J., Webster, G. and Rees, A. 1958. Notes and News - Magnetic Prospecting. *Antiquity*. 32(128). 270-271.

Barton, K. and Fenwick, J. 2005. Geophysical Investigations at the Ancient Royal Site of Rathcroghan, County Roscommon, Ireland. *Archaeological Prospection*. 12. 3-18.

Beglane, F., Bonsall, J., Nugent, L., Moran, J. & Meehan, H. 2018. An oasis in the desert: the early ecclesiastical site of Disert, Co. Donegal. *Journal of Irish Archaeology*. 27. 57-79.

Bonsall, J. & Dowd, M. 2017. Emerging from the waves: a Late Bronze Age intertidal saltwater fulacht fia at Coney Island, Co. Sligo. *Journal of Irish Archaeology*. 24. 79-95.

Bonsall, J. and Gaffney, C. 2016. Change is Good: Adapting Strategies for Archaeological Prospection in a Rapidly Changing Technological World. In F. Boschi (ed.) *Looking to the Future, Caring for the Past: Preventive Archaeology in Theory and Practice*. Bononia University Press. Bologna. 41-58.

Bonsall, J., Gaffney, C. and Armit, I. 2014a. A Decade of Ground Truthing: Reappraising Magnetometer Prospection Surveys on Linear Corridors in light of Excavation evidence 2001-2010. In Kamermans, H., Gojda M. and Posluschny, A.G. (eds.) *A Sense of the Past: Studies in current archaeological applications of remote sensing and non-invasive prospection methods*. BAR International Series 2588. Archaeopress, Oxford. 3-17.

Bonsall, J., Gaffney, C. and Armit, I. 2014b. *Preparing for the future: A reappraisal of archaeo-geophysical surveying on National Road Schemes 2001-2010*. University of Bradford report for the National Roads Authority. 1-151.

Byrne, M. 1995. *An introduction to the methods of Archaeological Prospection and a study of the theory and techniques of Electrical Resistivity*. MA Thesis. Department of Archaeology, University College Cork.

Cleary, R.M., Hurley, M.F. and Twohig E.A. 1987. *Archaeological Excavations on the Cork-Dublin Gas Pipeline (1981-1982)*. Cork Archaeological Studies No. 1. Department of Archaeology, University College Cork.

Cooney, G. 1990. Appendix 3 'Resistivity Survey' In G. Eogan, Ballynee Souterrains, County Meath. *The Journal of the Royal Society of Antiquaries of Ireland*. 120. 41-64

Corcoran, A. 2007. *An archaeological geophysical investigation of the ancient royal site of Dun Ailinne in Kilcullen, Co. Kildare*. MSc Thesis. Department of Geography, National University of Ireland Maynooth.

Doggart, R. 1983. The use of magnetic prospecting equipment in Northern Ireland. In Reeves-Smyth, T. and Hamond, F. (eds) *Landscape Archaeology in Ireland*. BAR Monograph British Series 116. Oxford. 35-46.

Dowd, M. and Carden, R.F. 2016. First evidence of a late Upper Palaeolithic human presence in Ireland. *Quaternary Science Reviews*. 139. 158-6.

Dowd, M., Beglane, F., Bonsall, J., Markley, S., Moore, S. and Read, C. Forthcoming. The past in the present: archaeology at IT Sligo. In Foran, J. (ed.) *IT Sligo 50th Anniversary Commemorative Volume*.

Gibson, P. 2007. Archaeo-magnetic investigation of the former monastic site at Kilskyre, Co. Meath. *Ríocht na Midhe*. 18. 33-38.

Gibson, P. and Breen, R. 2005. The internal structure of the Galtrim motte, County Meath, as revealed by ground-penetrating radar and electrical resistivity geophysical techniques. *Ríocht na Midhe*. 16. 23-28.

Gibson, P.J. and George, D.M. 2006. Geophysical Investigation of the Site of the Former Monastic Settlement, Clonard, County Meath, Ireland. *Archaeological Prospection*. 13. 45-56.

Johnston, S.A., Campana, D. and Crabtree, P. 2009. A Geophysical Survey at Dún Ailinne, County Kildare, Ireland. *Journal of Field Archaeology*. 34. 385-402.

Newman, C. 1997. *Tara: an archaeological survey*. Dublin: Royal Irish Academy for the Discovery Programme.

O'Rourke, T. and Gibson, P.J. 2009. Geophysical Investigation of the Environs of Rattin Castle Tower House, County Westmeath, Ireland. *Archaeological Prospection*. 16. 65-75.

Mitchell, G.F. 1978. Planning for Irish Archaeology in the Eighties (The Royal Irish Academy 1978 Presidential Address). *Irish Archaeological Research Forum*. 5. 1-14.

Monk, M. 1989. Archaeological Prospection Part 1: Geophysical Survey. *Archaeology Ireland*. 3. 30-33.

Napora, K.G., Bonsall, J., Rathbone, S. & Thompson, V.D. 2019. Geoarchaeological Analysis of a Dunefield Shell Midden Site in Carrowdough Townland, County Sligo, Ireland. *The Journal of Island & Coastal Archaeology*. 14(3). 394-410.

Ruffell, A., McCabe, A., Donnelly, C. and Sloan, B. 2009. Location and Assessment of an Historic (150–160 Years Old) Mass Grave Using Geographic and Ground Penetrating Radar Investigation, NW Ireland. *Journal of Forensic Science*. 54(2). 382-394.

Waddell, J., Fenwick, J. and Barton, K. 2009. *Rathcroghan: archaeological and geophysical survey in a ritual landscape*. Dublin. Wordwell.

Wailes, B. 1970. Excavations at Dun Ailinne, Co. Kildare 1968-9 Interim Report. *The Journal of the Royal Society of Antiquaries of Ireland*. 100(1). 79-90.

13th International Conference on Archaeological Prospection: Organisation Committee

- Dr James Bonsall, Department of Environmental Sciences, Faculty of Science, IT Sligo, Ireland
- Dr Kayt Armstrong, Magnitude Surveys Ltd., Bradford, UK
- Dr Immo Trinks, Ludwig Boltzmann Institute for Archaeological Prospection and Virtual Archaeology, Vienna, Austria
- Sam Moore, Department of Environmental Sciences, Faculty of Science, IT Sligo, Ireland
- Anthony Corns, The Discovery Programme: Centre for Archaeology and Innovation Ireland, Dublin, Ireland
- Fin Melia, Independent Archaeologist, Sligo, Ireland
- Oli Melia, IT Sligo Science Festival Organiser, Sligo, Ireland
- Greg Carew, Abbey Conference & Events, Dublin, Ireland
- Emma Wood, Abbey Conference & Events, Dublin, Ireland
- Julie Peacock, Abbey Conference & Events, Dublin, Ireland
- Iva Brankovic, Abbey Conference & Events, Dublin, Ireland

13th International Conference on Archaeological Prospection: Scientific Committee

Part One – Archaeological Prospection in Europe

Harbours from Antiquity to the Middle Ages: a Geophysical Panorama

Wolfgang Rabbel, Dennis Wilken, Tina Wunderlich, Annika Fediuk, Michaela Schwardt, Daniel Köhn, Clemens Mohr, SPP 1630 Working Groups
Kiel University, Kiel, Germany

Background

"Harbours from the Roman Period to the Middle Ages" were the theme of a recently completed Priority Program of the German Science Foundation ("SPP 1630" project). Its overarching aim was to study ancient harbours as a component of the system of past societies not only regarding constructional but also ecological, logistical, economic, social, and many other aspects. Geophysical investigations were incorporated in this project as the fundamental element for identifying ancient harbour sites and exploring their structure, infrastructure and geological setting (Rabbel *et al.* 2015).

The locations of ancient harbours and waterways prospected within the project extend from the Eastern Mediterranean to the North Atlantic, thereby covering more than 1000 years of infrastructural development. Concrete aims were to find harbour areas and waterways, to explore possible remains of harbour constructions, to determine the geometry of harbour basins and waterways and details of the construction and sedimentary infill, and to define environmental conditions in collaboration with archaeological and geoarchaeological working groups. We present a selection of data examples, covering this panoramic view from antiquity in the eastern Mediterranean to the Viking age in the North Atlantic, or concretely, from Miletos (Turkey) in the SE via Elaia, Ostia, Fossa Carolina, Rungholt, Leirovogur to Igaliku (Greenland) in the NW.

Methods

The variety of different on- and offshore environmental settings represented a special challenge and gave rise to ongoing methodical work. It included, *inter alia*, the construction of special instrument carriers (Fig. 1, right) and the development of new interpretation approaches and software tools. Highlights are the development of a dual-source marine seismic system enabling 3D seismic measurements in near-shore shallow waters (Fig. 1, left) (Wilken *et al.* 2019), a systemized approach to combine drilling and direct push technology with ERT (Electric Resistivity Tomography) to improve geoarchaeological profiling (Wunderlich *et al.* 2018), and the first successful application of seismic full waveform inversion (FWI) as an interpretational tool of archaeo-geophysics (Köhn *et al.* 2019). As verified by excavation, SH-wave FWI proved to be capable to image near-surface sediment structure at a dm-scale resolution. Special research efforts were dedicated to the suppression of multiples in shallow water seismics as well as to investigating the potential of ERT and Ground-penetrating radar (GPR) to image archaeological objects in near-shore areas.

Fig. 1. Dual-source seismic system "Ping-Pong" for 3D shallow water prospection (left) and vehicle for magnetic prospection in the Wadden Sea (right).

Results

The starting point of the geophysical work within the harbour project was the experience gathered beforehand mainly at classical Greek harbours and at the Viking harbour of Haithabu, all of which show clear evidence of harbour constructions in the geophysical records. Over the course of the SPP 1630 project, the initial ideas on how harbours and waterways may be identifiable needed to be revised and improved according to the varying natural environment and states of harbour development. Spatio-temporal examples on the side of the Roman period are the well-developed imperial harbour of Ostia antica (Fig. 2a), which turned out to be dual-phase, and the harbour of Kane (Fig. 2b) known from ancient sources to have hosted a Roman fleet in winter BC 191/190.

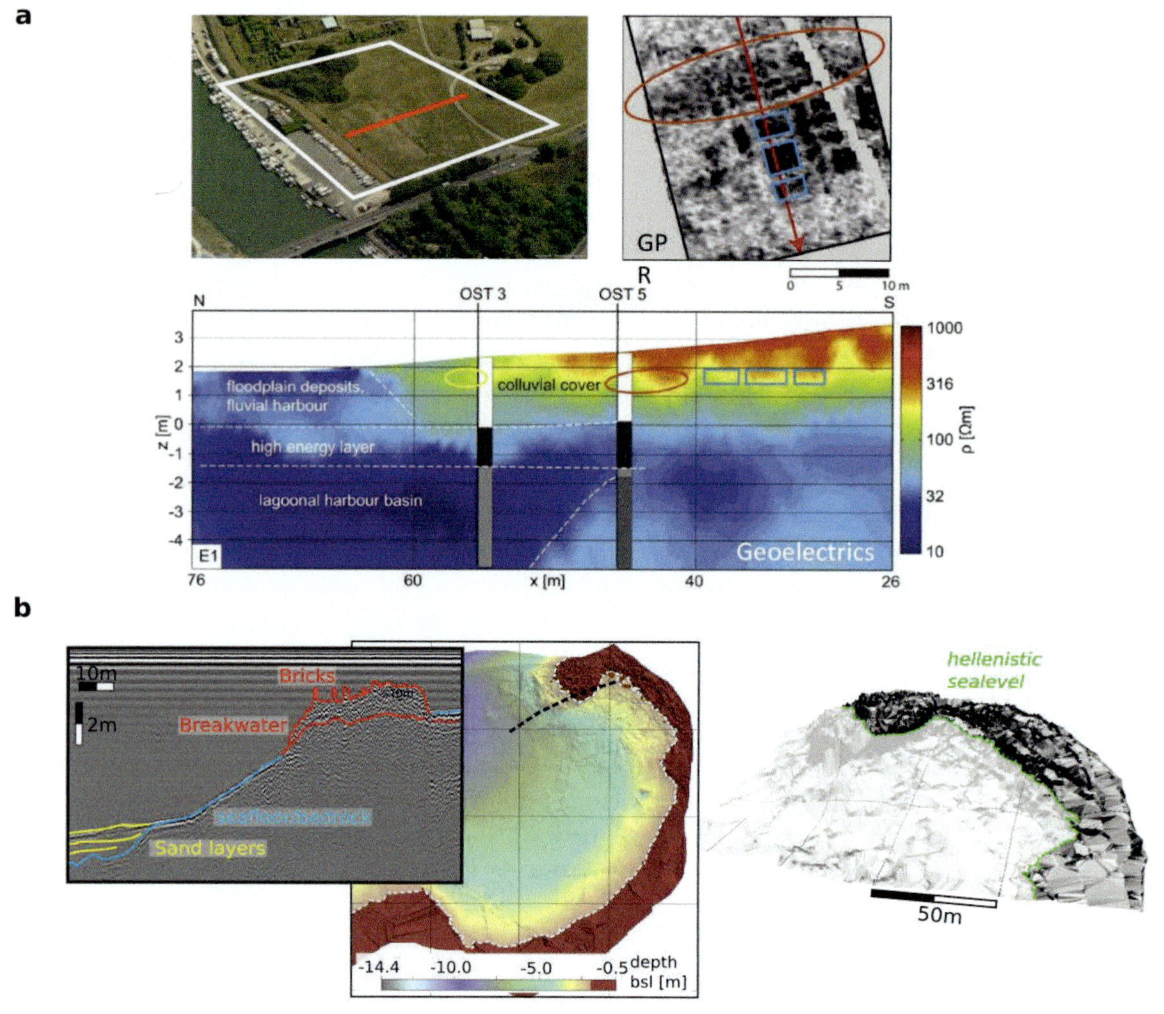

Fig. 2. Examples of the prospection of ancient harbours. (a) Fluvial harbour of Ostia Antica: Aerial view of the harbour and location of ERT profile (top left); ERT combined with drilling (bottom figure of a) revealed the two-phase nature of the harbour basin; GPR mapping shows the remains of houses and harbour construction (top right) (after Wunderlich *et al.* 2018). (b) Visualization of a Hellenistic/Roman sea harbour basin (right) on the Kane peninsula (western Turkey) based on seismic (left) and bathymetry (middle) data (after Fediuk *et al.* 2018).

An examples of the medieval harbours is the difficult to recognise Viking landing places in Greenland, such as the harbour of Garðar/Igaliku (Fig. 3a), which turned out to be a sheltered basin behind a peninsula after reconstructing the Viking age coastline. Other exploration highlights are from the tidal sand and mud flats of the Wadden Sea where the first geo-referenced remains of the major medieval settlement of Rungholt and related dike construction could be located by combined magnetic and seismic mapping (Fig. 3b), and from the Viking harbour of Hedeby, close to which a sedimented medieval ship wreck could be imaged in 3D (Fig. 3c).

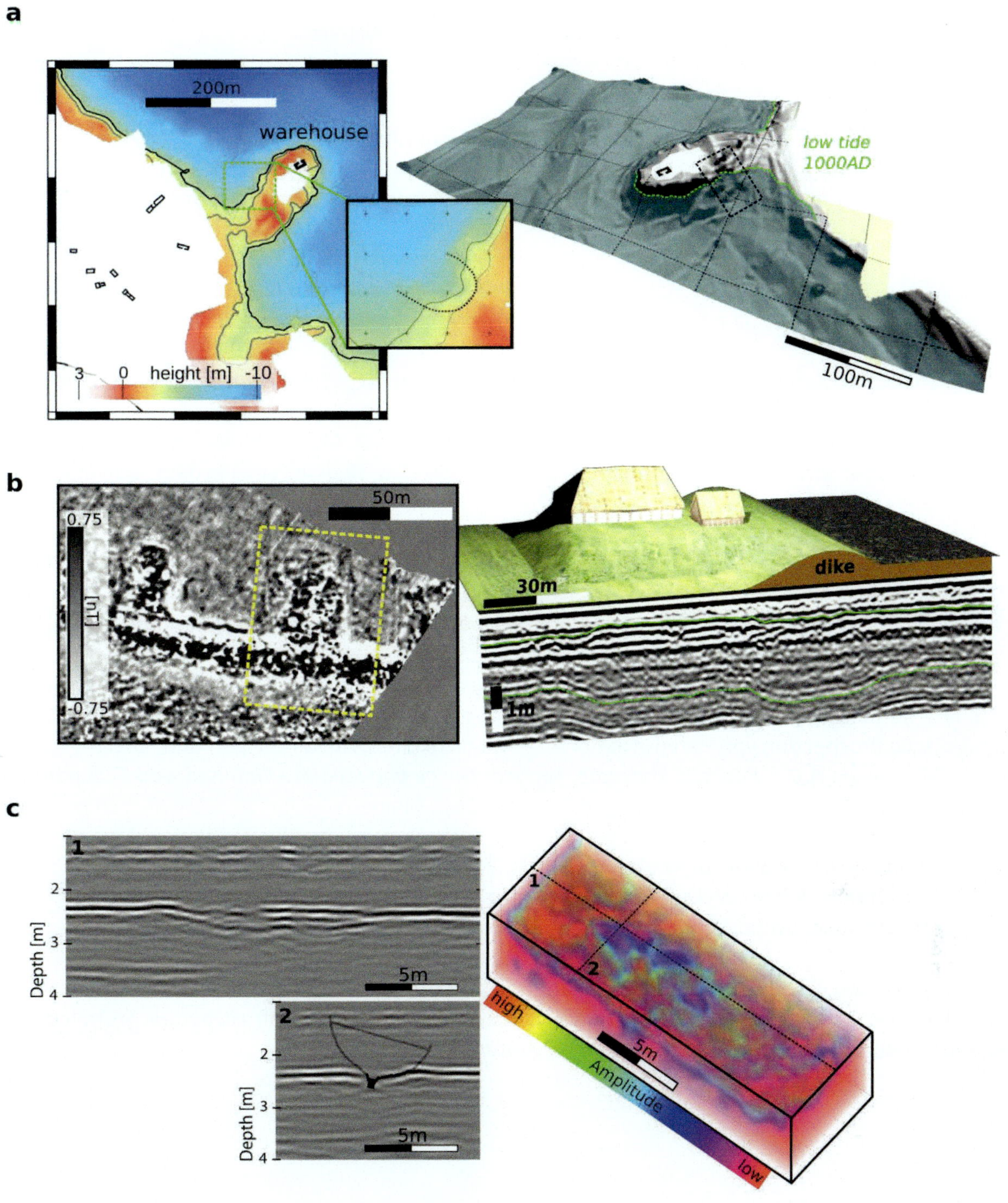

Fig. 3. Examples of the prospection of Medieval harbours. (a) Bathymetry based reconstruction of a Viking landing place on Greenland (after Coolen *et al.* 2017). (b) Magnetic and seismic based reconstruction of a medieval dike section with attached terp at the settlement of Rungholt, destroyed and "sunken" in the Wadden Sea of the German North Sea after a medieval storm flood (after Hadler *et al.* 2018). (c) 3D datacube of a submerged medieval wreck close to the Viking harbour of Hedeby (after Wilken *et al.* 2019).

Conclusions

The overview of geophysical prospections at ancient harbours performed in the SPP1630 project provides a panorama of antique and medieval harbour and waterway situations along a SE-to-NW oriented swath through Europe. The elaboration of these harbours shows a strong SE-NW decline too, according to the states of ancient civilizations, trade and marine traffic. Correspondingly, geophysical prospecting aims shift increasingly from detecting anthropogenic structure to exploring the geoarchaeological environment. In this context a systematic incorporation of drilling and downhole geophysics becomes especially important. Significant interpretational progress is to be expected from respective constrained inversion approaches, the waveform inversion of shear waves and small-scale HR 3D marine seismics.

References

Coolen, J., Mehler, N., Wilken, D., Weßling, R., Preston, J., Wunderlich, T. and Feldens, P. 2017. Archaeological prospection of Medieval harbours in the North Atlantic. In Jennings, B., Gaffney, C., Sparrow, T. and Gaffney, S. (eds.) *AP2017. 12th International Conference of Archaeological Prospection*. Archaeopress. Oxford. 138-139.

Fediuk, A., Wilken, D., Wunderlich, T., Rabbel, W., Seeliger, M., Laufer, E. and Pirson, F. 2018. Marine seismic investigation of the ancient Kane harbour bay, Turkey. *Quaternary International*. 511. 43-50.

Hadler, H., Fediuk, A., Fischer, P., Rabbel, W., Schwardt, M., Wilken, D., Willershäuser, T., Wunderlich, T. and Vött, A. 2018. Drowned by the Grote Mandrenke in 1362 AD - new geo archaeological research on the late medieval trading centre Rungholt (North Frisia). In Egberts, L., Schroor, M. (2018). *Waddenland Outstanding - History, Landscape and Cultural Heritage of the Wadden Sea Region*. Amsterdam University Press, Amsterdam: 239-252.

Köhn, D., Wilken, D., De Nil, D., Wunderlich, T., Rabbel, W., Werther, L., Schmidt, J., Zielhofer, C. and Linzen, S. 2019. Comparison of time-domain SH waveform inversion strategies based on sequential low and bandpass filtered data for improved resolution in near-surface prospecting. *Journal of Applied Geophysics.* 160. 69-83.

Rabbel W., Wilken D., Wunderlich T., Bödecker S., Brückner H., Byock J., von Carnap-Bornheim C., Kalmring S., Karle M., Kennecke H., Messal S., Schmidts T., Seeliger M., Segschneider M. and Zori D. 2015. Geophysikalische Prospektion von Hafensituationen – Möglichkeiten, Anwendungen und Forschungsbedarf. *Häfen im 1. Millennium A.D. – Bauliche Konzepte, herrschaftliche und religiöse Einflüsse*, eds. T. Schmidts and M. Vučetić. Romano-Germanic Central Museum Press (RGZM), Mainz.

Wilken, D., Wunderlich, T., Hollmann, H., Schwardt, M., Rabbel, W., Mohr, C., Schulte-Kortnack, D., Nakoinz, O. Enzmann, J., Jürgens, F. and Wilkes, F. 2019. Imaging a medieval shipwreck with the new PingPong 3D marine reflection seismic system. *Archaeological Prospection*. 2019. 1– 13. https://doi.org/10.1002/arp.1735

Wunderlich, T., Wilken, D., Erkul, E., Rabbel, W., Voett, A., Fischer, P., Hadler, H. and Heinzelmann, M. 2018. The river harbour of Ostia Antica - stratigraphy, extent and harbour infrastructure from combined geophysical measurements and drillings. *Quaternary International.* 473. 55-65.

High-resolution underwater archaeological prospection of Upper Austrian pile dwellings and lakes using multi-beam and sediment sonar

Immo Trinks[1], Wolfgang Neubauer[1,2], Timothy Taylor[2], Mario Wallner[1], Klaus Löcker[1,3], Jutta Leskovar[4]
[1]Ludwig Boltzmann Institute for Archaeological Prospection and Virtual Archaeology, Vienna, Austria, [2]Vienna Institute for Archaeological Science, University of Vienna, Vienna, Austria, [3]Central Institute for Meteorology and Geodynamics, Vienna, Austria, [4]Upper Austrian State Museum, Leonding, Austria

Background

The circum-alpine lakes have been home to important Neolithic and Bronze Age pile dwellings. In Upper Austria the four pile dwelling sites in Lake Mondsee and Abtsdorf I, Abtsdorf III and Litzlberg Süd in Attersee (Pohl 2016) constitute part of the UNESCO World Heritage Site "Prehistoric Lake Dwellings around the Alps". The goal of the "Archaeological Prospection of Upper Austrian Lake Dwellings" project is to provide new insights into these special settlement sites and their surroundings through the use of modern non-destructive archaeological prospection methods. While at some sites the remains of 4000-5000 year old wooden piles are still preserved in the shallow water above the lake floor, at other sites only the deposits of the prehistoric settlements have been preserved (Offenberger and Ruttkay 1997). The Ludwig Boltzmann Institute for Archaeological Prospection and Virtual Archaeology (LBI ArchPro) in cooperation with the Vienna Institute for Archaeological Science (VIAS) of the University of Vienna and the Direktion Kultur of the regional government of Upper Austria, have set up a new prospection system for efficient high-resolution underwater archaeological surveys with the aim of mapping and exploring the lake bed and lake subsurface of Lake Mondsee and Lake Attersee in detail. Until now, the water depth of these lakes was only known in from depth contour lines with 10m to 5m depth intervals (Behbehani 1987). The project presented here can best be compared with the project "Tiefenschärfe Bodensee" (Wessels *et al.* 2016), which successfully measured Lake Constance with a comparable research approach using modern echosounder and sediment sonar systems.

Methods

Modern sonar technology enables detailed surveying and mapping of lake bottoms as well as three-dimensional investigation of sediment layers and underground structures. For this project, state-of-the-art echosounder and sediment sonar systems with suitable positioning technology are used.

After test measurements with various high-resolution multi-beam sonar systems conducted at Lake Mondsee in May 2016, a high-resolution Teledyne Reson SeaBat T50-P multi-beam sonar system was acquired. This sonar permits the mapping of the sea floor with 512 beams (1° × 0.5° beam width @ 400kHz) recording up to 25,600 points and their reflection intensity per second. Reflections in the range of 0.5m to at maximum 575m can be detected. It is possible to record multiple detections as well as the entire water column for each beam in order to map complex structures in more detail. The specified depth resolution is 6mm. In 2m water depth the footprint of each beam is 3.5cm × 1.7cm, at the maximum depth of 68m of Lake Mondsee the footprint is 119cm × 59cm.

In addition to the multibeam echo sounder, the innovative four-channel parametric sediment sonar Innomar SES-2000 quattro was acquired for high-resolution three-dimensional measurements of the lakebed in extremely shallow water (Lowag *et al.* 2010). This sub-bottom profiler consists of transducer sounders and receivers mounted as a line array at a distance of only 25cm. It is possible to use the transmitters/receivers in single, dual, triple or quad beam mode, depending on the task and the required penetration depth or the desired horizontal resolution. The quad beam array can be used in a depth range from 0.5m to 30m.

The precise positioning of the measurement data is conducted with a dual antenna real-time Global Navigation Satellite System and a highly accurate inertial navigation system. A professional navigation and data recording solution is used for the survey planning and data acquisition. Profiles of the sound velocity in the water column are frequently measured.

As a surveying platform, a cabin motor boat with 6.35m length and only 35cm draught and separate power supply for the sonar systems was set up. The sonar heads are deployed over the bow of the boat via a swivelling mount.

Preliminary results and outlook

Since May 2018, the lake bottom of Lake Mondsee has been mapped almost completely with multi-beam echo-sounder measurements, with the exception of the extremely shallow coastal waters due to the low water level in summer 2018 (Fig. 1). The bathymetric data are of great interest not only from an archaeological and historical point of view, but also regarding lake morphology, sedimentology and geology. The multi beam sonar data swaths of Lake Mondsee, covering a total area of more than 13km^2, will be processed to generate an optimized three-dimensional model of the lake bed. In cooperation with the Vienna University of Technology, the software OPALS (Orientation and Processing of Airborne Laser Scanning data), developed for airborne laser scanning data processing and strip adaptation, will be adapted for application to multi-beam sonar data (Pfeifer *et al.* 2014). Multi-beam and sediment sonar data collection will continue at Lake Mondsee and Attersee in 2019. This project will generate essential basic data for the scientific exploration of these lakes, for environmental monitoring projects and targeted underwater archaeological investigations as well as geological and limnological research.

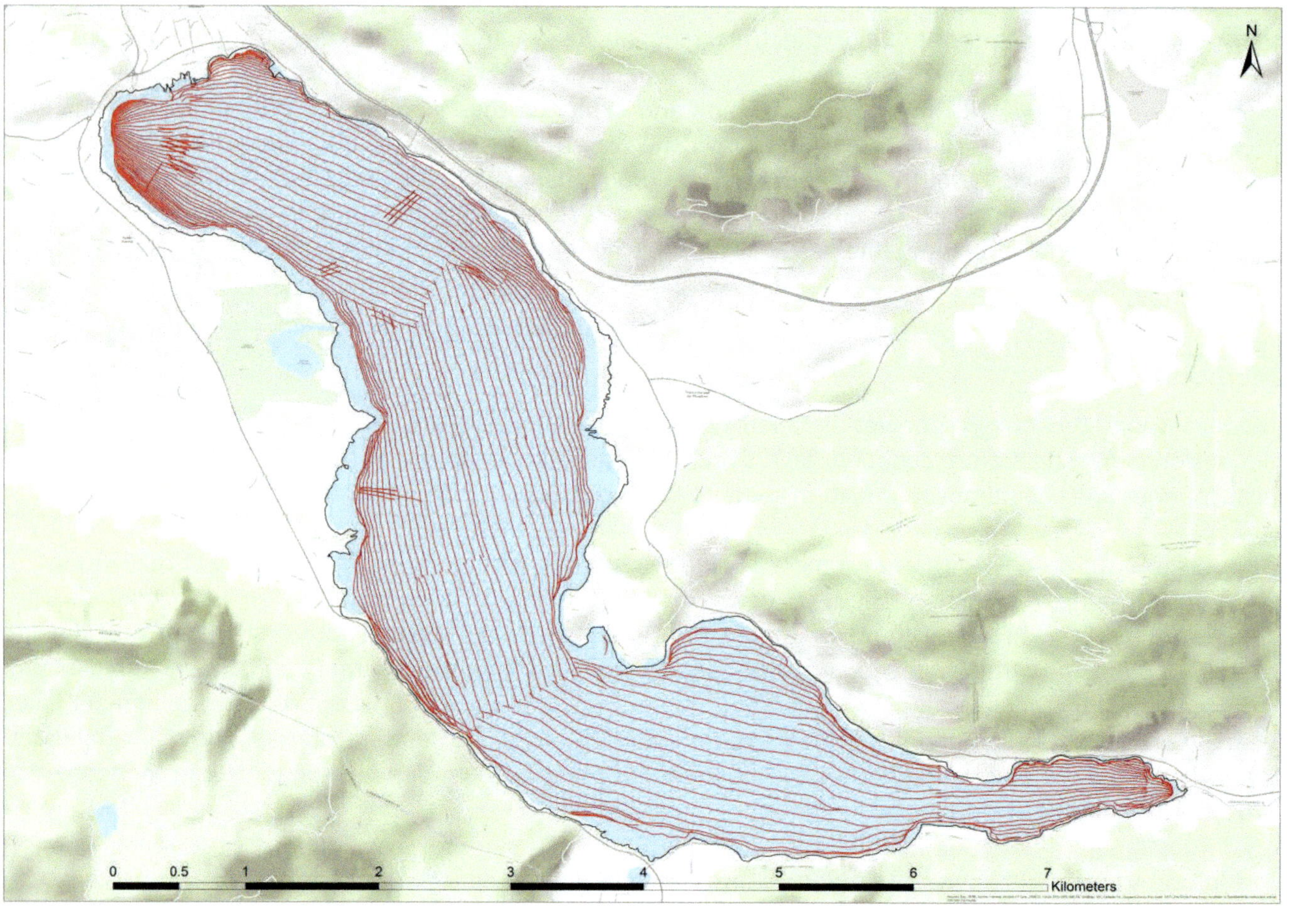

Fig. 1. Position and distribution of survey swaths of the multi-beam sonar measurements on Lake Mondsee (total length 304km).

Acknowledgements

We acknowledge the kind support of Mag. Nicoletta Waechter, the Institute of Limnology in Mondsee (Prof. Martin Hahn), Dipl.-Ing. Christian Klug (EPOSA), Riegl Laser Measurement Systems, Johannes Häupl, DI Jörg Brunken, Florian Hämmerling and Alexander Schmidt, Heinz Gruber, Gerald Egger.

The LBI ArchPro is based on an international cooperation of the Ludwig Boltzmann Society, the Office of the Provincial Government of Lower Austria, the University of Vienna, the Vienna University of Technology, the Central Institute for Meteorology and Geodynamics, Airborne Technologies, 7reasons, Römisch-Germanisches Zentralmuseum Mainz, Landschaftsverband Westfalen-Lippe - Archäologie für Westfalen, Norsk institutt for kulturminneforskning – NIKU, and Vestfold fylkeskommune - Kulturarv.

References

Behbehani, A.-R. 1987. Sedimentation and climate history of the Late and Postglacial in the Northern Limestone Alps (Salzkammergut lakes, Austria). *Göttingen work on geology and palaeontology*. 34th Geological Institutes of the Georg-August-University Göttingen.

Lowag, J., Wunderlich, J. and Huembs, P. 2010. Three-dimensional investigation of buried structures with multi-transducer parametric sub-bottom profiler as part of hydrographical applications. *Proc. Hydro 2010, Rostock-Warnemünde, Germany*, 02-05 November 2010.

Offenberger, J. and Ruttkay, E. 1997. Lake Dwelling Research in the Austrian Salzkammergut Lakes. In *Lake dwellings around the Alps. Archaeology in Germany, special edition*. 76-86.

Pfeifer, N., Mandelburger, G., Otepka, J. and Karel, W. 2014. OPALS - A framework for Airborne Laser Scanning data analysis. *Computers, Environment and Urban Systems*. 45. 125-136.

Pohl, H. 2016. Three years of underwater archaeological monitoring at the Austrian UNESCO World Heritage Sites. *Archaeology of Austria*. 27(1). 29-35.

Wessels, M., Anselmetti, F.S., Baran, R., Hilbe, M., Gessler, S. and Wintersteller, P. 2016. *Tiefenschärfe - Hochauflösende Vermessung Bodensee*. International Commission for the Protection of Lake Constance, Blue Series, 61.

Ground-penetrating radar study of the Asaviec 2 archaeological site, Belarus

Alina Tšugai-Tsyrulnikova[1], Jüri Plado[1], Maxim Charniauski[2], Irina Khrustaleva[3,4], Aivar Kriiska[4]
[1]Department of Geology, University of Tartu, Tartu, Estonia, [2]Department of Archaeology of Prehistory Societies, Institute of History of NAS of Belarus, Minsk, Belarus, [3]The State Hermitage Museum, St Petersburg, Russia, [4]Institute of History and Archaeology, University of Tartu, Tartu, Estonia

The Asaviec 2 archaeological site is situated in the Belarusian Paazierje (Lakeland) and is one of ten currently known Stone and Bronze Age Kryvina peatbog settlement sites. The occupation layer of Asaviec 2 contains abundant archaeological material and is very well preserved due to being covered by a thick peat layer (Charniauski and Kryvaltsevich 2011: 108). After being discovered in 1966, the site was repeatedly excavated (Cherniavskii 1967: 372–385). It is now one of the best investigated wetland settlement sites in Northern Belarus (Charniauski 1997: 202–204). In addition to archaeological excavations, stratigraphic and typological analyses, previous research of the area has included drilling, lithological, palynological analyses and radiocarbon dating, but geophysical methods have not been employed.

Kryvina peatbog is rich in settlement sites, well studied, and a promising area that needs further inverstigation via non-destructive survey methods. The application of such methods could reduce the amount of excavations of already known sites and help to discover new settlement sites. Ground-penetrating radar (GPR) is a relatively cheap, quick and mobile method. It creates an estimate of the subsurface remains in real time during fieldwork and can be used to correct plans, if needed.

For the current study, ground-penetrating radar survey was combined with hand drilling in August 2018. The method was applied to determine the extent of the occupation layer of the Asaviec 2 archaeological site. On a larger scale, however, it could be used to elaborate an algorithm for the determination of cultural material within peaty sediments. A Zond 12-e instrument, manufactured by Radar Systems Inc., was used, with centre frequencies of 300MHz and 900MHz. Antennas were pulled by hand at walking speed, making measurements using an odometer wheel every 2.5cm for the 300MHz antenna and every 1cm for the 900MHz antenna, with a time range of 200ns and 100ns, respectively. Data processing was performed by Prism2 software and included a band-pass filter, zero-point adjustment, background removal, and gain control, if needed. For converting time-to-depth scale the relative dielectric permittivity was found by analysing radar profiles and comparing them with drill-core data.

Three radar facies (Neal 2004) were distinguished and described (Fig. 1). The radar facies correspond to peat and revealed two types of patterns. The first includes long horizontal reflectors (RF1), the second is composed of chaotic often tilted reflections (RF2). The horizontal reflections of RF1, partly due to natural reflections, partly phantom, indicate accumulations in calm natural conditions or a horizontal humificated peat layer, undisturbed by human activity. Tilted chaotic reflections of RF2 are interpreted as peat layers with disturbed bedding due to the impact of human activity soon after its accumulation. In most cases, the presence of the cultural material recovered under or within facies RF2 of the peat layer corroborates this assumption. Radar facies 3 (RF3) corresponds to the mineral subsoil that is mostly presented by sands of various grain sizes.

The GPR results indicated that a boundary between the peat (RF1 and RF2) and the underlying mineral subsoil (RF3) is clearly distinguishable (Fig. 1). The boundary is associated with a strong continuous reflector due to the high contrast in the electrical properties between organic (the peat layer) and non-organic (mineral subsoil) material. The reflector allowed the peat thickness and the topography of underlying subsoil to be mapped (Fig. 2). Areas with low thickness (or high subsoil relief) are assumed to be potential places for human habitation.

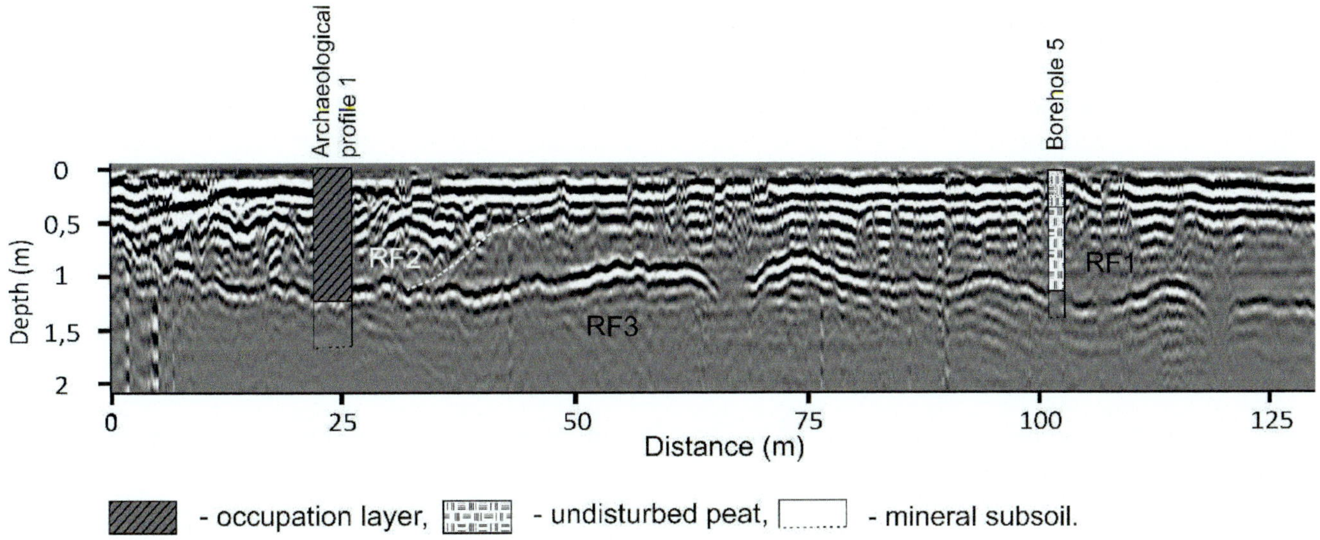

Fig. 1. Ground-penetrating radar image (with 300MHz centre frequency), demonstrating three characteristic radar facies: RF1 – long horizontal reflections corresponding to the undisturbed peat layer, RF2 – tilted chaotic reflections corresponding to the occupation layer, RF3 is separated from RF1 and RF2 by strong continuous reflector and corresponds to the mineral subsoil.

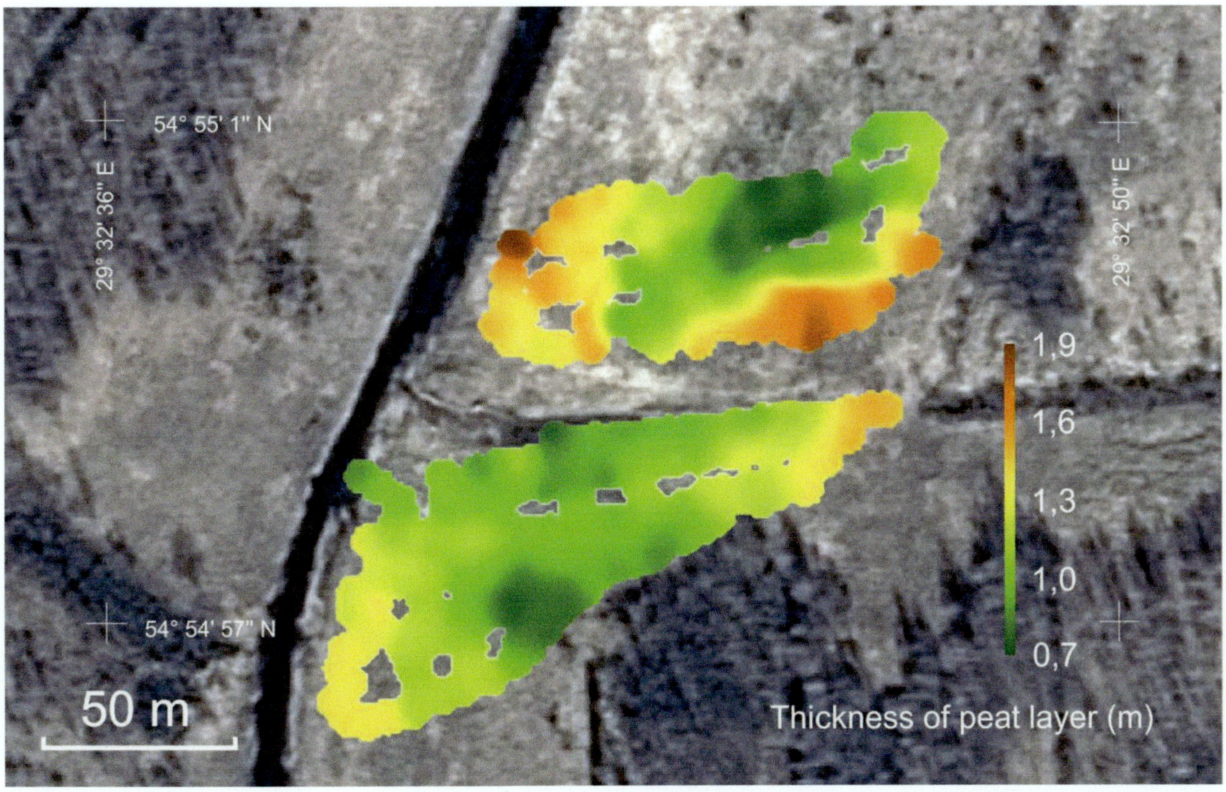

Fig. 2. Map of the peat layer's thickness.

References

Charniauski, M.M. 1997. Narvenskaja kultura (The Narva Culture). *Archealohija Bielarusi*. Tom 1. Kamienny I bronzavy viaki. 190–205. Minsk: Bielaruskaja navuka.

Charniauski, M. and Kryvaltsevich, M. 2011. Belarusian Wetland Settlements in Prehistory. In E. Pranckenaite (ed.), *Wetland settlements of the Baltic: a prehistoric perspective*, 105–122. Vilnius: Center of underwater archaeology.

Cherniavskii, M.M. 1967. Novoe neoliticheskoe poselenie na Krivinskom torfianike. In *Doklady K XI Konferencii Molodykh Uchenykh*, 372–385. BSSR.

Neal, A. 2004. Ground-penetrating radar and its use in sedimentology: principles, problems and progress. In *Earth-Science Review*. 66. 261–330.

The Challenge of Urban Archaeological Geophysics: the Example of Grand'Place in Brussels, Belgium

Michel Dabas[1], François Blary[2], Gianluca Catanzariti[3], Paulo Charruadas[2], Sébastien Flageul[4], Benjamin Van Nieuwenhoeve[2], Philippe Sosnowska[2], Alain Tabbagh[4]

[1]CNRS-ENS-PSL UMR8546, PARIS, FRANCE, [2]ULB - CReA Patrimoine, Bruxelles, Belgium, [3]3D Imaging, Torino, Italy, [4]Sorbonne Université - University P. et M. Curie, Paris, France

Introduction

The program "Voir sous les pavés" ("Seeing under cobblestones") launched for the study of the Grand'Place in Brussels, under the direction of François Blary (ULB) and Michel Dabas (CNRS-ENS), aimed to detect and map the archaeological structures under this square without performing excavations. If this goal seems very common for a geophysicist, it corresponds, in the case of Brussels, to a set of large-scale integrated non-destructive surveys that prefigure what could be undertaken in other downtown areas. This program has made it possible to design a new methodology for the study of archaeological remains buried in urban contexts that until now have been little studied because of their complexity. This complexity is linked to several well-known constraints that we have to cope with: a high footprint over the ground of urban furniture, cars and also tourists; high impact in the subsoil of utilities but also a complexity of archaeological multi-phased deposits and finally a disturbed electromagnetic spectrum.

Background

The current appearance of the Grand'Place is close to that of 1695, after bombardment by the French. Before 1695, the story remains to be written: the square was probably much smaller, texts mention buildings but their locations are unknown. One or more fountains existed and are depicted in some drawings. Also more ephemeral structures have existed: Gas street lamps, Christmas trees, a market place, and podium for public executions, remains that could eventually be detected by geophysics. Only a single very small archaeological excavation has occurred near the Town Hall and some occasional discoveries such as a water well or a cave are known.

Before the surveys took place, it was possible to get a high definition (5cm) aerial orthoimage and a map of the different utilities. No georeferencing of these utilities existed and one of the purposes of the survey was also to map all utilities under the main square and the streets around.

Considerations for the survey of a city centre

Studying a city centre needs to minimize the three above-mentioned constraints. We have found that the constraints situated over the ground surface can be minimized thanks to rigorous prior management. In the case of Brussels, this meant knowing the exact location of fixed and removable furniture, the schedules of all shops and their suppliers but also finding solutions to control the stream of tourists during measurement campaigns!

The subsoil constraints assume that the plan and depth of the known servitudes, possibly georeferenced, are available in advance, as well as the location of the visible sewer drains on the surface.

Finally, the applicability of geophysical methods to urban areas is limited directly by electromagnetic interferences and indirectly by the complexity of the structures to be detected: the usual situation of surveys of old city centres in open-fields is not applicable in our case. Moreover, we feel that the good results obtained over Roman towns for example are due to the conjunction of facts that are seldom encountered in todays city centres: low depth of structures (< 1m); existence of some destruction phase that has 'cleaned' the rubbles and made apparent walls or their negative counterpart; the existence of a single phase of construction of the urban layout. Our situation is closer for example to "tells" in the Middle East. This is clearly a challenge for the geophysicist.

Methods used

Three techniques were used simultaneously in the Grand'Place: ground-penetrating radar (GPR); the electrostatic method, also named CCR (Capacitively Coupled Resistivity); and micro-topography (Fig. 1).

Fig. 1. (left) MP3 prototype for measuring apparent resistivity and (right) GPR (StreamC 600MHz and StreamX 200MHz, IDS GeoRadar Corporation).

Several 3D-multi-channel radars were set up to find a compromise between penetration, spatial resolution and measurement speed. Simultaneously, a new prototype of electrostatic resistivity-meter "MP3", specifically developed for this project, was implemented. It allows the measurement of electrical resistivity simultaneously for three depths of investigation. A robotic total station allowed the positioning of all geophysical data in the same reference frame. The post-processing of these data made it possible to calculate a digital model of the soil surface (DSM) associated with a registration into the Belgian cartographic system (Lambert72) and the available high-resolution orthophotos.

The relatively simple processing and interpretation of the electrical data in a GIS has to be opposed to the very complex processing and interpretation of 3D-radar data. Two ways of dealing with spatial positioning of electrostatic resistivity data were tested and the corresponding maps were compared.

Radar maps in the form of time-slices, using a representation of the amplitude (phase) or the magnitude (envelope), taking into account - or not - the topography were tested. The final interpretation was made thanks to correlation with several historical representations of the Grand'Place.

Results

Resistivity (Fig. 2) was very successful and able to detect several resistive structures that can be interpreted as the remains of old dwellings, and despite the signature of some utilities, generally depicted as conductive elongated anomalies; distinctive walls of houses were detected all over the Grand'Place. These anomalies were detected with the most superficial quadrupole (a=0.5m), meaning that these anomalies are within the upper 50cm. They are best seen for the 1m dipole and have been confirmed with the different maps corresponding to the two orientations of the dipoles. The highest gradient corresponding to probably a massive building can be seen in the south-east corner (see the combined interpretation in Fig. 4). The density of resistive anomalies is less in the western part of the Grand'Place. We think that this does not necessarily imply a lower density of constructions in this area, but perhaps deeper or more erased structures.

The phase of destruction and reconstruction of the 17th century Grand'Place was clearly picked up especially in the south-eastern part: the widening of the place to the east has resulted in a re-alignment of the houses that resulted in the destruction of the former front of the houses (pushed approximately 10m to the north-east).

Among the known utilities, only one was clearly picked-up and corresponds to the trench associated with a water pipe. The other utilities were too small to be imaged by resistivity but were picked by GPR.

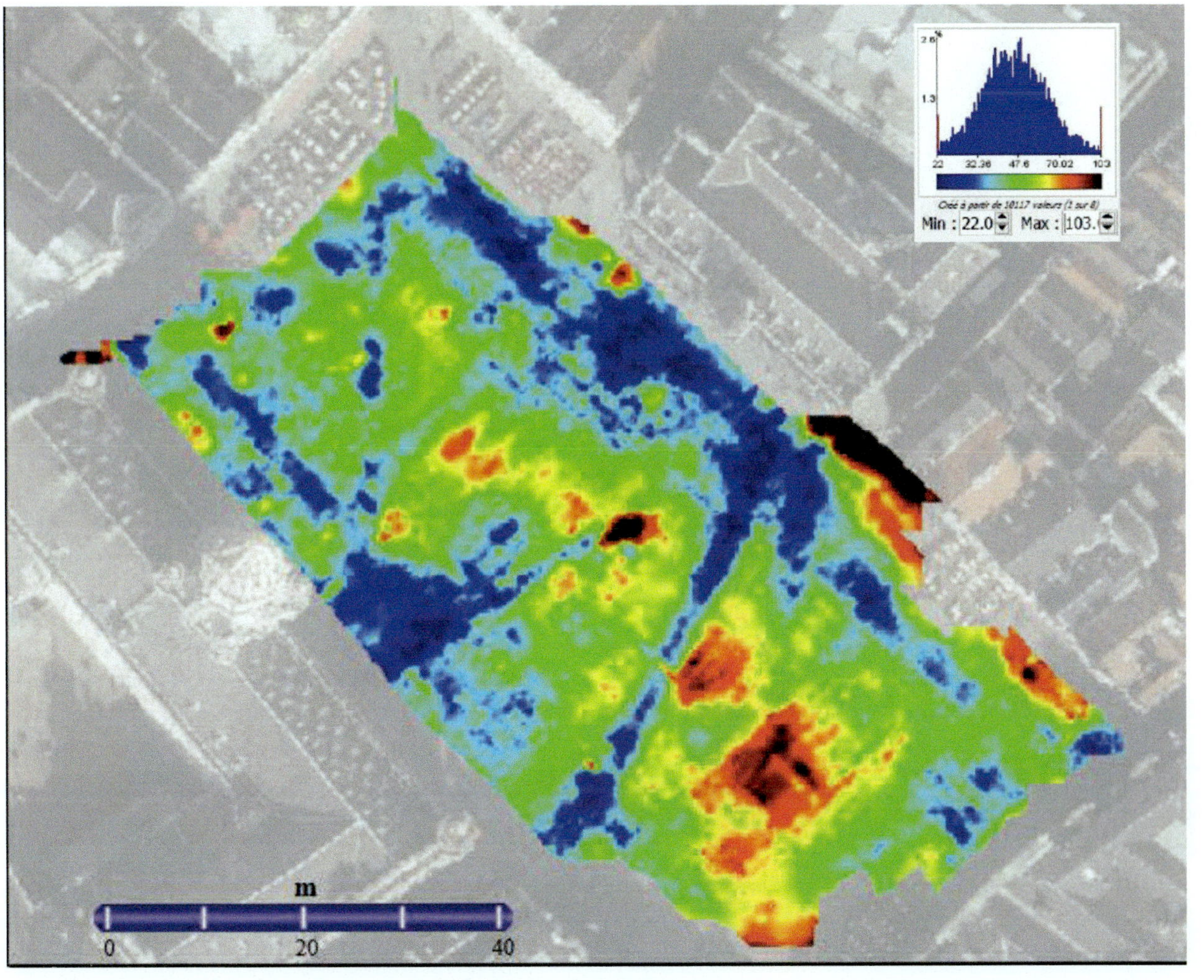

Fig. 2. Apparent resistivity data obtained with channel2 (0m to 1m) of the MP3 prototype (charge coupled resistivitimeter) overlaid on orthoimagery (June 2016, source: cirb.brussels, Open licence).

The GPR results - despite many maps corresponding to the different frequencies used, and for different processes, and even for different polarizations - were far more complicated to interpret. Moreover, the low electrical resistivity measured with CCR and confirmed by a single Vertical Electrical Sounding between the cobblestones has explained why the radar penetration depth, even at 200MHz, was so low (1m approximately). Nevertheless, the GPR results (Fig. 3) demonstrated, beside a very high number of unknown utilities, the existence of a platform for two fountains and probably three water wells (Fig. 4). The spatial resolution of GPR was necessary to detect these small-scale anomalies (compared to resistivity imaging). As an example of known structures, the location of the foundations for the Christmas tree and a former water well were discovered in the time-slice GPR maps, which were not known before our survey.

Surprisingly, the signature of the dwellings and walls discovered with resistivity imaging was not so clear. Only GPR amplitude time-slices (and not the envelope) - thanks to a difference in texture - was able to delineate the limits of these dwellings (but not the walls), and we should admit that a comparison with electrostatic resistivity maps was mandatory to secure our interpretation.

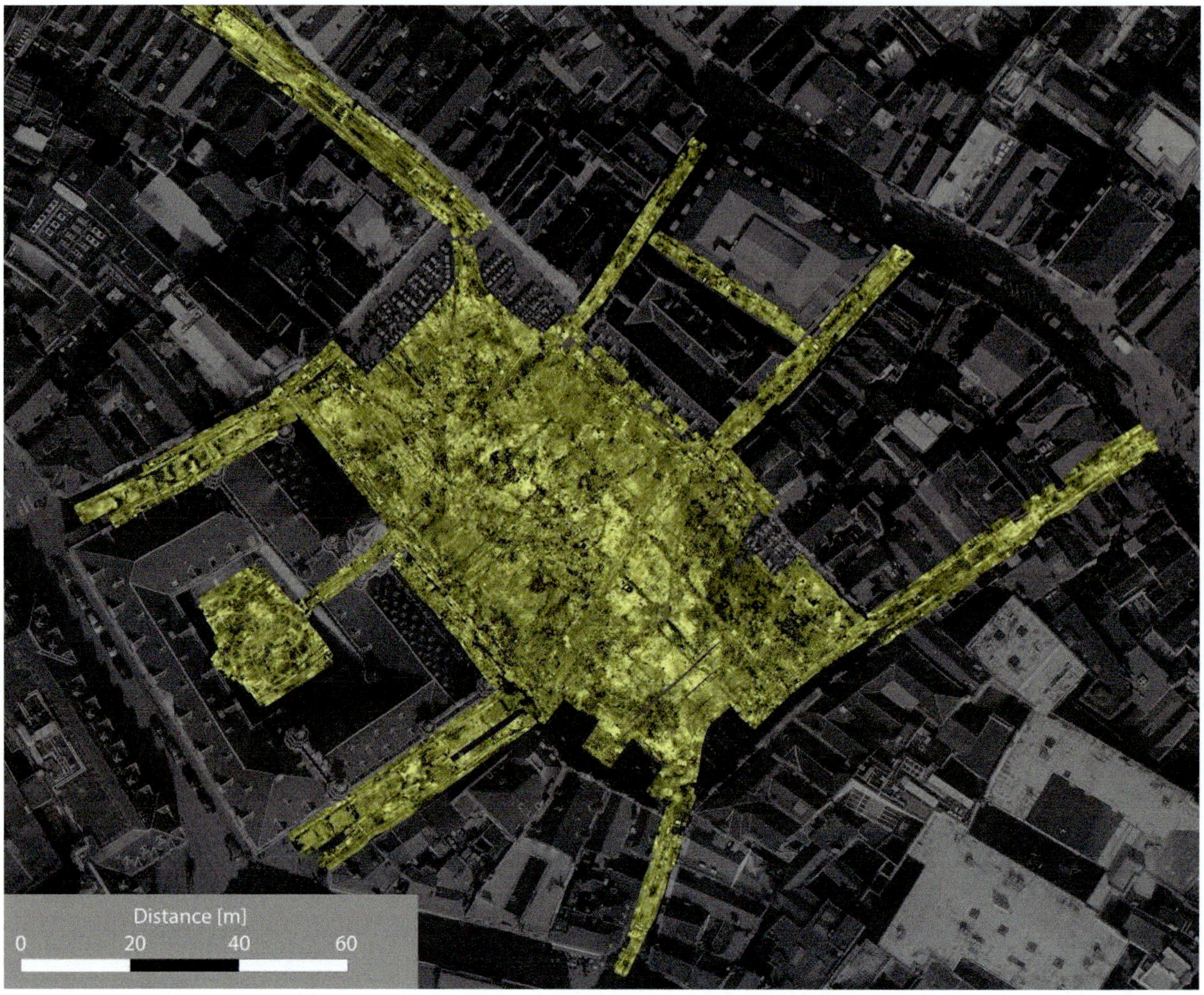

Fig. 3. 600MHz GPR phase time-slice centered on 0.25m overlaid on ortho-imagery (June 2016, source: cirb.brussels, Open licence).

Conclusions

The archaeological interpretation of geophysical data from Grand'Place is a challenge which is not finished: Our interpretation has changed a lot as new maps became available. In particular, new processes for GPR data were performed, that enable new archaeological structures to be visible. Confrontation with resistivity data was the most important in this situation. Despite the lack of spatial resolution, resistivity data were able to map quickly the main structures but also very conductive levels, probably caused by the swamp that extended in this area. GPR data proves to be the best for small scale structures and for utilities. Most importantly, the existence of old drawings and paintings has forced us to push further our first interpretations: without this information we will probably have missed most of the archaeological structures. We believe that the good achievements of this project are due to the integration of the archaeologists at every stage of the process. Thanks to a cross-analysis with historical documents, an important contribution to the knowledge of the genesis of this historic center of the Capital of Europe was attained.

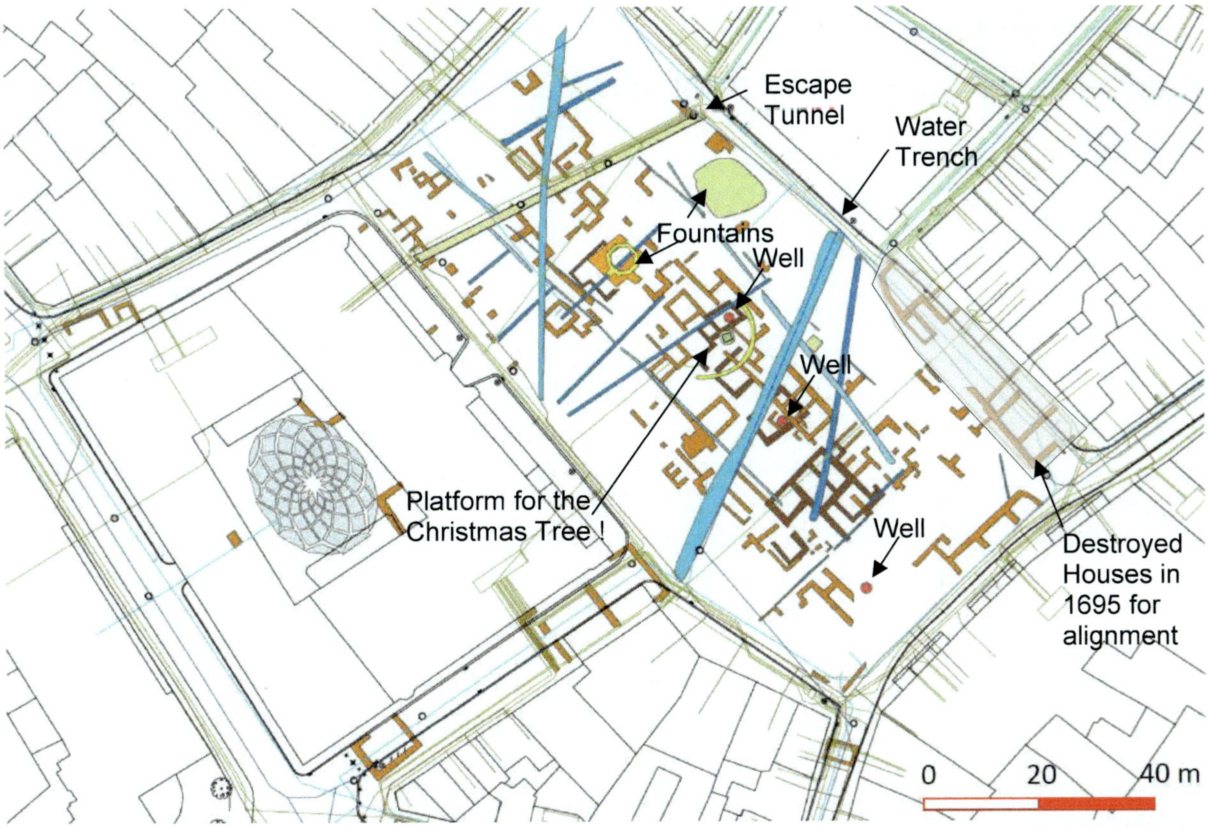

Fig. 4. First joint interpretation of resistivity maps and GPR time-lapse maps (March 2019 – F. Blary).

Acknowledgements

We would like to thank all the team involved in this project, representing more than 30 people. The GPR data were acquired by Geostudi Astier.

The Good, the Bad and the Ugly (Data): 100-year Discussion over Roman Fort in Herzegovina solved with shards of information

Michal Pisz[1], Tomasz Dziurdzik[2]
[1]*Faculty of Geology, University of Warsaw, Warszawa, Poland,* [2]*Institute of Archaeology, University of Warsaw, Warszawa, Poland*

Background

Research history and problem

The Roman site Gračine is dated to the 1st-3rd century AD and located on a plateau overlooking the Trebižat river in village Humac, Ljubuški municipality, West Herzegovina Canton, Bosnia and Herzegovina. Its function was discussed for over a century, since Carl Patsch stated it was a Roman auxiliary fort and identified it with ancient Bigeste (Patsch 1897). Excavations were conducted in 1977-1980 in the middle part of the site, but never fully published. The uncovered features (Bojanovski 1981) did not fit the expectations about the inner architecture of Roman forts, resulting in conflicting interpretations. Some proposed an attribution of different functions (Basler 1985: 22), others that the site was not the fort, but rather an annexe to it (Dodig 2011). The resolving of this problem is of special importance for the history of Roman Dalmatia, as the site had a crucial role in the defence of an important city, Narona, and was the southernmost installation of the so-called limes Delmaticus, the defensive system of the whole province (see an elaboration of the theory in Šašel 1974, recently Periša 2008 and Tončinić 2015).

Conditions on the site

Due to private ownership of the land plots, only part of the site is accessible for researchers. Of special importance is the geology of the area and the thickness of archaeological layers. While in some of the parts excavated in the 1970s it exceeded 1.5m, in other parts the archaeological structures are located just beneath - or even partially visible above - the modern surface. The bedrock is also very high, with outcrops in the immediate vicinity of the site.

Methods

Earth resistance was a principal method of prospection. Measurements were taken with a Geoscan Research RM85 meter. In October 2016 the northern part of the site was surveyed using a Wenner array (a = 0.5m) and a 1 × 1m sampling resolution. Measurements have been taken within two 40m × 40m grids, but *c.* 0.17ha of the plot was measured (Fig. 1). These were the very first geophysical prospection methods applied to Gračine. The area of prospection was limited by bushes, bedrock outcrops and contemporary infrastructure. The Wenner array had been chosen since it provides a relatively better mobility of the earth resistance meter and the field measurements can be operated by just two people. It is also considerably faster than surveying with remote probes in this environment, which was an important choice for such a short period of research conducted abroad.

In Spring 2017 a continuation of the earth resistance survey was performed at Gračine. During this campaign all remaining accessible ground was surveyed with the Wenner array, using electrodes with a 0.5m separation distance and taking measurements with a 1m profile interval and a 0.5m sample interval (Fig. 1). After first visualisation of the data, the survey was repeated over the most interesting part of the site using 0.5m × 0.5m resolution and again with 1m × 0.5m sampling and a 0.75m electrode separation distance. All the data were collected within 20m × 20m grids and covered 0.25ha and 0.1ha for the 1m × 0.5m and 0.5m × 0.5m sampling respectively.

Registered resistance values were normalized to obtain an apparent resistivity by applying a geometric factor. This was in order to compare the survey results of both the 0.5m and 0.75m electrodes separation distances and analyse changes of apparent resistivity distribution with increased depth of prospection.

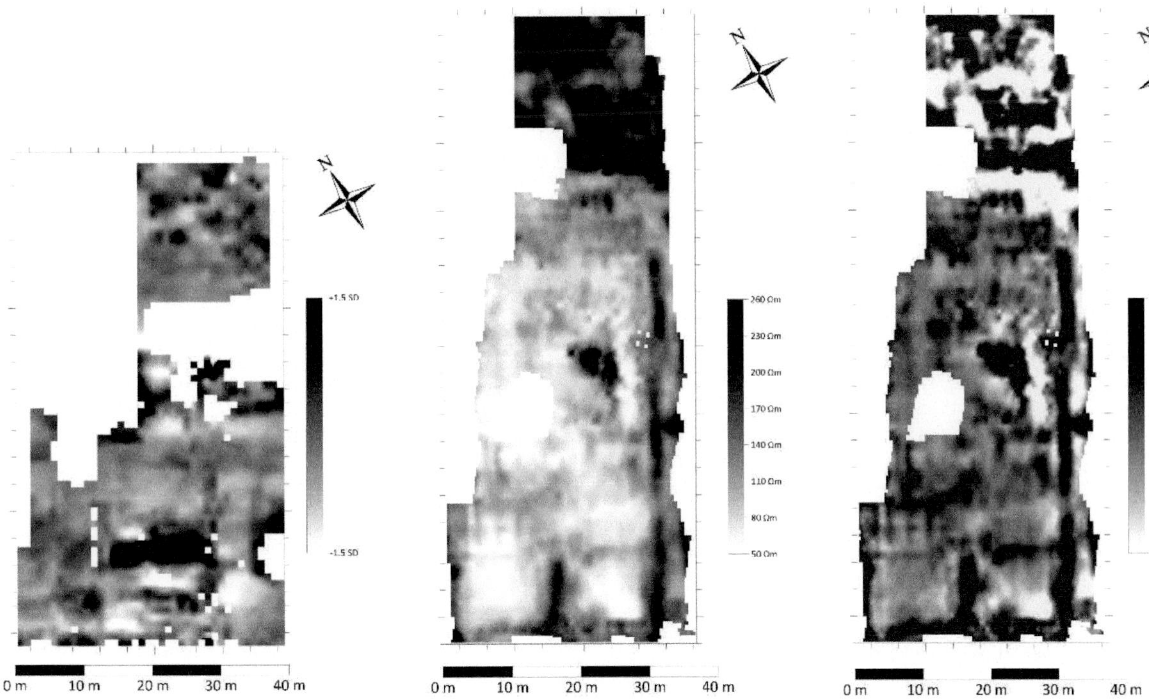

Fig. 1. Earth resistance results from Gračine. From the left: results from 2016 with despike, low pass and high pass filters applied; results from 2016 with despike and low pass filters applied; the same results with added high pass filter. All data are interpolated to 0.25m × 0.25m.

Results

The first approach to Gračine was not very promising. The data looked unclear and it was hard to distinguish any particular structures and interpret the data at this stage of research (Fig. 1). The image has been dominated by a very strong high resistance zone of anomalies of irregular shape. Registered apparent resistivity values rarely exceeded the range of 40-200 Ωm with the mean value of *c.* 90 Ωm.

The second campaign brought more satisfying results. At first sight it was possible to distinguish some linear high resistance anomalies of varying intensity. The highest measured resistivity values reached *c.* 500 Ωm, but the most of interesting features ranged from 70-130 Ωm with average background values 50-70 Ωm. All the data which might be regarded as unclear and far from ideal, when combined, provided a very interesting overview of the site and delivered a lot of important information (Fig. 2).

Archaeological interpretation

The shape of the anomalies is recognisable to anybody familiar with Roman military installations: they appear to be caused by the remains of two mirror barrack blocks, with rows of double rooms for the common soldiers and a larger house for the officers on the southern side (Fig. 3). They are separated from the buildings excavated in the 1970s by a strong anomaly which seems to be caused by the hard surface of a street. Similarly, on the eastern side *via sagularis*, a street running around the fort, separating the inner architecture from the defensive walls is expected. These assumptions were verified in 2017 and 2018 during excavations by a team from the University of Mostar, fully proving the geophysical interpretation.

Conclusions

The architecture is clearly that of a Roman auxiliary fort from the Principate, even though the middle part does not follow usual or expected patterns. This, however, is explained by a possibly early date of construction (as suggested by small finds from the excavation), before the design was standardised. Thus, geophysical survey allowed a final determination of the site's function despite the bad first impression of the data. Moreover, it opens new avenues of interpretation, including contributing to the discussion on the Empire-wide problem of the development of Roman military architecture.

Fig. 2. Earth resistance survey results on the drone orthophoto.

Fig. 3. The interpretation of earth resistance survey results.

Acknowledgements

The research was conducted within the framework of a Preludium 10 grant (number 2015/19/N/HS3/00886) awarded to Tomasz Dziurdzik by the National Science Centre, Poland. The authors would like to thank Mirko Rašić (University of Mostar) for his cooperation and information on the results of verification excavations.

This work was supported by an ITC Conference Grant from COST Action SAGA: The Soil Science & Archaeo-Geophysics Alliance - CA17131 (www.saga-cost.eu), supported by COST (European Cooperation in Science and Technology).

References

Basler, Đ. 1985. Nova postavka Muzeja Franjevačkog samostana Humac. In D. Vukojević (ed.), *100 godina Muzeja na Humcu*. 17-30. Ljubuški.

Bojanovski, I. 1981. Gračine, Ljubuški – rimski vojni logor. *Arheološki pregled.* 22. 63–66.

Dodig, R. 2011. Rimski kompleks u Gračinama. Vojni tabor ili...? *Izdanja Hrvatskog arheološkog društva*. 27. 327-345.

Patsch, C. 1897. Bigeste. In A. Pauly and G. Wissowa (eds.), *Paulys Realencyclopädie der classischen Altertumswissenschaft: neue Bearbeitung. Vol. II*. 428–469. Stuttgart.

Periša, D. 2008. Je li delmatsko područje presjekao rimski limes? *Archaeologia Adriatica*. 2(2). 507-517.

Šašel, J. 1974. Die Limes-Entwicklung in Illyricum. In D.M. Pippidi (ed.), *Actes du IXe Congrès International d'Etudes sur les Frontières Romaines*. 194-199. Bucureşti/Köln.

Tončinić, D. 2015. Der Donaulimes in Kroatien – von Augustus bis Claudius – von Dalmatien zur Donau. In L. Zerbini (ed.), *Culti e religiosita nelle province danubiane*. 335-345. Bologna.

Geophysical surveys of Eneolithic ditch enclosures in central Bohemia

Roman Křivánek
Institute of Archaeology, CAS, Prague, v.v.i., Prague, Czech Republic

Lowland areas along the main rivers and streams, with fertile soils, represent the most intensively settled areas, with common polycultural prehistoric, early medieval and medieval sites. These long-time agricultural areas are also those most intensively observed by aerial and geophysical prospection. The heaviest concentration of various ditch enclosures and hillforts is also typical in these same areas. Non-destructive methods have been employed here, to identify and verify the shape or extent of many Neolithic roundels, Bronze Age oval and La Téne quadrangular ditch enclosures. But new rescue archaeological results, together with accessible new sequences of ortho-photo aerial pictures, have, in the last two decades, initiated geophysical verifications of other ditch enclosure systems dated mainly to the Early Eneolithic (*c.* BC 4500 – 3800). These ditch enclosures are situated on lower terraces, not far from the flood plain areas of rivers, or on slightly elevated plateaus within lowland areas.

The collection of geophysically observed or verified Eneolithic ditch enclosures from Bohemian lowlands today has over 10 features (the total number of known and/or expected Eneolithic ditch enclosures in Bohemia is nearly 20 features). Some of these sites were surveyed only in particularly accessible areas as preliminary or subsequent prospection during different stages of rescue archaeological investigations (Dobeš *et al.* 2016; 2017). The other enclosures were later identified on aerial photographs and then systematically verified in the field by magnetometer prospection and field artefact collection (Křivánek 2004; 2006), (Fig. 1). These non-destructive results were verified by archaeological trenches (Foster 2004; Gojda *et al.* 2007) or new archaeological projects focused on the issue of Eneolithic ditch enclosures.

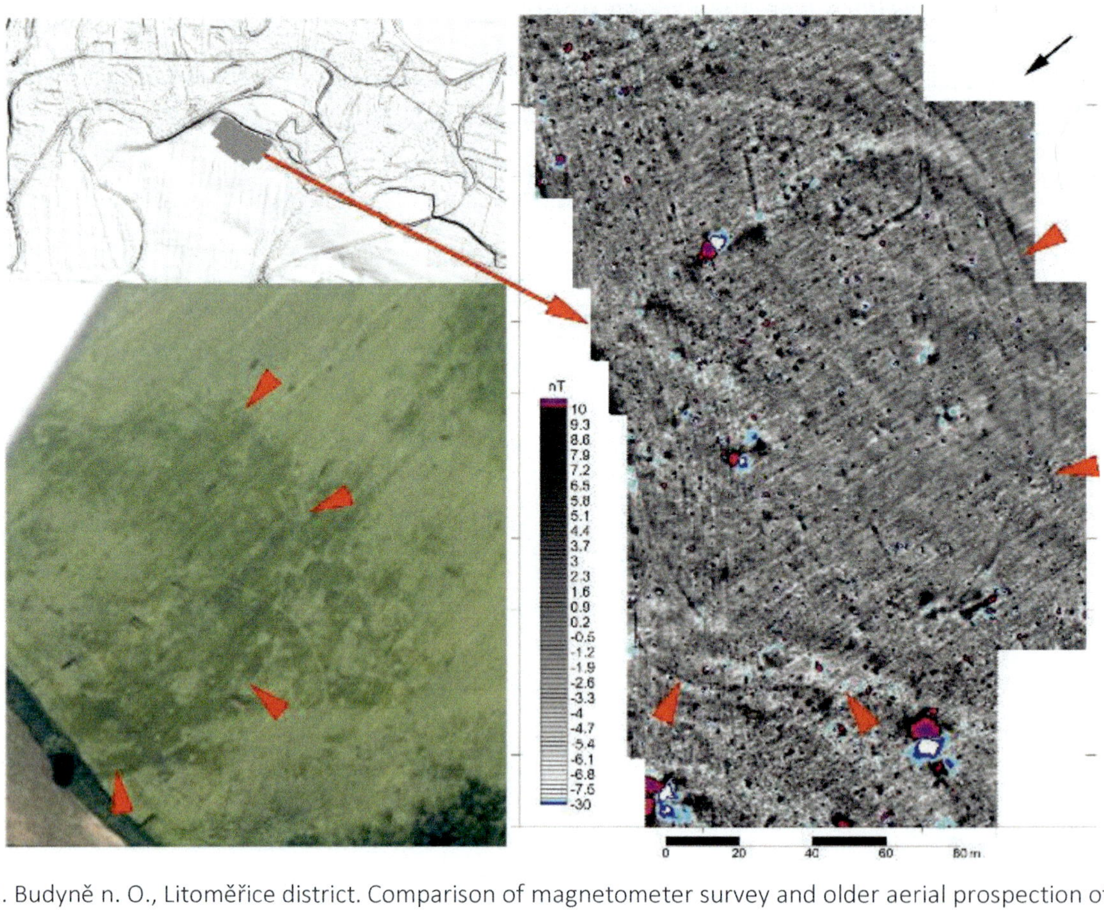

Fig. 1. Budyně n. O., Litoměřice district. Comparison of magnetometer survey and older aerial prospection of the polycultural site with Eneolithic enclosure and abandoned remains of the medieval village of Mileticko (source: Smrž; lidar: www.cuzk.cz; survey: Křivánek 2016; surveyed area: approx. 2.7ha; Sensys 5-channel push-cart fluxgate gradiometer system; sampling interval: 0.5m x 0.2m).

Budyně nad Ohří, Litoměřice district. The archaeological site, situated on the lower terrace above the southern bank of the Ohře River flood plain, has been known from old archaeological reports and aerial photographs for decades. The elevated flat terrain was settled in multiple prehistoric periods and later by the medieval village Mileticko (abandoned). Using aerial prospection, archaeologists also identified here the remains of an interrupted double ditch enclosure (Fig. 1). Magnetometer results confirm more remains of polycultural settlement (sunken features and small enclosures), but also a probable Eneolithic double ditch enclosure adhering to the edge of the lower terrace. All subsurface remains are very intensively ploughed and occupation layers are influenced by sand-gravel sediments of the Ohře River.

Chrášťany, Kolín district. The site was discovered during the rescue archaeological excavation in 2010. In the area of a new biogas plant, the remains of two ditches of an Early Eneolithic ditch enclosure (with ceramic finds from the Schussenried and Michelsberg cultures) were identified. New magnetometer surveys of the outer area of the biogas plant confirmed the presence of a larger double ditch enclosure (Fig. 2). Unfortunately, some parts of the original enclosure were on the sloped terrain of the plateau and were deeply eroded by long-time ploughing. Geophysical results also confirmed the continuation of an Early Eneolithic occupation area (sunken settlement features and the probable remains of a long barrow).

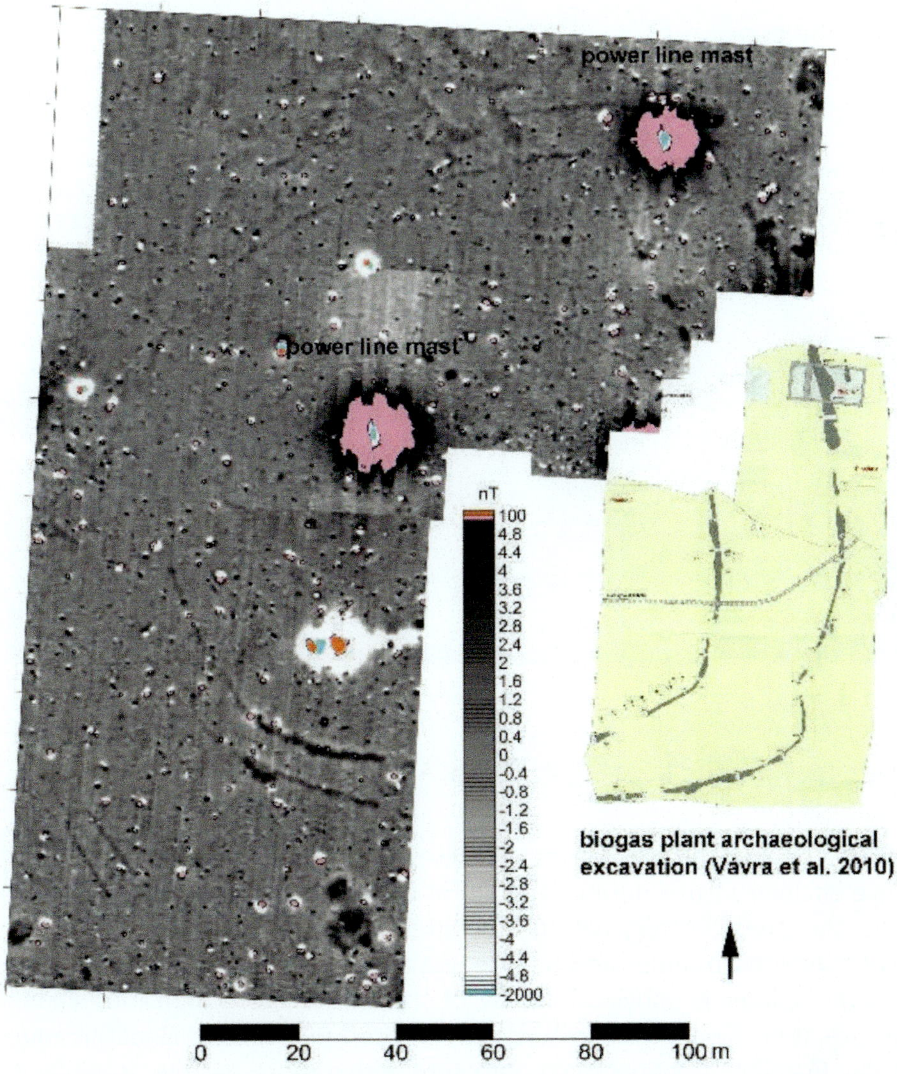

Fig. 2. Chrášťany, Kolín district. The result of the magnetometer survey combined with previous rescue archaeological investigation (source: plan from report of archaeological investigation by Vávra-Šťastný in 2010; survey: Křivánek 2018; surveyed area: approx. 2ha; Sensys 5-channel push-cart fluxgate gradiometer system; sampling interval: 0.5m x 0.2m).

Všechlapy, Nymburk district. The enclosed archaeological feature was first identified in 2012 due to publicly accessible sources of new aerial pictures (orthophoto maps from 2010 and some other years (Kos 2013)). A less irregular, closed ditch enclosure was later chosen for archaeological verification by trenching in 2016, during a new archaeological project of the Department of Archaeology of the University of West Bohemia in Pilsen. In the same year, magnetometer survey was launched in the vicinity of an archaeological trench with later verification of the whole older Eneolithic enclosure (Fig. 3). Geophysical results showed that this ditched enclosure had a less typical, irregular shape and interruptions; only three-quarters of it had double ditches, the rest was marked by only one ditch. It appears that the double ditch enclosure had not been completed. The remains of sunken settlement features were also identified inside and outside the enclosure (probably from polycultural prehistoric settlement of this slightly elevated area).

Fig. 3. Všechlapy, Nymburk district. Magnetometer survey combined with an aerial picture from 2010 (source: www.cuzk.cz; survey: Křivánek 2016 and 2018; surveyed area: approx. 3.7ha; Sensys 5-channel push-cart fluxgate gradiometer system; sampling interval: 0.5m x 0.2m).

Byseň, Kladno district. Czech archaeologists discovered and documented a quarter of a probable (semi)circular double ditch enclosure, during agricultural land-use changes of fields to new hopfields in 1952 (Knor 1954). Due to the Linear Pottery culture, the site was (preliminarily) classified at the time as the probable location of a Neolithic roundel, and the site was also monitored by (palaeo-) archaeo-astronomical observations. However, a later geophysical survey of the whole site significantly changed the previous archaeological interpretation. Geophysical results did not confirm a circular Neolithic roundel, but showed an elliptical (probably Eneolithic) double ditch enclosure within more narrow, irregular interruptions and also, nearby, the irregular shape of the whole elliptical bow (Křivánek 2018). The site has never been verified by any archaeological investigation and is still waiting for more precise dating.

The number of Early or later Eneolithic ditch enclosures in Bohemian archaeology has been increasing in recent years. It is no longer a unique and random situation in the Czech landscape. The combination of archaeological research and general geophysical surveys shows several similarities in the location and range of fortified areas. All enclosed features have a clear relationship with rivers and streams. In Bohemia, lowland Early Eneolithic settlement is conspicuously concentrated in the main lowland areas in central, northern and north-western Bohemia. The fortified area inside the closed enclosures is on average 2ha–4ha.

Systematic monitoring of various enclosed areas in agricultural lowland regions with long-time ploughing and soil erosion will also be necessary in the near feature. For the first time, this would be connected with an archaeo-geophysical project. The new project "Non-destructive geophysical research of endangered sites by agricultural activity with different types of ditch enclosures" is supported in 2019 by the programme of Regional Cooperation between the Regions and the Institutes of the Czech Academy of Sciences.

References

Dobeš, M., Stránská, P., Křivánek, R. and Limburský, P. 2016. Časně eneolitické ohrazeni ve Vliněvsi. Příspěvek k povaze kontaktu mezi jordanovskou a michelsberskou kulturou v Čechách. *Památky archeologické*. CVII. 51-115.

Dobeš, M., Křivánek, R., Kyselý, R. and Stolz, D. 2017. Časně eneolitické příkopovité ohrazení v Trubíně, okr, Beroun – Proto-Eneolithic causewayed enclosure in Trubín, central Bohemia. *Archeologické rozhledy*. LXX. 3-38.

Foster, P. 2004. 4.4 Large ditched enclosure at Chleby, (distr. Nymburk): a small-scale test excavation report. In M. Gojda *et al.* (eds.) *Ancient Landscape, Settlement Dynamics and Non-Destructive Archeology – Czech research project 1997-2002*. 214-227. Praha.

Gojda, M. (ed.), Dreslerová, D., Foster, P., Křivánek, R., Kuna, M., Vencl., S. and Zápotocký, M. 2002. Velké pravěké ohrazení v Klech (okr. Mělník). Využití nedestruktivních metod výzkumu k poznání nového typu areálu. *Archeologické rozhledy*. LIV-2002. 371-430.

Knor, A. 1954. Nálezy na chmelnicích v severozápadních Čechách. *Památky archeologické.* 45. 281–286.

Kos, L. 2013. Ohrazený areál u Všechlap, okr. Nymburk. Nové ohrazení časného eneolitu? *Archeologie ve středních Čechách.* 17. 377-386.

Křivánek, R. 2004. 2.3. Geophysical prospection. New perspectives for settlement studies in Bohemia. In M. Gojda *et al.* (ed.) *Ancient Landscape, Settlement Dynamics and Non-Destructive Archeology – Czech research project 1997-2002*. 39-71. Praha.

Křivánek, R. 2006. Magnetometric prospection of various types of large ditched enclosures in Bohemia. *Archaeological Prospection.* 13(1). 25-43.

Křivánek, R. 2018. Příkopové ohrazení na k. ú. Byseň, okr. Kladno: mýtus a realita z výsledku geofyzikálního měření - Ditch enclosure near Byseň, Kladno district – myth and reality from the results of geophysical measurement. *Archeologie ve středních Čechách.* 22(2). 865-872.

Vávra, M. and Šťastný, D. 2010. *Zpráva o záchranném archeologickém výzkumu v k. ú. Chrášťany u Českého Brodu, okr. Kolín, bioplynová stanice, v r. 2010. Archaeological report ÚAPPSČ*, Praha.

The application of the geophysical method in forested highland terrains of Bohemia

Roman Křivánek

Institute of Archaeology, CAS, Prague, v.v.i., Prague, Czech Republic

The typical landscape types for the Czech Republic vary between agricultural lowlands, terraces above streams and rivers, flat plateaus, geologically different highland regions and border mountains. The most intensive areas of settlement and activities were, of course, concentrated in lowland regions along the main rivers of the Czech water network, within the most fertile soils. Some very specific activities - such as separated production, mining, transport, burial activity and the settlement of strategic places - also occurred in the forested highland and mountain regions. The present forest cover of the Czech Republic is approximately 34%, but due to a combination of property changes (privatization), land use changes, climate-vegetation changes and different interests of owners and users, even the state of archaeological sites cannot be considered stabilized here. Deforestation and new reforestation within additional terrain modification are the main risks, not only for nature conservation but also for archaeology.

From the point of view of archaeo-geophysical prospection, forested highland regions represent variably sloped, less accessible terrains, with further limitations on the of application and/or interpretation of methods. Of course, we cannot apply large-scale magnetometer measurements in a multi-sensor configuration, or efficiently apply resistivity measurements during all seasons of the year in these areas. The spatial possibilities of the methods are limited, and the time and work demands are much higher than on fields in lowlands (Křivánek 2007). On the other hand, suitable combinations of geophysical methods, new Lidar data and other archaeological results could provide a variety of new results relating to the subsurface archaeological situation, before its deeper or definitive destruction.

Prehistoric barrow cemetery

An example of a combination of magnetometer and resistivity measurement in the northern area of the forested Bronze Age barrow cemetery near Plav, in the České Budějovice district (Chvojka and Kovář 2016), showed different possibilities for both prospection methods (Fig. 1). The results confirmed the different subsurface preservation of particular barrows, in some cases with preserved stones at the perimeter of barrows, with the nearly full destruction of the feature elsewhere (Křivánek 2018). This part of the forested area was subsequently investigated by archaeologists due to the planned corridor of a new motorway. Additional magnetometer measurements between barrows and on a meadow in the vicinity of the barrow cemetery did not confirm the continuation of the site or the presence of known sunken flat graves.

Medieval production areas

Geophysical prospection of forested abandoned medieval glass-working sites has rather a long tradition in the Czech Republic (Křivánek 1995; 1998; Černá 2016). One of the latest examples of a successful magnetometer survey of a forested abandoned medieval glass-works (13th or 14th century AD) in the Ore Mountains comes from Boč, in the Karlovy Vary district. The site was discovered by archaeologists from the Institute for Archaeological Heritage in NW Bohemia (Volf *et al.* 2017) and subsequently verified by detailed magnetometer measurement (Fig. 2). Two strong and identically-aligned magnetic anomalies confirmed the locations of two original glass furnaces in the flat sloped terrain, while scattered remains of inhomogeneous magnetic material near the deserted ravine road then confirmed the probable destruction of the original glass-waste heap.

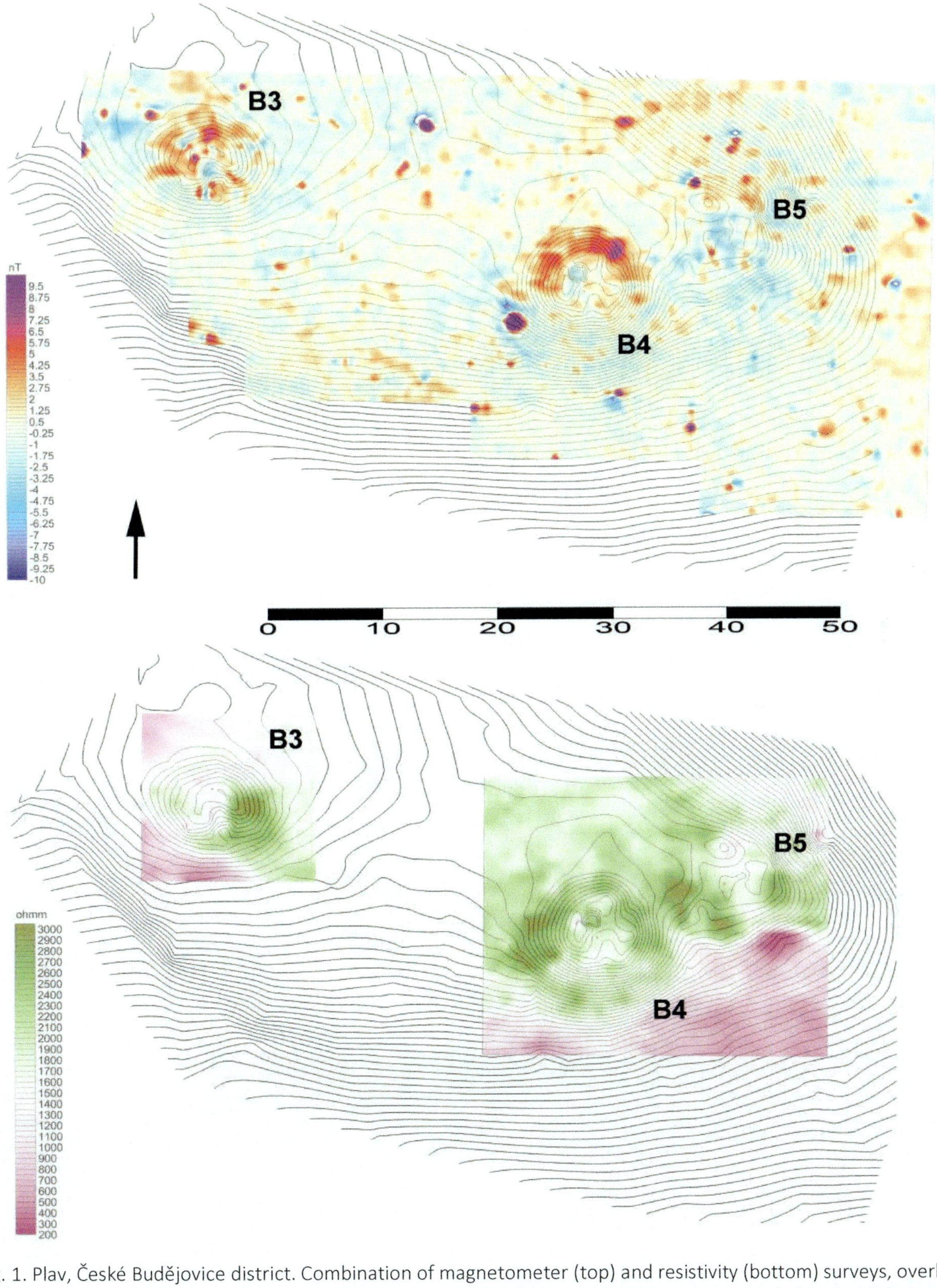

Fig. 1. Plav, České Budějovice district. Combination of magnetometer (top) and resistivity (bottom) surveys, overlain by a detailed contour line plan of area surveyed, forming part of a barrow cemetery (source of plan: South Bohemian University in Č. Budějovice; survey: Křivánek 2017; surveyed area: 0.237ha).

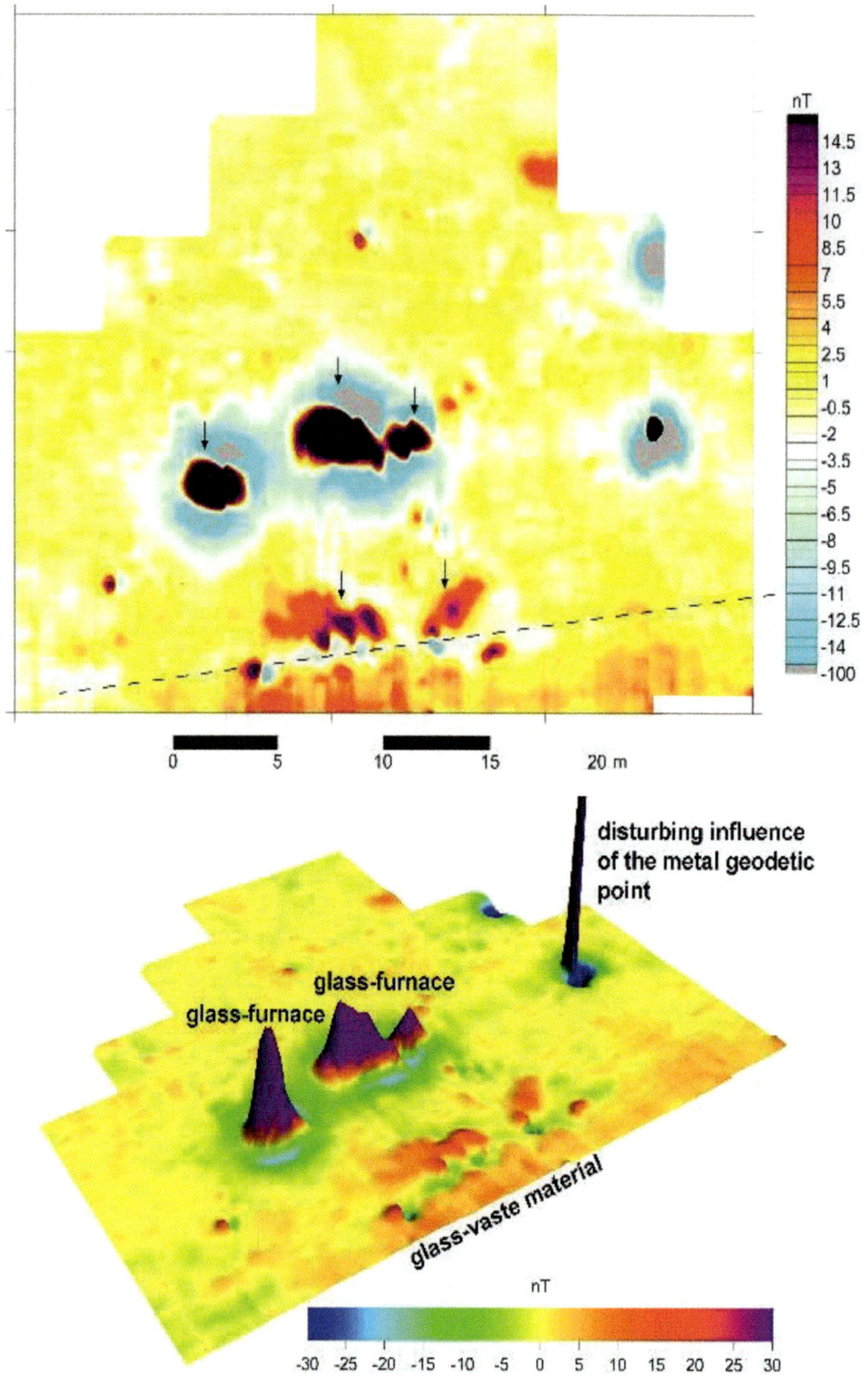

Fig. 2. Boč, Karlovy Vary district. The separation of two glass furnaces and the remains of a destroyed glass waste heap at the abandoned medieval glass-working site (dotted line – the edge of the deserted ravine road; survey: Křivánek 2017; surveyed area: approx. 25m x 20m).

Deserted mine area

In medieval and modern periods, the Ore Mountains were intensively mined regions with many local and specific mining sub-regions (iron, silver, tin, wolfram and other polymetallic ores). The montane micro-region near Pstruží, in the Karlovy Vary district, also includes areas with remains of waste iron-slag heaps, where the original iron processing locations are unknown. A magnetometer survey in the flat area of the mountain valley along the local stream, probably identified highly magnetic concentrations of the probable remains of an iron furnace. The initial processing of the extracted iron ore probably occurred in the mining area, near water sources.

Deserted medieval village

An example of a combination of magnetometer and resistivity measurement was used in the northern part of the forested and abandoned medieval (from 13th/14th to 15th century AD) village of Horní Zabělá, Plzeň-město district. The surveys documents different activities in different parts of the settled area (Sokol 2016). Remains of settlement features with magnetic anomalies and a concentration of stony remains on small platforms were identified in northern area. Some segments were also damaged by modern mechanical afforestation. A concentrated highly magnetic anomaly with scattered stones very probably confirms the production area (Fig. 3) separated inside the village area – the probable remains of a smithy.

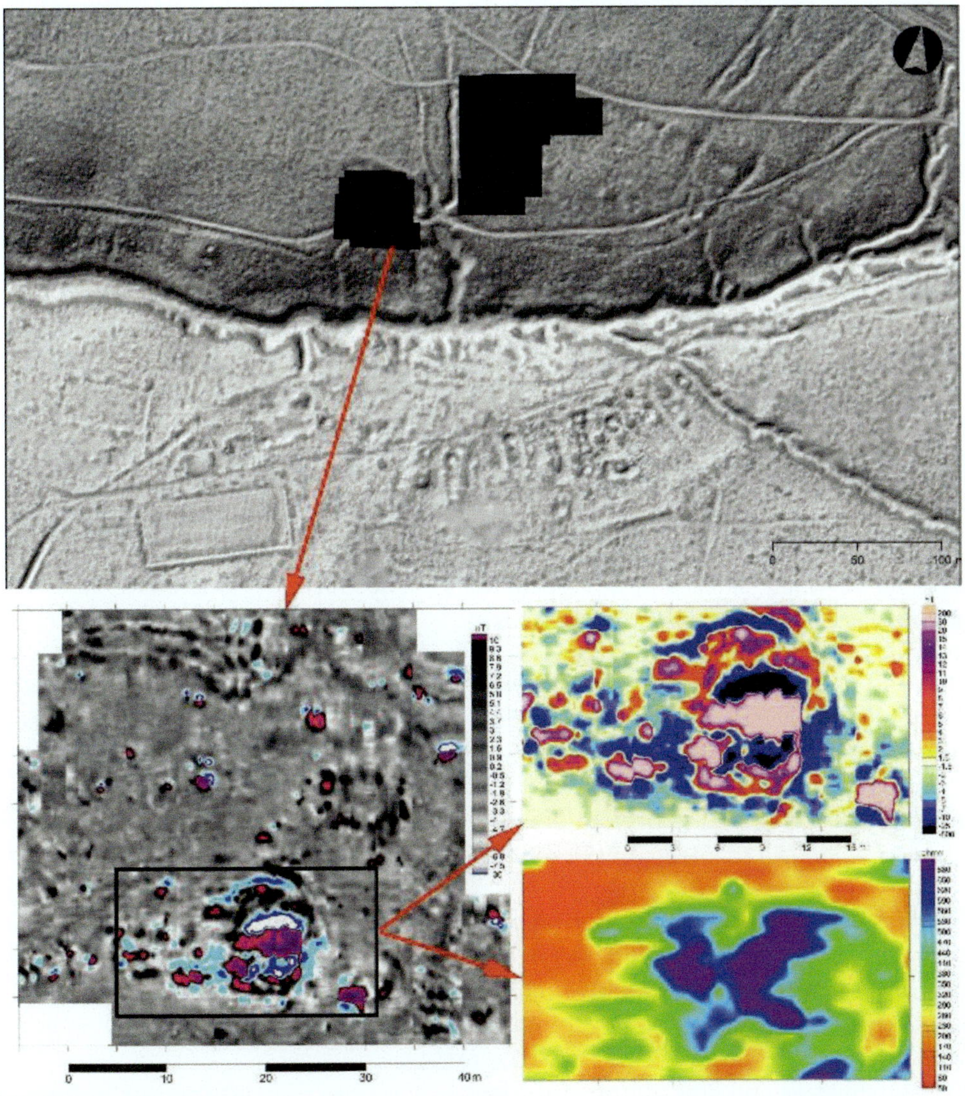

Fig. 3. Horní Zabělá, Plzeň-město district. Combination of magnetometer and resistivity surveys over remains of the smithy in the northern part of the abandoned medieval village, which is visible in the Lidar data (source of Lidar: www.cuzk.cz; survey: Křivánek 2016; surveyed area: 0.2ha).

Deserted medieval motte in the production area

The survey of a forested area near Strašice, in the Rokycany district, demonstrates the application of different geophysical methods for the detection of different archaeological deposits (Křivánek 2010). The small medieval fortified site comprises a motte inside a larger area of pitch-production, with many deserted lines of ravine roads (14th/15th century AD). In the heavily forested area, two stony accumulations from the probable original stone features built-up inside the motte were identified by resistivity measurement. More pitch-furnace locations were confirmed by magnetometer survey of the more accessible outer areas with many small elevations along systems of ravine roads.

Conclusions

The application of various geophysical methods to the survey of archaeological sites in forested highland terrains is possible (and irreplaceable) and could contribute to uncovering many archaeological sites. In Bohemia, between 60-80 archaeological sites were surveyed by different geophysical methods (at least half were verified by archaeological trenches). However, their efficiency is often dependant on more specific conditions that during a year (climate, precipitation, humidity, vegetation time) or in space (local geology, slope or land use changes). Non-vegetation time is, for example, the most suitable period for magnetometer surveys. Magnetometry is the most successful method for the verification of various production and mine areas, funeral and fortified settlement places. In the case of Bohemia, drier periods in the summer or autumn seem to be good for resistivity measurements. This method helps at many barrow cemetery sites, and over medieval or modern settlement with remains of stone architecture. Local flat and accessible terrains seem to be suitable areas for potential electromagnetic and radar surveys. Methods are applied, as in mine areas, for the detection of cavities in various medieval and modern sites. For all geophysical methods or combinations chosen for a specific site, it is always prudent to avoid the big influences of steep or dramatically variable slope, or geological changes in the background. If this is not possible, the survey of small separate areas with similar conditions is often better than the survey of one complete area with these great influences or recent landscape changes. The quality of the survey interpretation is also dependant on how the speed of collected data varies in the case of continuous measurements for some methods and instruments. But with most of them, the quality of prospection in forested highland terrains is connected with the quality of the subsurface preservation of the archaeological situation still *in situ*. We are not always in time at sites, and irreversible afforestation changes do not always respect these sites, however the potential for archaeo-geophysical prospection in forested highland terrains exists and is not negligible.

References

Černá, E. 2016. *Středověké sklárny v severozápadních Čechách = Mittelalterliche Glashütten in Nordwestböhmen*. ÚAPPSZČ v Mostě, ARÚ Praha. Most.

Chvojka, O. and Kovář, O. 2016. Dvojí výročí jihočeské archeologie: první odborný archeologický výzkum v Plavu a první archeologická výstava v Českých Budějovicích před 150 lety. *Výběr: časopis pro historii a vlastivědu jižních Čech.* 53(3). 190-194. České Budějovice.

Křivánek, R. 1995. Příklady výsledků geofyziky při vyhledávání středověkých sklářských hutí v Krušných a Lužických horách. *Archeologické rozhledy*. 47(3). 486-500. Praha.

Křivánek, R. 1998. Ergebnisse geophysikalischer Messungen von mittelalterlichen Glashütten im Erzgebirge. In Materialhefte zur Archäologie Band 41, *Unsichtbares sichtbar machen - Geophysikalische Prospektionsmethoden in der Archäologie, Kolloquium* vom 27. October 1994 in Leipzig, Landesdenkmalamt Baden-Württemberg, 147-159. Stuttgart.

Křivánek, R. 2007. Possibilities and limitations of surveys by caesium magnetometers in forested terrains of archaeological sites. In I. Kuzma (ed.) - *Topics and Abstracts, 7th International Conference on Archaeological Prospection*, 11-15 September 11-15, 2007, Nitra, Slovakia, Archeologický ústav SAV Nitra, 202-204.

Křivánek, R. 2010. Přehled využití geofyzikálních měření v průběhu dílčích podprojektů a výzkumů v rámci výzkumného záměru Opomíjená archeologie v letech 2007-2008. In Krištuf, P. and Vařeka, P. (eds.) *Opomíjená archeologie 2007-2008 – Neglected Archaeology 2007-2008*, 160-171. Plzeň.

Křivánek, R. 2018. Archeogeofyzikální průzkumy Archeologického ústavu AV ČR Praha v jižních Čechách v letech 2014-2017. *Archeologické výzkumy v jižních Čechách* 31, Č. Budějovice, 345-361.

Sokol, P. 2016. Zaniklá středověká ves v poloze Horní Zábělá (okr. Plzeň-město). *Zprávy České archeologické společnosti, Suppl.* 101. 46-47.

Volf. M., Volf, V. and Opicová, K. 2017. *Zpráva o lokalizaci neznámé středověké sklárny na katastrálním území Boč*. Archaeological report ÚAPPSZČ. Most.

Non-destructive survey of early medieval ramparts in the Czech Republic and Slovakia

Peter Milo, Igor Murín, Michaela Prišťáková, Tomáš Tencer, Michal Vágner
Department of Archaeology and Museology, Masaryk University, Brno, Czech Republic

The investigation of early medieval centres is one of the principal tasks of Central European archaeological research. Despite the great importance of this issue there are still plenty of unanswered questions. One of these questions is concerned with the nature and development of fortification systems. In this task geophysical research could bring very significant results. As part of a scientific project solution, geophysical surveys have been conducted across 10 early medieval hillforts in the Czech Republic and Slovakia.

The main goal of these geophysical surveys was to determine the internal structure of the ramparts in the selected hillforts. The other aim was to check the efficiency of ground-penetrating radar (GPR), electrical resistivity tomography (ERT) and magnetometry on sites with early medieval fortifications. The combination of various geophysical methods will allow us to follow a wider spectrum of physical properties of individual ramparts. All geophysical surveys on ramparts have been carried out close to the area that was archaeologically explored in the past. This fact is an important advantage that enables us to compare the results from geophysical survey with results from previous archaeological excavations. The last aim of the project was to evaluate the individual method used on these surveyed sites. Two aspects are important: the natural environment and the current state of preservation. For comparison, we have hillforts situated in a hilly environment, whose ramparts are rich in stone material and at the same time they are relatively well preserved. On the other hand, there are hillforts situated in a lowland environment, whose ramparts consist mostly of a wood-clay construction with a smaller amount of stones. Due to agricultural activity, these sites tend to be, in principle, less preserved.

Some of the hillforts, especially those with an adjacent bailey, are surrounded by multiple, more complex, systems of fortification. Therefore the survey occurred on, at least, two different places for each site. For the purposes of the GPR survey, a single channel apparatus X3M Ramac (Geoscience Malå) with two shielded antennas with a central frequency of 500MHz and 250MHz was used. Where possible, we conducted a survey in polygon, where not, only single profiles were measured. The collected data were processed using the software RAMAC GroundVision (Geoscience Malå). We applied 2D ERT measurements using three types of configuration: Wenner, Schlumberger and dipole-dipole, with electrodes at a distance of 0.5m or 1m. The used lengths of profiles were 36m, 47m and 73m, depending on the demand of each site to cover by measurement the whole rampart. The measurements were carried out using an ARES resisitivity meter (GF Instruments Brno) and the collected data were processed with Res2Dinv software (Geomoto Inc.). The magnetic survey was carried out using a LEA MAX magnetometer (Eastern Atlas) and LEAD2 software. Topographic information for each survey was gathered using RTK GPS or Total Station.

Two presented case studies - from the hillforts Bíňa and Svätý Jur Neštich - show surveys from sites of two different characters. The site Bíňa is located in a flat area of lowlands. The fortification was built without using stone material. The GPR profiles show a relatively homogenous picture. Interesting results were also brought by the ERT survey. The results (Fig. 1) show that the rampart has higher values of electrical resistivity in the lower central part, which is covered by a layer of lower values above it. This is a result of construction phases, when the lower central part of the bank was made from top soil from the area of the ditch and covered by a fill of clay soil from the lower part of the ditch. The top part of the banks is made by loose and drained organic soil.

The hillfort Neštich is located in a hilly area and for the construction of the fortification, stone material from the immediate surroundings was used to a great extent. Both the ERT and GPR survey results (Fig. 2) show the inner structure of the rampart, with the remnants of stone destruction and the original bottom of the filled outer ditch in front of the rampart. The results also show that the surveyed ditch reached the bedrock. The rampart is made from material of the upper sediment, lower bedrock and possibly some clay fill.

The range of results achieved on individual ramparts is well above expectations. By combining individual methods, several structural elements were identified in detail for several ramparts. With the methods of ERT and GPR, the original levels of the terrain, the stone and clay cores of the ramparts were identified as well as the extent of the destruction of the stone walls, the character and depth of the outer ditches. Magnetometry in turn pointed to the presence of burnt parts of the ramparts as well as the differences between the ditch fills and the surroundings.

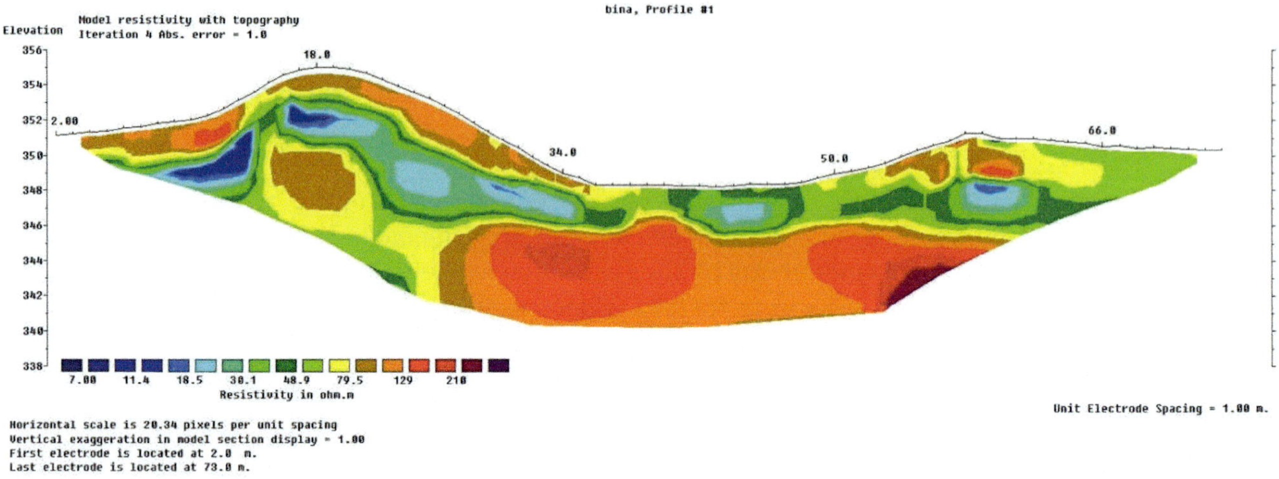

Fig. 1. Case study of ERT survey (Schlumberger configuration) on the early medieval hillfort Bíňa, located in the south-west region of Slovakia. The rampart has higher values of electrical resistivity in the lower central part covered by a layer of lower values above it.

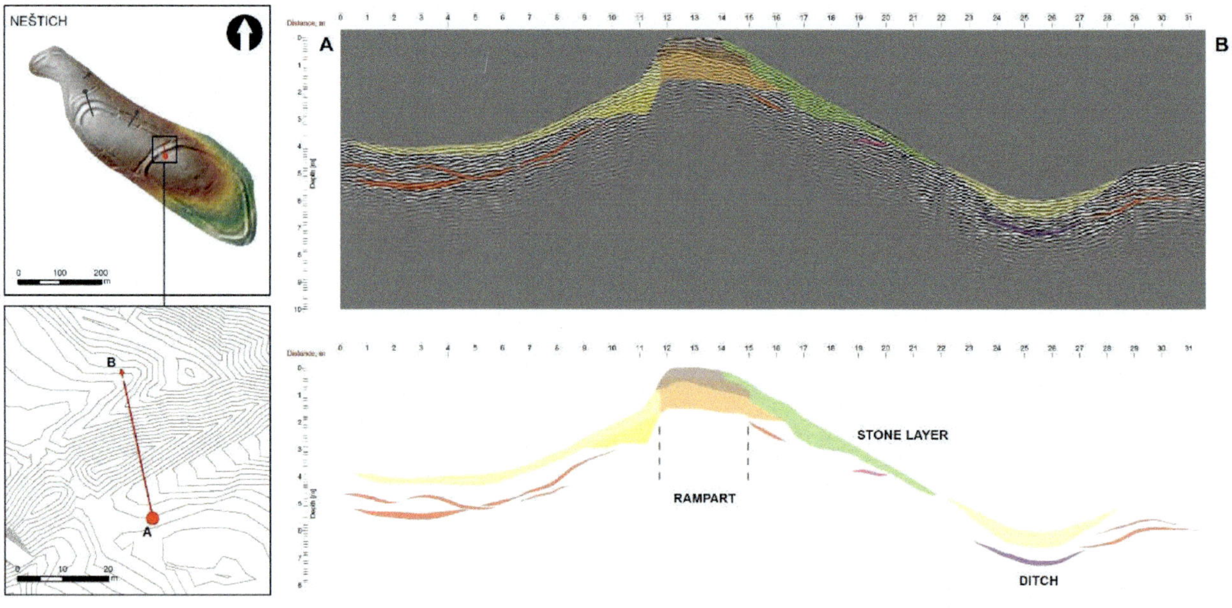

Fig. 2. GPR survey on early medieval hillfort Neštich, located in the south-west region of Slovakia. Results shows the inner structure of the rampart with the remnants of stone destruction and the original bottom of the filled outer ditch in front of the rampart.

Looking for military remains of the Battle of Gergovia: Benefits of a towed multi-frequency EMI survey

François-Xavier Simon[1], Guillaume Hulin[2], Yann Deberge[3], Marion Dacko[4]

[1]Inrap, UMR 6249 Chrono-Environnement, Besançon, France, [2]Inrap, UMR 7619 Metis, Paris, France, [3]Inrap, UMR 8546 Aoroc, Clermont-Ferrand, France, [4]University Clermont Auvergne, MSH, Intelespace, Clermont-Ferrand, France

Background

This study is part of a project focusing on the Gergovia plateau where Caesar besieged the Gallic army led by Vercingetorix. During the battle, which was one of the greatest of the Gallic Wars (BC 58-52) the oppidum of Gergovia was surrounded by two major Roman military camps (named the "Small camp" and the "Big camp") and other military ditches. Since the 19th century several archaeological excavations were conducted on the "Small camp" settled on a hill in front of the oppidum (Deberge *et al.* 2018). This site is located on the Limagne des Buttes area and is surrounded by low limestone mounts covered by basaltic deposits. The La Roche Blanche mount, where the camp is located, is made of limestone and silty clay formations with a few exogenous basaltic rocks occurring in the topsoil.

Eugène Stoffel first conducted a field evaluation in 1862 at the instigation of Napoleon III. In 1995-1996 archaeological excavations confirmed the discovery made in the 19th century, with a V-shaped ditch enclosing the top of the hill which was partially filled with a black organic earth deposit. In these ditches, several Roman republican weapons were found: ballista balls, catapult bolts and legionary shoe nails.

A geophysical survey was implemented on the "Small camp" in order to explore all the cultivated land of the site by mapping ditches partially excavated by the previous archaeological teams. Here we discuss the interest of electromagnetic induction (EMI) measurements compared to the magnetic and resistivity survey (Tabbagh 1984).

Methodology

A first geophysical campaign was performed in autumn 2017 in order to optimize the design of further survey for these geological contexts and archaeological remains. Magnetic, electrical resistivity, Ground-penetrating Radar (GPR) and EMI were then tested on the site. The most promising results came from the magnetometry and EMI techniques. Electrical resistivity data gave completely different results showing some resistive anomalies not directly connected to the known ditches. GPR results were considerably limited due to the high conductivity of the soil.

Based on this first trial, an extensive magnetic survey (Sensys MXPDA) was planned in 2018 over 13ha with some focused areas surveyed with EMI (multi-frequency Geophex GEM2). Both methods were towed by an ATV mounted on a cart for the magnetic device and a sledge for the EMI sensor (Fig. 1). All the geophysical measurements were positioned with a GNSS RTK system. Electrical resistivity measurements (Geoscan Research RM85) were also carried out along the main ditches. Magnetic, EMI and electrical resistivity data were processed with QGIS through AGT-Archaeological Geophysics Toolbox.

Results

All the methods used in 2018 confirmed the location of the ditches found in the 19th century by the Napoleonic team (Fig. 2, Fig. 3). New discoveries were also made from these surveys. Each geophysical dataset shows a 12m wide interruption along the main ditch combined with several anomalies. This could be interpreted as a door system similar to those known on the site of Alésia and other military camps (*clavicula*). Behind the ditch, a resistive and linear anomaly was clearly detected with the RM85. It could correspond to a limestone bank or a defensive embankment. In front of the ditch, several pits were mapped with the magnetic and EMI methods. A direct correlation between the military camp and these pits is not obvious and further ground investigations would be necessary to establish a clear relationship between them.

Fig. 1. Towed GEM2 system.

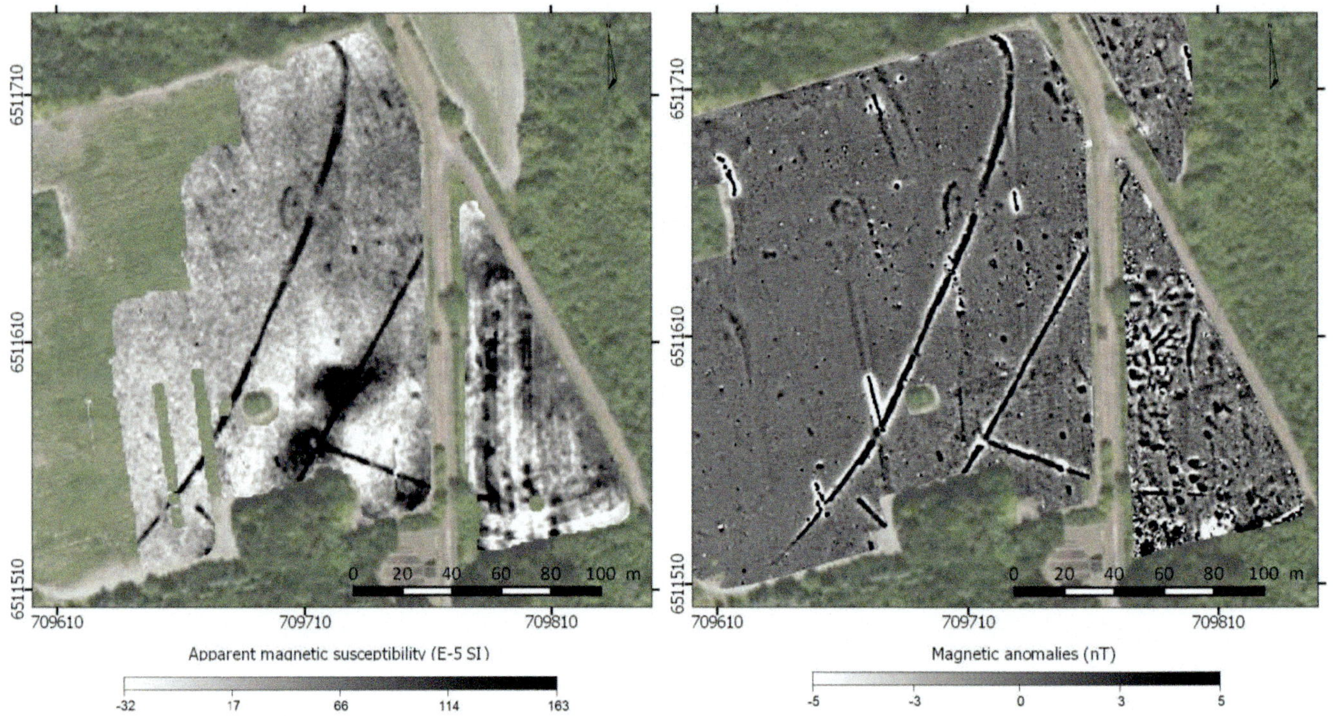

Fig. 2. Apparent magnetic susceptibility (left) and magnetic data (right) map of the central area of the site.

Discussion

For EMI measurements, data quality clearly depends on the stability and clearance of the device. Distortion of the frame, rolling and pitching of the sensors and ground clearance variations can affect the measurement. Use of a sledge clearly reduced the noise as observed between the data collected in 2017 with a handheld system and this collected with the sledge in 2018 (Fig. 1).

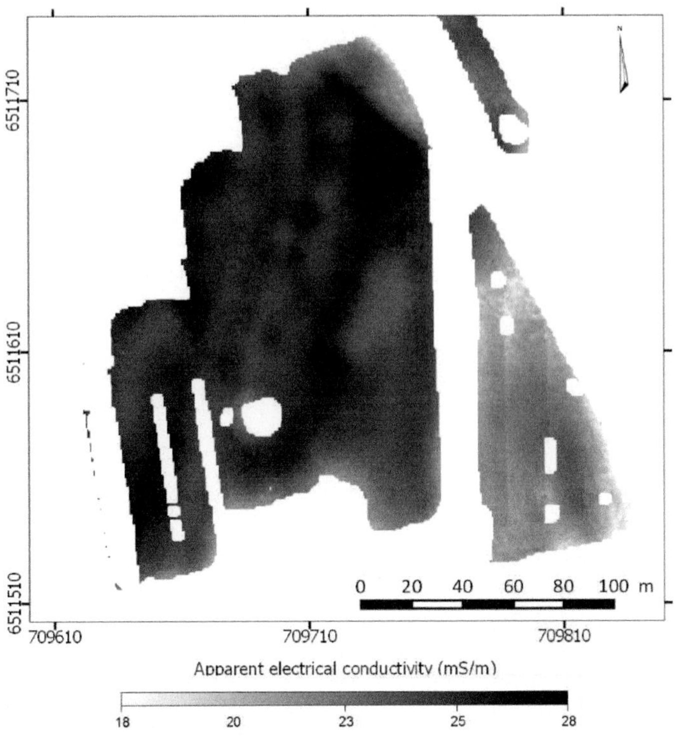

Fig. 3. Apparent electrical conductivity map of the central area of the site.

Based on this more suitable EMI data, it is possible to compare magnetometry and both the EMI-derived magnetic susceptibility and electrical conductivity maps. Several differences can be noticed. The first one is the detection, on magnetic susceptibility maps, of large anomalies probably corresponding to soil layers, and undetected by differential magnetometry as known by the results of modelling (Fig. 4). This difference comes from the very low sensitivity of magnetic methods to vertical contrasts contrary to EMI sensors, which are more sensitive to the filling. The second one is the difference of geophysical signature between both EMI and magnetometry surveys to detect magnetic features. Due to their shape or depth, archaeological pits with a magnetic infill can be clearly detected or not by one or both methods. The third one is the detection of electrical resistivity anomalies on the highest frequency (89kHz) corresponding probably to archaeological pits that are not visible on the magnetometry or magnetic susceptibility maps. We suspect here a very low magnetic contrast between the infilling and its surrounding but the existence of an electrical conductivity contrast. This last point reminds us, that pits and ditches are not necessarily magnetic even on a site offering several features with strong magnetization and low magnetic noise at the site scale.

Finally, the results from preliminary modelling demonstrate that negative values of apparent magnetic susceptibility could exist if the interpretation is based on a homogeneous half space approximation (the only one possible with single coil separation devices). This significant effect greatly reduces the possibility to proceed to a joint interpretation of the magnetic and EMI measurements for the GEM2 (Benech *et al.* 2002).

Conclusion

Despite some difficulties, resulting from the non-standard coils geometry of the GEM2 (Simon *et al.* 2015) (actually a *dHCP* for weighted differential HCP) some original information results from the comparison with magnetometry and resistivity. The shape of some magnetic features was better defined with EMI. Moreover, several features can be only detected with the electrical conductivity as they don't have any magnetic contrast. Consequently, EMI surveys help to better characterize some archaeological remains (for lenticular shapes, features with contrasting electric conductivity) and should be considered more often for the characterization of magnetic and electrical targets. Nevertheless, these benefits result from the use of a non-magnetic sledge with a high accuracy positioning system. The upgraded EMI data quality, associated with other geophysical methods, offers a more exhaustive view of the archaeological remains occurring on this site.

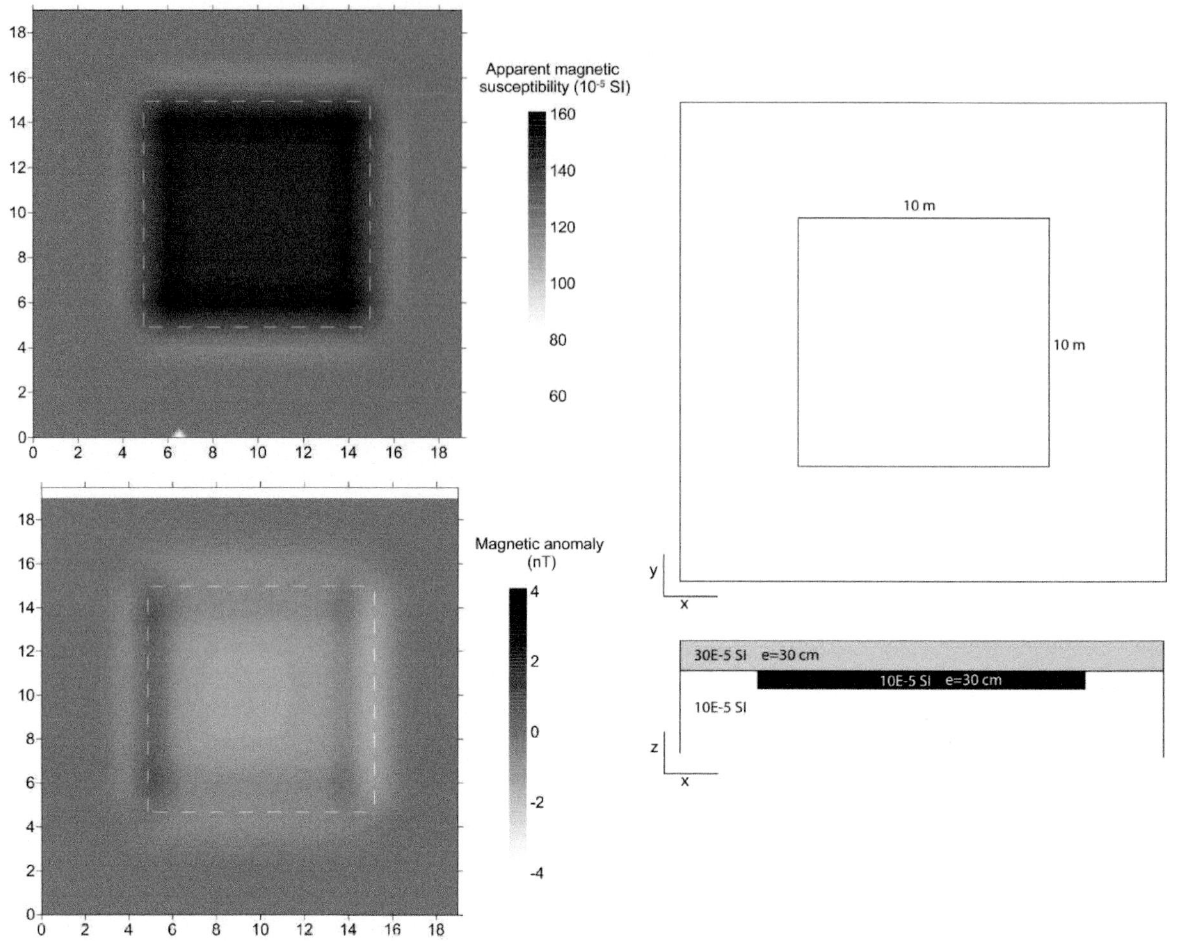

Fig. 4. Modelling of a magnetic thin plate of 0.3m width and 100E-5 SI magnetic susceptibility in a two layer model of homogeneous electrical resistivity at 30 Ohm.m. First modelling for a GEM2 working at 21,235Hz with 0.3m ground clearance and the second one for a fluxgate magnetometer (0.65cm length) at 0.3m ground clearance.

References

Benech, C., Tabbagh, A. and Desvignes, G. 2002. Joint inversion of EM and magnetic data for near surface studies. *Geophysics*. 67. 1729–1739.

Deberge, Y., Foucras, S., Pasty, J.F., Hulin, G., Jud, P., Leguet, D., Millet, M. and Simon, F-X. 2018. Présences militaires romains dans le bassin clermontois, In M. Reddé, *Les armées romaines en Gaule à l'opoque républicaine. Nouveaux témoignages archéologiques*, collection Bibracte. 28. 46-73.

Simon, F.-X., Sarris, A., Thiesson, J. and Tabbagh, A. 2015. Mapping of quadrature magnetic susceptibility / magnetic viscosity of soils by using multi-frequency EMI. *Journal of Applied Geophysics*. 120. 36–47.

Tabbagh, A. 1984. On the comparison between magnetic and electromagnetic prospection methods for magnetic features detection. *Archaeometry*. 26. 171–182.

First 3D reconstruction of the palaeoenvironment at the Mesolithic site of Duvensee, Germany, using geophysics and geoarchaeology

Erica Corradini[1], Dennis Wilken[1], Marco Zanon[2], Daniel Groß[3], Harald Lübke[3], Diana Panning[1], Katharina Rusch[1], Rebekka Mecking[1], Ercan Erkul[1], Natalie Pickartz[1], Ingo Feeser[2], Wolfgang Rabbel[1]
[1]*Institute of Geoscience, Christian-Albrechts-University, Kiel, Germany,* [2]*Institute of Pre-and Protohistoric Archaeology, Christian-Albrechts-University, Kiel, Germany,* [3]*Centre for Baltic and Scandinavian Archaeology (ZBSA), Schleswig-Holstein State Museums Foundation Schloss Gottorf, Schleswig, Germany*

Introduction

The ancient lake Duvensee is one of the best known archaeological sites of the Mesolithic in Germany. The ancient lake bed was formed during the Older Dryas as a dead ice hole; the lake reached its largest extent during the early Preboreal but over the course of the Holocene it gradually silted up and was overgrown by peat (Averdieck 1986: 407-410). Archaeological sites on the former lakeshore provide a vivid illustration of early Mesolithic life, with bark mats including evidence for the extensive use of hazelnuts (Groß *et al.* 2019). Today 23 different camps of Mesolithic hunter-gatherer groups are recorded in the Duvensee bog and 17 of them have been (partly) excavated. Radiocarbon dating sets the establishment of the first known Mesolithic camps within the Preboreal (9000 cal. BC), whereas the youngest Mesolithic occupation is dated to the early Atlantic period (6500 cal. BC). In this study we present the first reconstruction of the palaeoenvironment during the Mesolithic with the joint application of geophysical methods and stratigraphical information. The aim is to identify the location and extension of five former islands with Mesolithic camps already hypothesized (Groß *et al.* 2019, Bokelmann 2012: 369-380) using a large scale Ground-Penetrating Radar (GPR) Survey.

Methods

A GSSI GPR Antenna with 200MHz frequency was used to perform common offset data acquisition in the Duvensee area. A coarse rectangular grid of GPR profiles was established to investigate the palaeolandscape (Fig. 1).

A velocity analysis on diffraction hyperbolas throughout the area was carried out and migration was performed with a resulting average velocity of 4.5cm/ns for the entire survey. A further step was to recognize and pick the reflections associated with the main layers visible in several cores (Peat on top, Gyttja in between and Clay/Sand as bottom of the lake). This procedure has been done using the Kingdom IHS Software. After interpolation of the picked horizons, a map showing the depth of each layer was created.

Results

After the detection of the reflection coming from the main sediments a map showing the bottom of each layer was created. Fig. 2 illustrates the transition between clay and sand associated with the bottom of the lake. If we compare the map with the dwelling sites we see a very good match between the supposed islands (called 'islets', Bokelmann 2012: 369-380) and the archaeological evidence. The Mesolithic camps are indeed concentrated in the brown areas (islet 1 to 5) in accordance with the archaeological field study (Groß *et al.* 2019). From the GPR data we also created a 3D model of the lake area based on a study of the velocity of the radar waves with depth (Fig. 3). The 3D model includes spatial information of the main facies with time information from drillings and excavations together with a hypothetical regression of the water between the Late Preboreal and the Subboreal (Neolithic).

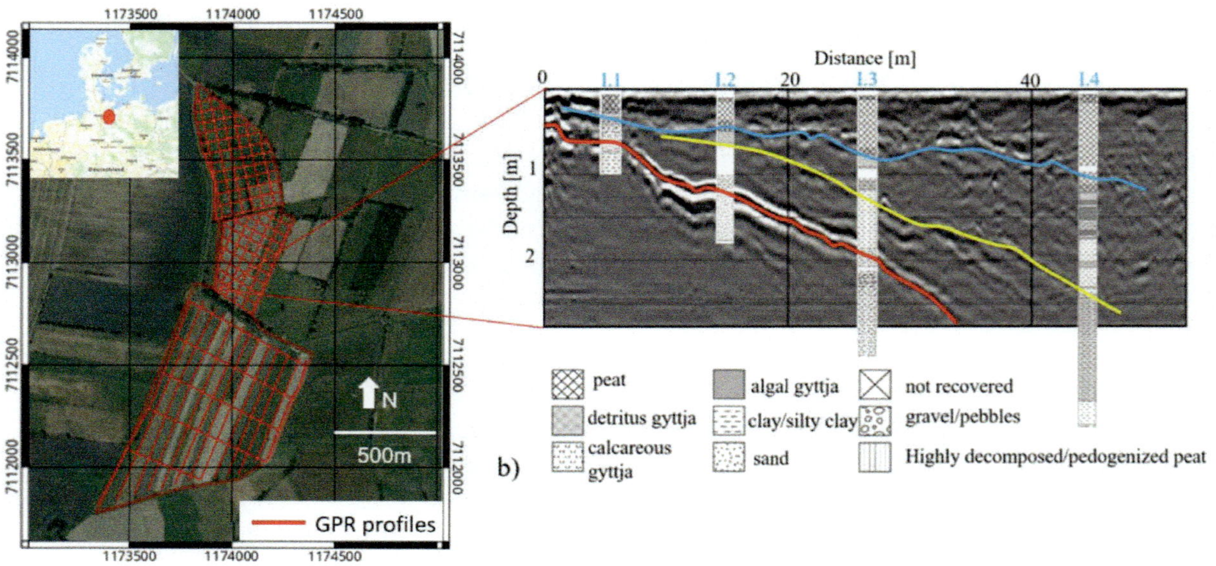

Fig. 1. (a) The investigated area of Duvensee with red lines indicating the GPR profiles. (b) Comparison between a radargram and the stratigraphy from the corings I.1, I.2, I.3 and I.4. The red line indicates the reflection coming from the bottom of the lake visible in the cores with the transition between clay and sand, yellow and blue lines indicate the clay layer and the peat on top.

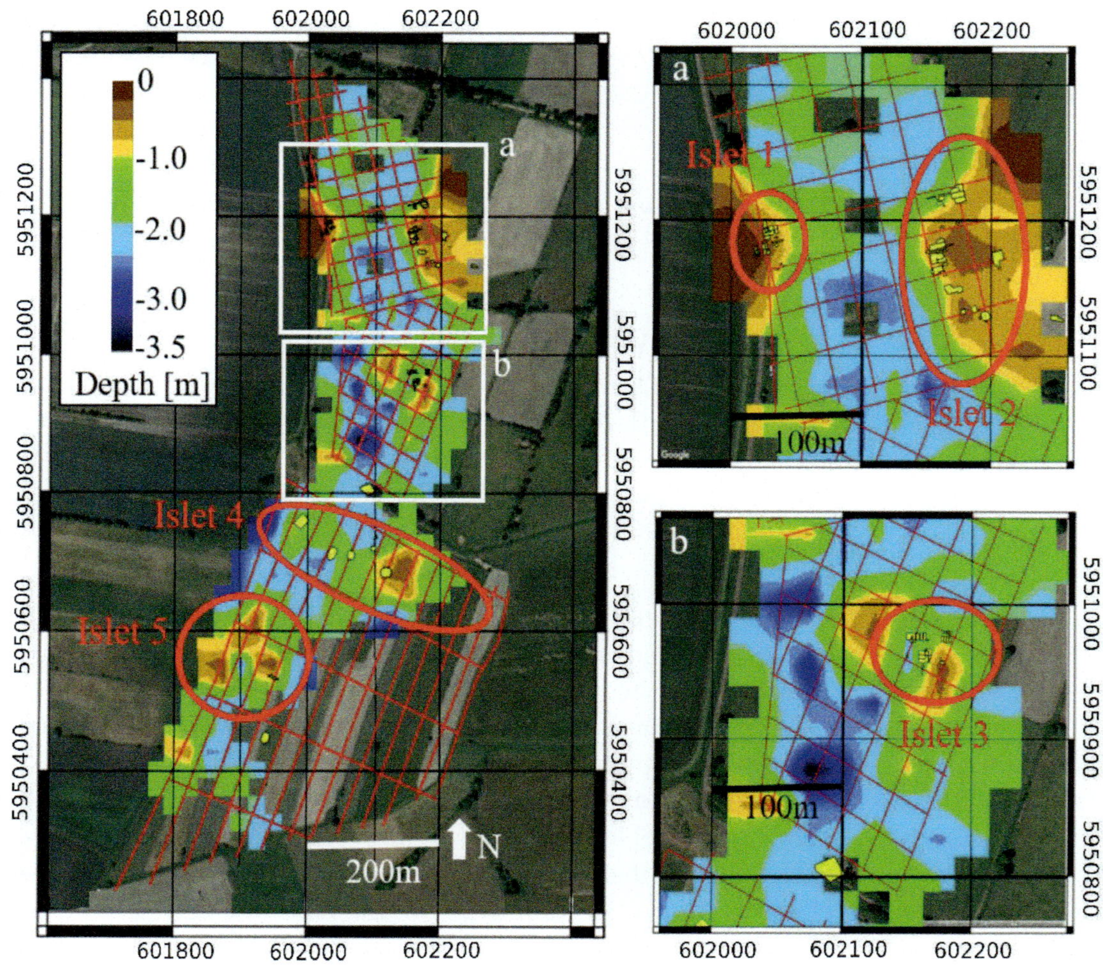

Fig. 2. The map of the surface related to the bottom of the lake. This illustrates a match between the brown areas (elevated palaeosurfaces, associated with the islets) and the Mesolithic camps (the yellow squares).

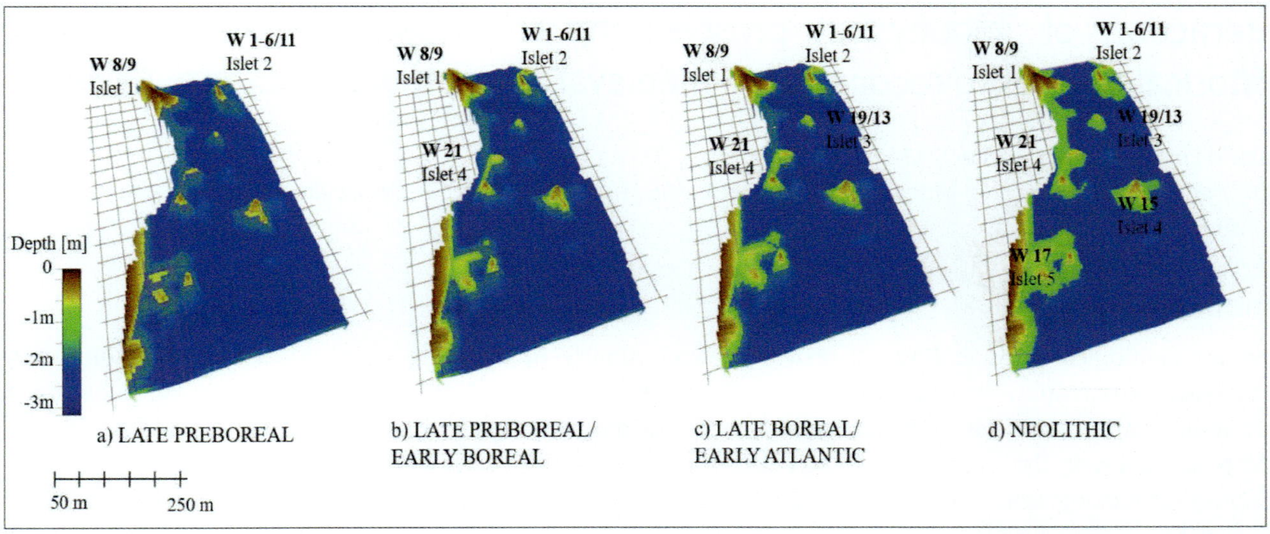

Fig. 3. Preliminary 3D reconstruction of the Duvensee area with a regression of the water level and the occupation of the islands by hunter-gathers ('W' indicates the Mesolithic camps: Wohnplätze).

Conclusions

The GPR survey gave good results for the detection and recognition of the islets already hypothesized from the archaeological research. These results allow us to apply the procedure in the future to unknown areas to enlarge the research at Duvensee for discovering new islands. As we can see from Fig. 3, the oldest Mesolithic camps are concentrated in the northern part of the ancient lake. A preliminary 3D reconstruction of the lake bottom also allows us to see that islets 1 and 2 were probably the first outside the water, which might be the reason for the first colonization during the lowering of the water level in the Preboreal. With the regression of the water, the other islets have been considered as a place for roasting sites at a later period.

Acknowledgements

We are grateful to the German Research Foundation (DFG) grant 2901391021 (SFB 1266) for funding the presented research.

References

Averdieck, F.-R. 1986. Palynological investigations in sediments of ancient Lake Duvensee, Schleswig-Holstein (North Germany). *Hydrobiologica.* 143. 407-410.

Bokelmann, K. 1971. Duvensee, ein Wohnplatz des Mesolithikums in Schleswig-Holstein, und die Duvenseegruppe. *Offa.* 28. 5-26.

Bokelmann, K. 2012. Spade paddling on a Mesolithic lake - Remarks on Preboreal and Boreal sites from Duvensee (Northern Germany), In Niekus, M.J.L.T., Barton, R.N.E., Street, M., Terberger, T. (eds.), *A mind set on flint. Studies in honour of Dick Stapert*. Groningen Archaeological Studies, Groningen. 369-380.

Groß, D., Lübke, H., Schmölcke, U. and Zanon, M. 2019. Early Mesolithic activities at ancient Lake Duvensee, Germany. *The Holocene.* 29(2). 197-208.

Neal, A. 2004. Ground-penetrating radar and its use in sedimentology: principles, problems and progress. *Earth-Science Review*. 66(3). 261e330.

Interaction of geophysical prospection, archaeological excavation and historical sources to reconstruct a medieval monastery in Southern Bavaria

Roland Linck[1], Florian Becker[1], Mario Hölzl[2]
[1]*Bavarian State Department of Monuments and Sites (BLfD), Munich, Germany,* [2]*X-Cavate Archaeology PG, Munich, Germany*

Introduction

Few archaeological sites in Bavaria provide the possibility of verifying geophysical survey results with subsequent excavation results. The reason is that normally the BLfD tries to preserve the subsurface archaeological remains without 'destroying' them via excavations. On the other hand, if there is planned construction work, the owner has to pay for the excavation anyway. Hence, often there is no preceding geophysical survey. Here we present one example of positive interaction results between geophysics and excavation.

Location and Historical Background of Survey Area

The region *Pfaffenwinkel* is located in Southern Bavaria and derives its name from the 41 monasteries founded there in the 8th century AD in the dukedom of *Baiern*. One of them is the Benedictine monastery of Schlehdorf, which is located on the shore of the Kochelsee (Hölzl 2018). The buildings of the first construction phase were destroyed in the 9th century AD, during the Hungarian invasion, and nowadays nothing is known about their exact location. Augustine friars built a second monastery in the middle of the 12th century AD. Again, the exact location is not recorded and the only hints come from 17th century copper engravings. Due to its position on the lakeshore, structural instabilities soon developed. Hence, the monastery re-located to a nearby hill in the early 18th century AD (Paula and Wegener-Hüssen 1994, Reichling 2011).

Survey Methodology

From the Baroque copper engravings by Wening and others, the location of the medieval building phase can be approximately assumed to be directly west of and partly under a modern rest home in the middle of the village of Schlehdorf. To verify this thesis, the BLfD executed a radar survey of all accessible areas in 2012 (Linck 2013a). Due to extensive technical disturbances near the survey area inside the modern settlement, and the expected stone remains in the subsurface, we chose a GPR measurement executed with a GSSI SIR-3000 and a 400MHz antenna.

As the water table of the Kochelsee is at the same height level than the survey area, the survey is strongly affected by groundwater. Thus, a Time-Domain-Reflectometry measurement of the soil moisture during the radar survey revealed quite a high value of 50% volume. Normally, the geological background of the late Würm glacial stage gravel could be expected to show a much lower value at the time of the survey in July 2012 (Linck 2013a).

Some years later, in 2017, a modern building replaced the old rest home. Therefore, the examination of the 'missing' eastern part of the monastery by archaeological excavations was also possible. In the central section of the site, a small area of the excavation overlaps with the GPR survey.

Results of GPR

The GPR depth slices in Fig. 1 show the archaeological remains at depths of 40cm-140cm below the modern surface. As the north-western most room of the monastery reveals an area of strong reflectivity between 80 and 100cm depth that can be interpreted as a preserved medieval floor, the anomalies depict preserved walls that are 40cm high (i.e. between 40cm and 80cm depth) with foundations from 100cm depth downwards. In fact, the footings must have originally reached much deeper to support a massive building like the monastery. However, the high soil moisture content did not allow a greater penetration depth of the electromagnetic waves (Linck 2013a).

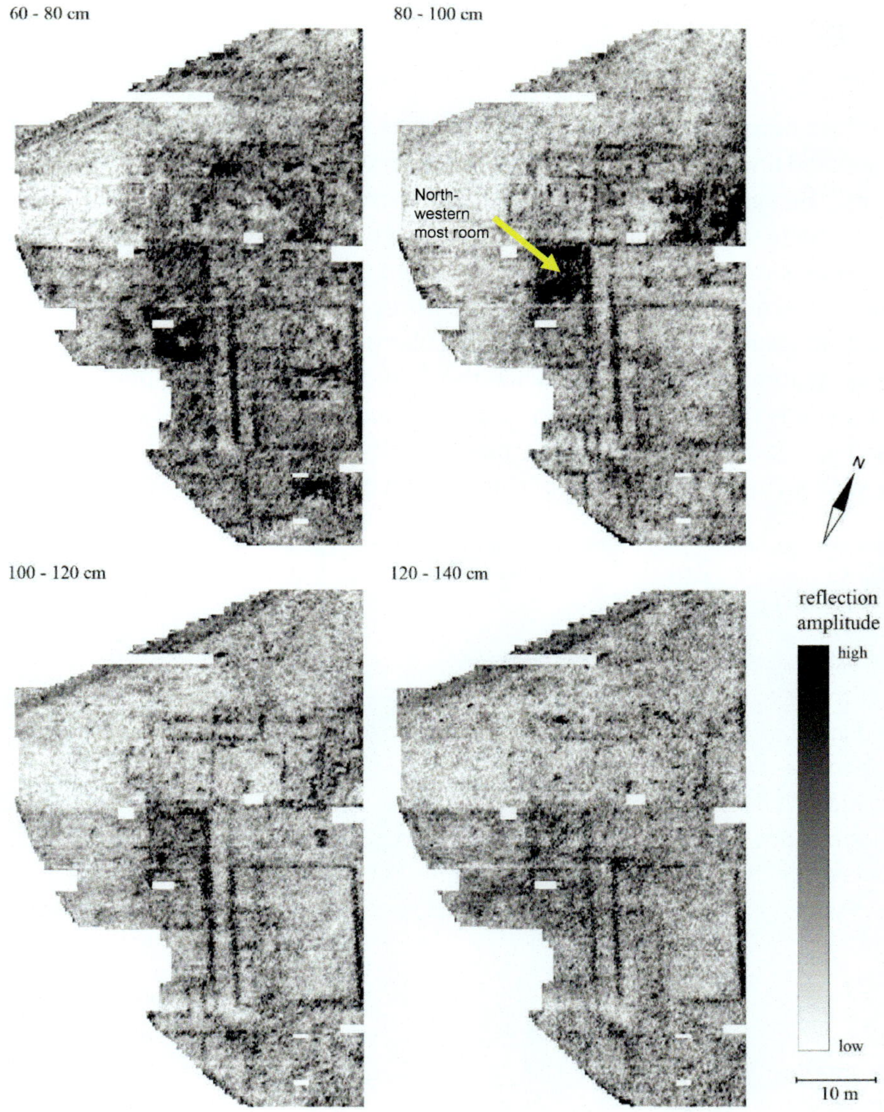

Fig. 1. Selection of depth slices of the GPR survey. GSSI SIR-3000 with 400MHz antenna, sample density: 2cm x 25cm, grid size: 67m x 46 m. The gaps in the depth slices are caused by trees.

The linear anomalies in the northern part of the survey area belong to a former Gothic church that was destroyed by fire in the late 18th century. It is nearly oriented in the typical east-west direction and has a small rectangular entrance building to the west (Fig. 1 and 2). The nave has a size of approximately 13m x 28m and may have been three-aisled, as several strong reflective anomalies in its interior can be interpreted as two rows of pillars. Another possible interpretation of these rectangular anomalies could be the northern continuation of an early medieval cemetery detected in the excavation further to the southeast. Other anomalies in the church's interior could belong to a crypt or to a predecessor of the monastery, as they lay below the supposed medieval floor level of 80cm depth. The results of the adjacent excavation also support the latter hypothesis. Outside the church, traces of the enclosure wall of the monastery as well as a small garden with a fountain, as shown in the copper engravings, are visible.

The actual monastery buildings are located to the south of the remains of the church. The radar survey covered the western and southern wings, which show a large number of small rooms, as well as two exterior bays: a rectangular one and a polygonal one. Within the monastery, a courtyard can be identified due to the lack of reflective anomalies in this area. Under the assumption that all wings around the central courtyard should have had approximately the same length, a total size of the monastery of 40m x 40m can be reconstructed.

Excavation results

Although the modern rest home disturbs the remaining parts of the monastery, the excavations executed after its demolition revealed unexpected and well-preserved archaeological remains in the areas not affected by modern installations. The eastern wing of the Augustine monastery was uncovered. Its preserved walls fit quite well with those mapped with GPR and almost no misalignment is detected (Fig. 2). The archaeological excavation exposed remains of shale paving tiles that caused the reflective anomaly in the corresponding depth slices. Additionally, several previously unknown features were found: in the south-eastern part of the planum, the remains of another small church with a circular apse (destroyed in 1684) are visible (already removed before the excavations seen in Fig. 2, since this figure shows a deeper phase of the excavations). The most surprising finding is the discovery of an octagonal building of possibly Carolingian date below the medieval monastery (Fig. 2); a detailed scientific dating of the corresponding findings is still ongoing. It measures 12m x 8m, with an internal diameter of the octagon of 5.5m (Hölzl 2018).

Fig. 2. Interpretation of the GPR survey (on the left side; red lines plotted onto modern surface) projected onto an UAV RGB-orthophoto of the excavation results (on the right side).

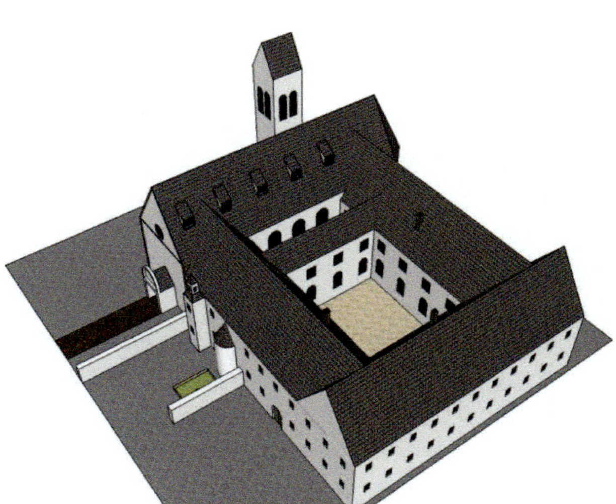

Fig. 3. Virtual reconstruction of the medieval monastery of Schlehdorf based on GPR, excavation results and copper engravings. View from the south.

Based on the results of the geophysical survey combined with a study of preserved copper engravings, a detailed digital reconstruction of the medieval monastery can be drawn (Fig. 3) (Linck 2013b). In this reconstruction, the layout of the buildings is based on GPR results, whereas the shape of the raised features, including roofing, windows, etc., was taken from the engravings.

Conclusion

A comparison of the results of the GPR survey with the excavated walls demonstrates that both methods fit together without significant misalignment. The small sections of masonry that cannot be identified in the GPR results, but which were uncovered by the excavation, lies within a part of the survey area that could not be covered due to modern installations. Hence, it is established that the geophysical survey of the site generally provided a useful image of the buried remains without the need to excavate each archaeological site. Thus, it is possible to accommodate the need for non-destructive heritage protection. In these cases, further geophysical methods have to be applied as an integrative survey to cover all types of archaeological anomalies, e.g. graves and/or ditches and pits, and not only stone walls.

Taking into account that the geophysical survey, as well as the excavation, revealed building structures below the floor of the medieval monastery, it can be assumed that the second phase of construction of the monastery has now been documented. In addition, an earlier, possibly Carolingian, set of structures has been discovered, for which no evidence had previously been found.

References

Hölzl, M. 2018. Ein Oktogon mit Rechteckapsis: dem frühmittelalterlichen Kloster von Schlehdorf auf der Spur. *Das Archäologische Jahr in Bayern 2017*. 87-91.

Linck, R. 2013a. *Methodische Untersuchungen zur Weiterentwicklung der Boden- und Satellitenradarprospektion in der Archäologie*. 109-114. Aachen.

Linck, R. 2013b. Virtual reconstruction of a medieval monastery in Bavaria based on geophysical prospection and old copper engravings. *ISAP News*. 37. 5-7.

Paula, G. and Wegener-Hüssen, A. 1994. *Denkmäler in Bayern I.5, Oberbayern. Landkreis Bad Tölz-Wolfratshausen*. 474. Munich.

Reichling, A. 2011. Die Orgeln der Pfarrei Schlehdorf und der Pfarrei Ohlstadt und ihre Geschichte. *Veröffentlichung der Gesellschaft der Orgelfreunde.* 253. 5.

The sunken trade centre of Rungholt – Geophysical investigations in the German North Frisian Wadden Sea

Michaela Schwardt[1], Dennis Wilken[1], Hanna Hadler[2], Annika Fediuk[1], Tina Wunderlich[1], Andreas Vött[2], Wolfgang Rabbel[1]

[1]*Institute of Geosciences, Christian-Albrechts-University Kiel, Kiel, Germany,* [2]*Institute of Geography, Johannes Gutenberg University Mainz, Mainz, Germany*

Introduction

The coastal landscape of North Frisia has been exposed to ever-changing environmental conditions and anthropogenic impacts. Newly reclaimed marshland, obtained by extensive dike building in the beginning of the 12th century, offered the opportunity for new settlements beyond the Geest margins. However, this dike building had major consequences on the coastal area: first, as the formerly flooding plains were now restricted, the tidal range was increased, and second, the dewatering of the cultivated land caused a subsidence of the ground below the mean high tide. During the Middle Ages, the coastal regions were affected by extreme storm surges and large parts of the formerly settled and cultivated area were taken by the sea and turned into tidal flats (Hadler *et al.* 2018).

In the 12th and 13th centuries, the newly reclaimed marshland around today's Hallig Südfall, between the islands of Pellworm and Nordstrand, belonged to the historical administrative district of Edomsharde, with the main settlement of Rungholt (Fig. 1). During the 1st Grote Mandrenke in 1362, large areas of the low-lying marshland, including the settlement of Rungholt, were flooded and turned from cultivated land to tidal flats (Hadler *et al.* 2018a). Although only a little is known about Rungholt and the medieval cultivated landscape, some remains are still visible in the Wadden Sea to the present day.

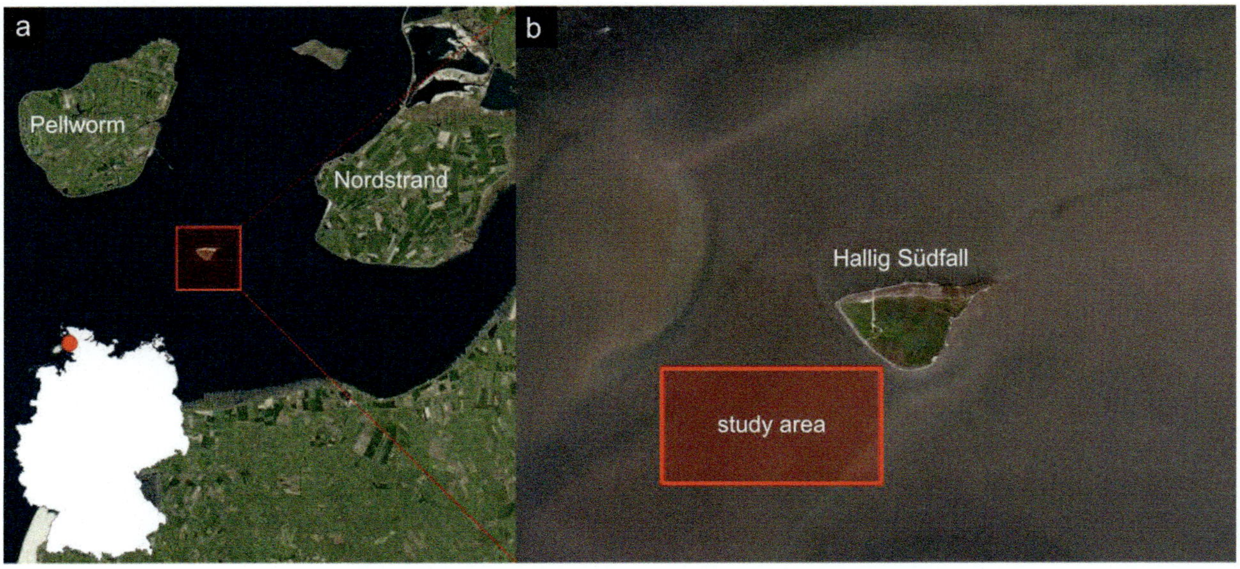

Fig. 1. Areas of investigation. (a) Aerial image of the coastal area between the islands of Pellworm, Nordstrand and Eiderstedt. (b) Aerial image of Hallig Südfall and the study area.

Based on previous surveys, aerial photographs, and historical maps, an area of the Wadden Sea south of Hallig Südfall was geophysically investigated (Fig. 1). The objectives of the investigations are three-fold:

1. Identification of possible remains in the Wadden Sea.
2. The reconstruction of the palaeolandscape of the former cultivated area.
3. Preservation of information about the medieval landscape in a constantly changing environment dominated by tidal-dynamics.

Methods

Geophysical investigations were carried out on the tidal flats south and north of Hallig Südfall by means of magnetic gradiometry during low tide and high-resolution marine reflection seismics during high tide.

Magnetic gradiometry

For the magnetic measurements, we used an array of four to six Foerster fluxgate differential vertical component magnetometers with an internal sensor spacing of 0.65m and a horizontal spacing of 0.5m. Positioning was achieved using a RTK-DGPS (real time kinematic differential global positioning system, Leica 500), which yields an accuracy of 0.01m–0.02m. The mean of the measured magnetic data is subtracted from the data measured by each sensor for every profile, followed by an interpolation of the whole dataset onto a 0.2m x 0.2m grid. In total, 13ha south of the Hallig Südfall were investigated.

High-resolution marine reflection seismics

To obtain cross-sections of the detected magnetic anomalies and take a closer look at the geomorphology, high-resolution marine reflection seismics were applied. We used one acoustic source (ELAC NAUTIC TL-444) with a frequency of 2kHz-8kHz firing eight times per second and two hydrophones (INNOMAR) to record the signals, both mounted on a frame in front of a small boat. Positioning was achieved with a RTK-DGPS (Leica 500). In total, reflection seismic profiles with a total length of 50km were acquired. The data were bandpass filtered, followed by a coherency filter to smooth the seafloor. Then, the seafloor and signals above were muted. Lastly, a dip filter was applied.

Results

The resulting magnetic map (Fig. 2) is characterized by anomalies of very low amplitudes (±0.5nT). When compared to historical maps, the observed magnetic anomalies can be associated with the remains of a medieval dike and former rectangular dwelling mounds (terps), either attached to the inner side of the dike or located separately inside the diked land, connected to channel or sewer systems. Furthermore, two anomalies might indicate the location of two tidal gates (Busch 1923, Hadler *et al.* 2018b).

The reflection seismic profiles show cross-sections of the magnetic anomalies and provide a close look at the geomorphology of the site. South-west of Hallig Südfall, the profiles show a former tidal creek, which eroded most of the cultural traces, also visible as a decrease in the strength of the magnetic anomalies. Further to the east, remains of the medieval dike and dwelling mounds can be recognized in several seismic profiles. These remains are characterized by trough-shaped reflectors, representing deformed and compacted sediment layers due to the load of the dike. The deformed layers, which are the imprint of the dike, could be identified in the whole area, although the remains of the dike itself were eroded away by the tidal activities over the past 600 years.

Conclusions

The application of geophysical investigations in the Wadden Sea helped to identify and map several medieval remains of the cultivated area. Based on the information gained by the magnetic and seismic investigations, the palaeolandscape was reconstructed. It is shown that large areas of the present-day tidal flats were formerly cultivated and settled land and that after more than 600 years some remnants are still preserved. Furthermore, evidence for a growing endangerment of the cultural remains by the constantly changing environment dominated by tidal-dynamics was found, showing the need for further investigations and preservation of information about this area.

Acknowledgements

This work was funded by the German Research Foundation (DFG) within the frame of the Priority Program 1630 "Harbours from the Roman Period to the Middle Ages" (von Carnap-Bornheim and Kalmring 2011).

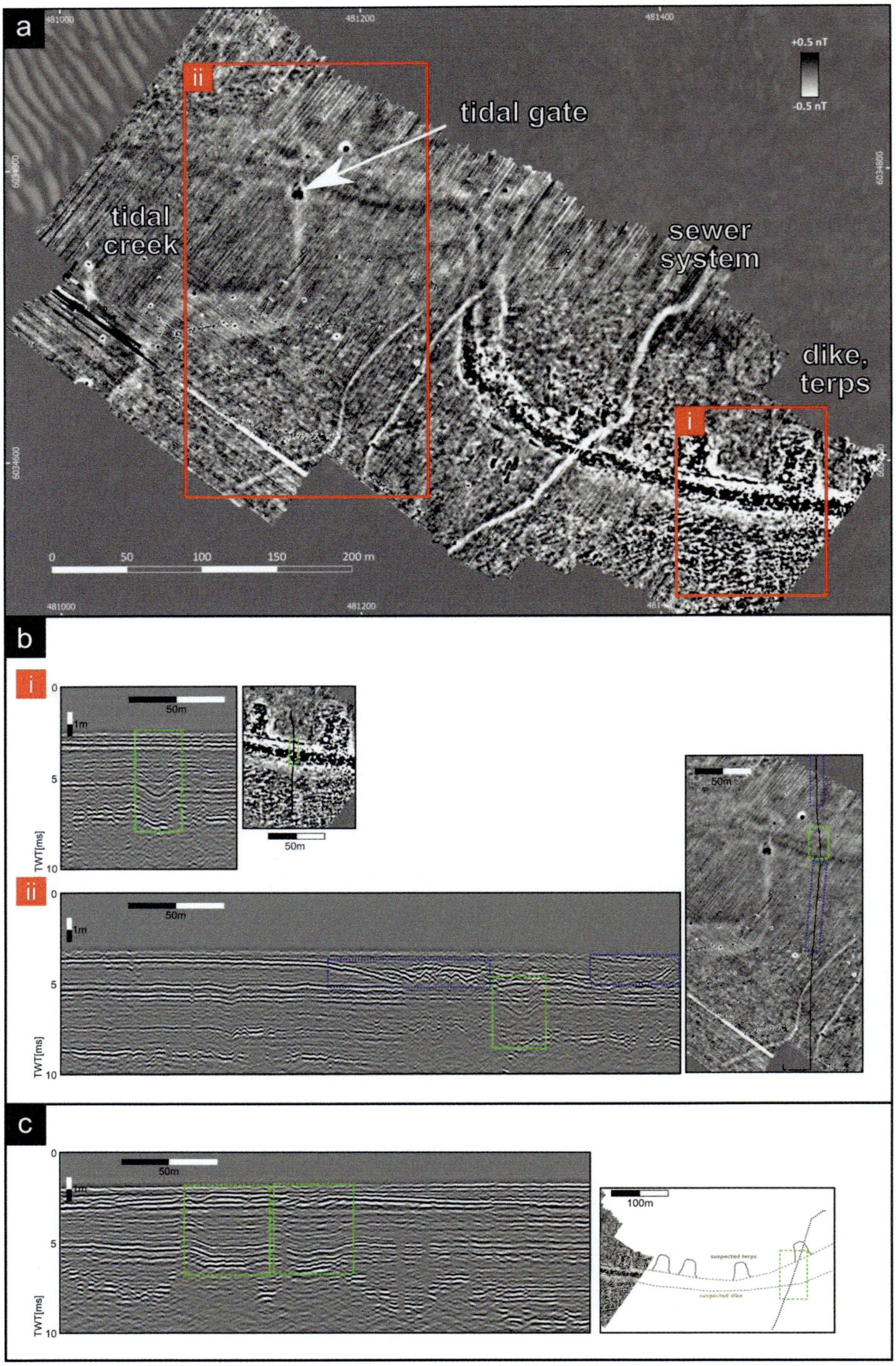

Fig.2. (a) Magnetic anomaly map of study area 1. Several features like the dike, terps, the tidal gate and sewer systems can be identified. (b) Close-up of two selected areas and comparison with seismic profiles. In close-up (i) the imprint of the former dike is seen as trough-shaped reflector, marked in green. In close-up (ii) the seismic profile shows the former tidal creek (blue) and the dike's imprint. (c) A seismic profile further to the east crossing a dike and terp, characterized again as trough shaped reflectors.

References

Busch, A. 1923. Die Entdeckung der letzten Spuren Rungholts. *Jahrbuch des nordfriesischen Vereins für Heimatkunde.* 10. 1-32.

Hadler, H., Vött, A., Newig, J., Emde, K., Finkler, C., Fischer, P. and Willershäuser, T. 2018a. Geoarchaeological evidence of marshland destruction in the area of Rungholt, present-day Wadden Sea around Hallig Südfall (North Frisia, Germany), by the Grote Mandrenke in 1362 AD. *Quaternary International.* 473. 37-54.

Hadler, H., Wilken, D., Wunderlich, T., Fediuk, A., Fischer, P., Schwardt, M., Willershäuser, T., Rabbel, W. and Vött, A. 2018b. Drowned by the Grote Mandrenke in 1362 AD - new geo-archaeological research on the late medieval trading centre Rungholt (North Frisia). In L. Egberts and M. Schroor (eds.), *Waddenland Outstanding. History, landscape and cultural heritage of the Wadden Sea Region*. Amsterdam University Press, Landscape and Heritage Studies, 239-251.

3-D Resistivity Imaging of Rock-Cut Chamber Tombs: the case of the Mycenaean Cemetery in Prosilio, Greece

Nikos Papadopoulos, Dimitris Oikonomou, Gianluca Cantoro, Nikos Andronikidis, Apostolos Sarris
GeoSat ReSeArch Lab, IMS-FORTH, Rethymno, Greece

Introduction

The excavation activities undertaken in the summer of 2017 at Prosilio in Boeotia, Greece between the Ephorate of Antiquities of Boeotia and the British School at Athens/University of Cambridge brought to light a monumental Mycenaean rock-cut chamber tomb, which is one of the largest of its kind ever discovered in Greece. A rock-cut passageway (*dromos*), 20m in length, leads to a monumental façade 5.40m in depth. The façade gives access, through a doorway (*stomion*), to the burial chamber, which has an area of 42 square metres, making this example the ninth largest out of *c.* 4000 Mycenaean chamber tombs that have been excavated in Greece in the last 150 years. Furthermore, another rock-cut passageway, with similar size, was found at a distance of 15m from the above-mentioned one leading to a tomb with similar dimensions to the already excavated one.

Based on the above significant archaeological evidence and findings, a geophysical prospection survey was planned and completed in autumn 2017 in an effort to investigate the possibility for the existence of other tombs in the area. The geophysical method of 3-D Electrical Resistivity Tomography (ERT) was considered as the optimum technique to meet the specific goals of the project considering the nature and dimensions of the targets being mapped in combination with their burial depth. The strategy used was to collect a dense network of parallel 2-D lines and subsequently apply 3-D resistivity inversion to reconstruct the subsurface resistivity structure in a 3-D context, based on the approach described by Papadopoulos *et al.* (2007). A total area of almost 10,000 square metres was covered with ERT along 86 lines with a cumulative length of 5.8km (Fig. 1).

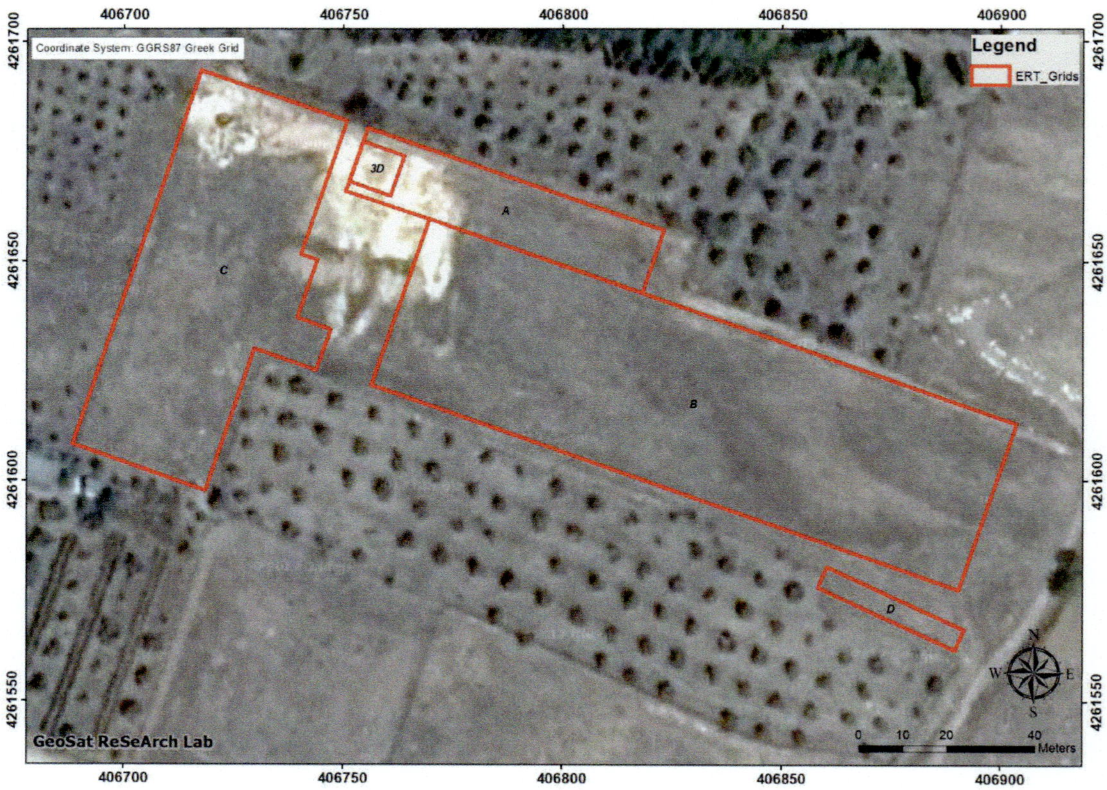

Fig. 1. Outline of the four grids which were laid out in Prosilio to complete the 3-D ERT survey. Axes units in metres. The above area was scanned with parallel lines oriented along the NW-SE direction.

Methodology

The elevation of the site was recorded by moving the rover GNSS unit along almost parallel NW-SE tracks mapping a total area of more than 12,500 square metres with an average horizontal spatial resolution of 2m. The interpolated digital elevation model ranged between 123m above mean sea level at the southern part of the site and 144m towards the northern part, with an average elevation of 136m.

The dipole-dipole electrode configuration was used to complete the ERT survey along west-east individual and parallel transects in different Grids (A, B, C and D). The unit electrode distance, a, along the lines was 1m or 2m and the N separation (the ratio of the distance between A-B dipole to M-N dipole length) ranged from 1 to 8. In order to increase the investigation depth and the recorded signal, additional data with unit electrode spacing "2a", "3a" and "4a" were collected.

A systematic method for data pre-processing was followed for all the individual 2-D survey lines. After de-spiking outlier values (due to high geometric factors, small potential values or values collected with insufficient current) the data were exported as resistance values. Afterwards the 2-D lines were combined to a single data file and a full 3-D inversion program (Loke and Barker 1996) was used to reconstruct the subsurface resistivity variation imposing robust constraints in the inversion routine (Loke *et al.* 2003). The topography along each line was extracted from the digital elevation model and incorporated in the inversion procedure to correct the topography effect.

Results

The depth slice of 0.5-1.0m from Grid A (Fig. 2, left) shows six conductive linear anomalies that all have the same S-N orientation and common width (~1.5m). These are correlated with the already known roads that are clearly visible in the respective orthophoto. Likewise, similar conductive features outline the location of the S-N roads in the northern section of Grid C (Fig. 2, right). It is evident that ERT is highly effective in mapping these buried roads as conductive targets due to the fine moisturized soil that fills the rock cut roads. Within the deeper slices (Fig. 3), the resistivity image reveals the resistive region T2 that is attributed to the chamber of the excavated tomb. The horizontal dimensions of the chamber based on the ERT results are 3.6m by 6.0m. The road leading to the chamber of the tomb is characterized by high resistivity values since the filling material had been removed by the time the ERT data were collected. At a distance of about 14m west of the known tomb, the ERT survey crosses the location of a second tomb (T3) showing a resistive structure with dimensions ~4m by 4m within the depth range of 2.0m to 4.5m. The strong resistive signature of the respective anomaly is comparable to the anomaly of the known tomb, indicating the relatively good preservation level of the buried target.

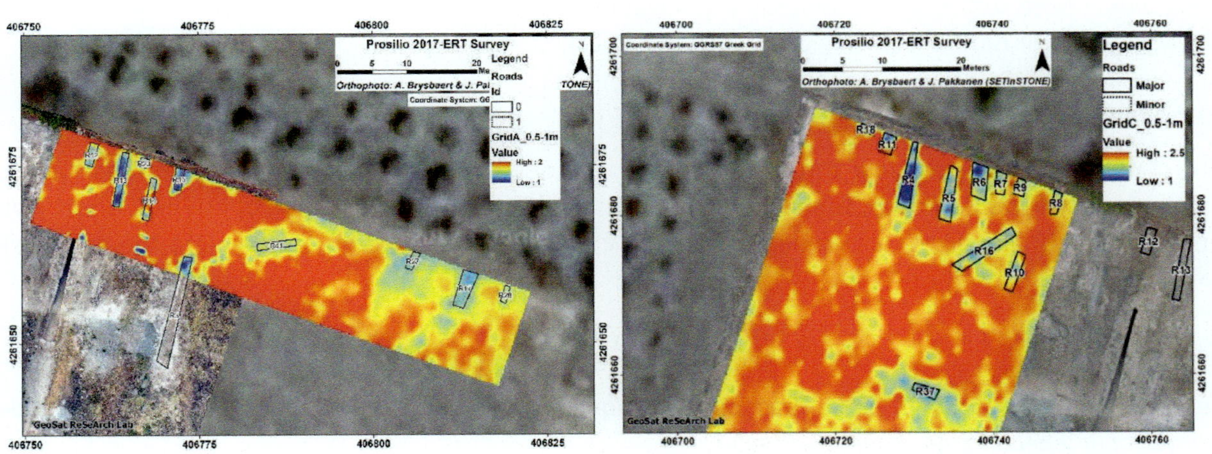

Fig. 2. Resistivity depth slice of 0.5-1.0m from Grid A (left) and the northern part of Grid C (right). Axes units in metres and resistivity in Ohm.m.

The space between tombs T2 and T3 is occupied by the resistive anomaly T4 that appears in similar depth below the ground but with smaller dimensions (2 x 2 x 2m). However, the shape and its resistive signature gave a possible indication of another smaller tomb (Fig. 3, right).

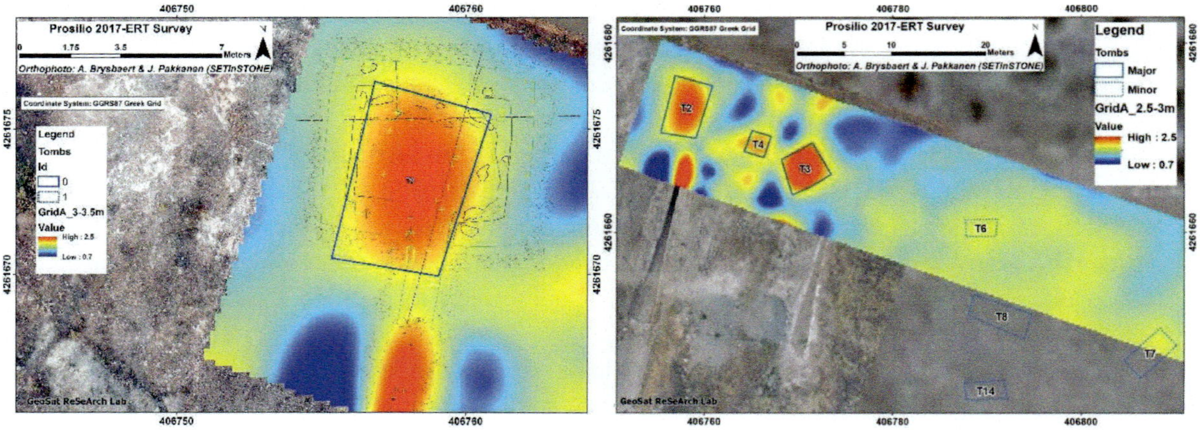

Fig. 3. (left) Resistivity depth slice of 3-3.5m over the known excavated tomb T2. (right) Resistivity depth slice of 2.5-3m from Grid A. Axes units in metres and resistivity in Ohm.m.

Conclusions

In archaeological geophysics, tombs comprise the most common subterranean human-made cavities and in cases where looting has not taken place, they may contain important findings of great historical and economical value. Several successful case studies employing various geophysical methods in the detection of tombs have been reported in the literature (e.g Nyari and Kanli, 2007). However, most of the reported examples refer to structures buried at small depths in relation to their dimensions, relatively flat terrain and two-dimensional approaches (Testone *et al.* 2018). The approach of the geophysical strategy for the effective mapping of tombs within a fully 3-D context was been investigated thoroughly by Papadopoulos *et al.* (2010) and Tsourlos *et al.* (2014).

The 3-D ERT survey in Prosilio was able to indicate a number of candidate targets, thus completing to a certain degree the picture for the burial site. The ERT depth slices up to 1m below the ground floor outlined mainly linear conductive anomalies related to the soil material of the roads leading to known or potential chamber tombs. The results verified the existence of another tomb with comparable dimensions, burial depth and preservation level with the excavated one. Furthermore, the space between these tombs seem to be occupied by a smaller unknown tomb. The remaining area is also characterized by a number of resistive targets, which need further archaeological investigation, either with test trenches or with small drillings, to verify whether their nature is geological or archaeological.

Acknowledgements

The research was made possible through the funding allocated by the University of Cambridge and the assistance of Dr Yannis Galanakis, Senior Lecturer in Classics (Classical Art & Archaeology). The Archaeological Ephorate of Boeotia is highly acknowledged for the granting the permission to complete the geophysical survey in Prosilio.

References

Loke, M.H. and Barker, R.D. 1996. Practical Techniques for 3D Resistivity Surveys and Data Inversion. *Geophysical Prospecting.* 44. 499-523.

Loke, M.H., Acworth, I. and Dahlin, T. 2003. A comparison of smooth and blocky inversion methods in 2D electrical imaging surveys. *Exploration Geophysics.* 34. 182-187.

Nyari, Z. and Kanli, A.I. 2007. Imaging of Buried 3D Objects by using Electrical Profiling Methods with GPR and 3D Geoelectrical Measurements. *Journal of Geophysics and Engineering*. 4. 83-93.

Papadopoulos, N.G, Tsourlos, P., Tsokas, G.N. and Sarris, A. 2007. Efficient ERT Measuring and Inversion Strategies for 3D Imaging of Buried Antiquities. *Near Surface Geophysics.* 5(6). 349-362.

Papadopoulos, N.G, Yi, M.Y., Kim, J-H., Tsokas, G.N. and Tsourlos, P. 2010. Geophysical Investigation of Tumuli by means of Surface 3D Electrical Resistivity Tomography. *Journal of Applied Geophysics.* 70. 192–205.

Testone, V., Longo, V., Mameli, P. and Rovina, D. 2018. Geophysical prospection pilot in rock-cut tombs: The case study of Anghelu Ruju necropolis (Sardinia). *Archaeological Prospection*. 25. 271– 277.

Tsourlos, P., Papadopoulos, N., Yi, M-J., Kim, J-H. and Tsokas, G. 2014. Comparison of measuring strategies for the 3-D electrical resistivity imaging of tumuli. *Journal of Applied Geophysics.* 101. 77-85.

Manifestation of the urban design of ancient cities in northern Greece by archaeological prospection

Gregory N. Tsokas, Panagiotis Tsourlos, George Vargemezis
Exploration Geophysics Lab., Aristotle University of Thessaloniki, Thessaloniki, Greece

Introduction

Geophysical prospecting is widely used for archaeological research all over the globe. It comprises an integral part of archaeological survey and consists of an ensemble of different methods, each of which are based on the contrast of physical properties between the buried ancient structures and vestiges with respect to the soil that covers them. The discernible physical properties create anomalies that allow for the detection and mapping of buried antiquities and in some cases imaging, if specific conditions are fulfilled and appropriate methods have been used.

This process allows for an assessment of the identity, dimensions and the burial depth of subsurface structures. Consequently, these methods are capable of imaging buried vestiges which are usually found in the remains of foundation walls. Therefore, the urban plan of ancient cities can be revealed by the layout of the vestiges.

This paper does not aim to compare geophysical methods nor assess their attributes regarding resolution, depth penetration, technical merits, drawbacks, etc. This paper will clearly assess the ability of geophysics to detect structures which were important parts of the life of ancient cities (e.g. public buildings, theatres, temples agorae, etc.). By locating and studying the ruins of such features, important parameters of the life at ancient cities can be revealed.

Geophysical prospection as a tool for the manifestation of the urban design of ancient cities

It is possible, under favourable conditions, to obtain an image which resembles the result that would have been drawn if an excavation had taken place (Scollar *et al.* 1986). An example is given in Fig. 1 where the ground view of the buried remnants of ancient foundation walls is easily recognized. The example (Tsokas *et al.* 1994) is drawn from the geophysical prospection at the site of ancient Europos in Kilkis Prefecture (Region of Macedonia, northern Greece). It is immediately evident that the ancient urban complex is revealed in this map. It is further evident that two construction phases are seen, presumably reflecting the last two phases of city life. The use of such images allows archaeological characterization of the imaged subsurface structures (e.g. function of the buildings, identification of domestic or craft areas, empty spaces, etc.).

The city of Europos was the homeland of Seleucus I Nicator who founded the Dura-Europos in Syria, presumably honouring his origin. The urban design of that city was revealed by the survey leaded by Christophe Benech (2007). Generally, studies such as these lead to valuable conclusions about the spatial organization of ancient urbanization (Donati *et al.* 2017). Further, the swift and relatively cheap attribute of the geophysical survey add to the advantages of the approach.

Electrical Resistivity Tomography (ERT) can be used but this kind of survey needs more effort and time than classical resistivity and magnetic (both total field and gradiometry) mapping techniques. Of course, the ERT survey is superior since it can provide depth slices and 3-D subsurface imaging. This is evident in the next example.

Fig. 2 (Bonias *et al.* 2017) shows the distribution of the resistivity at a 1m depth level at ancient Argilos (Region of Macedonia, northern Greece, at the mouth of river Strymon). The distribution was inferred from the 3D inversion of the data. The high resistivities form very clear patterns. In fact, these patterns are linear, and they tend to form rectangular shapes. Therefore, they are attributed to subsurface resistive features such as the ruins of ancient foundation walls or other antique structures.

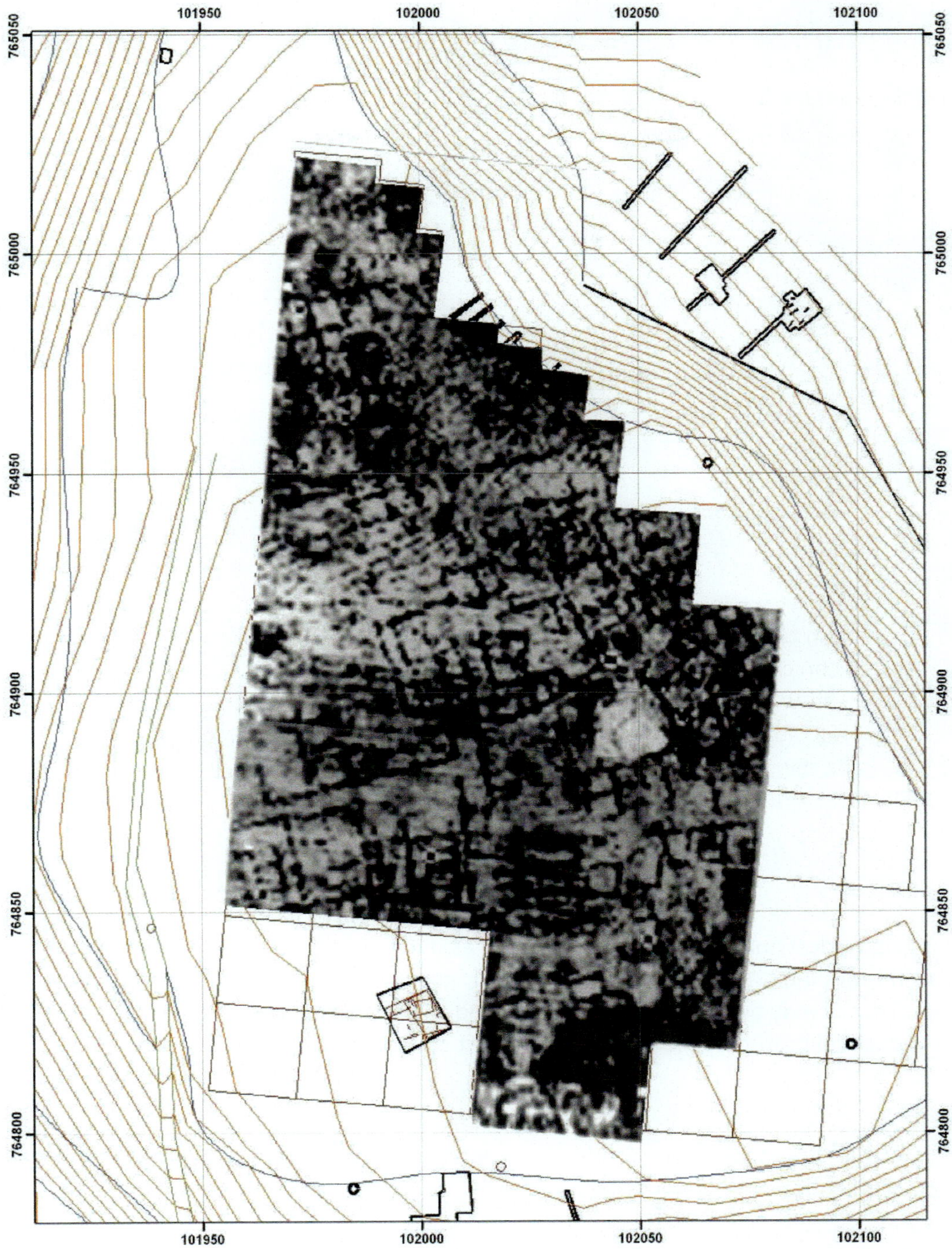

Fig. 1. The ancient ruins comprise resistive structures buried in a relatively conductive environment, creating high resistance anomalies which are mapped by geophysical prospecting. The subsurface resistance has been mapped and its distribution is depicted in a greyscale. The darker tones indicate areas of high resistance which are presumably caused by the concealed ancient ruins. The example has been drawn from the exploration of the site hosting the ruins of the ancient city "Europos" in northern Greece (Tsokas *et al.* 1994). The closed rectangular anomalies, reflecting the presence of remnants of foundation walls, form a plan view showing the shape of the ancient urban complex.

Conclusions

Images which show elongated features forming geometrical shapes are usually the outcome of any archaeological prospection survey, often attributed to the remains of foundation walls and therefore outlining the spatial distribution of ancient buildings and urban networks. In this context they reveal the urban planning of the cities that do not exist anymore. Further, important centres of the life at the ancient cities can be identified.

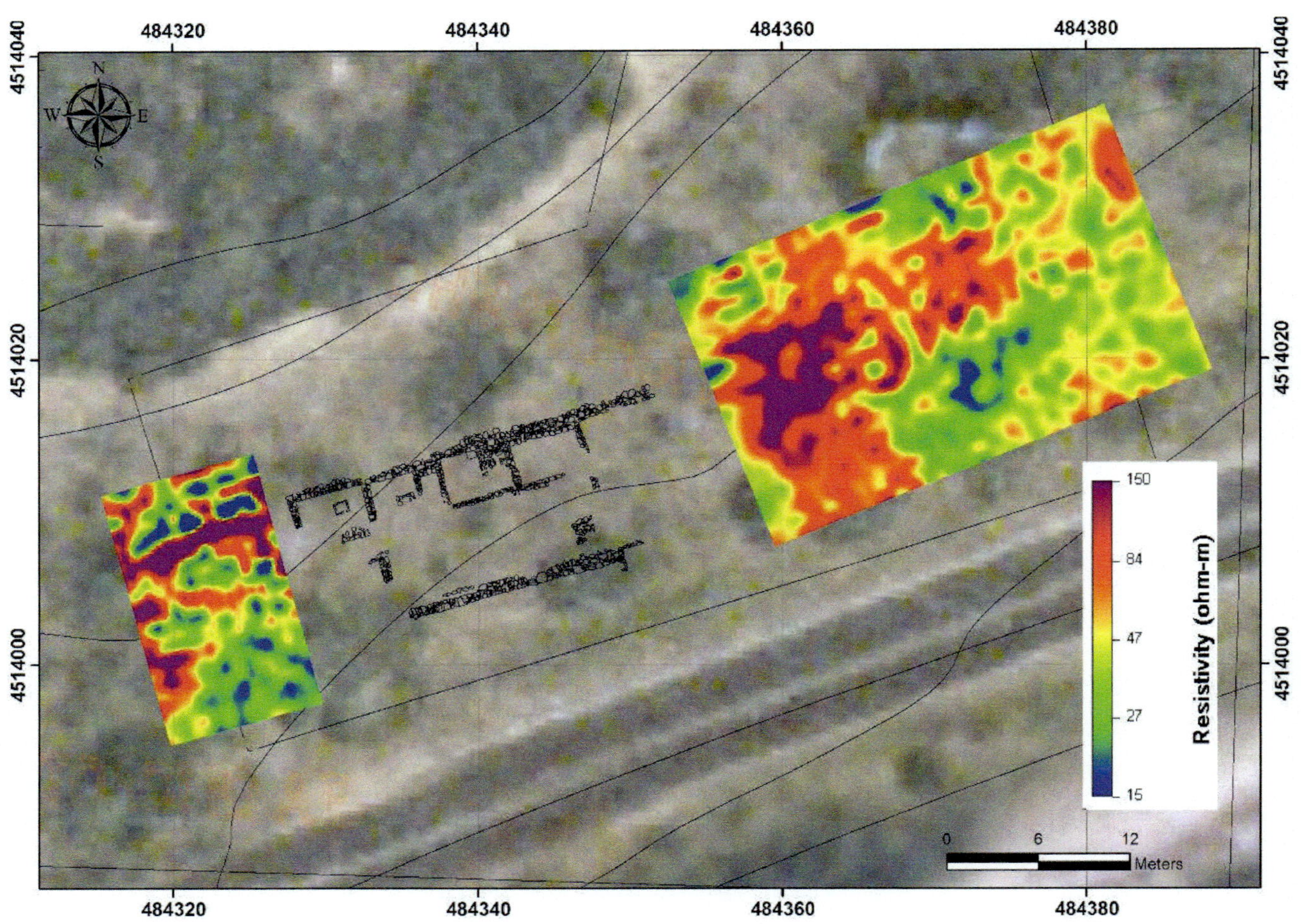

Fig. 2. Resistivity distribution (slice) at the depth of about 1.0m in ancient Argilos (Bonias *et al.* 2017).

However, the geophysical images must be considered as dynamic elements. Their interpretation is heavily dependent on the available archaeological, geological, geomorphological information and sometimes on historical clues. Thus, it may be altered or complemented if new aspects are added. They may change and be improved if they are calibrated by the findings in excavation trenches.

References

Benech, C. 2007. New Approach to the Study of City Planning and Domestic Dwellings in the Ancient Near East. *Archaeological Prospection*. 14. 87–103.

Bonias, Z., Perreault, J. and Tsokas, G.N. 2017. Argilos 2012: Preliminary results of the archaeological research and geophysical Survey. *The Archaeological Work in Macedonia and Thrace, Thessaloniki, Greece, March 20-22, 2013. Volume of Proceedings*, 513–520.

Donati, J.C., Sarris, A., Papadopoulos, N., Kalaycı, T., Simon, F.-X., Manataki, M., Moffat, I. and Cuenca-García, C. 2017. A Regional Approach to Ancient Urban Studies in Greece Through Multi-Settlement Geophysical Survey. *Journal of Field Archaeology*. 42. 450–467.

Scollar, I., Weidner, B. and Segeth, K. 1986. Display of archaeological magnetic data. *Geophysics*. 51(3). 623-633.

Tsokas, G.N., Giannopoulos, A., Tsourlos, P., Vargemezis, G., Tealby, J.M., Sarris, A., Papazachos, C.B. and Savopoulou, T. 1994. A large-scale geophysical survey in the archaeological site of Europos (northern Greece). *Journal of Applied Geophysics.* 32. 85-98.

Ground-Penetrating radar (GPR) for non-destructive testing of monument walls

Gregory Tsokas[1], Nectaria Diamanti[1], Dimitrios Angelis[2], Panagiotis Tsourlos[1], George Vargemezis[1]
[1]*Department of Geophysics, Aristotle University of Thessaloniki, Thessaloniki, Greece,* [2]*Northumbria University, Newcastle upon Tyne, UK*

Introduction

The use of geophysics to assist with the detection and mapping of subsurface archaeological structures is common practice. Ground-penetrating radar (GPR) – a geophysical method that uses electromagnetic pulses to image the subsurface (Annan 2005) – is very popular amongst other geophysical techniques due to its non-destructive nature and because it can provide in a timely manner, real-time subsurface images of the surveyed area.

GPR data are typically displayed in cross-sectional views. A GPR unit sends out radio waves that are reflected when they encounter electrically different materials, and the GPR system measures (a) how long it takes for a reflection to return and (b) the strength of that signal. In the context of locating archaeological features, soil disturbances and/or hyperbolas indicate the presence of subterranean objects and by marking several locations of a similar target, it is possible to demarcate ancient remains and help archaeologists plan an excavation. In many cases, depending on the goal of the survey, it may be necessary to create a plan map in order to shed light on the underground conditions.

In this paper, we present cases where we have used GPR to study the interior of walls at ancient monuments located in Greece. There are many more examples and we show only a few to demonstrate that GPR is a powerful tool that can locate and map in detail walls' internal structural features (such as pipes, etc.) as well as wall identify damage due to voids and/or water ingress.

Examples of GPR surveys for monument wall investigation

The first example that is presented relates to GPR data collected at the Eupalinian Aqueduct, a 2.5km long water feeding system on the island of Samos, Greece. This is one of the most impressive ancient Greek technical constructions because it was excavated from both ends having a meeting point at about the centre, and the measured mismatch (Kienast 1995) is very small even to present-day standards. We employed both electrical resistivity tomography (ERT) and GPR methods to investigate the condition of the tunnel wall lining (Tsokas *et al.* 2015). With GPR, we collected a number of profiles constituting a total length of 772m (Fig. 1). Fig. 2 shows four GPR profiles. A clear continuous reflection (red line) very close to the wall's surface (~0.3 m–0.5m) is attributed to the interface between the lining's building stones and the backfill material. At the locations where there was visible water dripping from the walls, the GPR signal disappears.

We also investigated the walls of the Megali Panagia church in Northern Greece. We used GPR to obtain information on the structural condition of the church's walls (voids, water ingress, etc.). Fig. 3. shows a GPR cross section collected at a height equal to 1.70m above the church floor. The red rectangle annotates a reflection possibly caused by a vertical internal surface within the wall structure. This reflection is seen in all the cross sectional GPR images.

Fig. 1. View of the start of the archaic lining from an undressed portion of the Eupalinian tunnel (left). GPR data collection using a pulseEKKO 1000 GPR system by Sensors & Software Inc (right).

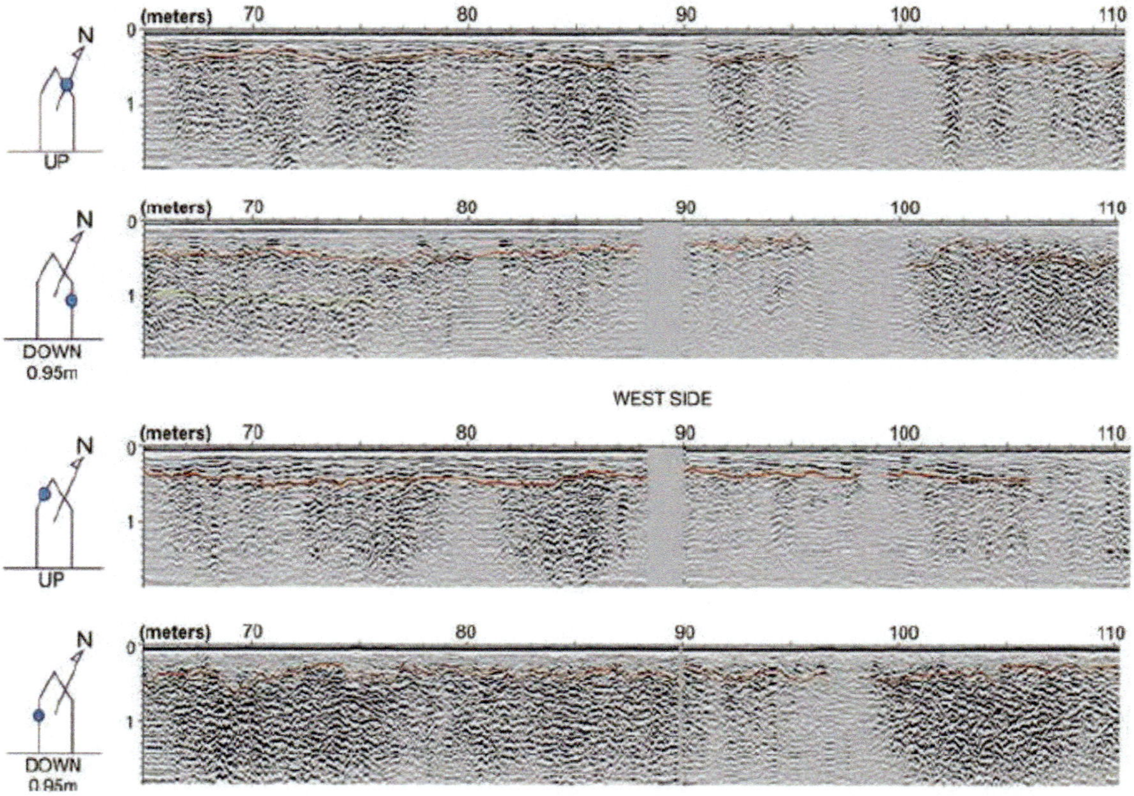

Fig. 2. Processed GPR sections on the sidewalls of the lining of Roman age. The small sketches on the left depict the direction of the profiles, and the dots mark the levels where they were carried out. The red line annotates a clear continuous reflection very close to the wall's surface at a distance varying ~0.3 m–0.5m. This reflection corresponds to the interface between the lining's building stones and the backfill material.

In another case study, GPR was used to study the walls of Heptapyrgion, a Byzantine fortress overlooking the city of Thessaloniki in Northern Greece (Angelis *et al.* 2018). The fortress is comprised by thick walls and defensive towers that suffer from structural and moisture problems that are not always visible as in Fig. 4. Hence, a GPR campaign was scheduled to study the walls' internal structure and detect areas with possible high moisture problems. Apart from GPR measurements, GPR numerical modelling was also employed to help understand better the wave scattering mechanisms, evaluate and optimize the survey measuring parameters and aid with the interpretation of recorded field data. Fig. 5 shows some synthetic and field data collected on one of the Heptapyrgion walls. Different types of interesting reflections were identified that could be attributed to different construction wall phases, water damage and voids.

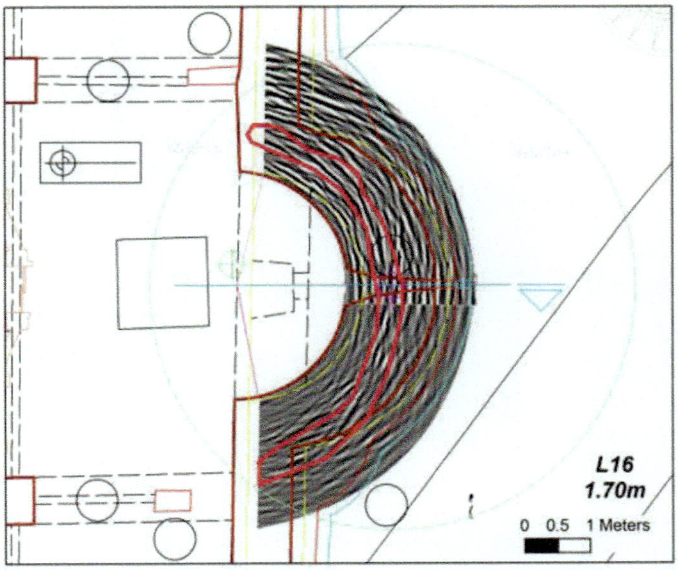

Fig. 3. GPR cross section on a semi-circular church wall at a height equal to 1.70m above the church floor. The red rectangle annotates a reflection possibly caused by a vertical internal surface within the wall structure. This reflection is seen in all the cross sectional GPR images collected on this wall.

Fig. 4. Visible structural damages (left) and areas of high moisture (right) at various locations of the Heptapyrgion fortress walls.

Conclusions

Apart from structural elements within the wall, we also identified problematic areas where voids and damage due to water ingress occurred. In most instances, we could identify problems associated with the structural wall elements in real time in the field, while looking at raw data. However, many situations were too complicated to work out in the field and required post-processing.

GPR is a very valuable tool when used with experience and site knowledge to understand the complex structures under study. When used in conjunction with construction knowledge and visual clues, it will provide archaeologists with a powerful tool to assist in imaging monuments walls as well as the subsurface, in order to map underlying archaeological features.

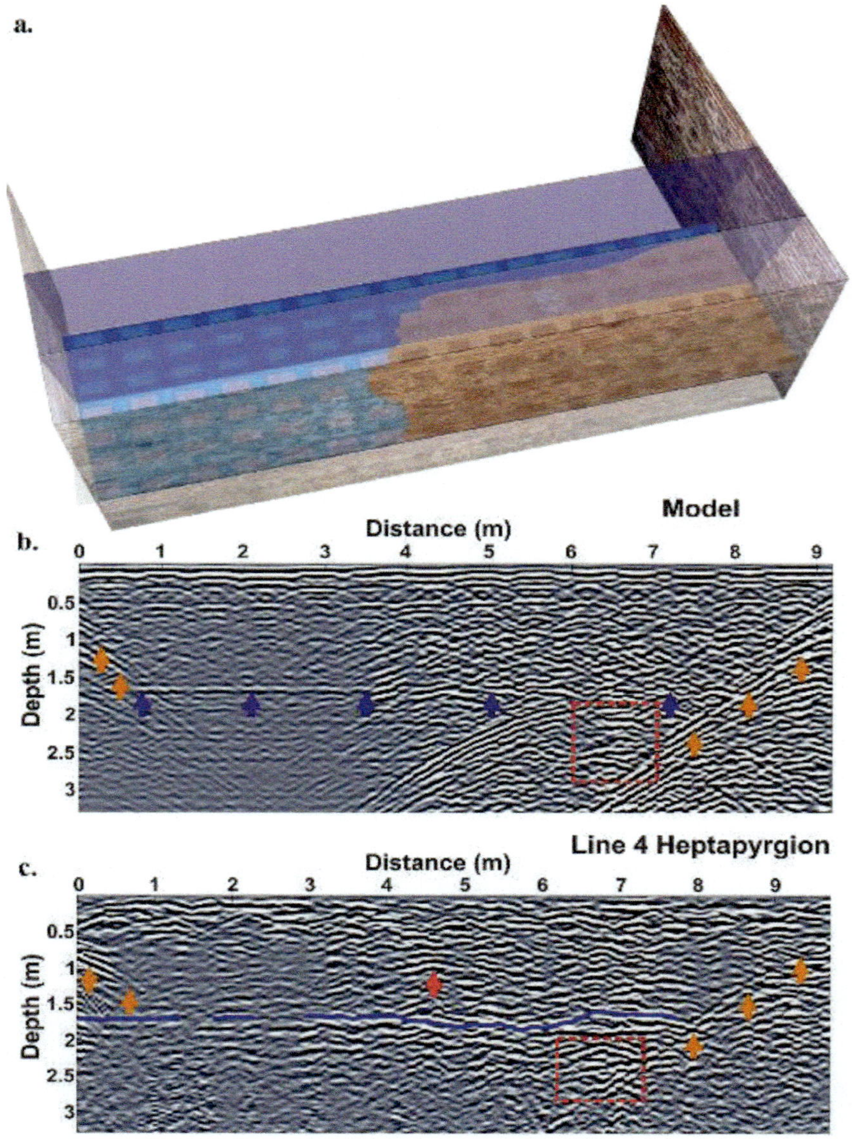

Fig. 5. a) A simulated 2D wall model fused with a 3D representation of the surveyed wall. b) Processed synthetic cross section of the wall model. Blue and orange arrows and the red rectangle indicate the different construction wall phases, lateral air reflections and distorted areas, respectively. c) Processed GPR cross section collected on one of the Heptapyrgion walls. Indicated are: the second construction phase of the wall (blue line), possible voids (red arrow), distorted region (red rectangle), lateral air reflections (orange arrows).

References

Angelis, D., Tsourlos, P.I., Tsokas, G.N., Vargemezis, G., Zacharopoulou, G. and Power C. 2018. Combined application of GPR and ERT for the assessment of a wall structure at the Heptapyrgion fortress (Thessaloniki, Greece). *Journal of Applied Geophysics.* 152. 208-220.

Annan, A.P. 2005. Ground-Penetrating Radar. In *Near-Surface Geophysics: Society of Exploration Geophysicists.* D.K. Butler (ed.), No. 13, Chapter 11, 357–438.

Tsokas, G.N., Kim, J.H., Tsourlos, P.I., Angistalis, G., Vargemezis, G., Stampolidis, A. and Diamanti N. 2015. Investigating behind the lining of the tunnel of Eupalinus in Samos (Greece) using ERT and GPR. *Near Surface Geophysics.* 13(6). 571-583.

From Roman Villas to 19th century gardens: case studies of geophysical surveys for built heritage in Hungary

László Nagy
Várkapitányság Nonprofit Zrt., Budapest, Hungary

In recent years, there has been a growing demand in Hungarian archaeology for the geophysical survey of archaeological sites before (or instead of) excavations. Our research group, consisting of archaeologists and geophysicists, is carrying out such investigations throughout Hungary for pre-investment projects and for exclusively scientific purposes as well, with the aim of mapping anomalies referring to archaeological deposits. As part of this, we have been involved in a number of projects aimed at surveying specifically deserted buildings. Our goal was always to get the most information about the built heritage of sites with all the methods available, in the shortest possible time.

These surveys were conducted by using magnetometer, ground-penetrating radar, and electrical resistivity tomography. To locate structures – possibly – made from stone or brick, we mainly used ground-penetrating radar (GPR) and electrical resistivity tomography, but in several cases magnetometry proved to also be successful. Where it was possible – and made sense, depending on the characteristics of the site - we used integrated research by the joint application of these methods.

The case study of the abandoned medieval village Ság (next to Sárszentmihály, Fejér county) and its parish church is a good example for this integrated survey method: the site was investigated by a SENSYS MXPDA 5-channel magnetometer and a MALA GX GPR with a 450MHz antenna. The magnetometer results first revealed the exact location of the church, so we were able to narrow down the investigation area for the GPR to a 30m x 30m square (Fig. 1).

By using both methods, we gained a very accurate ground plan: the church had a single-nave with 8.5m x 6m inner size and a semi-circular apse with 5m length and 4m width. On the western part of the nave the probable foundation of a pillar can be seen, it may have held a choir. Around the church, the magnetometer data shows the wall around the churchyard and numerous other anomalies, which can be mostly related to the deserted village of Ság.

Larger ramparts, earth-built fortifications, were typically investigated by electrical resistivity tomography, such as the case study of the medieval castle of Sárospatak (Borsod-Abaúj-Zemplén county), where we had to provide information about the inner structure of the fortification around the castle before excavation. Cross-sections were measured through the rampart using an ARES RES5 – 5CH device, the distance between the electrodes was 1m. According to the results of these cross-sections, the main structural layers of the rampart were identified, and these were confirmed by a later excavation.

The raw data obtained from these measurements were processed, then placed on a map after georeferencing, after which geophysicists and archaeologists interpreted it together. In a few cases (like in Sárospatak), excavations have also taken place after the surveys, which provided opportunities for comparing the unearthed walls and the measurement results, that helped to draw valuable conclusions about the archaeological sites.

Based on the results, we were, in most cases, able to successfully determine the full or partial ground plan or the inner structure of the buildings that were being searched for. Compared to the conventional archaeological methods, these techniques were much quicker, much cheaper, and most importantly, non-destructive, thus providing important new data for the archaeological and historical research of built heritage.

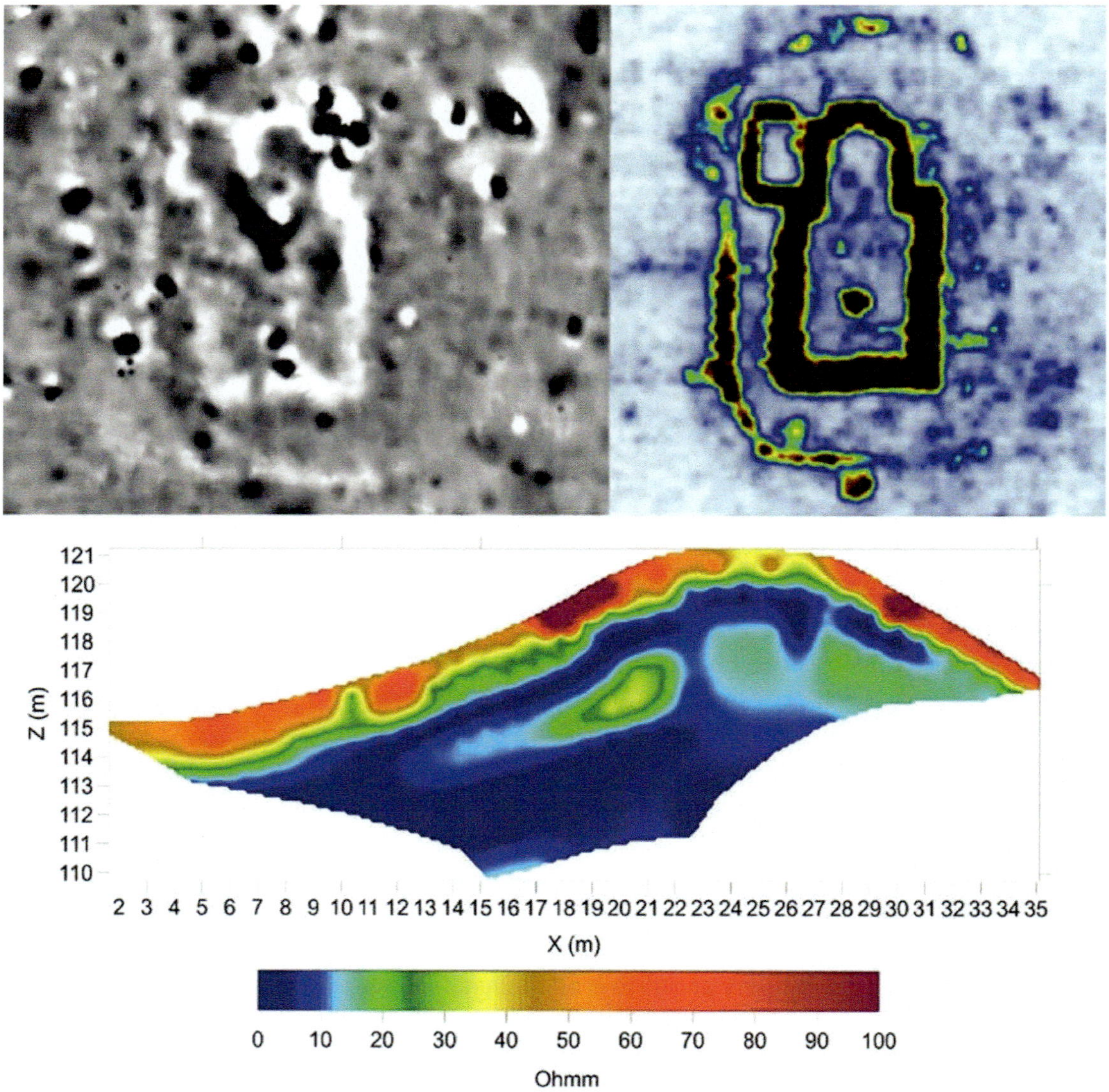

Fig. 1. Magnetometer (upper left) and GPR (upper right) results of the medieval church of Ság, and a cross-section from the electrical resistivity tomography survey of Sárospatak (below)

Hidden Depths and Empty Spaces: the contribution of archaeological prospection to the study of early medieval Ireland

Susan Curran
UCD School of Archaeology, Dublin, Ireland

Ireland is home to perhaps the richest and best-preserved early medieval settlement archaeology in Europe (O'Sullivan 2011, ix), but this wealth of existing evidence is only part of the story. From the arrival of Christianity and the development of literacy, to the establishment of Ireland's first towns, the early medieval period (*c.* AD 400-1100) is one of the most dynamic and complex periods of Ireland's past. The changes brought about during this period not only had a significant impact on society, but also left an indelible imprint on the Irish landscape. More than 1000 years on, many of these early medieval ecclesiastical and secular sites are still clearly visible on the Irish landscape, marking the locations of long abandoned settlements. Many monuments are still quite prominent and visible to the naked eye, the most common being the *rath*, of which more than 43,000 examples have been recorded in the Republic of Ireland. However, the advent of archaeological survey techniques such as lidar has demonstrated that there is even more to the early medieval landscape than meets the eye.

This paper presents the results of PhD research which comprised the application of lidar and geophysical survey techniques to the exploration of the Irish early medieval landscape. The research focused on two case study areas, the first in parts of counties Leitrim and Roscommon (also the subject of M.A. research, see Curran 2012; Curran 2013), and the second in the north-east of County Monaghan. Prospection for 'new' monuments was the first - and indeed a key - step in this process, and lidar analysis proved extremely successful in this regard. Prior to this, both case study areas had attracted little attention by way of archaeological excavation (or even focused research) and are, for the most part, somewhat sparsely populated. This was of immense benefit to the application of a remote sensing approach as it meant that these relatively untouched landscapes provided an optimum base for investigations using lidar. The early medieval population have been kind to those attempting to investigate the landscape they inhabited, as their settlements (both secular and ecclesiastical), which were defined by concentric banks and ditches, are by and large, archaeologically distinctive and particularly well-suited to identification by non-invasive techniques. This was evidenced not only by the number of monuments which were newly discovered over the course of the analysis, but also the way in which existing monuments could be visualised and their possible relationship with their counterparts examined. Preliminary lidar investigations identified 149 potential 'new' *raths* within the Leitrim/Roscommon study area, constituting an increase of 44% in the number of early medieval settlements within this catchment. Similar analysis of the Monaghan study area led to an increase of 11% in the number of early medieval *raths*.

However, prospection and monument-identification by remote sensing is not always infallible, and there are several challenges to overcome in order to achieve the best possible interpretation. First and foremost, the experience, knowledge, and skill of the researcher are paramount (e.g. Cowley 2013, 24; Palmer 2013, 76-77); without these, the foundations upon which an interpretation can be built are somewhat shaky. This was most evident in the revision of the number of discoveries in Leitrim/Roscommon between the M.A. and PhD research, a direct reflection of the skills and confidence gained in line with the building of experience. Re-investigation of the preliminary analysis saw the initial identification of 149 potential *raths* revised down to 90; although this represents a significant drop, the classification of the 'new' monuments is more solid, and this still constitutes an overall increase of 21% to the number of previously recorded monuments.

Lidar is not a tool which can be used in isolation. Investigations must be undertaken in conjunction with other techniques and resources, including historic mapping, aerial imagery, and documentary research in order to maximise the identification and interpretation of new discoveries. While lidar can make significant contributions to our knowledge of early medieval settlement, particularly in relation to the identification of enclosing elements, there are many features that it simply cannot reach. Archaeological features without topographic expression, and those more subtle structural elements of early medieval settlement (e.g. houses, field boundaries, entrances, etc.), are not especially suited to this technique. As such, geophysical survey (magnetic gradiometry and earth resistance) were employed at selected sites in order to enhance the lidar findings and to further our knowledge of those monuments. The identification of key diagnostic features such as enclosures at three of the ecclesiastical sites points to their early medieval origins, of particular importance in relation to the chronology of the sites and settlement in the area. The earth resistance survey at the newly discovered (by lidar analysis) enclosure at Mullaghmore, Co. Roscommon (Fig. 1) revealed a circular internal structure which is consistent with the morphology of houses of the early medieval period. This additional layer of information helped to classify this enclosure more definitively as an early medieval *rath* rather than a similar enclosure from another period.

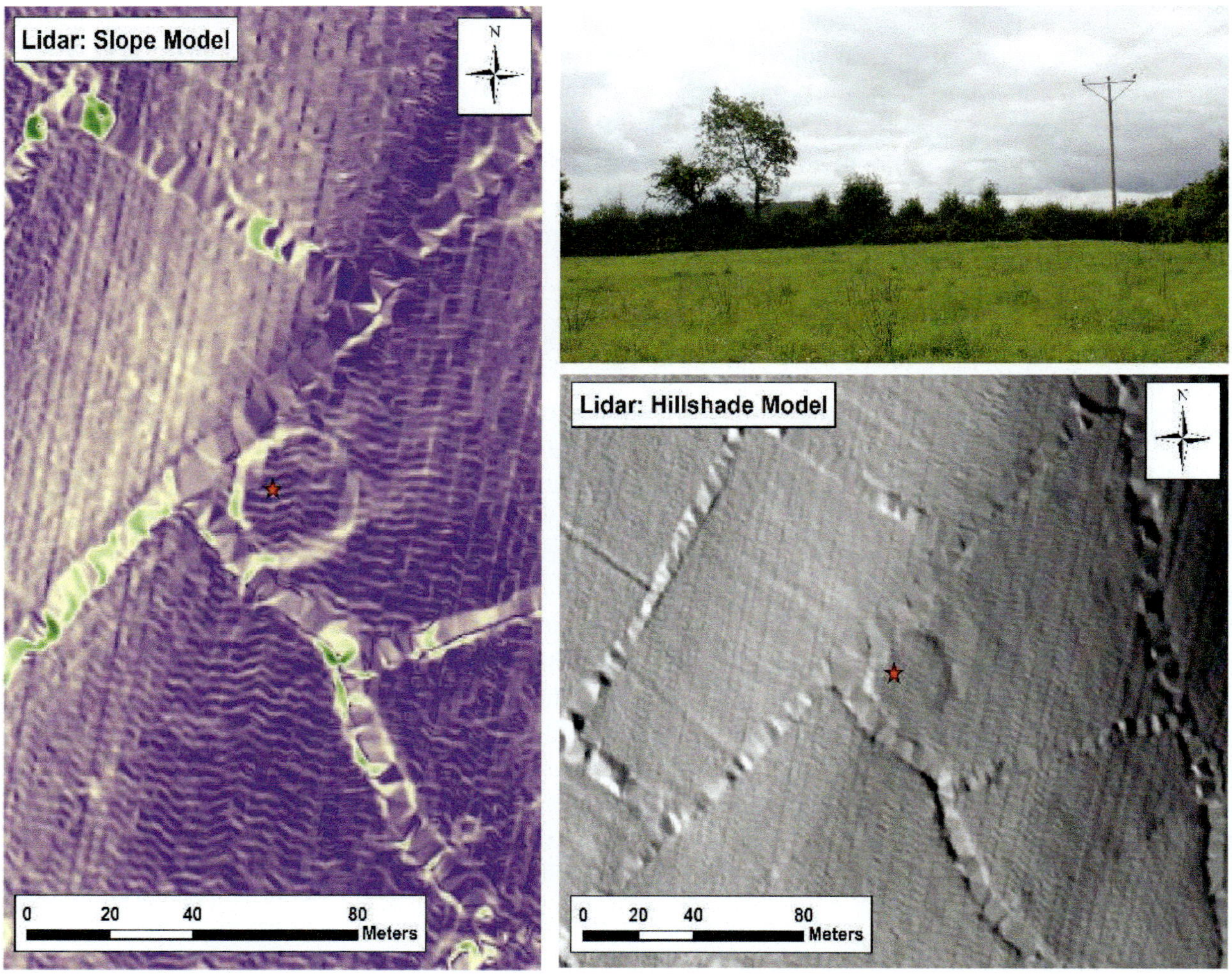

Fig. 1. The newly discovered *rath* at Mullaghmore, Co. Roscommon

However, prospection was only the first step in the project, the ultimate goal of the research was to interpret the findings and build upon them in order to achieve a more comprehensive understanding of the early medieval landscape. The appropriate identification and interpretation of new monuments and the integration of these findings with those already known aspects of the early medieval landscape are crucial parts of this process; without them, an informed study of the early medieval landscape is futile. By filling in some of the empty spaces, and by exploring the hidden depths of the early medieval landscape, these techniques have made a substantial contribution towards developing an improved picture of early medieval society. Moreover, the ability to visualise the landscape using lidar (and GIS) allows us to refocus our attention on small-scale settlement patterns, moving away from a previous concentration on national overviews with arguably little regard for understanding how the landscape actually worked on a day-to-day basis at a community level.

Acknowledgements

Sincere thanks to Transport Infrastructure Ireland, Leitrim County Council and Monaghan County Council for their permission to use the lidar datasets for the purposes of my research. Thanks also to Roscommon County Council for the awarding of a bursary with which to purchase a higher resolution lidar dataset from Ordnance Survey Ireland. I am very grateful to the landowners who allowed me to undertake the geophysical surveys on their land, and indeed, to the many survey volunteers who assisted me. Sincere gratitude to UCD School of Archaeology and the Discovery Programme for their continued support throughout my research, and to the Irish Research Council who funded the project.

References

Cowley, D. 2013. In with the new, out with the old? Digital workflows and auto-extraction in remote sensing archaeology. In D. Hookk (ed.), *Papers of the First International Conference on Virtual Archaeology*, St Petersburg 2012. 18-30. St. Petersburg.

Curran, S. 2012. *Early Medieval Settlement in Leitrim and Roscommon: A LiDAR and GIS Initiative*. Unpublished M.A. Thesis, University College Dublin.

Curran, S. 2013. LiDAR and Early Medieval Settlement in Counties Roscommon and Leitrim: a New Angle on Early Medieval Landscape Studies. *Co. Roscommon Historical and Archaeological Society Journal, 30th Anniversary Edition*. 97-101.

O'Sullivan, A. 2011. Foreword. In C. Corlett and M. Potterton (eds.), *Settlement in Early Medieval Ireland in the light of recent archaeological excavations*. ix-xi. Dublin.

Palmer, R. 2013. Reading aerial images. In R. Opitz and D. Cowley (eds.), *Interpreting Archaeological Topography: Airborne Laser Scanning, 3D Data and Ground Observation*. 76-87. Oxford.

Visualizing the Village: A Comparative Assessment of Remote Sensing Methods on Inishark, Co. Galway, Ireland

Sean Field, Ian Kuijt
University of Notre Dame, South Bend, United States

Over the last ten years researchers have made remarkable advancements in the use of aerial remote sensing methods. Among these methods, light detection and ranging (LiDAR) has generally been viewed as the 'gold-standard' due to its effectiveness in detecting archaeological sites in high-vegetation contexts and over large landscapes (Chase *et al.* 2012, Grammer *et al.* 2017, McNeary 2011, Opitz and Cowley 2013, Risbøl and Gustavsen 2018). During the last five years, however, research using photogrammetric modelling has also advanced in significant ways, partially due to its cost-effectiveness and ability to produce high-resolution imagery of individual sites (Armstrong *et al.* 2018, Campana 2017, Daponte *et al.* 2017, Kullman 2017, Quartermaine *et al.* 2014, Verhoeven 2017). To better understand the documentation capabilities of these different methods, we present a comparison of LiDAR and photogrammetry within a single low-vegetation archaeological landscape on the island of Inishark, Ireland.

The island of Inishark, County Galway, has been occupied at least from the Early Bronze Age, with major occupations from the 7th-14th centuries AD, and then again from the 1780s onward. This has resulted in a complex array of archaeological features in variable conditions across the island. To better understand the documentation capabilities of aerial based remote sensing for the period AD 1780 onward, we conducted a LiDAR and aerial photography survey to produce two separate digital terrain models (DTM) of the eastern portion of the island and the historic village of Inishark; LiDAR was conducted over the entire island (~1.3 square km) and photogrammetry over a portion of the island with the greatest concentration of archaeological features (~0.3 square km). LiDAR produced a DTM with an average resolution of 10 pts./m^2 (low resolution between 4-9 pts./m^2 and a high resolution between 20-15 pts./m^2). Photogrammetry produced a point cloud of 782 pts./m^2 that when converted to a DTM had an average resolution of 516.5 pts./m^2. While the resolution differences are significant, we wanted to assess the extent to which the physical characteristics of different archaeological features can be documented under the same conditions with the two methods. Specifically, we compared the plan view (Fig. 1) and profiles (Fig. 2) produced from these DTMs on four buildings and two depressed surface features. To produce profiles, we first manually drew a series of closely spaced (sub 2cm) shapefile lines across each feature in QGIS. Then, using a short script in R we extracted the value of every cell that was intersected by the transect lines, and then plotted these values recursively to visualize a comprehensive profile of the feature.

These data demonstrate that on average a DTM produced by photogrammetry has an effective resolution thirty times greater than the DTM produced by LiDAR. Under conditions of minimal vegetation, photogrammetry provides a more accurate documentation of standing architecture, including wall height and wall thickness, and more subtle aspects like the overall shape and scale of architectural features on floor and wall surfaces. Further, the documenting advantages provided by photogrammetry occur for both large and moderate sized architectural features. LiDAR, however, outperformed photogrammetry in documenting certain deflated surface features, as photogrammetry would at times distort berms or other similar features due to surface vegetation. On the other hand, photogrammetry provided a better rendering of certain surface features like deflated walls and foundations. This suggests that in the case of more complex surface features, photogrammetry would be more likely to correctly identify the number, size, and shape of unique rooms or features than LiDAR. Together, these results demonstrate that in conditions of low vegetation, photogrammetry out-performs LiDAR when documenting architectural structures, and in some cases, small scale features. Additionally, these data also demonstrate the growing ability of aerial based photogrammetry to document larger landscapes.

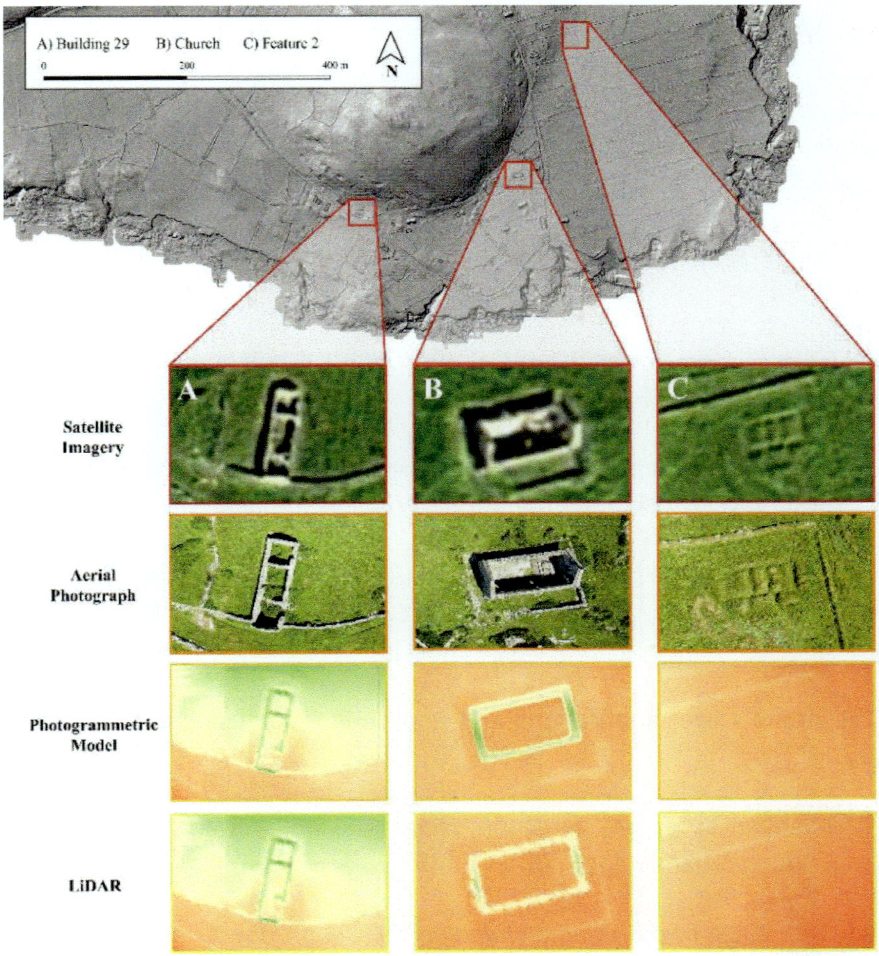

Fig. 1. An example of plan view comparisons of Building 29, the Church, and Feature 2 via satellite imagery, aerial photography, photogrammetry and LiDAR.

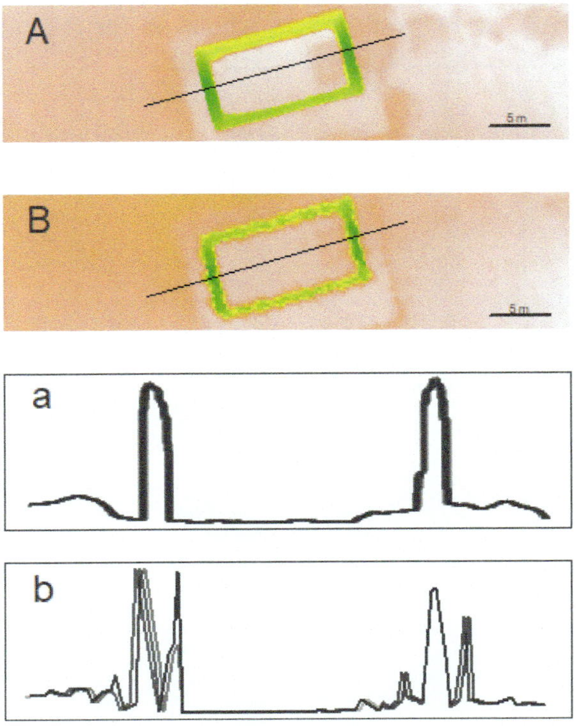

Fig. 2. An example of profile comparisons. Upper: Comparison of (A) photogrammetry and (B) LiDAR of St. Leo's Church, Inishark, Co. Galway, Ireland. Lower: west-east cross section profile of St. Leo's Church developed with (a) photogrammetry and (b) LiDAR.

References

Armstrong, B.J., Blackwood, A.F., Penzo-Kajewski, P., Menter, C.G. and Herries, A.I.R. 2018. Terrestrial laser scanning and photogrammetry techniques for documenting fossil-bearing palaeokarst with an example from the Drimolen Palaeocave System, South Africa. *Archaeological Prospection*. 25. 45-58.

Campana, S. 2017. Drones in Archaeology. State-of-the-art and Future Perspectives. *Archaeological Prospection*. 24. 275-296.

Chase, A.F., Chase D., Fisher, C., Leisz, S. and Weishampel, J.F. 2012. *Geospatial revolution and remote sensing LiDAR in Mesoamerican archaeology*. 1–6.

Daponte, P., Vito, L.D., Mazzilli, G., Picariello, F. and Rapuano, S. 2017. A height measurement uncertainty model for archaeological surveys by aerial photogrammetry. *Measurement.* 98. 192–198.

Grammer, B., Draganits, E., Gretscher, M. and Muss, U. 2017. LiDAR-guided Archaeological Survey of a Mediterranean Landscape: Lessons from the Ancient Greek Polis of Kolophon (Ionia, Western Anatolia). *Archaeological Prospection.* 24. 311– 333.

Kullman, K. 2017. The drone's eye: applications and implications for landscape architecture. *Landscape Research*. 43. 906-921.

McNeary, R. 2011. Archaeological feature detection using airborne laser scanning at Shane's Castle Park, County Antrim. *Ulster Journal of Archaeology*. 70. 111-125.

Opitz, R.S. and Cowley, D.C. 2013. *Interpreting archaeological topography*. Oxbow Books. Oxford, UK.

Quartermaine, J., Olson, B.R. and Killebrew, A.E. 2014. Image-Based Modeling Approaches to 2D and 3D Digital Drafting in Archaeology at Tel Akko and Qasrin Two Case Studies. *Journal of Eastern Mediterranean Archaeology & Heritage Studies*. 2(2). 110–127.

Risbøl, O. and Gustavsen, L. 2018. LiDAR from drones employed for mapping archaeology – Potential, benefits and challenges. *Archaeological Prospection*. 25. 329-338.

Verhoeven, G. 2017. DRDF and its Impact on Aerial Archaeological Photography. *Archaeological Prospection*. 24. 133-140.

Unusual monuments, Unusual molecules: geochemical processes at work in County Limerick, Ireland

Heather Gimson, Cian Hogan, Ursula Garner
Earthsound Geophysics Ltd., Claremorris, Ireland

Introduction

Kilfinane is a small town located in the south-west of Ireland, near the border of Counties Limerick and Cork, sited on rolling grasslands at the foothills of the Ballyhoura Mountains. Overlooking the town is the impressive monument of 'The Moat' – a large motte surrounded by three incomplete banks and ditches (Fig. 1). This enigmatic monument seems to comprise several periods of activity, with the banks and ditches representing an earlier enclosure, into which was inserted the motte at a later period.

Fig. 1. Aerial view of the motte and banks at Kilfinane (Source: N. Jackman, Abarta Heritage).

The site at Kilfinane is mentioned in historical texts. The *Leabhar na gCeart* (Book of Rights) has written versions dating to AD 1390 and AD 1418, though it is likely that the poems that it is comprised of date earlier still. The site of *Tréada-na-Riġ*, thought to refer to Kilfinane, is mentioned twice in the text (O'Donovan 1847: 89, 93) and is interpreted as a fort belonging to the King of Cashel.

In 2017, the Kilfinane Community Council (KCC) took on The Moat as a research project through the 'Adopt a Monument' scheme run by the Heritage Council. The KCC are helping in the interpretation and conservation of the town's heritage. As part of that project, Earthsound were commissioned to undertake a series of geophysical surveys on lands surrounding the motte.

The Geophysical Surveys

Two methods of investigation were deployed. A magnetometer survey was undertaken at a resolution of 0.5m x 0.25m over 6.2Ha, using an Eastern Atlas LEA MAX system. The data were processed using data normalisation and drift correction prior to being gridded and interpolated.

Further targeted surveys were undertaken using a GF Instruments CMD-Mini Explorer, in the Vertical Coplanar Coil configuration. This electromagnetic induction survey, at 1m x 0.25m resolution, collected both in-phase and quadrature data and was carried out over targeted areas totalling 2.18Ha. Data were processed using a moving filter drift correction, despike, low pass gaussian filter and interpolation.

The Results

The surveys at Kilfinane have revealed a dense archaeological landscape, including previously unknown archaeological features and monuments (Fig. 2). The extent of the earthwork banks and ditches around the motte were confirmed as well as the identification of a possible entrance, outer bailey, external sub-enclosures or annexes and a number of possibly related pits.

In addition to these principal features, evidence was also detected for burgage plots or enclosures associated with the historic town of Kilfinane. A previously unknown early medieval bivallate enclosure complex was also detected containing habitation and industrial activity. A prominent natural palaeochannel may have been utilised as a delimiter between the large earthwork and this feature.

While the presence of a motte within - and the reuse of - an earlier multi-vallatted earthwork is not unique, it is the geochemical processes at work across a large extent of the site which made it unusual from a geophysical perspective. The magnetic data revealed the presence of anomalies that ran counter to the typical form. Extensive cut features were detected across the site which contained negative magnetism, while other cut features, presumably from a different historic period(s), contained the expected positive magnetism. These geochemical processes are particularly strong within the magnetometer data, but have also had some effect on the apparent magnetic susceptibility portion of the electromagnetic data, with the excavated (moisture retaining) nature of many of these anomalies confirmed through the use of EMI quadrature data.

The magnetometer anomalies appear to have formed across the site in two distinct ways; cut features which follow the 'expected' principle and have produced a higher magnetic susceptibility within their fills than the surrounding soil. A large number of anomalies however were also detected which contain a lower magnetic susceptibility than the surrounding soils.

Two explanations exist for the lowering of magnetic susceptibility readings; either the leaching of magnetic iron-oxides from upper layers and depositing them at depth (Cunningham *et al.* 2001) or the destruction of the susceptibility-bearing iron oxides via redoxic conditions (Weston 2002). Both explanations are associated with water.

Although periods of extensive rain and waterlogging are not unusual within the west of Ireland, the geophysical surveys at Kilfinane have revealed a localised example of extreme soil, anaerobic or water processes. This suggests that the site experienced unusual activity at some point in its history. The full extent of the process is undetermined, which, along with a lack of geochemical investigation, hinders a complete understanding of the process.

Cut features open within the landscape during this period would have suffered from unusually extensive waterlogging leading to features being permanently filled with stagnant water. The palaeochannel which runs through the site also appears as a broad negatively magnetic feature. It seems to be incorporated into some of the numerous archaeological ditches and features and may have played a part in the waterlogging process of these features. This waterlogging could have caused the magnetic iron-oxides to leach out from the upper layers and deposit them as iron-pan above the deeper layers. Alternatively, the waterlogged conditions may have combined with specific anaerobic conditions and 'destroyed' the susceptibility-bearing iron oxides (Weston 2002, Thompson and Oldfield 1986, Kattenberg and Aalbersberg 2004, Schmidt 2017).

Fig. 2. Results of the magnetometer survey.

This process has been mapped in features before, but what is unusual at Kilfinane is the extent of the effect detected, something not seen in Ireland to date. Along with the detection of negative cut features a further portion of cut features remained positive, indicating that the wet climatic conditions on site were limited.

Conclusion

A number of further geophysical and geochemical surveys have been proposed for the motte and the surrounding lands. These aim to target the previously unknown archaeology, in particular several of the negative magnetometer anomalies to assess the true nature of the archaeological and geochemical processes which have affected the site and the causes for these. Unfortunately, at this point in time, due to lack of secure funding further investigations have been put on hold, but we are hopeful that it will be available at a future date.

Acknowledgements

Earthsound Geophysics would like to thank The Heritage Council, Kilfinane Community Council and the local community, Abarta Heritage and GeodataWIZ.

References

Cunningham, D.A., Collins, J.F. and Cummins, T. 2001. Anthropogenically-triggered iron pan formation in some Irish soils over various time spans. *Catena.* 43(3). 167-176.

Kattenberg, A. E. and Aalbersberg, G. 2004. Archaeological prospection of the Dutch perimarine landscape by means of magnetic methods. *Archaeological Prospection.* 11(4). 227-235.

O'Donovan, J. 1847. *The Book of Rights: with translation and notes*. Dublin. Dublin University Press.

Schmidt, A. 2017. Comments on Magnetometer Data – Kilfinane, Co. Limerick. GeodataWIZ: unpublished report.

Thompson, R. and Oldfield, F. 1986. *Environmental Magnetism*. London. Allen and Unwin.

Weston, D.G. 2002. Soil and susceptibility: aspects of thermally induced magnetism within the dynamic pedological system. *Archaeological Prospection.* 9(4). 207-215.

Newgrange, New Monuments and New Perspectives

Joanna Leigh[1], Geraldine Stout[2], Matthew Stout[3]
[1]J. M. Leigh Surveys Ltd, Leixlip, County Kildare, Ireland, [2]National Monuments Section, Dublin, Ireland, [3]Dublin City University, Dublin, Ireland

New discoveries from aerial drone footage made international headlines in 2018 after a prolonged heatwave revealed new monuments in the fields surrounding the World Heritage megalithic site of Newgrange. The images released of known and newly discovered monuments significantly added to the knowledge and understanding of the archaeological environment of the famous 'bend in the Boyne' landscape. However, the story of the discoveries begins a little earlier, in 2015, with a geophysical survey on a steep south facing slope, to the south of the monument.

In 2015 The Office of Public Works (OPW) commissioned an archaeological survey in the north-east of a field to the immediate south of the Neolithic Newgrange passage tomb, as part of a planning application to re-build the onsite facilities for the visitor centre. A detailed gradiometer and resistance survey were conducted in a small area immediately adjacent to the existing interpretive centre buildings (Leigh 2015). The gradiometer survey was conducted with a Bartington Grad 601 instrument, sample interval 0.25m x 1.0m. The resistance survey was conducted with a Geoscan RM15 with a sample interval of 1.0m x 1.0m.

Despite the small size of the original survey area, the results identified a series of curious pit or post-hole type responses. At first it was thought this may represent a former farmers fence and interpretation was far from clear. The survey was extended to further investigate the responses. The results presented a very clear, rectilinear feature, measuring *c.*45m in width and extending at least 110m in length. Within the rectilinear feature was a series of both large and small pit-type responses. These features were very clear in both the gradiometer and resistance surveys (Fig. 1).

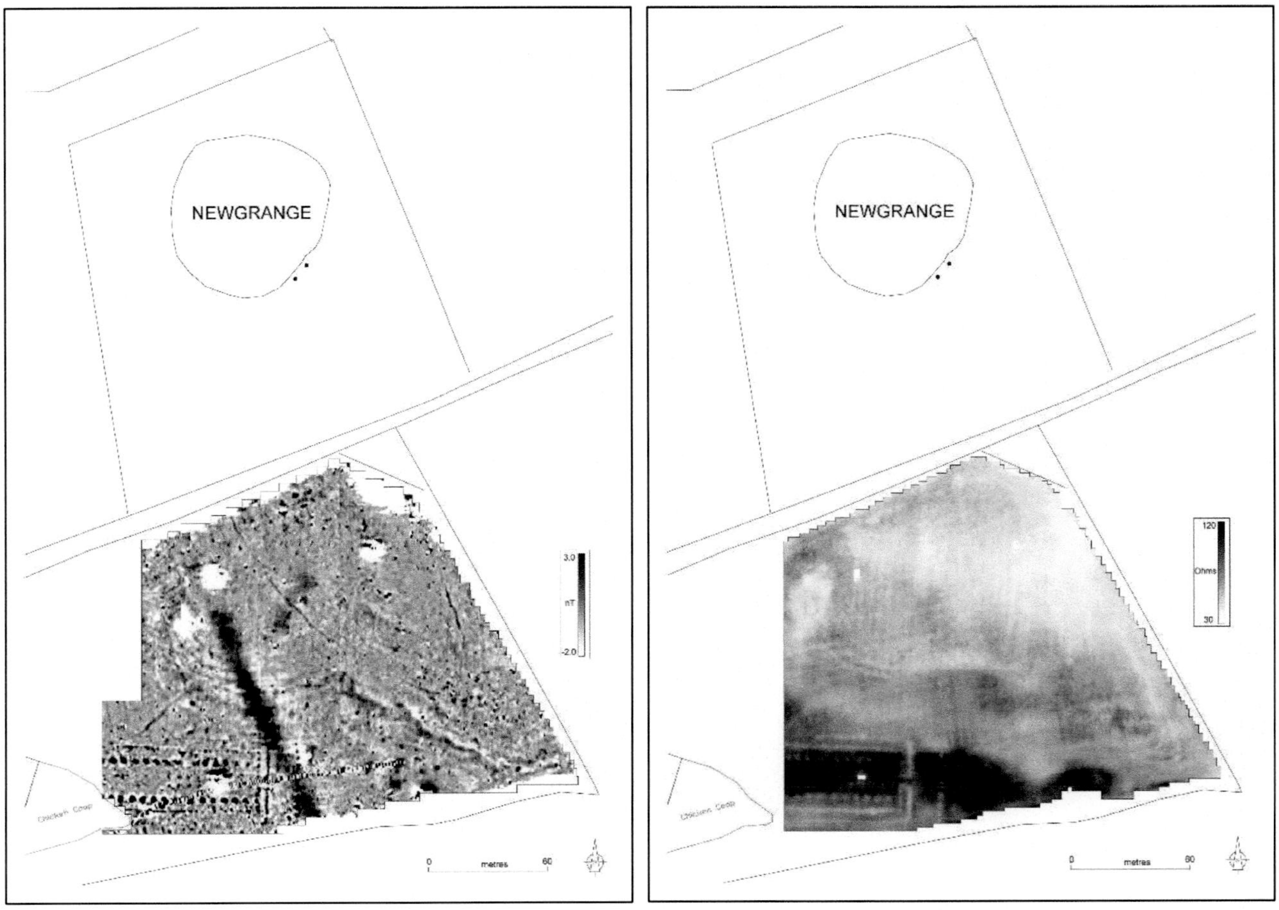

Fig. 1. (left) Magnetometer data and (right) earth resistance data, south of Newgrange passage tomb.

However, the interpretation was less straightforward. Could this be a monument relate to the use Newgrange passage tomb itself? The clearly defined linear form of the responses was suggestive of a possible building or structure. However, the size of the rectilinear feature was inconsistent with a building. There was an absence of ferrous responses which would be inidvciatve of a post-Iron Age date, caused by nails and door fixings. However, this was not evident in the gradiometer results. This data set was relatively clean, except for the usual background modern litter responses. Certainly not what would be expected for a building of this size. So, a prehistoric interpretation was preferred.

In 2018 we conducted a small research excavation that extended across one of the large pits and the outer fragmented ditched feature to investigate the origin of the geophysical responses. Excavation confirmed a fragmented outer ditch cut into re-deposited gravels. The inner ditch measured 1.5m-2m in width and 1.2m deep and ran the width of the excavation trench. In the southern half of the excavation, a cluster of four small shallow pits and a 'great pit' corresponded with the responses identified in the geophysical survey. Of clear interest was one of the very broad (*c.* 3m wide) responses which was targeted. This corresponded with what became known as 'the great pit'. This was 4.8m wide and 1.65m deep and was cut into redeposited gravels. The outer lip was defined by a baked clay ledge. A concentration of fire-reddened boulders was located at the base of the large pit with charcoal and red burnt clay overlying the boulders. Within this deposit were concentrations of burnt animal bone with cut marks consistent with butchering. A flue-like feature opened into the pit from the east.

Charcoal taken from the outer ditch produced a radiocarbon date of 2632-2472 cal. BC, placing this in the late Neolithic period and confirming its status as a prehistoric site. From the excavation results we surmise that the site represents a unique late Neolithic monument type, combining some of the elements of a cursus monument with those of a pit alignment (Leigh *et al.* 2018). The geophysical surveys (Leigh 2015) suggests the response associated with 'the great pit' appears to be repeated along two parallel linear alignments, both with possibly 13 fire pits or cooking sites. The western extent of the site could not be determined through due to access restrictions, however, it is speculated that the monument continues westerly and may extend for some distance.

The combination of geophysical survey and targeted excavation has allowed the discovery of a new monument type, adding to the knowledge of the Brú na Bóinne landscape. The clear images provided by the geophysical results facilitated targeted excavation, which in turn greatly added to the interpretation of the site. It is hoped that further excavations will take place across the eastern extent of the pit alignments to provide additional information about this remarkable monument.

References

Leigh, J., Stout, G. and Stout, M. 2018. A pathway to the Cosmos at Newgrange Farm. *Archaeology Ireland* Winter 2018. 32(4). Wordwell Ltd.

Leigh, J. 2015. *Geophysics Report, Newgrange Farm, Newgrange, County Meath*. Unpublished geophysics report J. M. Leigh Surveys Ltd.

Microgravimetry for cavity detection–an example from pilot measurements on Newgrange passage tomb (Brú na Bóinne World Heritage Site, Ireland)

Roman Pasteka[1], Pavol Zahorec[2], Juraj Papco[3], Conor Brady[4], Kevin Barton[5]

[1]Comenius University, Department of Applied and Environmental Geophysics, Bratislava, Slovakia, [2]Division of Geophysics, Earth Science Institute, Slovak Academy of Sciences, Bratislava, Slovakia, [3]Department of Theoretical Geodesy, Faculty of Civil Engineering, Slovak University of Technology, Bratislava, Slovakia, [4]Department of Humanities, Dundalk Institute of Technology, Dundalk, Ireland, [5]Landscape and Geophysical Services, Claremorris, Ireland

Introduction

The detection of cavities/voids in non-invasive archaeological prospection plays an important part in the search for tombs, chambers, crypts, cellars, etc. Among the different geophysical methods available, the best results are usually obtained from GPR, ERT and microgravity methods (e.g. Panisova *et al.* 2013: 163–174). The microgravity method (microgravimetry) is based on the precise measurement of anomalies in the Earth's gravity field with the objective of detecting the responses of shallow, small-scale, density inhomogeneities. Relative gravity meters can measure changes in the Earth's gravity field with a precision of approximately ±5 microGals (1 microGal = 10^{-8} m·s^{-2}), so cavities with an anomalous gravitational effect above 10 microGals (in absolute value) can be detected. The potential for detecting gravity anomalies using the microgravity method is dependent on the depth of the target; cavities with a volume of several m^3 can be recognized at a few metres depth (depth to top of cavity). There have been several successful archaeological applications of microgravity methods in the detection of cavities/voids (Pašteka and Zahorec 2000: 373–387, Abad *et al.* 2007: 197–201). Microgravity was used in the search for hidden cavities inside the Great Pyramid at Giza in Egypt (Lakshmanan and Montlucon 1987: 10–17).

There has been little research into the use of microgravity in the detection of chambers in mounds due to their generally small dimensions. The situation in the large Neolithic mounds at Brú na Bóinne, is different. Some of the chambers beneath the Newgrange, Knowth and Dowth mounds have relatively large dimensions, and are ideal targets for microgravity. As it is possible that some of chambers remain undiscovered, there is potential in using this approach. A pilot study, consisting of both computer simulations (modelling) and field measurements were carried out in October 2011. Preliminary results were published by Barton *et al.* (2011: 11–12), here we present the re-processed final results.

Results

The first step involved carrying out computer simulations (3D density modelling). Based on preliminary geodetic work within the Newgrange chamber, the shape and dimensions of the chamber (and passageway) were approximated by several polyhedral bodies and their gravity effect was calculated (Fig. 1). The target is made up of two elements: the passage and the chamber. The dimensions of the passageway are *c.* 20m long, *c.* 2m high, *c.* 2m wide (maximum). The chamber is cross-shaped and maximum dimensions are *c.* 6m long, *c.* 6m wide and 6m high.

This simulation showed that the calculated effect on gravity of the chamber volume would produce a measurable negative gravity anomaly with a magnitude of minus 25 microGals (detectable with state-of-the-art gravity meters). Subsequently, measurements were taken on-site (Fig. 2), using Scintrex CG-3M and CG-5 gravity meters. Field measurements were made over the known chamber along lines with a 1m station separation. The horizontal and vertical position of each of the 149 measurement points was determined using a combination of differential GPS and laser tachymetry. Strong winds at the time of survey required special wooden windshields to be used to protect the instruments from gusts (Fig. 2) and each measurement had to be verified by repeat readings. The average error from repeated and independently controlled measurement points was ±15 microGals. This is a relatively high, but still acceptable value.

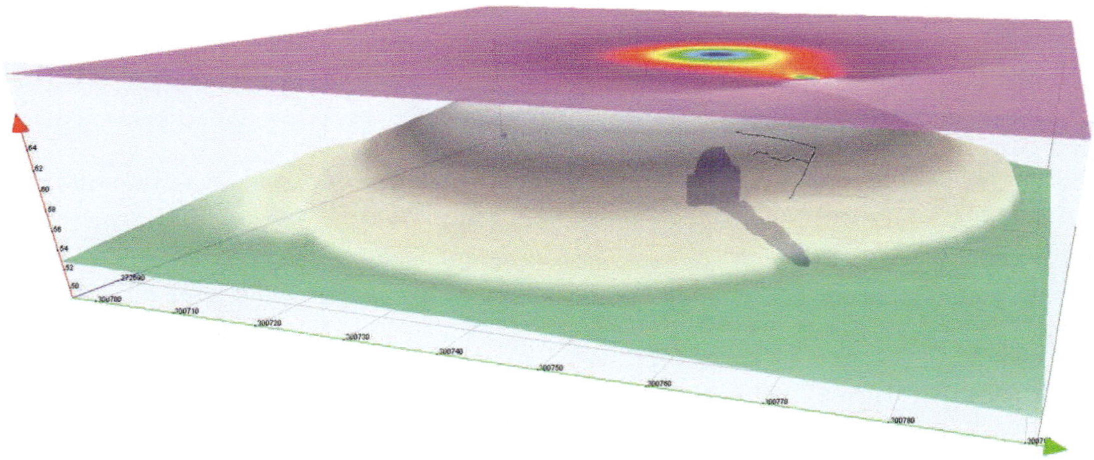

Fig. 1. Results from the synthetic density modelling - a 3D view of the Newgrange passage tomb together with the modelled passageway/chamber structure and its simulated gravitational anomaly for measurements that would be made on the top surface of the mound.

Fig. 2. Gravity data acquisition using a Scintrex CG-3M and CG-5 on the surface of the Newgrange mound – over the chamber and a part of the passageway.

Acquired data were processed to remove the effects of elevation, tidal variation, topography, latitude, and instrument drift. The removal of the effect of surrounding topography on the measurements was greatly facilitated by the availability of high-resolution (1m x 1m) LiDAR digital elevation data. The processing output was a gravity anomaly map (Bouguer anomalies) for the selected correction density 2200kg·m^{-3} (Fig. 3). This map shows a strong negative anomaly over the centre of the chamber – with a magnitude of approximately minus 70 microGals. The size of the anomaly at its centre is several times larger than the precision of the instruments. The passageway leading to the chamber could not be detected - its dimensions are too small and depth too great to be detected. The magnitude of the main anomaly over the chamber was several times larger than its modelled value. During the reconstruction of the Newgrange mound, about 100 lengths of hollow concrete pipe were used to embed a structure over the chamber (to protect it from the weight of the overlying cairn) and these objects contributed significantly to the final anomalous gravity signal by lowering it (later verified by additional density modelling). It is thus necessary to test the method in the future at a site with an intact unreconstructed chamber.

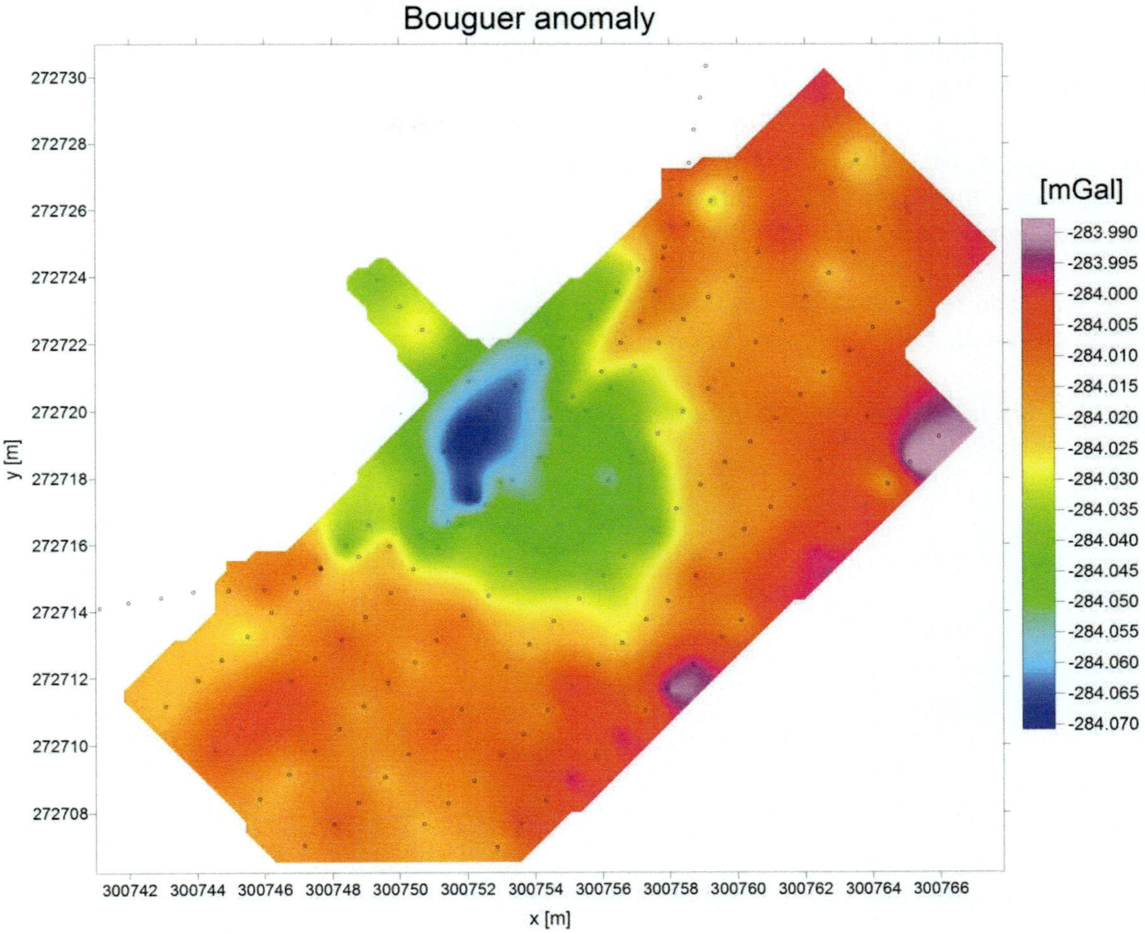

Fig. 3. Main result of microgravity data acquisition – the Bouguer anomaly map (correction density: 2200kg·m^{-3}) on the surface of the mound. Dots represent positions of measured gravity points. Units: milliGals (1 milliGal = 10^{-5} m·s^{-2}).

Conclusions

Results obtained from this project are very encouraging. Some of the chambers in the large mounds in the Brú na Bóinne WHS are quite large and are ideal targets for microgravity. We would like to continue with this research, because there is still a chance that chambers remain undiscovered beneath the three great mounds. This aim is in line with the research objective of exploring the structural sequence, phasing and interpretation of the passage tombs as set out in the *Brú na Bóinne World Heritage Site Research Framework* (Smyth 2009).

Acknowledgements

This research was carried out by scientists from Comenius University, the Slovakian Academy of Sciences and the University of Technology, Bratislava, Slovakia, in collaboration with Dundalk Institute of Technology (DkIT), Landscape and Geophysical Services (LGS), the National Monuments Service of the Department of Arts Heritage and the Gaeltacht (NMS) facilitated by the Office of Public Works (OPW). First author would like to acknowledge projects VEGA 1/0462/16 and COST action CA17131 "SAGA". LiDAR data courtesy of Meath County Council and the Discovery Programme.

References

Abad, I.R., García, F.G, Abad, I.R., Blanco, M.R., Conesa, J.L.M, Marco, J.B. and Liadro, R.C. 2007. Non-destructive assessment of a buried rainwater cistern at the Carthusian Monastery 'Vall de Crist' (Spain, 14th century) derived by microgravimetric 2D modelling. *Journal of Cultural Heritage.* 8. 197–201.

Barton, K., Pašteka, R., Zahorec, P., Papčo J. and Brady, C. 2011. In search of hidden chambers at Newgrange Passage Tomb. *Archaeology Ireland.* 25(4). 11–12.

Lakshmanan, J. and Montlucon, J. 1987. Microgravity probes the Great Pyramid. *Geophysics: The Leading Edge of Exploration.* 6(1). 10–17.

Panisova, J., Fraštia, M., Wunderlich, T., Pašteka, R. and Kušnirák D. 2013. Microgravity and Ground-penetrating Radar Investigations of Subsurface Features at the St. Catherine's Monastery, Slovakia. *Archaeological Prospection.* 20(3). 163–174.

Pašteka, R. and Zahorec, P. 2000. Interpretation of microgravimetrical anomalies in the region of the former church of St. Catherine, Dechtice. *Contribution to Geophysics and Geodesy.* 30(4). 373–387.

Smyth, J. 2009. *Brú na Bóinne World Heritage Site Research Framework.* Kilkenny. The Heritage Council.

When the Norsemen return: Complementary GPR surveys at the Viking Age site of Woodstown, County Waterford, Ireland

Petra Schneidhofer[1], Christer Tonning[1], James Eogan[2], Erich Nau[3], Alois Hinterleitner[4], Ian Russell[5], Terje Gansum[1], Wolfgang Neubauer[4,6]

[1]Vestfold Fylkeskommune, 3126 Tønsberg, Norway, [2]Transport Infrastructure Ireland, Tramore, Ireland, [3]Norwegian Institute for Cultural Heritage Research, Oslo, Norway, [4]Ludwig Boltzmann Institute for Archaeological Prospection and Virtual Archaoelogy, Vienna, Austria, [5]Archaeological Consultancy Services Unit, Drogheda, Ireland, [6]Vienna Institute for Archaeological Science, Vienna, Austria

Introduction

Woodstown (Fig. 1) is located on the southern bank of the River Suir, *c.* 6km to the west of Waterford in County Waterford, Ireland. The site was discovered in 2003 over the course of archaeological investigations prior to the construction of the N25 Waterford City Bypass carried out by Archaeological Consultancy Services Ltd. Test trenches as well as excavations revealed a large number of archaeological features including postholes, pits and two deep double ditches in the designated corridor. To gain further knowledge about the site's layout within the proposed corridor, a magnetic survey was conducted by Earthsound Archaeological Geophysics in 2004. The survey was conducted in between the already opened test trenches, as well as to the north and the south-east of the excavation area, resulting in a patchwork of prospection data. In 2007, a supplementary research project (SRP) aimed - among other things - at establishing the extent of the site, conducted further magnetic surveys and excavations in an area outside the original road corridor. The combined results of the excavations and both surveys characterised the site as a fortified settlement at the edge of the River Suir. Human activities seemed to be concentrated to the inside of the two enclosures and consisted of pits and postholes, potential hearths and/or kilns and at least two structures interpreted as houses (Russell and Hurley 2014). Finds, in particular coins, hack silver, lead weights and weapons supported an interpretation of Woodstown as a Viking Age trading post, dated to around AD 850–950. The discovery of a burial with a typical sword, spear, axe, shield and whetstone outside Enclosure 1 confirmed the presence of Vikings in Woodstown. Based on these results, the site was declared a National Monument in 2005, leading to the road corridor being moved to preserve the archaeological remains *in situ*. Plans for further investigations and the opening of the site to the public, however, had to be shelved in 2007 as Ireland was struck hard by the financial crisis; for the next eleven years, the site sank back into a slumber.

Ground-penetrating radar survey

At the initiative of the Norwegian embassy in Ireland, the Vestfold County Council (VFK) was invited in 2018 by the Waterford Museum of Treasures to perform a motorized, high-resolution ground-penetrating radar (GPR) survey in Woodstown. In the years since 2010, VFK has gained extensive experience with non-invasive investigations of Viking Age sites through their research collaboration with the Ludwig Boltzmann Institute for Archaeological Prospection and Virtual Archaeology (LBI ArchPro) and the Norwegian Institute for Cultural Heritage Research (NIKU) (Trinks *et al.* 2018). The GPR survey at Woodstown was primarily aimed at gaining further information on the extent and layout of the Viking Age settlement by providing a complementary dataset to the magnetic measurements acquired in 2004 and 2007. Data were collected using a 16 channel 400MHz MALÅ Geoscience Imaging Radar Array (MIRA) with a sampling resolution of 8cm x 4cm between the 24th-30th of May 2018 by Petra Schneidhofer and Christer Tonning, with logistic support from Mario Wallner (LBI ArchPro), James Eogan (Transport Infrastructure Ireland) and the Survey Department of the Tramore House RDO. Of the proposed 13.8ha, 10.7ha were accessible and could be surveyed over six working days. The survey area was divided into six fields (A-F) based on terrain and infrastructure. Data were processed using ApRadar and interpreted based on on georeferenced composite depth slices using specific tools developed by the LBI ArchPro within a Geographic Information System (GIS).

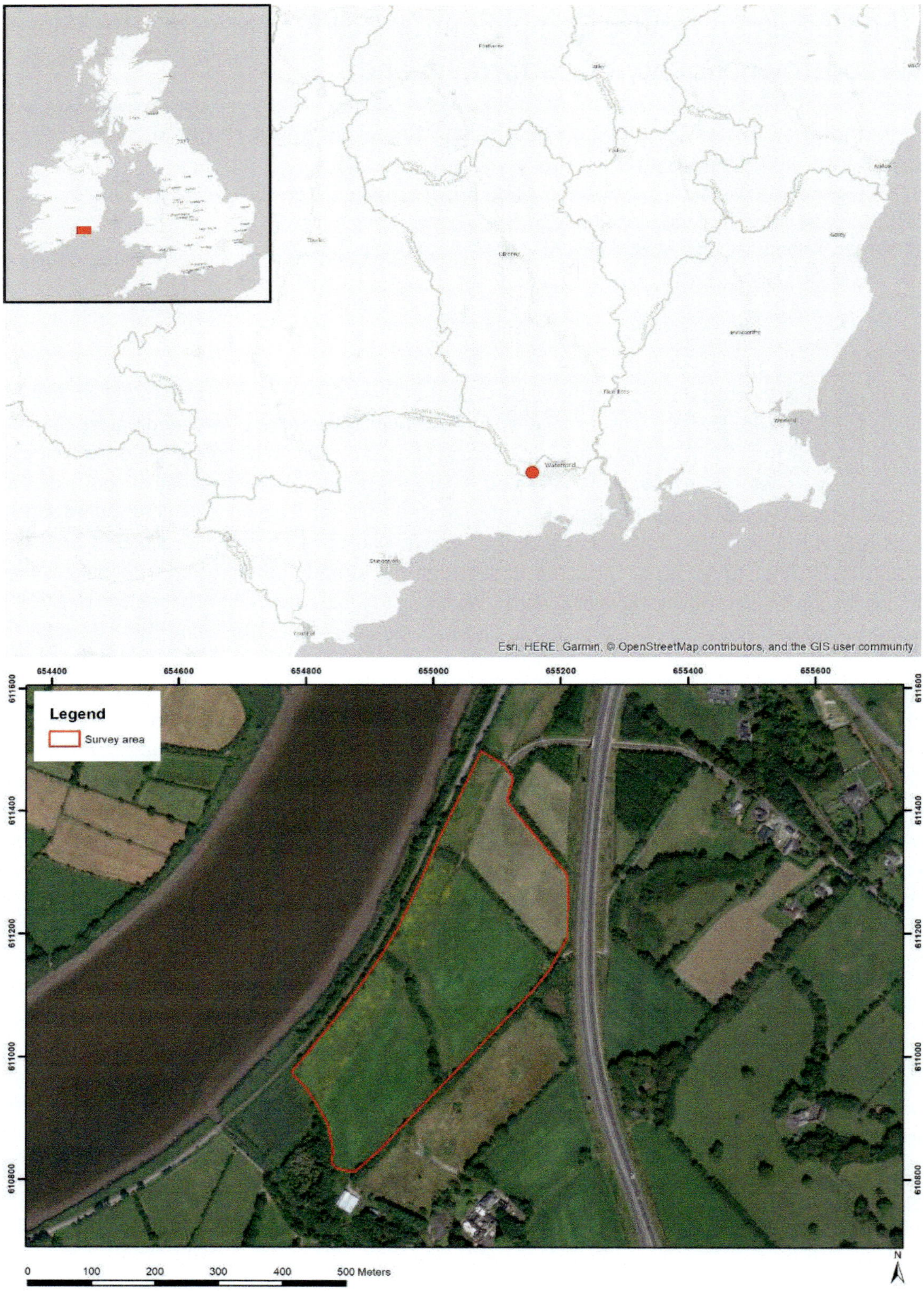

Fig. 1. Location of the site at Woodstown and delineation of the survey area.

Results and Discussions

A preliminary data interpretation revealed different types of archaeological structures, many of them previously unknown. The complementary, non-invasive approach was able to further delineate the site and provide a more detailed layout of the settlement.

Enclosures 1 and 2, the site's most defining elements which were previously detected in magnetic data and aerial images as field patterns and partially investigated through excavation, could also be identified in the

GPR data (Fig. 2). Enclosure 1 can be traced as a slightly curved double ditch of *c.* 164m length that seems to be largely consistent with the results of the magnetic surveys. The same applies to Enclosure 2, likewise consisting of two parallel ditches, which extends over *c.* 260m on a slightly curved alignment. Based on the GPR data it was possible to further trace Enclosure 2 towards its south-western limit; an area not covered by the magnetic survey. Both enclosures, but in particular Enclosure 2, feature linear structures that connect the pairs of parallel ditches at irregular intervals, seemingly forming compartments; these linear structures can also be retrospectively recognised in the magnetic data of 2007, albeit only vaguely.

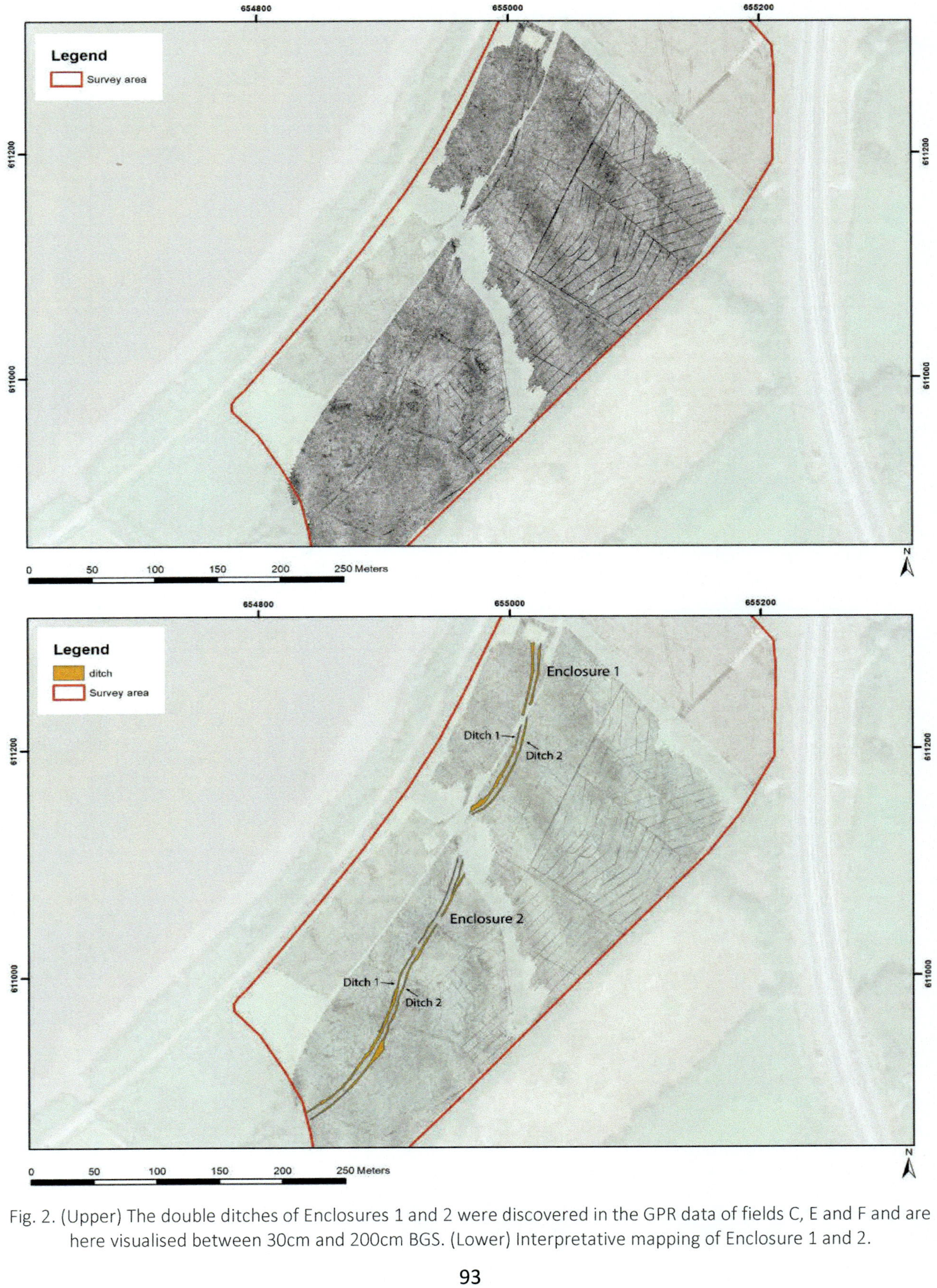

Fig. 2. (Upper) The double ditches of Enclosures 1 and 2 were discovered in the GPR data of fields C, E and F and are here visualised between 30cm and 200cm BGS. (Lower) Interpretative mapping of Enclosure 1 and 2.

Towards its north-eastern end, Enclosure 2 shows an area of faintly reflective features, perpendicularly located both inside and outside of the double ditched enclosure. Interpreted as ditches, they form roughly rectangular parcels, which are also partially visible in the magnetic data of 2007. Similar features were discovered in the GPR data at the Viking Age trading post Heimdaljordet in Gokstad, Norway and were subsequently excavated (Bill and Rødsrud 2013).

The remains of at least four buildings (5, 6, 12 and 13) could be clearly identified in the GPR data. Buildings 12 and 13 were first detected by the magnetic surveys, building 13 was excavated during the SRP in 2007. GPR data interpretation added ten further structures that can tentatively be interpreted as potential buildings, almost all of which are located within the enclosed areas (Fig. 3).

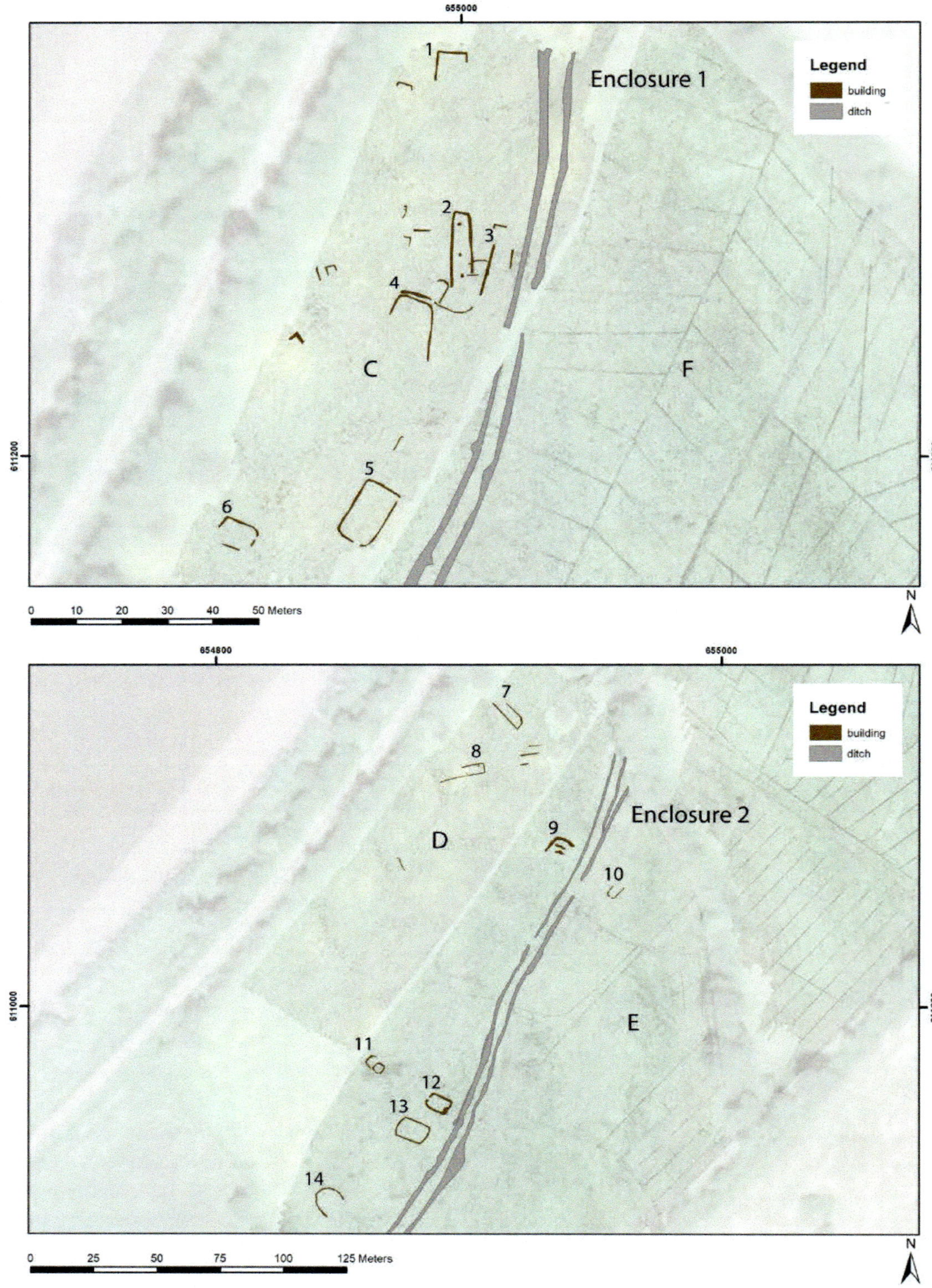

Fig. 3. Details of buildings and potential buildings (upper) inside Enclosure 1 and (lower) around Enclosure 2.

In the south-western part of field A, close to the location of a Viking Age burial discovered in 2004 (Russell and Hurley 2014), faint, linear structures as well as features resembling narrow ditches might be connected to burial structures. However, the sparse traces in the data do not yet allow a conclusive interpretation.

Conclusions

Large-scale, high-resolution GPR was successfully applied at the Viking Age site of Woodstown. The preliminary data interpretation yielded considerable contributions to the archaeological database based on the complementary nature of the method. It is hoped that this endeavour presents a new starting point for further interdisciplinary and multi-method research at Woodstown.

References

Bill, J. and Rødsrud, L. 2013. En ny markeds- og produksjonsplass ved Gokstad i Vestfold. *Nicolay*. 120. 5-12.

Russell, I. and Hurley, M.F. 2014. *Woodstown: A Viking-Age settlement in Co. Waterford*. Dublin.

Trinks, I., Hinterleitner, A., Neubauer, W. *et al.* 2018. Large-area high-resolution ground-penetrating radar measurements for archaeological prospection. *Archaeological Prospection*. 25. 171-195.

Revisiting the Segesta and the Monreale Survey Sites - The Benefits and Possibilities of Digitising Analog Archaeological Spatial Data

Doris Jetzinger, Christopher Sevara

University of Vienna, Department of Prehistoric and Historical Archaeology, Franz-Klein-Gasse 1-III, 1190 Vienna, Austria

Introduction and Research Background

As the applications of digital methods for archaeological research continue to evolve, it has become increasingly apparent that these methods can not only be used to help generate new data, but also to further process and use existing analogue and archival material. In particular, high-resolution topographic data generated from Airborne Laser Scanning (ALS) systems can provide a valuable reference source when converting analogue spatial data to digital form. As part of research related to the Prospecting Boundaries project (https://mazaro.univie.ac.at/, see also, Sevara *et al.*, this volume) at the Department of Prehistoric and Historical Archaeology of the University of Vienna, a study was conducted regarding digitising archival spatial archaeological data from western Sicily with the aid of such data and assessing the digitisation results. The study's main aim was to find useful, transferrable methods for digitising these analogue datasets as a basis for future digitising efforts. The source material consists of two maps and associated data describing the locations and contexts of archaeological sites published in two articles from the Italian conference proceedings "Giornate Internazionali di Studi sull'Area Elima". The first of these articles focuses on categorising different types of sites in the surrounding area of Segesta during the Hellenistic period (Cambi 2003). The second article presents the results of a survey conducted in a 72km^2 area in the S. Maria di Monreale territory and analyses the development of the settlement system (Johns 1992).

Approaches and Methods used in the Study

The study comprised two main parts: georeferencing and digitising the Segesta and the Monreale maps, and assessing the results. All steps were carried out in ArcMap v10.4 (ESRI 2019). The initial ArcMap project setup contained several sets of background data, which were used as a basis for georeferencing, digitising, and validation of the results. A 2m spatial resolution hillshade visualisation of an ALS-derived terrain model was employed as a basemap (Regione Siciliana 2019). Further map data, including the hydrographic network and a 1:25,000 scale map of Sicily, was obtained from the Italian National online geoportal (Geoportale Nazionale 2019) via Web Map Service (WMS).

In addition to the site locations, the Segesta and the Monreale maps only depict river courses and some basic terrain information in the form of contour lines (Fig. 1). Due to this, control points typically used for georeferencing were missing in both maps. Therefore, they were aligned and georeferenced using river forks visible in both the maps and the background data. The most important references for this process were provided by the hillshade visualisation and by the hydrographic network map. The georeferencing process proved challenging for both datasets, and the workflow had to be adapted for each of them to fit their individual characteristics. This involved multiple iterations of experimentation, result evaluation and method refinement to ensure the best possible map alignments with the control points at hand.

One of the most noteworthy features of the source maps are the markers indicating the site positions. In both maps, these site markers are of a relatively large size (see Fig. 1), although in the Segesta map they vary in size according to site type. Measurements performed with the measurement tool in ArcMap showed the Monreale site markers to be 360m in diameter, and the largest Segesta site markers to measure 300m in diameter. The size of these markers and the scale of the base map strongly influence the position uncertainty of the digitisation. Therefore, the aim of the study for the digitisations was to achieve position uncertainties smaller than the diameters of the largest site markers in the maps, i.e. smaller than 300m for the Segesta- and smaller than 360m for the Monreale digitisations. Due to the general nature of the site placement and the scale of the original maps, other types of distortion, such as scanner induced distortions and source material warping, were not deemed necessary to attempt to compensate for, although their importance in other contexts is recognised.

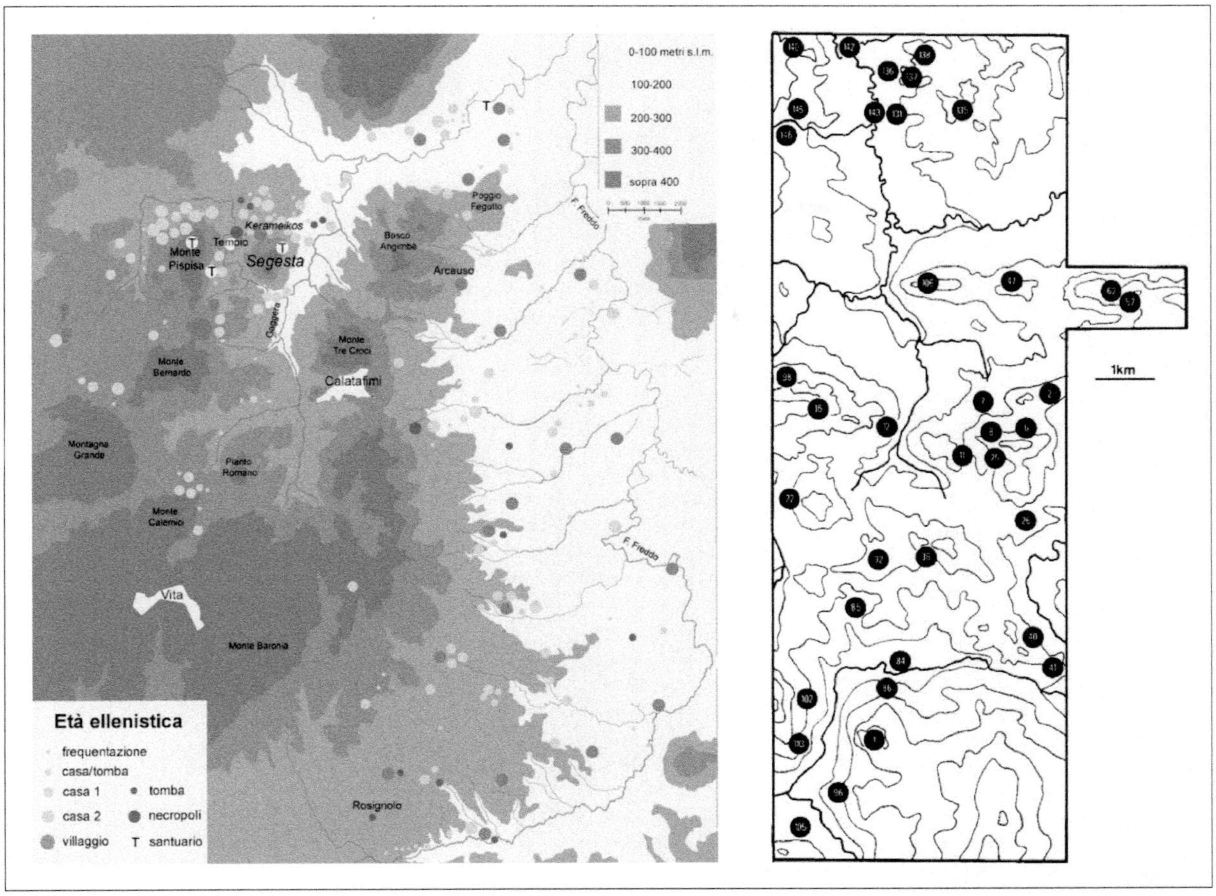

Fig. 1. The base data used in the project: the Segesta site map (left; Cambi 2003, Tab. XIX), and one of the Monreale site maps, showing the Hellenistic period sites (right; Johns 1992, Tab. XLVIII), with the respective site markers.

Segesta and Monreale digitisations and their estimated accuracy

Both datasets could be successfully digitised, with their position uncertainties ranging below the set limits (300m for Segesta and 360m for Monreale). It can therefore be assumed with reasonable confidence that the digitisations generated in the course of this study are at least as accurate as the original site locations depicted on the maps. To achieve this degree of accuracy and to estimate it with some confidence, it was crucial to mitigate causes of imprecision throughout the process. During the georeferencing, any control points featuring residual errors higher than 100m were deactivated to keep distortions as small as possible. Great care was taken to incorporate remaining map distortions into the site digitisations. This was achieved by using differences in the river course alignments between the maps and modern geospatial datasets as a reference for how far from a site marker's centre a point needed to be placed in order to incorporate any alignment inaccuracies into the site digitisation (see Fig. 2). The accuracy of the point placement was checked at regular intervals by measuring the distances between the centres of the site markers on the maps and the matching points in the digitisation shapefiles, and comparing these measurements using a threshold of 100m for acceptable accuracy.

Conclusions

The outcome of this study shows that it is possible to digitise older, analogue spatial datasets with an acceptable effort and expenditure of time, even when the source material initially lacks background data, features a poor resolution, or proves otherwise challenging. Key to the success of this is the use of modern, large area prospection data and reliable cartographic sources as a spatial reference. Digitising older, low quality analogue spatial archaeological data such as these maps provides a considerable increase in information as well as further benefits for modern GIS-based analysis. Furthermore, estimation of their general level of accuracy provides an indication of the reliability of such information and the acceptable scales at which it can be used. Moreover, the methods used here are broadly applicable. Newly created, or rather, re-created, digital datasets like these present an important possibility of adding reliable spatial information to the general pool of digital archaeological spatial data.

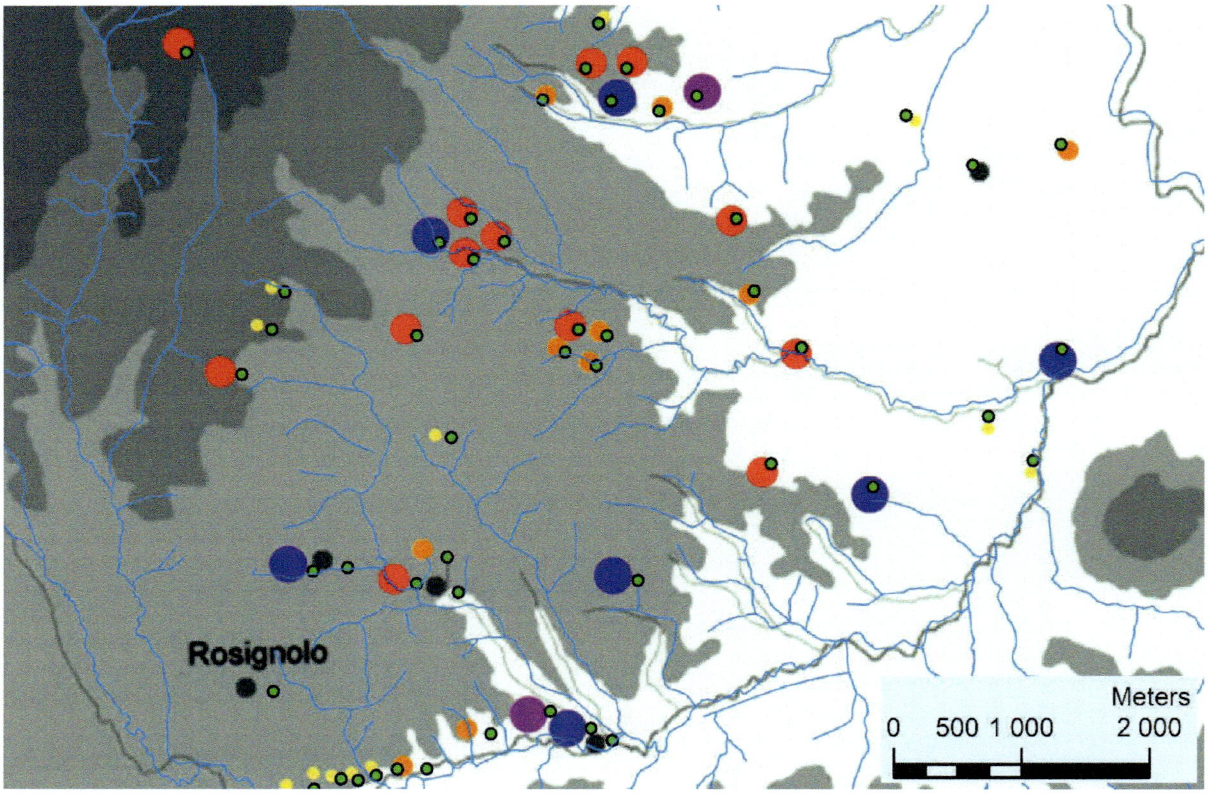

Fig. 2. Remaining distortions after georeferencing in the right bottom half of the Segesta map, visible through the differences between the river course alignments. These alignment inaccuracies have been incorporated into the site digitisations (depicted as green markers).

Acknowledgements

The Prospecting Boundaries project is funded by a grant from the Austrian Science Fund (Project ID: P-28410) and is carried out in cooperation with the Soprintendenza per I Beni Culturali ed Ambientali di Trapani. Doris Jetzinger built the research design and conducted the digitisation and assessment of the spatial data. Christopher Sevara provided the basic research concept, support for spatial data processing and background material.

References

Cambi, F. 2003. Insediamenti Ellenistici nella Sicilia Occidentale. Il caso Segestano. *Quarte Giornate Internazionali di Studi sull'area Elima, Erice, 1-4 Dicembre 2000, Atti I.* 135-169. Pisa.

ESRI. 2019. *ArcMap*. Available at: http://desktop.arcgis.com/en/arcmap/ (accessed 07 April 2019).

Geoportale Nazionale 2019. *The Italian National Geoportal.* WMS Services. Available at: http://www.pcn.minambiente.it/mattm/en/view-service-wms/ (accessed 07 April 2019).

Johns, J. 1992. Monreale Survey. *L'insediamento umano nell'alto Belice dall'età Paleolitica al 1250 D.C. Giornate Internazionali di Studi sull'area Elima, Gibellina, 19-22 Settembre 1991, Atti.* 407-420. Pisa-Gibellina.

Regione Siciliana. 2019. *Geoportale Regione Siciliana*. Available at:
http://www.sitr.regione.sicilia.it/geoportale/it/ (accessed 07 April 2019).

Integrating geophysical and geoarchaeological surveys for the reconstruction of a Roman Port infrastructure: the Claudian Harbour at Portus

Stephen Kay[1,2], Elena Pomar[1], Simon Keay[1,2], Kristian Strutt[2], Stoil Chapkanski[3], Jean-Philippe Goiran[4]
[1]British School at Rome, Rome, Italy, [2]University of Southampton, Southampton, UK, [3]University of Paris 1 Panthéon-Sorbonne, Paris, France, [4]CNRS - Université de Lyon UMR 5133, Lyon, France

Introduction

The site of Portus, the Imperial port of Rome, is located 23km to the west of Rome on the Tyrrhenian coast. The construction of the artificial harbour was initiated by the Emperor Claudius in AD 42 and it was completed sometime around AD 64, as it was commemorated on coinage around this time (Fig. 1). There followed a second major building phase under the Emperor Trajan who added a unique hexagonal inner basin connected to the Claudian harbour through an artificial channel (Fig. 2).

Fig. 1. A sestertius from the reign of Nero showing the Claudian harbour. Image Christophe Jacquand.

The importance of Portus is now well documented and the configuration and functionality of the Trajanic harbour has become better understood through targeted excavation, geophysical surveys and cores (Keay *et al.* 2005, Keay and Paroli 2011, Keay 2012). In contrast, several aspects regarding the organization of the Claudian harbour are still unclear as major parts now lay under modern infrastructure and buildings. Since the 17th century, the area of Portus started to become incorporated inland due to coastal advancement (Giraudi 2011: 21), with the result that the *Portico di Claudio* (Fig. 2) now lies 2.5km from the coastline. This progressive modification of the landscape is reflected in the discrepancy of antiquarian representations of the site, which when combined with insufficient archaeological evidence has led to various proposals for the form and location of the moles of the Claudian harbour (for an overview see Giuliani 1996, Quilici 2017).

A methodology was developed to better understand the landscape, through a combination of geophysical survey, analysis of historical cartography and the drilling of mechanical boreholes. The aim was to accurately locate the course of the northern mole (on the right of the coin in Fig. 1) and the possible location of the lighthouse.

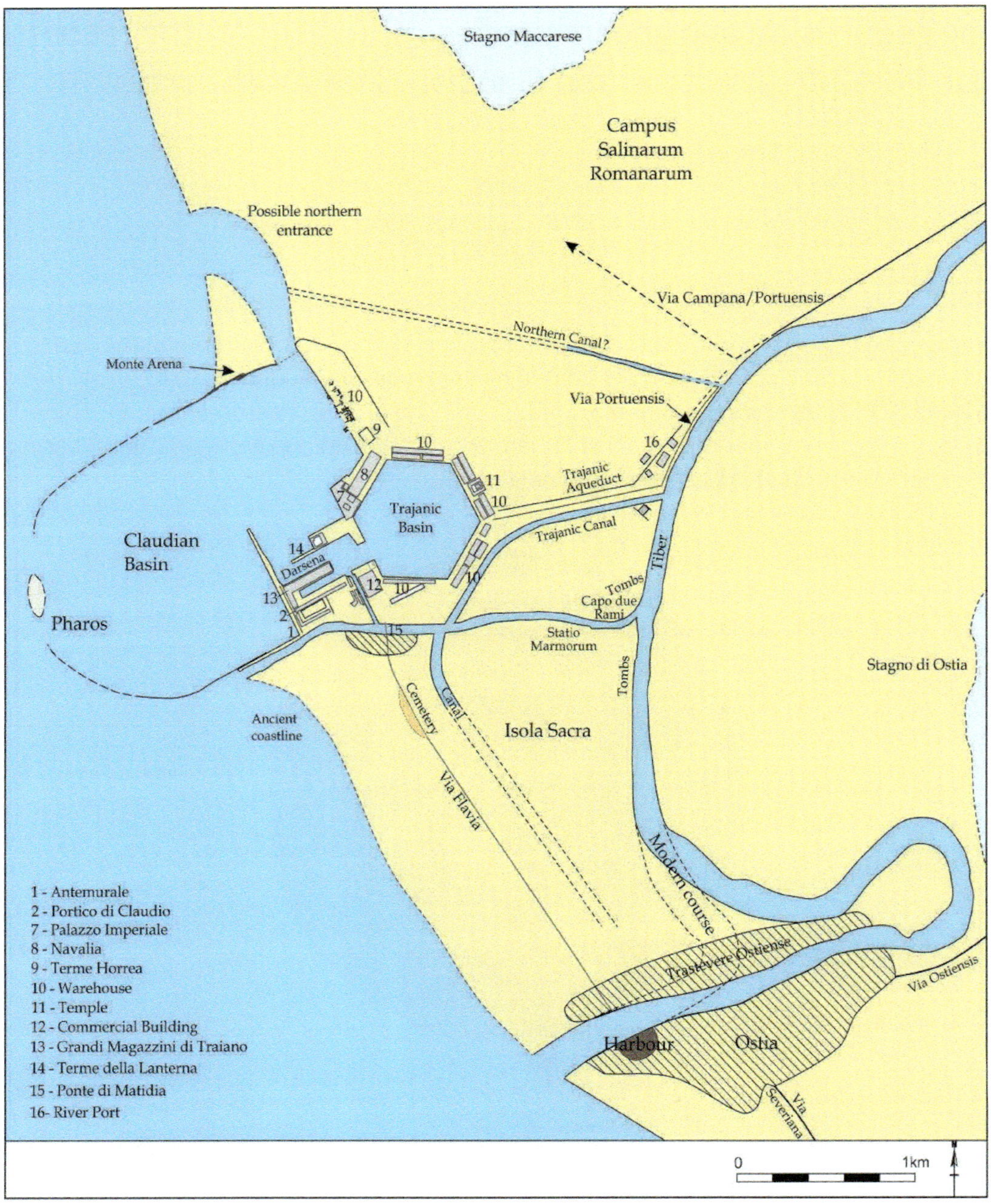

Fig. 2. The harbour complex at Portus.

Geophysical surveys of the northern mole of the Claudian harbour

The central section of the northern mole (Fig. 2, Monte Arena) was partially excavated by Testaguzza (1970) during construction for Leonardo Da Vinci Fiumicino airport. In 1998, at the request of the *Soprintendenza per i Beni Archeologici di Ostia*, magnetometry survey was undertaken (Keay *et al.* 2005: 71-75) alongside the mole and extended eastward towards Monte Giulio (Fig. 2, 10). In 2009 Ground-Penetrating Radar (GPR) was used by the Portus Project to explore parts of the isthmus between the Claudian and Trajanic harbours. Between 2006 and 2007 the archaeological authority drilled 33 mechanical cores to the west of the airport, in an area where it was expected the northern mole would continue southwest. Of these cores, 26 intercepted archaeological deposits at depths between 4.3m-15.5m below modern ground level (Morelli *et al.* 2011: 56-58).

In 2016 a preliminary GPR survey was undertaken with a 200MHz antenna to test the potential of mapping the spread of material recorded by the 2006-2007 cores. However, the shallowness of the water table in this area led to unsatisfactory results, therefore in 2017 and in subsequent seasons Electrical Resistivity Tomography (ERT) was used to assess the location and direction of the spread of material (Keay and Kay 2018).

Methodology

The ERT survey (with a configuration of 64 probes using an Allied Associates Tigre Resistivity Meter) extended across an area of 2.43ha which was surveyed in April 2017 and May 2018. The electrodes were placed at regular 2m intervals for a total length of 128m. The traverses were oriented north-east/south-west with a 4m separation for a total of 54 parallel profiles. Data were collected in ImagerPro software and processed with Res2Dinv. The processed data were georeferenced in a Geographic Information System (GIS) together with the local topography, lidar data and the results of the previous surveys. The creation facilitated the visualization of results, the display of the horizontal distribution of recorded features and increased the capability of depth correlation between features. The following stage was the drilling of 3 new mechanical boreholes to test the results of the ERT survey.

If a correct correlation could be made between the ERT profiles and the cores, the geophysics could then be used to further map the length of the mole, its depth and the hypothesised location of the lighthouse.

Results

The most prominent feature in the ERT results is a high resistivity anomaly of *c.* 100-110 Ωm, which traverses the survey area in a north-east/south-west direction (Fig. 3). The feature is recorded in continuous depth between 5m-15m below ground level, measures 220m in length and has a maximum width of approximately 50m. The significant dimensions as well as the orientation and depth suggest that the feature indicates the position of the northern Claudian mole. In order to confirm the interpretation, three mechanical cores were drilled in the survey area:

- Core CL31 was drilled north of the anomaly, potentially outside the Claudian harbour, in the sea (Fig. 3, CL31).
- Core CL30 was drilled directly above the anomaly (Fig. 3, CL30).
- Core CL32 was drilled to the east, potentially inside the harbour basin (Fig. 3, CL32).

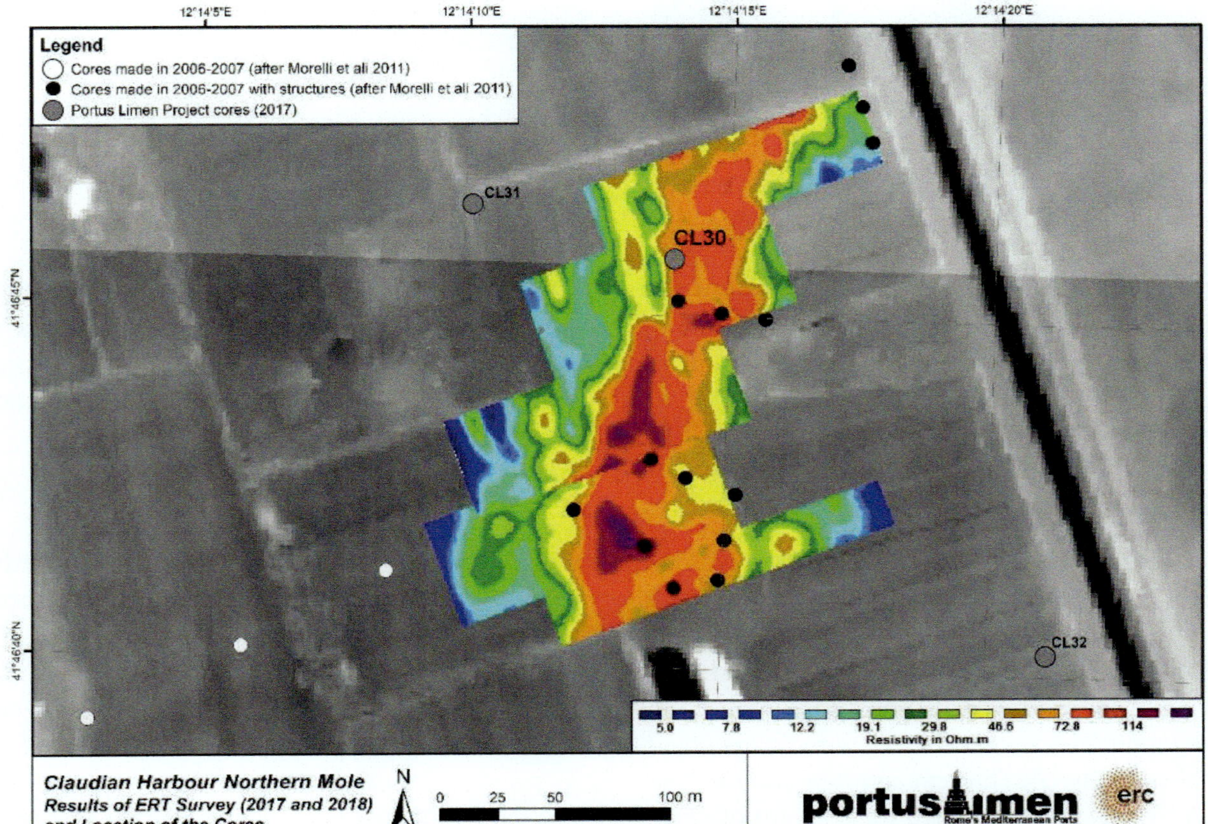

Fig. 3. Results of the ERT survey, the location of the 2006-2007 cores and the new 2017 cores.

The results of the core CL30 (Fig. 4), which was drilled through the strong anomaly, have shown that, at a depth between 4.3m-15.5m, there is a deep deposit of pozzolana, basalt, tuff and some Roman pottery.

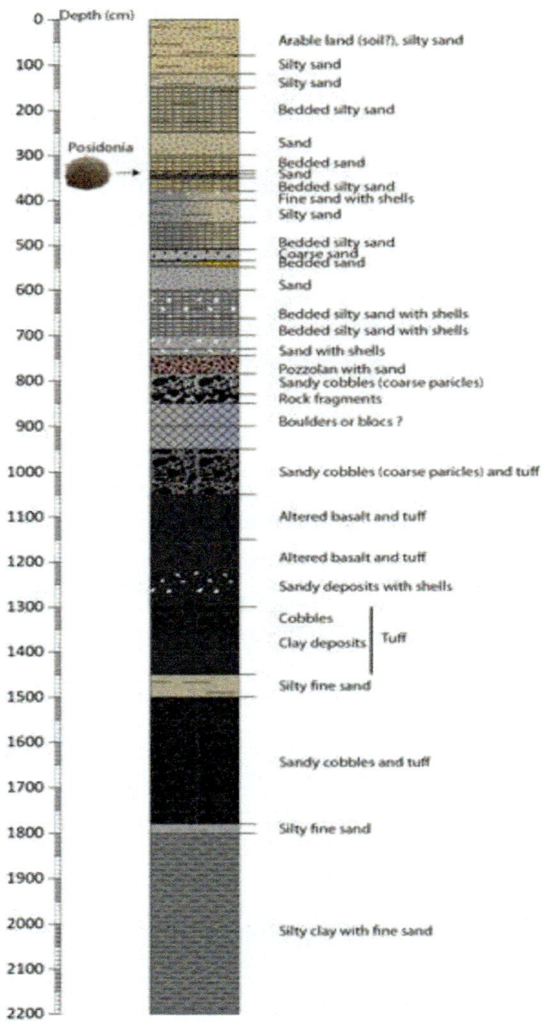

Fig. 4. Composition of core CL30.

The next stage was the mineralogical analysis of the materials used to construct the mole in order to identify the sources of raw materials used in its construction. Samples of mole materials, collected at different depths, were compared with tuff and pozzolana from different outcrops in the vicinity of Rome using Mid-Infrared Spectroscopy. The first results have suggested a choice was made to use a Tuff Lionato to build the foundation of the northern mole of the harbour.

Conclusion

The extensive ERT survey combined with targeted cores within the Claudian harbour has provided a better understanding of the location, construction and functioning of the port. Core CL32, located inside the harbour (Fig. 3) provided important information concerning the marine environment inside the port and how the mole affected the movement of water and sediment. Likewise, the information from the ERT surveys has shown that the mole was built on a continuous foundation, with a base width of approximately 50m. The size of the foundation suggests that the mole may have supported an upper structure, as perhaps indicated by the coinage (Fig. 1). Through an integration of the resistivity and coring data, the location of the northern mole can now be established for a length of *c.* 1704m. The methodology of combining ERT with targeted cores and subsequent petrographic analysis provides a useful framework for the combination of these techniques to study a port environment.

Acknowledgements

The research leading to these results has received funding from the European Research Council under the European Union's Seventh Framework Programme (FP7/2007-2013)/ ERC grant agreement no. 339123. The project gratefully acknowledges the support of Dott.ssa Mariarosaria Barbara, Dott. Luigi Gambaro and Dott.ssa Marina Lo Blundo (Parco Archeologico di Ostia Antica), Dott. Claudio Marangio (Fondazione Gerini), the British School at Rome and the University of Southampton.

References

Giraudi, C. 2011. La geologia dell'agro portuense nell'ambito dell'evoluzione del delta del Tevere. In S. Keay and L. Paroli (eds.), *Portus and its Hinterland. Archaeological Monograph of the British School at Rome* 18, 21-30. London.

Giuliani, C.F. 1996. Note sulla topografia di Portus. In V. Mannucci (ed.), *Il parco archeologico naturalistico del Porto di traiano: metodo e progetto*. 29-44. Roma.

Keay, S. and Kay, S. 2018. The Roman Ports Project, in *Papers of the British School at Rome*. 86. 317-320.

Keay, S., Millet, M., Paroli, L. and Strutt, K. (eds.) 2005. *Portus. An archaeological survey of the port of Imperial Rome.* Archaeological Monograph of the British School at Rome 15. London.

Keay, S. and Paroli L. (eds.) 2011. *Portus and its Hinterland.* Archaeological Monograph of the British School at Rome. 18. London.

Keay S. (ed.) 2012. *Rome, Portus and the Mediterranean*. Archaeological Monograph of the British School at Rome. 21. London.

Morelli, C. Marinucci, A. and Arnoldus-Huyzendveld, A. 2011. Il porto di Claudio: nuove scoperte. In S. Keay and L. Paroli (eds.), *Portus and its Hinterland*. Archaeological Monograph of the British School at Rome. 18. London. 47-65.

Quilici, L. 2017. Sul faro di Portus e una nota in margine al Porto di Traiano, in *Atlante Tematico di Topografia Antica*. 27. 45-64.

Testaguzza, O. 1970. *Portus: illustrazioni dei Porti di Claudio e Traiano e della città di Porto a Fiumicino*. Roma.

Extensive Geophysical Investigations to study the Archaeological Site of Norba (Norma, Central Italy)

Salvatore Piro[1], Stefania Quilici Gigli[2], Enrico Papale[1], Daniela Zamuner[1]
[1]ITABC - National Research Council, Monterotondo Scalo, Italy, [2]Università della Campania, Luigi Vanvitelli, Caserta, Italy, Caserta, Italy

Background

This study is part of the "Norba Project", jointly developed between the University of Campania region "Luigi Vanvitelli", the Municipality of Norma (Latina) and the Institute of Technologies Applied to Cultural Heritage (ITABC-CNR).

The archaeological site of Norba is located in the Latium Region, about 90km south of Rome, in Italy. The ancient town of Norba rises on a high plateau overlooking the Pontine plain. As stated in previous studies, the town has been founded in the archaic age and its most important period was between BC 450 and BC 81 when it was destroyed. During the Middle Ages, the ancient town was partially inhabited and only for a short period of time. This is one of the reasons for which the archaeological site is still so well preserved today. The city represents one of the best examples of urban town planning, with a regular layout dating back to an ancient age. During the last ten years, the site has undergone many studies, followed by archaeological excavations, which allowed several buildings and other important archaeological features to be brought to light. The conservation of the polygonal walls and other structures, has attracted the attention of the historians and archaeologists to Norba, since the beginning of the eighteenth century; it has also been used for the experimentation of the first examples of aerophotogrammetric restitution. Thanks to the work of Latium Region and the Municipality of Norma, now the whole site is part of an archaeological Park. The archaeological excavations made in a small portion of the town allowed for the recognition of a regular urban layout, marked by terraces in polygonal work.

The Project started in 2017 with the acquisition and processing of new extensive geophysical surveys to investigate unexcavated portions of the archaeological site with the aim to enhance the knowledge of the urban plan of the ancient town. Ground-penetrating Radar and the Gradiometric methods have been applied to investigate this site. In particular, GPR system SIR 3000 (GSSI), equipped with a 400MHz antenna with constant offset, was employed to survey 18 different areas close to some of the unearthed structures. Furthermore, differential magnetic surveys were carried out using the Geoscan FM256 in some areas investigated with GPR.

The results obtained from the geophysical surveys were interpreted together with the archaeologists to define the meaning of all the identified anomalies and to enhance the knowledge of the ancient town's layout and mapping.

Methods

The geophysical surveys were carried out using the Ground-penetrating Radar and the gradiometric methods. For the measurements a SIR3000 GPR system (GSSI) equipped with a 400MHz (GSSI) bistatic antenna with constant offset was employed. The horizontal spacing between parallel profiles at the site was 0.25m and 0.50m. In the investigated areas, a total of 1199 adjacent profiles across the site were collected alternatively in forward and reverse directions, employing the GSSI cart system equipped with odometer. All radar reflections within the 90 ns (two-way-travel) time window were recorded in the field as 16-bit data and 512 samples per radar scan.

Part of the area surveyed with GPR was also investigated using the gradiometric method. The surface was divided in squares of 20m x 20m and of 10m x 10m. The vertical gradient of the magnetic field was measured using a fluxgate gradiometer FM256 (Geoscan Research, UK) along parallel profiles with a horizontal spacing of 1m and with a sampling interval of 0.5m.

Results

All the GPR data were processed with GPR-SLICE v7.0 Ground Penetrating Radar Imaging Software (Goodman 2017). The basic radargram signal processing steps included: (i) post-processing pulse regaining; (ii) DC drift removal; (iii) data resampling, (iv) band pass filtering, (v) background filter and (vi) migration. With the aim of obtaining a planimetric vision of all possible anomalous bodies, the time-slice representation technique was applied using all processed profiles (Goodman and Piro 2013). Time slices for a depth of about 0.90m are represented in Fig. 1.

Fig. 1. GPR time slices at the estimated depth of 0.90 m.

The magnetic data were processed with Geoplot 3.0 software (Geoscan Research). After de-spiking, filtering and rearranging processes, the data were assembled in a contour of the vertical gradient of the total magnetic field (Fig. 2). The magnetic data were processed using the 2D cross-correlation technique in order to enhance the S/N ratio and to better define the spatial location and orientation of the possible targets (Piro *et al.* 2009). This method is a measure of the similarity between the raw data and calculated synthetic anomalies.

With the aim to have a better understanding of the subsurface, qualitative and quantitative integration of the results was employed in some of the investigated areas. For the integration process the following techniques were applied: map overlays and RGB colour composites (graphical integration), binary data analysis and cluster analysis (discrete data integration) and data sum, data product and principal component analysis (continuous data integration).

Fig. 2. Gradiometric contour map of the investigated area. Range -300 to +300nT/m.

Conclusions

The detailed analysis of the GPR survey results allowed us to recognize the organization of the sectors of the town. Taking into account the environmental conditions and the characteristics of the searched structures, the intrinsic high-resolution of the employed method has allowed the identification and recognition of weak anomalies in the internal part of the buildings. This project is still in progress and new integrated field surveys are planned over the next few months.

References

Goodman, D. and Piro, S. 2013. *GPR Remote sensing in Archaeology*. Springer. Berlin.

Piro, S. and Gabrielli, R. 2009. Multimethodological Approach to investigate Chamber Tombs in the Sabine Necropolis at Colle del Forno (CNR, Rome, Italy). *Archaeological Prospection.* 16. 111-124.

Piro, S., Papale, E., Zamuner, D. and Kuculdemirci, M. 2018. Multimethodological approach to investigate urban and suburban archaeological sites. In (eds.) Persico, R., Piro, S. and Linford N. *Innovation in Near Surface Geophysics. Instrumentation, application and data processing methods*. 461–504. ScienceDirect.

Piro, S. and Quilici Gigli, S. 2018. Indagini geofisiche per l'integrazione dei dati di scavo a Norba. In *Norba, scavi e ricerche*, Atlante Tematico di Topografia Antica, XXII Supplemento 2018. L'Erma di Bretschneider. 15-27.

Quilici Gigli, S. and Quilici, L. 2000. Sulle mura di Norba. *Atlante Tematico di Topografia Antica*. IX, 181-244.

Innovating Archaeological Investigations in Mediterranean Landscapes: Contributions from the Prospecting Boundaries Project

Christopher Sevara[1], Michael Doneus[1, 2], Erich Draganits[3], Roderick B. Salisbury[1], Cipriano Frazzetta[1], Doris Jetzinger[1], Sheba-Celina Schilk[1], Sebastiano Tusa[4, 5]

[1]University of Vienna, Department of Prehistoric and Historical Archaeology, Vienna, Austria, [2]LBI for Archaeological Prospection, Vienna, Austria, [3]University of Vienna, Department of Geodynamics and Sedimentology, Vienna, Austria, [4]Departimento dei Beni Culturale e dell'Identità Siciliana, Palermo, Italy, [5]Deceased (March 2019)

Introduction

The Prospecting Boundaries project (http://mazaro.univie.ac.at, see also, Jetzinger and Sevara, this volume) is an archaeological research project focused on understanding land use and social change in western Sicily. Between 2016 and 2019, the project explored and documented the Mazaro River corridor using integrated archaeological prospection approaches, building on prior reconnaissance work in the area (Calafato *et al.* 2001, Doneus 2007). This included the use of data obtained from archival sources, historic and modern remote sensing, directed airborne laser scanning (ALS), geoarchaeological evaluation and geophysical prospection campaigns. Intensive field surveys and artefact analysis were carried out to obtain datable material to corroborate information obtained from prospection data. Combined with the large-grain categorization of a historic landscape characterisation to represent the general time depth of the landscape, this project represents a diachronic, area-based approach to documenting and understanding the changing cultural and environmental landscapes of western Sicily.

Prospecting Boundaries: Methodological and Archaeological Developments

As part of the project research, new approaches to the processing and interpretation of historic and modern data were developed. Among these was a process for recovery of historic topographic data from archival aerial imagery in order to examine how modern landscape development has affected visibility of archaeological remains (Sevara *et al.* 2018). Using a sequence of aerial photographs from as early as the 1940s, a series of historic terrain models was generated and used to calculate changes in land use. The results showed the major impacts of change on survivability of prior iterations of land use (Fig. 1). Additionally, the project developed new tools for processing of ALS data, including specific approaches for separating archaeologically relevant information from scrub vegetation typical to the region, and a workflow to calibrate radiometric values in ALS data, allowing for the generation of true undistorted orthoimagery in the infrared spectrum (Sevara *et al.* 2019a).

Information obtained through application of these approaches yielded new discoveries regarding the archaeology of the Mazaro River corridor. One of the most significant was a previously undocumented ditched settlement on the bank of the river, where integrated prospection yielded evidence of an occupation sequence stretching from the mid-second to first millennium BC, and provided evidence of an extensive indigenous settlement during the 6th century BC (Sevara *et al.* 2017). Previously, it was thought that by that time period indigenous populations had retreated inland as colonial settlements advanced from the coasts into the hinterlands. This major new discovery has the potential to help significantly revise old hypotheses about movement and occupation of various cultural groups in western Sicily during the mid-first millennium BC.

Other significant outcomes include new discoveries at the settlement of Mokarta, where *c.* 40 new structures contemporary to the excavated portion of the settlement were identified, significantly expanding our view of the spatial layout of one of the most important Late Bronze Age settlements so far discovered in western Sicily (Sevara *et al.* 2019b). Additionally, the historic landscape characterisation approach has reiterated that while the modern structure of the landscape appears to conform largely to post-ecclesiastical land reform in the 1860s when looking at cadastral data and modern land use, there is still an abundance of surviving pre-reform land use and infrastructure. The extent of these features, in the form of extensive networks of animal-drawn vehicle tracks, relict terraces and agricultural surfaces, and quarries, was previously unknown. High-resolution ALS and orthoimagery data provided us with the ability to track and quantify the extent of these resources, and provided 3D data about cross-cutting of features for relative dating (Fig. 2). Fieldwork has confirmed a number of these interpretations. This has been critical to refining our view of the importance of pre-reform infrastructure on the development of the modern landscape and forms a basis for the next phase of project development, which will focus on understanding these changes through spatial and environmental analyses, and further investigation of relict features.

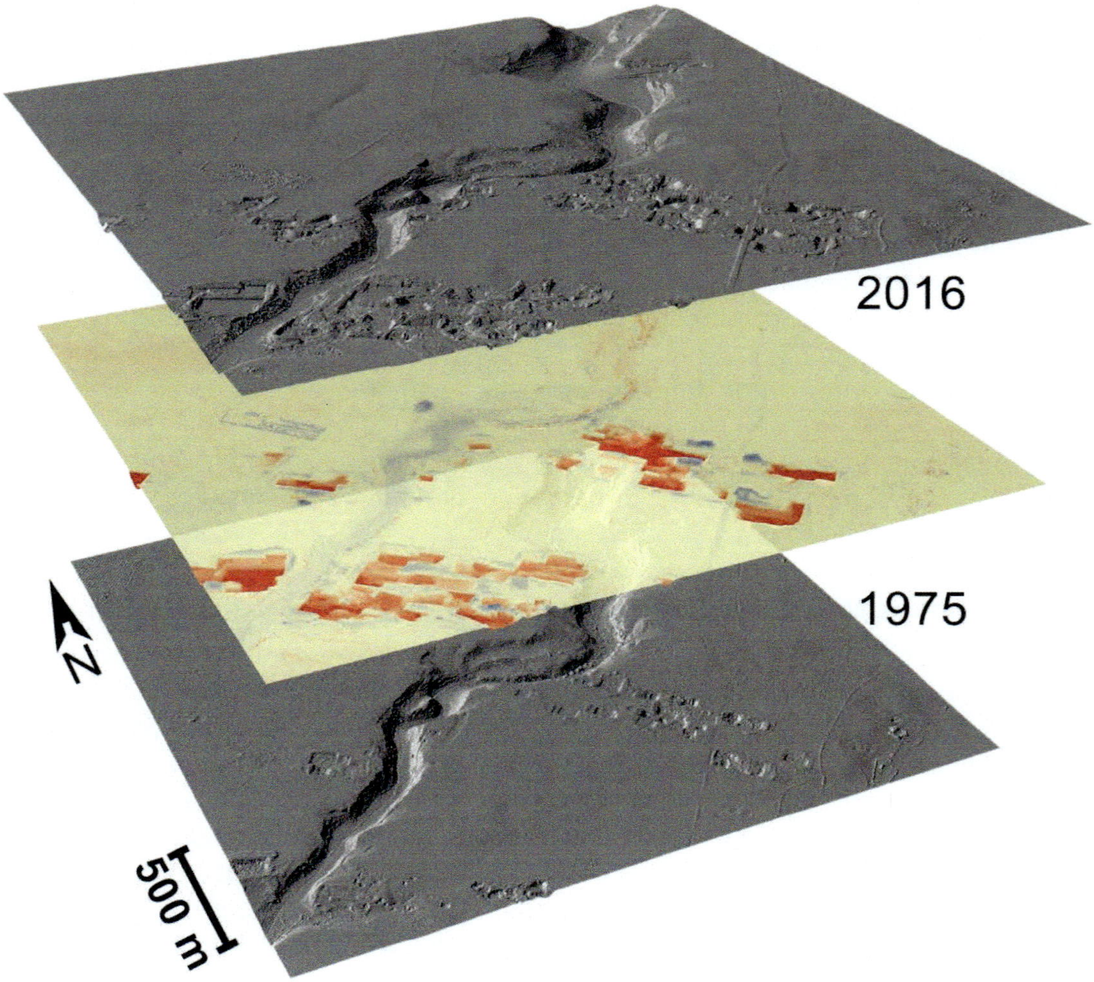

Fig. 1. Hillshade representation of an ALS elevation model from 2016 (top) and an image-based elevation model from 1975 (bottom), with difference calculation in the middle. Red indicates surface removal (loss), blue indicates deposition (gain). Corresponding surfaces are indicated in yellow. Image source: Sevara *et al.* (2018: 627).

Fig. 2. Relict vehicle track networks in the project area. Left: Track outlines visible in ALS-derived relief model visualization. Right: Representative image of a relict track set.

Conclusions

The Prospecting Boundaries Project has applied a conceptually and methodologically new integrative approach to examine the landscape of western Sicily in order to acquire new information about landscape change, past land use, infrastructure development and sociocultural interaction. In order to achieve this, several methodological approaches have been developed, including the use of historic photographs to reconstruct earlier versions of the terrain and land use, and specific data processing approaches focused on improving the extraction of archaeologically and environmentally relevant information from airborne laser scanning data. These have helped to both directly identify previously unknown archaeological resources, such as the above mentioned land use infrastructure and settlement activity, and to indirectly identify areas where we may no longer be able to see traces of earlier human activity due to extensive modification of the landscape. Furthermore, this research has produced a wealth of new archaeological information, particularly regarding activities in the mid-first millennium BC. Previously, it was thought that there were no indigenous settlements from this period in our project area, and that such groups may have retreated inland as Greek and Phoenician colonists occupied more land and moved inland from the coasts. Our research has shown that this is not the case, and that land use, occupation and interaction between indigenous and colonial groups during the first millennium BC were likely more complex than previously thought.

Acknowledgements

The Prospecting Boundaries project is funded by a grant from the Austrian Science Fund (Project ID: P-28410) and is carried out in cooperation with the Soprintendenza per I Beni Culturali ed Ambientali di Trapani. We would like to thank Dot.ssa Rossella Giglio, formerly Soprintendenza per I Beni Archeologici, Trapani Region for her continued support of our project.

References

Calafato, B., Tusa, S. and Mammina, G. 2001. *Uomo e ambiente nella storia di Mazara del Vallo - Indagine topografica nell'agro mazarese*. Grafistampa Palermo, Palermo.

Doneus, M. 2007. Aerial archaeology in the area of Mazara del Vallo, Sicily. *Report of the reconnaissance seasons 2003-2004. Archeologia Aerea, Studi di Aerotopografia Aerea.* 2. 269-284.

Sevara, C., Doneus, M., Draganits, E., Cusumano, R., Frazzetta, C., Palermo, B., Pisciotta, F., Stallone, R., Totschnig, R., Tusa, S. and Valenti, A. 2017. Testing Boundaries: Integrated Prospection from Site to Landscape in Western Sicily, In Jennings, B., Gaffney, C., Sparrow, T. and Gaffney, S. (eds.) *AP2017. 12th International Conference of Archaeological Prospection*. Archaeopress. Oxford. 230–232.

Sevara, C., Verhoeven, G., Doneus, M. and Draganits, E. 2018. Surfaces from the Visual Past: Recovering High-Resolution Terrain Data from Historic Aerial Imagery for Multitemporal Landscape Analysis. *Journal of Archaeological Method and Theory.* 25(2). 611–642.

Sevara, C., Wieser, M., Doneus, M. and Pfeifer, N. 2019a. Relative Radiometric Calibration of Airborne LiDAR Data for Archaeological Applications. *Remote Sensing.* 11(8). 945.

Sevara, C., Salisbury, R.B., Totschnig, R., Doneus, M., Löcker, K. and Tusa, S. 2019b New Discoveries at Mokarta, a Bronze Age Hilltop Settlement in Western Sicily. Forthcoming.

Looking for Etruscan harbours: geophysical survey of the ancient site of Pyrgi

Quentin Vitale[1,2], Christophe Benech[2], Jean-Philippe Goiran[2]
[1]*Éveha International, Lyon, France,* [2]*Université de Lyon, CNRS, Archeorient, UMR 5133, Maison de l'Orient et de la Méditerranée, France*

Introduction

The Etruscans were one of the first important civilization on the Italian peninsula. They emerged around the Late Bronze Age (12th–10th centuries BC) and the early Iron Age (Villanovian period, 10th–8th centuries BC) and, by the 9th century BC, they controlled territories around the Tyrrhenian coast extending from modern day Tuscany, Lazio, and Umbria. They benefited from the many mineral resources available on their territory to establish an economic and military power, based on overseas trade through a network of inland urban city centres and coastal trading posts (*emporia*). According to ancient texts, the Etruscans were both merchants and pirates forming a "thalassocracy" relying on both commercial harbour basins and military naval bases. Despite the importance of the Etruscan sea power and the numerous archaeological studies along the Tyrrhenian coastline, no evidences of Etruscan harbour activity have been discovered yet (Enei 2008).

In the 1990s, based on aerial photographs of the region, Frau (1990) assumed that the Etruscan settlement of Pyrgi (modern-day Santa Severa, Fig. 1) could have used former inland lagoons instead of large maritime harbour basins (Fig. 2.B). Etruscan port structures would hence be sought inland and not offshore (Frau 1990).

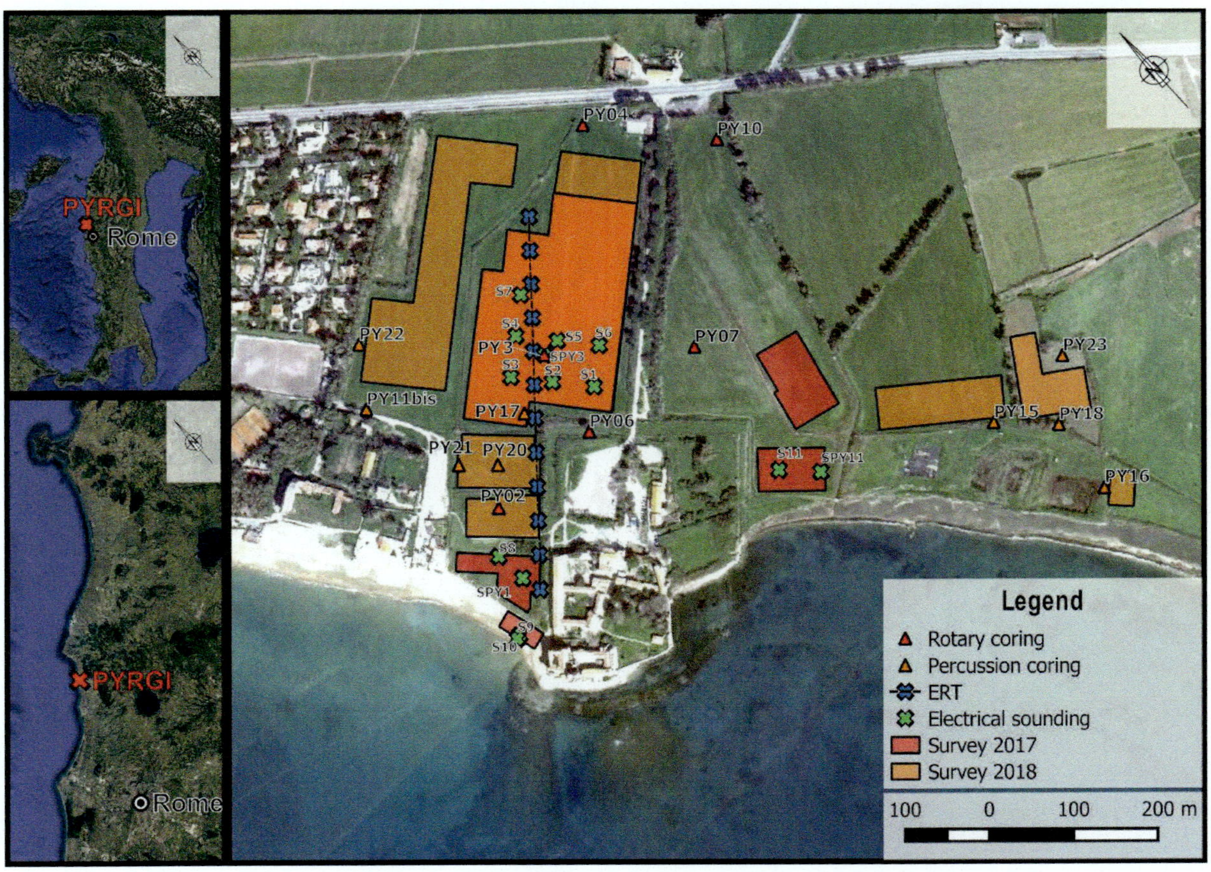

Fig. 1. Map of the archaeological site of Pyrgi, with the location of the different geophysical and coring methods used.

In order to test this hypothesis, two expeditions were conducted during the last two years (2017-2018) at the site of Pyrgi, using a multidisciplinary approach involving geophysical surveys and sedimentological analysis (coring) to find evidence of an ancient inland lagoon that could hold the traces of the Etruscan port Pyrgi.

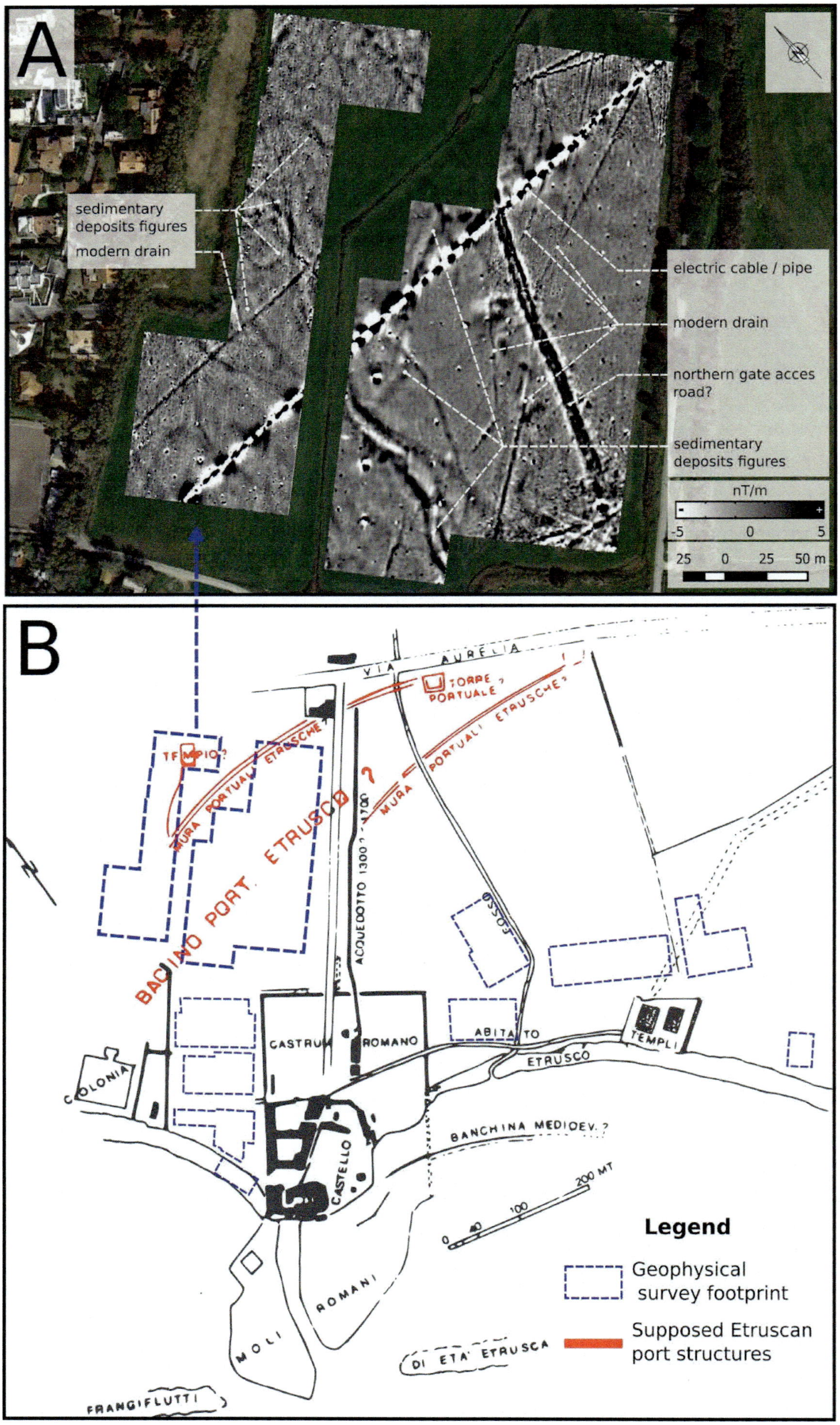

Fig. 2. A: Magnetic map of the North part of the site of Pyrgi. B: Comparison of the geophysical surveys' footprints and the expected Etruscan port structures' location (based on Frau 1990).

Methodological approach

To assess the geomorphological background of the site, sedimentary cores (rotary coring) were drilled at six different locations distributed on the site. In combination, electromagnetic surveys, electrical sounding and an electrical resistivity tomography (ERT) profile were carried out in the fields around the ancient city walls of Pyrgi. In order to look for archaeological structures, magnetic surveys were also carried out. The magnetic data were processed with the open-source software WuMapPy (Marty *et al.* 2015, see also Vitale *et al.* this volume).

The locations of the sedimentary cores and the geophysical surveys are shown on Fig. 1. The surveyed areas are concentrated on the North-West part of the site where most of the lagoon is expected to be, according to Frau's hypothesis (Fig. 2).

Preliminary results

The preliminary results focus on the sedimentary core analysis as wells as the results of the electrical sounding and the magnetic survey interpretation in the north-west part of the site.

The sedimentary description of the cores revealed that the major stratigraphic layers are silt-clay deposits overlying a sandstone bedrock (Fig. 3). Silt-clay sediments are characteristic of a low energy depositional environment compatible with a lagoon or a protected harbour.

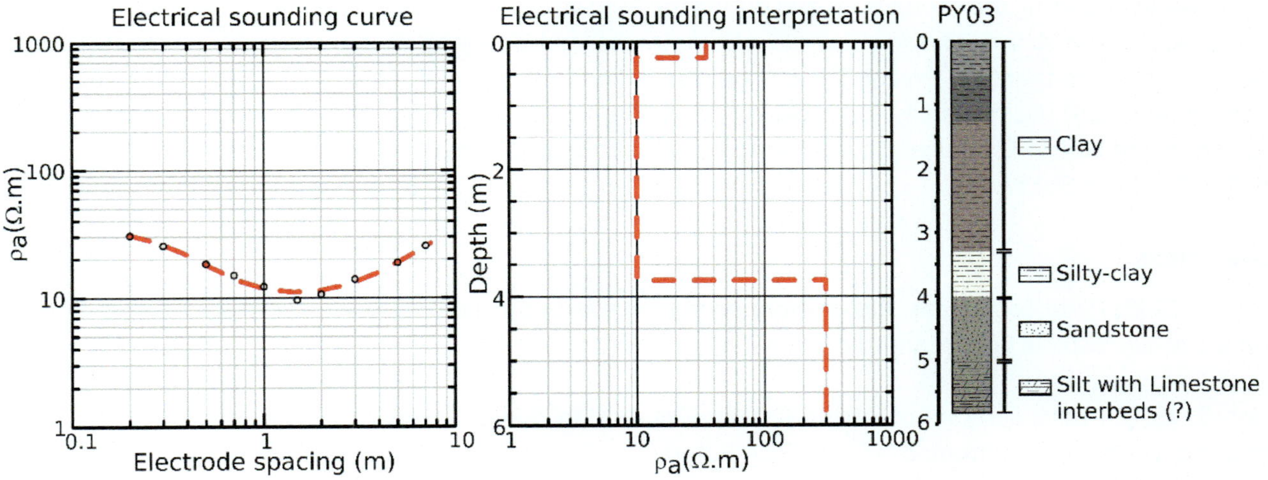

Fig. 3. Electrical sounding data and interpretation at the PY3 core location highlighting the interface between the silt-clay deposits and the overlying sandstone bedrock.

The contrasting electrical resistivities of those two sedimentary units make it possible to follow the clay-sandstone interface and estimate the clay thickness using the different electrical soundings. The estimated clay thickness is a maximum of 4m.

The magnetic survey highlighted structures with the characteristics of sedimentary deposits similar to a palaeochannel meander, road elements (Fig. 2.A) and buildings feature within the proximity of the ancient city walls (not shown), confirming that the ancient urban network continues beyond known city limits along a main road passing close from the temple area. No other archaeological structure appears on the map.

Discussion and conclusion

After correlating the cores, the sandstone bedrock in this area appears to be higher than relative sea-level during the Roman epoch (Rovere *et al.* 2011), which conflicts with the existence of an ancient lagoon located north-west of Pyrgi. In addition, the sedimentary structure recalling a meander in the magnetic map is coherent with an alluvial-origin for the clay deposits (flood plain clay) rather than a lagoon deposit. Furthermore, if associated with a lagoon, the maximum clay thickness of 4m, which gives an insight of the lagoon's navigability, would be insufficient for port activities. Finally however, there are no traces of harbour structures (a port wall) or other archaeological structures (a temple) proposed by Frau in any of the geophysical maps so far (Fig. 2). As such, these preliminary results disagree with Frau's hypothesis regarding an inland lagoon related to Etruscan harbour settlements. These findings, however, are not conclusive as it only covers one area of Pyrgi. Future research is needed to establish a higher spatial resolution for a more detailed palaeoenvironmental reconstruction of the area.

Although the geophysical coverage at the site contains several gaps and is yet to be completed, it has revealed interesting preliminary results. Thus, an access road to the northern city gate, as well as what appears to be an eastern neighbourhood (not shown), have been identified on the magnetic map.

The geophysical survey, however, is not fully completed at this stage. The preliminary results presented here reveal promising information concerning the urban network and the ancient landscape of Pyrgi. It will be possible to better reconstruct the geomorphological landscape with the completion of the geophysical surveys and the interpretation of the electromagnetic surveys and the electrical resistivity tomography cross-section.

Acknowledgements

The authors would like to deeply thank Dr Flavio Enei, Director of the *Museo del Mare e della Navigazione Antica* at Santa Severa, and his team. The help and assistance were much appreciated and facilitated every aspect while preparing the expeditions and during the fieldwork. Dr Rossella Zaccagnin of the *Soprintendenza Archeologia, Belle Arti e Paesaggio per l'area metropolitana di Roma, la provincia di Viterbo e l'Etruria meridionale* is also thanked. Finally, the authors would like to thank all the people that actively participated in the fieldwork of this research, and especially Stoil Chapkanski, Yves Bière, Vincent Gaertner, Emmanuelle Régagnon and Solène Chevalier. This work was partly performed under the "EHUES-PYRGI" project, funded by the MSH Lyon St-Etienne.

References

Enei, F. 2008. *Pyrgi sommersa. Ricognizioni archeologiche subacquee nel porto dell'antica Caere, Pyrgi – Santa Severa*. Historia Editore Viterbo.

Frau, B. 1990. I porti ceretani di Pyrgi e Castrum Novum. In Maffei A. and Nastasi F. (eds.), *Caere e il suo territorio*. Da Agylla a Centumcellae. 319-327. Rome.

Marty, P., Darras, L., Tabbagh, J., Benech, C., Simon, F-X. and Thiesson, J. 2015. WuMapPy – an open-source software for geophysical prospection data processing. *Archaeologia Polona*. 53. 563-566.

Rovere, A., Antonioli, F., Enei, F. and Giorgi, S. 2011. Relative sea level change at the archaeological site of Pyrgi (Santa Severa, Rome) during the last seven millennia. *Quaternary International*. 232(1-2). 82-91.

Investigating a tumulus in the Etruscan necropolis of Banditaccia – Applying multiple non-invasive prospection methods on a World Heritage Site

Mario Wallner[1], Immo Trinks[1], Ingrid Schlögel[2], Alois Hinterleitner[1,2], Hannes Schiel[1,2], Tanja Trausmuth[1], Fabio Colivicchi[3], Maurizio Forte[4], Wolfgang Neubauer[1]

[1]Ludwig Boltzmann Institute for Archaeological Prospection and Virtual Archaeology, Vienna, Austria, [2]Zentralanstalt für Meteorologie und Geodynamik, Vienna, Austria, [3]Queen's University, Classics Department, Kingston, Canada, [4]Duke University, Department of Art, Art History and Visual Studies, Durham, USA

Background

The Etruscan city of Cerveteri (Latin *Caere*) was, from the 9th century BC onwards until its incorporation into the Roman political system, one of the most important settlements in the Mediterranean world. The city was founded on a plateau surrounded by necropolises. The most famous and monumental necropolis is that of Banditaccia. The high-quality architecture and enormous extension of these burial grounds granted Cerveteri, in the year 2004, the status of a World Heritage Site. The architectonical symbol of the necropolises of Cerveteri is the burial mound (*tumulus*), a structure that consists of a circular trench dug into the tuff bedrock and accumulating a large conical barrow of earth. Each tumulus may contain one or more chamber tombs which are accessible through a corridor (*dromos*).

The burial grounds of Caere have been known and excavated for centuries, with a concentration of effort on the central sector of the Banditaccia cemetery. However, very large sectors of the necropolis lie outside the best-known area and have been only partially explored and sparsely published (Tartara 2018). Our research project was therefore focused to go beyond the information extracted from the aerial photos, looking deeper into the ground to map the internal structures of the tumuli. The case study is a burial mound located North of the Recinto, in a sector that has seen little systematic work.

Methods

Aerial imagery

The vast extent of the Etruscan necropolises surrounding Cerveteri is best recognisable on various open source aerial photographs. The visible vegetation marks show in particular those areas with a thinner top soil layer above the tumuli and their entrance areas. The continuation of the densely occupied necropolis outside the fenced off area of Banditaccia is clearly visible.

Ground-Penetrating radar (GPR)

To gain detailed information on the still unexcavated burial structures, the use of GPR antennae with different operating frequencies was tested in a field with clearly expressed vegetation marks. Thus it was ensured that actual burial structures were present beneath the surface. In total an area of 4ha was surveyed with a motorised six-channel 500MHz Sensors & Software SPIDAR GPR array. The central part of the area of interest was re-mapped using a shielded single-channel 160MHz MALÅ GX HD GPR antenna with a cross-line spacing of 50cm.

Electrical resistivity tomography

As a complementary near-surface geophysical prospection method, 2D electrical resistivity tomography (ERT) measurements were conducted with an ABEM Terrameter SAS 1000 with one-meter electrode spacing.

Image based modelling

By coincidence, during the geophysical prospection surveys, a hole in the ground was discovered, that appears to have opened over a partially collapsed roof of a burial chamber. Thus, it became possible to digitise this burial chamber using image-based modelling (IBM). The mapped 3D volume had also been covered with the different GPR instruments as well as the ERT measurements.

Data processing

The processing of the GPR data was conducted using the software ApRadar. Data interpolation, channel balancing, adaptive time-zero adjustments, background removal and frequency filtering as well as the Kirchhoff migration were applied to improve the imaging of the buried structures (Trinks *et al.* 2018). Furthermore, a depth-variable velocity model has been used for time-to-depth conversion and the dataset was topographically corrected. From the 3D data-volume, greyscale GPR depth-slice images were generated with various thicknesses, as well as a 3D XYZ+amplitude point-cloud.

Data analysis

The visualisation of the reflection amplitudes of GPR data in form of a 3D point-cloud (Kamp *et al.* 2014) avoids the loss of spatial information that comes along with the reduction of the volume data to 2D raster visualisations. The traditional and common approach in analysing archaeological prospection GPR datasets is the computation of 2D data representations, the subsequent interpretation of the archaeological relevant structures, and finally the reintroduction of the third dimension by extruding interpretation polygons to 2½ presentations (Trinks *et al.* 2018, Torrejón *et al.* 2016).

GPR data is by its very nature three-dimensional. Therefore, the generation of a 3D point-cloud visualisation is suited for highly accurate representations and visualisation of such data. Other promising novel 3D visualisation approaches for GPR data comprise the application of medical 3D imaging techniques for advanced iso-surface renderings (Bornik *et al.* 2018, see also Neubauer *et al.*, this volume).

With the increasing computational and visualisation possibilities offered by modern workstations it has become possible to load dense point-clouds into open source visualisation software, such as CloudCompare (CloudCompare). Within the three-dimensional environments we are then able to apply suitable filters to edit the point-cloud for an improved visualisation of the archaeological structures of interest (Fig. 1).

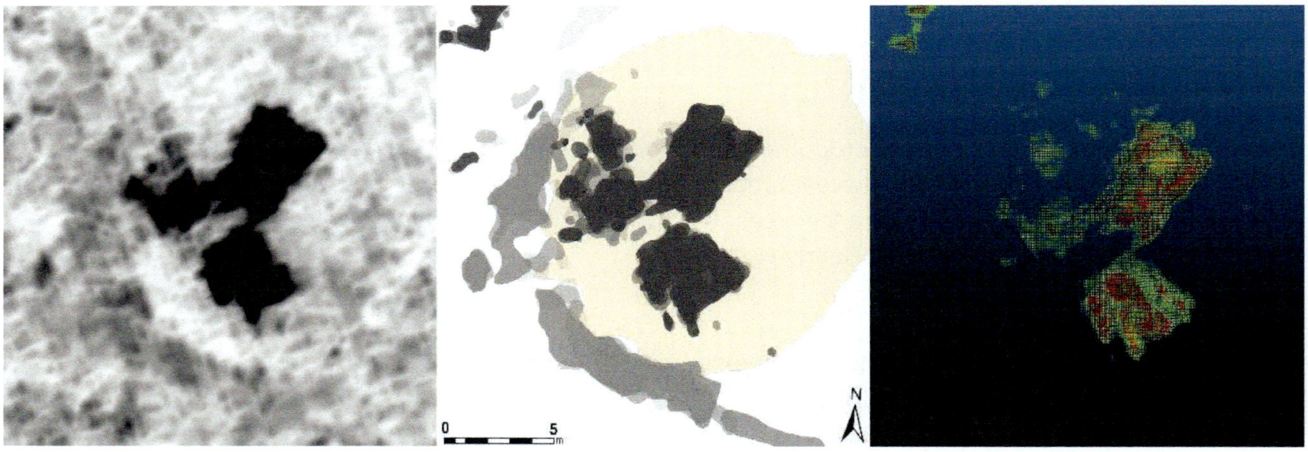

Fig. 1. Left: 160MHz GPR depth-slice 40cm to 250cm. Centre: interpretative mapping of GPR data (grey=path, black=burial chamber, ochre=tumuli). Right: 3D point-cloud amplitude rendering derived from the GPR dataset.

Results

The integrated analysis of vegetation marks, ERT measurements and high-resolution GPR data offered the opportunity to investigate an Etruscan necropolis at the modern town of Cerveteri in great detail, revealing valuable information on the state of preservation of the buried archaeology.

By integrating the information derived from the different survey methods it became clear that the strong GPR amplitudes actually were not caused by stones but the cavities of the individual burial chambers. Using 3D point-cloud visualisations of IBM and GPR data, it was possible to map and model the vault of the burial chambers to such a detail that it became possible to understand even internal details (Fig. 2, Fig. 3).

Fig. 2. Combined visualisation of the 3D point-clouds derived from IBM and GPR data, with the GPR reflection amplitude expressed by different colours (blue=25, red=78).

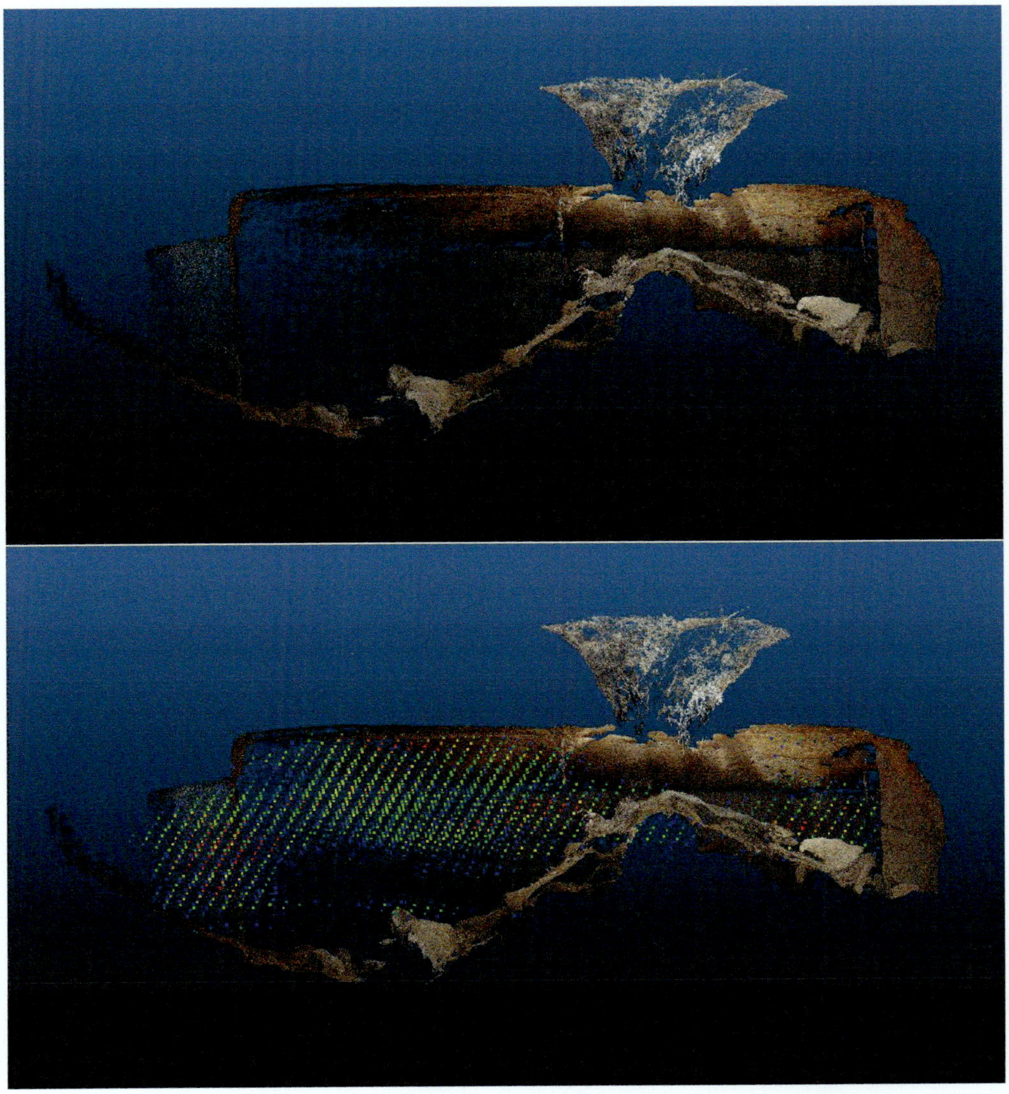

Fig. 3. Top: Section through the IBM point-cloud of the burial chamber. Bottom: Combination of the IBM and GPR point-clouds.

Conclusions

This study demonstrates the great potential offered by the integration of different archaeological prospection and documentation methods. It was possible not only to map the extent of several tumuli, but also to identify and image their burial chambers with GPR and ERT. The high level of detail of the data revealed individual chambers and showed extraordinary details regarding architectural elements, such as ceiling inclination, funeral niches and entrances.

The computation of a 3D point-cloud visualization of the GPR data illustrates the inherent information loss associated with time-consuming interpretative mapping based on 2D GPR depth-slices. The case study presented here proved once more the ability of GPR to visualise archaeological details of still buried remains even under challenging geological conditions.

References

Bornik, A., Wallner, M., Hinterleitner, A., Verhoeven, G. and Neubauer, W. 2018. Integrated Volume Visualisation of Archaeological Ground Penetrating Radar Data, In *Proceedings of the 16th Eurographics Workshop on Graphics and Cultural Heritage (GCH 2018)*, 231–234. Aire-la-Ville. The Eurographics Association.

CloudCompare: http://www.cloudcompare.org/

Kamp, N., Russ, S., Sass, O., Tiefengraber, G. and Tiefengraber, S. 2014. A Fusion of GPR- and LiDAR-Data for Surveying and Visualisation of Archaeological Structures – a case example of an archaeological site in Strettweg, District of Murtal, Austria. *EGU General Assembly Conference Abstracts*. 16. 12621.

Tartara, P. 2018. Territorio di Caere: viabilità e distribuzione delle necropoli attraverso la fotografia aerea, In A. Naso and M. Botto (eds.) *Caere Orientalizzante. Nuovi studi su città e necropoli (Studia Caeretana 1)* Roma C.N.R. 123-153.

Torrejón Valdelomar, J., Wallner, M., Trinks, I., Kucera, M., Luznik, N., Löcker, K. and Neubauer, W. 2016. Big Data in Landscape Archaeological Prospection, *ARQUEOLÓGICA 2.0 - 8th International Congress on Archaeology, Computer Graphics, Cultural Heritage and Innovation*, Valencia, Spain, September 05-07, 2016.

Trinks, I., Hinterleitner, A., Neubauer, W. *et al.* 2018. Large-area high-resolution ground-penetrating radar measurements for archaeological prospection. *Archaeological Prospection*. 25. 171–195.

Prospecting Mesolithic buried landscapes and sites. Two case studies from two different types of landscapes in the southern Netherlands

Eelco Rensink
Cultural Heritage Agency of the Netherlands, Amersfoort, Netherlands

Prospecting early prehistoric (Palaeolithic, Mesolithic and Early Neolithic) sites has a long tradition in the Pleistocene sandy areas and in the hilly loess area of the Netherlands (Rensink *et al.* 2006, Peeters *et al.* 2017). In these areas many surveys of ploughed fields and visual inspections of non-archaeological digging works have led to the discovery and (far more sporadically) excavation of numerous sites.

In the 1980s, new methods and techniques of field work (boring, trial squares) and new procedures regarding the collection of finds (sieving of sediment) were introduced, aimed at detecting and evaluating lithic scatters. In the current field of Dutch early prehistoric archaeology, the use of standardized procedures has become increasingly important for this category of sites in order to make well-founded decisions in the context of archaeological heritage management. For a long time, the focus of archaeological fieldwork was on individual early prehistoric sites. Many spatial developments in the Netherlands however, do not threaten one individual site, but larger landscape entities consisting of several or even numerous early prehistoric sites in addition to all kinds of 'off-site' phenomena representing integral components of the early prehistoric archaeological record. Therefore, detection and evaluation must also consider the characteristics of larger areas or landscape zones and the intrinsic and physical archaeological qualities of these landscapes (Rensink *et al.* 2006: 214-220). But how can we prospect and assess the value of these early prehistoric landscapes in the most accurate way?

Two case studies from the southern Netherlands are presented. The first one deals with the project area of Weerterbergen near Eindhoven in the Pleistocene sandy landscape. Because of the planned construction of a wildlife crossing, archaeological prospection was carried out by RAAP Archaeological Consultancy in various stages (Van Dijk 2010). In the first stage of field work boring was carried out in a grid of 40m x 50m to collect information on the intactness of the soil profile. The first borings with a diameter of 15cm in a grid of 15m x 20m yielded some Mesolithic stone artefacts. Using this information, zones were selected for additional borings in a more dense grid (10m x 5m and 5m x 5 m). Working this way, at least four Mesolithic concentrations were detected. Flint artefacts were found embedded in almost complete soils and showing no - or only limited - evidence for erosion or human disturbance in recent times. Subsequently, small trenches with a width of 1m yielded not only high numbers of flint artefacts, but also burned hazelnuts. This showed that the vertical dispersion of artefacts was relatively limited; for instance at location 3, 95% of the flint material was found between 30cm and 55cm below the present surface.

The second project area of Well-Aijen is located in the province of Limburg in the south-eastern part of the Netherlands. Since 1995 numerous archaeological investigations have been undertaken there along the river Meuse as part of the major 'Maaswerken' infrastructural project. To improve flood risk management, several areas of land covering a total of almost 2000ha have been (or in the near future, will be) excavated to a great depth. In anticipation of this, archaeological research was undertaken for the purposes of recording and documenting archaeological remains in the most important areas and locations (Stoepker *et al.* 2004, Rensink 2017). From 1995 to 2015 the Cultural Heritage Agency of the Netherlands (Amersfoort) was in charge of the investigations, and acted as adviser to national public works agency *Rijkswaterstaat.*

Initially the project area of Well-Aijen was investigated by means of large-scale borehole surveys and using 40m x 30m grids by RAAP Archaeological Consultancy (Van Dijk 2003). Though many finds from different archaeological periods were collected, it became clear that these surveys were of rather limited value for the identification and assessment of individual sites in the Holocene valley bottom of the Meuse. It appeared that the results of the borehole survey were difficult to translate in discrete temporally and spatially defined archaeological sites independent of their age or size. Even more importantly, the borehole surveys yielded almost no flint artefacts. As became clear later in the project, this observation was not in tune with the high

potential of the project area for the presence of buried early prehistoric sites, i.e. Mesolithic and Early Neolithic find concentrations associated with early Holocene point bars.
Rather unexpectedly, the first flint artefacts pointing to a Mesolithic concentration came to light in 2004 in one of the trial trenches that were dug by ADC ArcheoProjecten (Tichelman 2005) for the detection of archaeological features and remains from the Neolithic and possibly also younger periods. After the recovery of the first flint artefacts, the methods and strategies of field work were adapted to this new find situation, and sediments collected from small squares of 50cm x 50cm were wet sieved over a screen of 3mm. In the subsequent period of field work, and following the same procedures, high numbers of Mesolithic artefacts were recovered from several locations and associated with (at least) two stratigraphic levels about 1m-2m below the present surface and underlying a Neolithic layer (Müller *et al.* 2018). For the Netherlands this is a unique find situation, with well-preserved sites occurring in fine-grained, fluviatile (point bar) sediments dating from the Early and Middle Holocene.

Conclusions

The two case studies presented above differ in the rate of success with regard to the use of borehole surveys for detecting and evaluating Mesolithic buried landscapes and sites. Methodological evaluations, like the ones carried out for Weerterbergen and Well-Aijen, should be a standard practice in Dutch archaeology. This will improve the practice and quality of archaeological prospection and to ensure a full role of early prehistoric landscapes and sites in the cycle of archaeological heritage management in the Netherlands.

References

Dijk, X.C.C. van 2003. *Project Zandmaas, Deelgebied Well-Aijen, een Aanvullende Archeologische Inventarisatie*. RAAP-rapport 799. Amsterdam.

Dijk, X.C.C. van 2010. *Plangebied Weerterbergen, gemeente Weert en Nederweert; archeologisch vooronderzoek: een inventariserend veldonderzoek (proefsleuven)*. RAAP-rapport 1993. Weesp.

Müller, A., Tebbens, L. and Flamman, J. (eds.) 2018. *Van jager-verzamelaars naar vroege boeren langs een migrerende Maas. Onderzoek van veranderend landschap en prehistorische bewoning tussen Well en Aijen. Opgravingen in de werkvakken 2 en 4 van Hoogwatergeul Well-Aijen, gemeente Bergen (L.)*. Amersfoort.

Peeters, J., Raemaekers, D., Devriendt, I.I.J.A.L.M., Hoebe, P.W., Niekus, M., Nobles, G.R. and Schepers, M. 2017. *Paradise lost? Insights into the early prehistory of the Netherlands from development-led archaeology,* Nederlandse Archeologische Rapporten 62. Amersfoort.

Rensink, E., Deeben, J. and Peeters, H. 2006. Early prehistoric sites and landscapes in the Netherlands. Building frameworks for archaeological heritage management. In E. Rensink and J.H.M. Peeters (eds.), *Preserving the Early Past. Investigation, selection and preservation of Palaeolithic and Mesolithic sites and landscapes*, Nederlandse Archeologische Rapporten 31, 201-226. Amersfoort.

Rensink, E. 2017. Archaeological heritage management in the Meuse valley (Limburg, the Netherlands) from a national perspective: aims, methods and results. *Netherlands Journal of Geosciences – Geologie en Mijnbouw.* 96(2). 197-209.

Stoepker, H., Rensink, E. and Drenth, E. 2004. Behoud en onderzoek van archeologische waarden in het Maasdal in het kader van De Maaswerken en de Via Limburg. Wetenschappelijk beleidsplan. *Rapportage Archeologische Monumentenzorg* 111. Amersfoort.

Tichelman, G. 2005. *Archeologisch onderzoek in het kader van De Maaswerken: Inventariserend veldonderzoek (IVO), waarderende fase Well-Aijen.* ADC-rapport 404. Amersfoort.

"Where animals were equal to humans". Surprising results of complementary geophysical survey on gallows in Lower Silesia (Poland)

Michal Pisz[1], Magdalena Majorek[4], Bartosz Świątkowski[2], Daniel Wojtucki[3], Karolina Wojtucka[3], Krzysztof Narloch[5]
[1]Faculty of Geology, University of Warsaw, Warszawa, Poland, [2]University of Gdańsk, Gdańsk, Poland, [3]University of Wrocław, Wrocław, Poland, [4]University of Łódź, Łódź, Poland, [5]University of Warsaw, Warsaw, Poland

Background

Research

The survey in Chełmsko Śląskie (Lower Silesia, Poland) was carried out in April 2018 as a part of the project "Where animals were equal to humans. The former places of execution in Silesia in an interdisciplinary perspective". One of the activities within the project is archaeological research aimed at verifying places of execution, the initial locations of which were estimated based on historical and cartographic sources. The aim of the research in Chełmsko Śląskie was to determine the exact location of the former gallows (Wojtucki 2009: 25, 46, 271, 523).

Researchers intend to reveal the appearance of the place of execution and various types of archaeological finds and equipment functioning there. The research focuses on finding out how corpses of executed criminals have been treated and presented to the public.

Site

Based on archival research, it has been determined that the brick gallows construction in Chełmsko Śląskie was erected in 1593, probably in the vicinity of an older wooden gallows. A thief, Hans Schneutter, was the first to be hanged on one of its beams. This gallows was a brick construction on top of which a wooden structure was constructed with posts and beams for execution. The building was disassembled in 1826 and the construction material was probably used for raising some features of a nearby shooting range.

The approximate location of the gallows in Chełmsko Śląskie was established based on historical sources and cartographic materials. Historical maps have been georeferenced using GIS. During the first field reconnaissance in 2017 there was no visible evidence of gallows in the designated area. It was assumed that such a state of affairs could be the result of a reasonably thorough demolition, mentioned in historical sources, as well as intensive agricultural field cultivation in the recent past.

Conditions on the site

The site is located on a hill, west of the town of Chełmsko Śląskie. The hills in this area are made of conglomerate, arkose sandstones and siltstones, which are not highly magnetic. Nowadays the field is used as a meadow, which provides favourable conditions for geophysical prospection.

Methods

Geophysical prospection was planned and performed based on EAC guidelines (Schmidt *et al.* 2015). In accordance with a Level II survey (Gaffney and Gater 2003: 88-91), two complementary methods were used: magnetic gradiometry and earth resistance. Magnetic measurements have been taken with a SENSYS MX V3 multichannel gradiometer, first with a 0.5m sensor spacing, which covered an area of 1.36ha, and then with a 0.25m sensor spacing to increase the resolution of detected features. The position of the instrument was located in real time with an RTK GPS.

Earth resistance (ER) measurements were collected with a Geoscan Research RM85 meter with 1m × 1m sampling interval and a Wenner α electrode array. ER measurements were first taken with an electrode spacing a = 0.5m and repeated using a = 0.75m. Corners of the survey grids were measured with RTK GPS.

Results

The sought features were expected to provide a positive magnetic anomaly and a high resistance anomaly if foundations were preserved. Ring shapes appeared in both datasets, as expected. However, it turned out that they are displaced one from another and the registered signals must come from two different features.

The magnetic measurements provided quite a clear image of a round structure with a diameter of *c.* 8m (Fig. 1). The magnitude of the anomaly was quite low: it barely exceeded +5nT. However, the earth resistance results provided an indication of another round structure of *c.* 9m in diameter, located a dozen metres north from the magnetic anomaly (Fig. 2). This high resistance circular feature was at first interpreted as another gallows, but both anomalies were subsequently investigated via test trenches.

Archaeological verification

To verify both anomalies, four test trenches were located over the circular anomalies. The first trench examined the southern edge of the earth resistance anomaly. The excavations revealed a clearly inhomogeneous soil structure with a concentration of small sandstones, which probably caused the high resistance anomaly.

The other three test trenches examined the magnetic anomaly. In each of those trenches the occurrence of small fragments of brick and coal was recorded in the topsoil. Below the topsoil, a ceiling of the brick masonry layer was unveiled with numerous small pieces of brick and small coal, most likely a relic of the foundation of the dismantled gallows. Full excavations are intended to take place in the near future, in order to capture an overview of the structure.

Conclusions

The example of the survey in Chełmsko Śląskie points out how important complementarity is in the workflow of archaeo-geophysical research. Earth resistance survey results provided an image of a deceptively similar structure, but did not register any disturbance from the remains of the actual gallows. Test trenches provided an explanation of this result. The remains were just small pieces of brick and they are clearly not detectable with earth resistance. However, magnetic traces of the fired bricks were detectable.

Thanks to the use of non-invasive methods it was possible to confirm the location of the gallows, as recognised from the analysis of spatially referenced historic sources. The importance of this discovery is even higher because there was no other surface or remote sensing traces of the structure. Moreover, this case study shows how important an invasive verification of the geophysical prospection results might be in order to obtain a reliable interpretation of the data.

Acknowledgements

These studies are part of the project "Where animals were equal to humans. The former places of execution in Silesia in an interdisciplinary perspective" (2016/22 / E / HS3 / 00373), which is implemented as part of the SONATA-BIS grant financed by the National Centre of Science. Dr Daniel Wojtucki is the principal investigator of the project.

This work was supported by an ITC Conference Grant from COST Action SAGA: The Soil Science & Archaeo-Geophysics Alliance - CA17131 (www.saga-cost.eu), supported by COST (European Cooperation in Science and Technology).

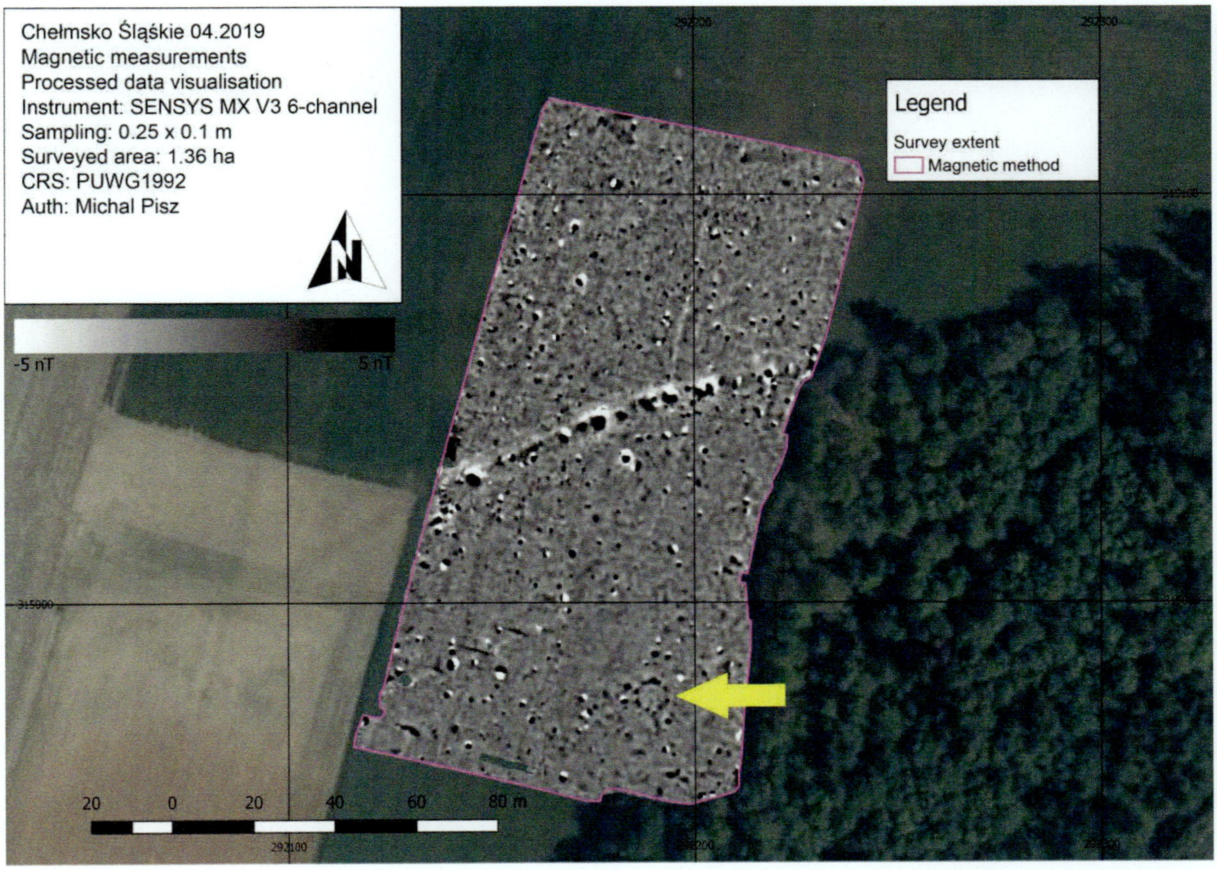

Fig. 1. Magnetic measurements. Magnetic anomaly interpreted as the remains of gallows is marked by a yellow arrow.

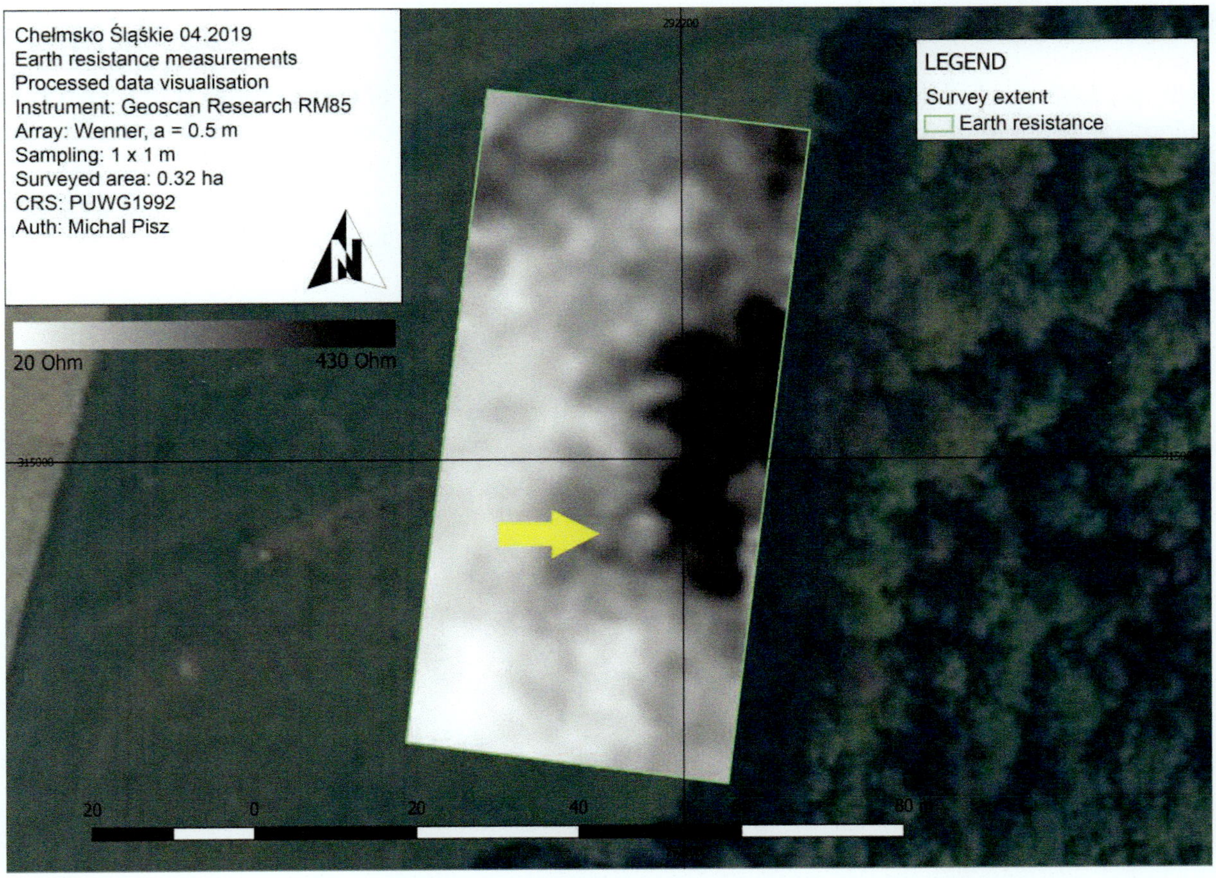

Fig. 2. Earth resistance survey. A round high resistance anomaly (marked by a yellow arrow) was initally interpreted as the possible remains of a gallows.

Fig. 3. The best-preserved remains of a gallows – a single brick and pottery shards found in one of test trenches

References

Gaffney, C.F. and Gater, J. 2003. *Revealing the Buried Past: Geophysics for Archaeologists*. Tempus.

Schmidt, A., Linford, P., Linford, N., David, A., Gaffney, C., Sarris, A. and Fassbinder, J.W.E. 2015. *EAC Guidelines for the Use of Geophysics in Archaeology: Questions to Ask and Points to Consider*. Belgium, Europae Archaeologia Consilium.

Wojtucki, D. 2009. *Publiczne miejsca straceń na Dolnym Śląsku od XV do połowy XIX wieku*. Katowice.

Back to the roots. Remote sensing techniques for rediscovering the Chalcolithic eponymous settlement of Cucuteni culture, Romania

Andrei Asandulesei, Felix-Adrian Tencariu, Radu-Ștefan Balaur, Mihaela Asandulesei
Interdisciplinary Research Institute, "Alexandru Ioan Cuza" University, Iași, Romania

Introduction and background

In 1884, located 50km from Iaşi, in north-east of Romania (Fig. 1a, 1b), the archaeological site of Cucuteni was discovered, a site which was to become one of the eponymous settlements of the most renowned prehistoric civilizations in Europe – Cucuteni-Trypillia, whose area of spread exceeded 350,000km^2 over modern Romania, Ukraine and the Republic of Moldova (Lazarovici *et al.* 2009: 15). 135 years after the discovery, and over 50 years since the last research there, the eponymous site of the Cucuteni culture (*Cetățuie* archaeological site) continues to surprise. Previous research, reflected in dozens of articles, studies and monographs and considered to be quasi-ended, is, as the latest investigations prove, far from finished. Older and newer observations suggested the existence of a prolongation of habitation or the existence of some satellite like settlements around the well-known site located on the *Cetățuie* promontory, but suppositions that were not confirmed by previous research. Having this in mind, a team of archaeologists from the "Alexandru Ioan Cuza" University of Iaşi made a series of investigations across an area of about 4ha, situated in the natural extension of the site to the southwest.

Regional settings

The eponymous site of the Cucuteni culture is located in the northwest of the Băiceni village, Iaşi county (Fig. 1c), on an outlier east of the wide Laiu plateau, bordered on three sides (south, east, and north) by steep slopes (Fig. 1d-f). As a geographical location, the area is located at the boundary between the Moldavian Plain and Suceava Plateau, within the Moldavian Plateau unit, in a hilly area with low altitudes to the east and south (under 150m), and higher to the west and north (above 300-350m). Hydrographically, the site belongs to the Oii Valley basin, the final tributary of the Bahluieţ River (Fig. 1c) before it flows into the Bahlui River (Petrescu-Dîmbovița and Văleanu 2004: 39-40).

Milestones of archaeological research

As stated above, the Cucuteni settlement was discovered in 1884, thanks to the ethnologist Theodor Burada, who, aware of the importance of the ancient remains of *Cetățuie*, stopped the destruction caused by the site's rock exploitation work. The first systematic research took place after 1888. The next period of intensive research was due to archaeologist Hubert Schmidt from the Ethnographic Museum of Berlin, which carried out two vast excavation campaigns in the years 1909 and 1910. The systematic research was resumed between 1961-1966 by a team headed by Mircea Petrescu-Dîmboviţa.

The settlement (as it was delimited initially) lies over about 1.5ha, mostly investigated over time. Both the initial and the subsequent research in the 20th century, focused only on the prominent part of the terrain (the *Cetățuie* itself), naturally defended on three sides by steep slopes, and on the fourth (westward) - through a system of ditches and ramparts (Fig. 2). Previous surveys have found that the habitation has expanded beyond this system, on the western plateau, to the forest that existed there. Thus, it is noted that the extension of the settlement was postulated and even partially documented, but a thorough investigation of the problem has never been undertaken. In this context, the main purpose of our field evaluation was to open a much broader perspective on living near the most famous settlement of the Cucutenian civilization, by way of the modern possibilities of investigation in the Arheoinvest Platform (archaeological topography, magnetometry, aerial photography, etc.). Surface surveys and magnetic prospection focused on an area of about 4ha, located southwest of the *Cetățuie*.

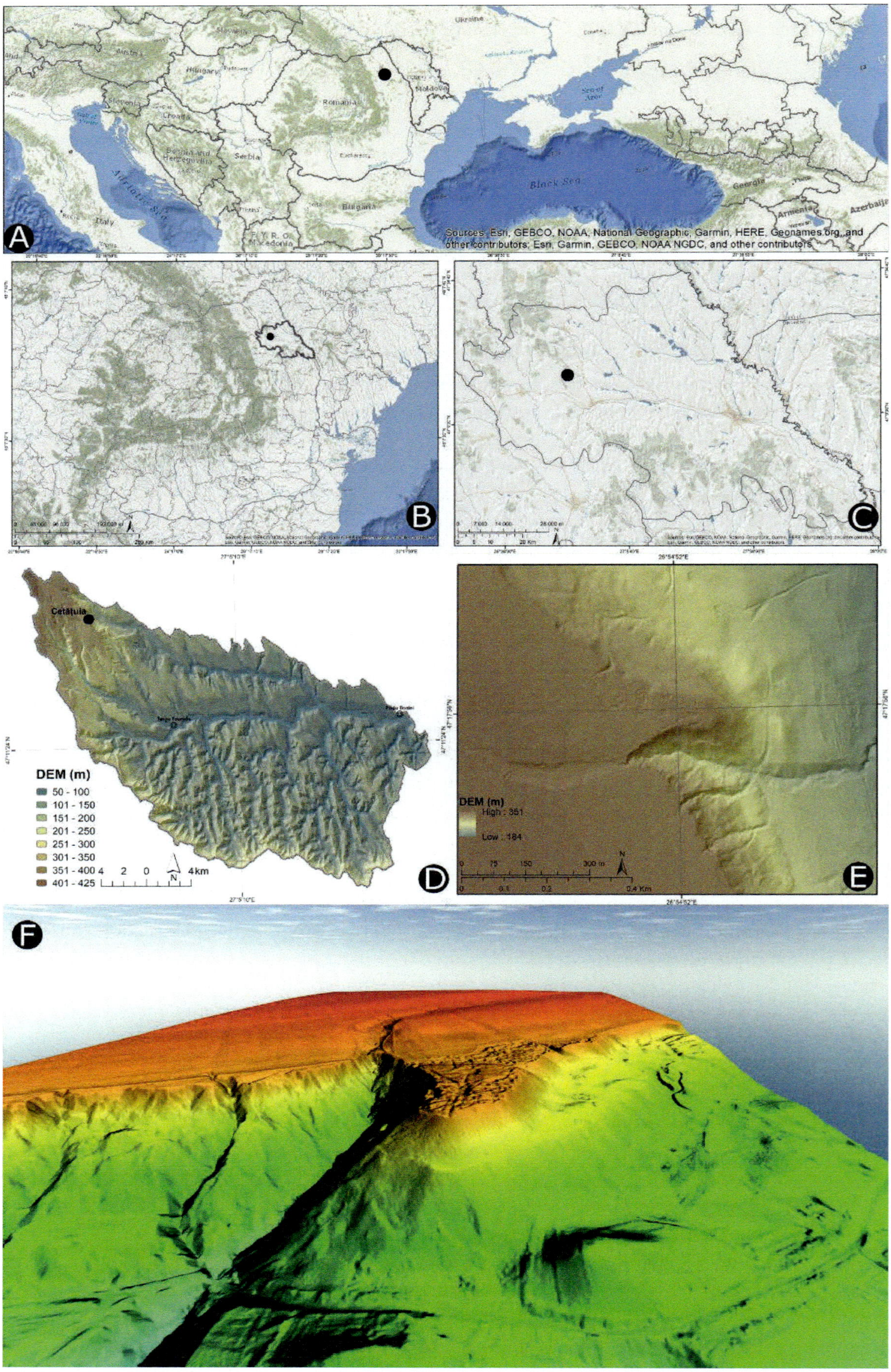

Fig. 1. The location of Cucuteni-*Cetățuie* archaeological site in the frame of Europe (a), Romania (b) Iaşi county (c) and Bahluieț river catchment (d-f).

Fig. 2. Oblique aerial image of the *Cetăţuie* archaeological site taken from the north.

Methods and results

The interpretation of vertical gradient magnetometry data (Fig. 3), Lidar and oblique aerial images for the large plateau highlighted the existence of a much different planimetry of the settlement than what was known to date. Although excavations have taken place outside the known ditches, it seems that misfortune has made the planned sections fall into the free space between two rows of dwellings. This area without constructions present, is probably one of the two access ways in the settlement, which separates the rows of dwellings. Only the southern row remains entirely today, but the layout of disturbed structures towards the centre and north of the external habitation suggests the presence of three alignments of chalcolithic dwellings.

It seems that the tip of the promontory constituted the initial nucleus of the settlement, which probably developed to the west in several stages, in the context of demographic growth. Several stages of evolution of the Cucutenian site can be distinguished, for which we cannot define chronological intervals, for the present research. The initial settlement was fortified with two defensive ditches, confirmed by the archaeological excavation (we express doubts about their chronological framing). These were followed by another very distinct ditch, with an average width of about 2.5m, the fill of which has, in certain sectors, a high degree of magnetic susceptibility.

The southern row consists of at least 12 heavily burned dwellings (with values that may even exceed 100nT), some of which are visibly disturbed. Most seem to be North-north-west to South-south-east oriented, with a few exceptions. The latter may be a category of special constructions. Here too we can include the anomaly at the western end of the row which, although disturbed, seems to have a surface that exceeds 300m^2. Structures on both sides of the access ways could also be classified as "bastions" or observation points.

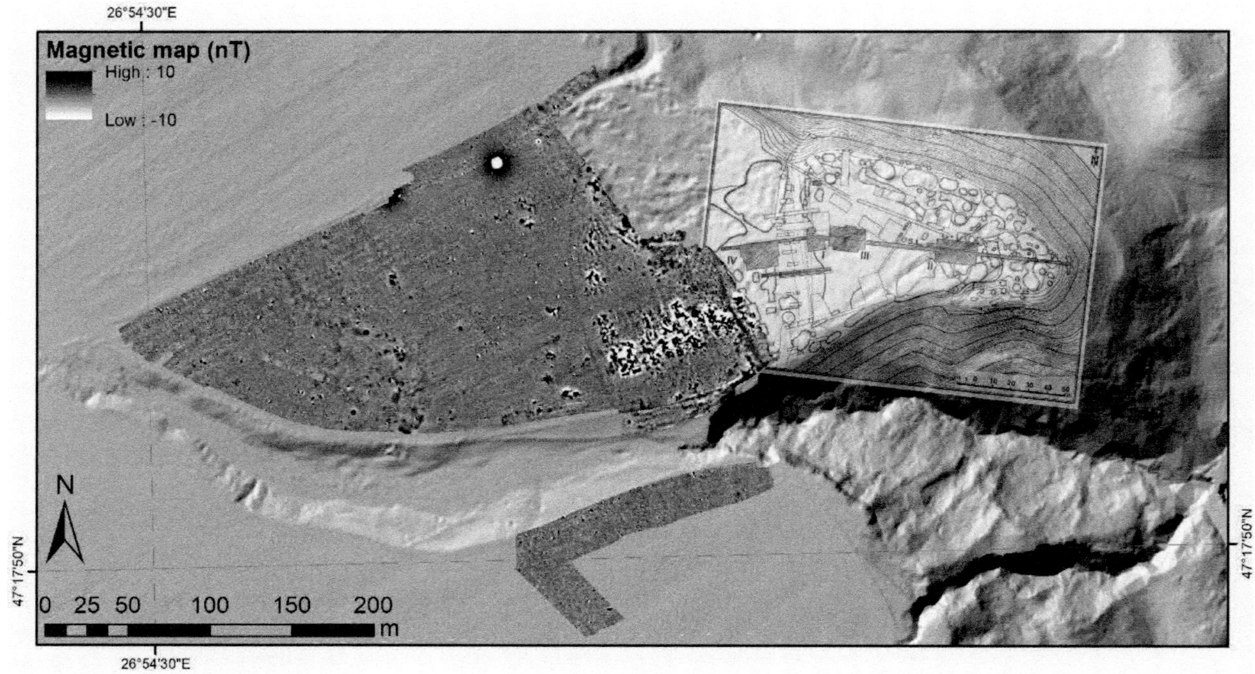

Fig. 3. Magnetic map (-10+10nT / white-black) and old excavation plan overlaid on Lidar DEM.

The settlement also has a system of trenches arranged in a circular arc, narrow (not more than 1.5m width) and not very deep, which delimits the western side. The interruptions on their route, alongside the space without dwellings, suggest access areas in the settlement. There is the suggestion of a palisade system.

Positive anomalies with relatively high magnetic signal strength, are visible outside the western trenches in the vicinity, but without a clear arrangement. Further on, to the west, there is a set of many positive, burned or unburned anomalies, possibly pits, are still visible. It is difficult to tell whether they have any connection to the Cucuteni habitation. There are also two circular anomalies in this sector, which exhibit a thermoremanent magnetism (signal intensity of about 55-60nT), which may suggest the presence of kilns.

Conclusions

As it can be seen, the recent research from Cucuteni *Cetățuie* provided new interesting data worthy of a thorough future investigation. It demonstrates the existence of habitation over a much larger area than the one that was researched 50 years ago, with many new interesting archaeological features (double fortifications, dwellings, possible kilns). Beyond the archaeological potential of the site, there is, of course, a symbolic meaning of our approach, given that we are talking about the eponymous site, with a long history of research.

Acknowledgements

This work was supported by a grant of the "Alexandru Ioan Cuza" University of Iasi, within the Research Grants program, Grant UAIC, code GI-UAIC-2017-04.

References

Lazarovici C-M., Lazarovici, G-C. and Ţurcanu, S. 2009. *Cucuteni. A Great Civilization of the Prehistoric World*. Iaşi.

Petrescu-Dîmbovița, M. and Văleanu, M.C. 2004. *Cucuteni-Cetățuie. Monografie arheologică*. Piatra-Neamț.

Understanding the Anomaly: Multi-Method Geoscientific Research Applied on a Roman Fort in Pojejena

Michal Pisz[1], Radoslaw Mieszkowski[1], Emil Jęczmienowski[2]
[1]*Faculty of Geology, University of Warsaw, Warszawa, Poland,* [2]*Institute of Archaeology, University of Warsaw, Warszawa, Poland*

Understanding The Anomaly – aim of the research

For more than 60 years geophysical methods have been commonly applied in archaeological prospection. During this time both field instrumentation and data processing tools have been strongly developed and adjusted to the needs of archaeologists. Instruments are easier to operate and nowadays some practitioners with very little training or instruction, can carry out measurements and create a basic visualisation of the results.

However, this progress has its dark sides as well. According to Meyer (2016: 42-43), a disturbing tendency might be observed: many archaeo-geophysical surveys are performed by semi-skilled freelancers, aiming at obtaining the most "spectacular" images, which leads to a downgrade of both the goals and the potential scientific contribution of the research. The lack of knowledge about the principles of the methods could lead to misinterpretations of the data. At the research verification stage, when the results do not match the reality, the entire blame is attributed to geophysics as a method. The "magic wand", as geophysics is still considered by many archaeologists, does not work.

The Understanding The Anomaly (UTA) Project aims at the testing and assessment of the responsivity of various complementary geophysical methods, together with further geoscientific studies over the registered anomalies and subsurface conditions. For this purpose, a group of characteristic and representative sites have been chosen for the research: Roman forts in the Romanian Banat region. There are several reasons for such a choice. First of all, Roman forts have a similar layout and construction. Second of all, they are buried in specific environmental conditions, which differ in the Banat region. There are examples of structures made of igneous, metamorphic and sedimentary rocks; surrounding soil may indicate differing magnetic susceptibility. Last but not least, most of the selected sites have not been thoroughly researched, hence the outcome of the Project should impact on the state of knowledge about the chain of Roman forts in Western Dacia.

Pilot UTA study in Pojejena

The Roman fort in Pojejena, located in Romanian Banat on the eastern bank of the Danube, was partially excavated in the 1970s by Romanian archaeologists (Gudea 1975). Only small sections of the stone fortifications were excavated which made it possible to estimate the general size of the fort. A few small trenches were opened in the centre of the fort, but they returned inconclusive results. The area around the fort was not researched.

The Polish-Romanian non-destructive investigations started in 2017 and aimed to evaluate the fort and to localize and determine the extent of a civil settlement in its surroundings by using fieldwalking and geophysical methods. The fort had been surveyed in 2017-2018 with magnetometry and small-scale earth resistance (ER) measurements. These results provided a general overview of the site and indicated the most interesting areas around the fort.

In 2019 a large scale ER survey was performed, with the use of a multi-depth twin-probe array (Fig. 1). A total number of 114 grids (20m × 20m) were measured, covering the accessible area of 4.15 ha: 3.23ha inside the fort and 0.92ha in its southern vicinity. Based on combined results of earth resistance and magnetometry, two polygons were selected for further studies with Ground-penetrating Radar (GPR), Electrical Resistivity Tomography (ERT) and Seismic Refraction Tomography (SRT) methods. Polygon 1 was placed in the area of the previously unknown eastern gate of the fort, while Polygon 2 was located over the fort's *principia*. Low resolution magnetic susceptibility survey was performed inside the fort, and in its northern vicinity. 12 drillings up to 3m depth have been undertaken, and soil samples of (at least) each 0.25m layer were taken for laboratory analysis, together with stone and brick samples.

Results

The applied methods have shown varied effectiveness, but in general all of them provided a positive response. Before the March 2019 campaign, the location of the eastern gate was estimated based on a quite faint negative linear magnetic anomaly. ER measurements provided a clear image of linear and broad high-resistance anomalies, the shape of which could be easily interpreted as a gate (Fig. 2). GPR measurements registered a number of locally occurring anomalies, corresponding to the results of the ER (Fig. 3). That was quite surprising, considering the silty-clayey subsurface was characterized by a very low resistivity (*c.* 30-60 Ωm). It was expected that the electromagnetic wave would be suppressed. However, while the maximum depth of penetration was not big, the positive result might be an effect of high contrast between the permittivity of the objects and their surroundings.

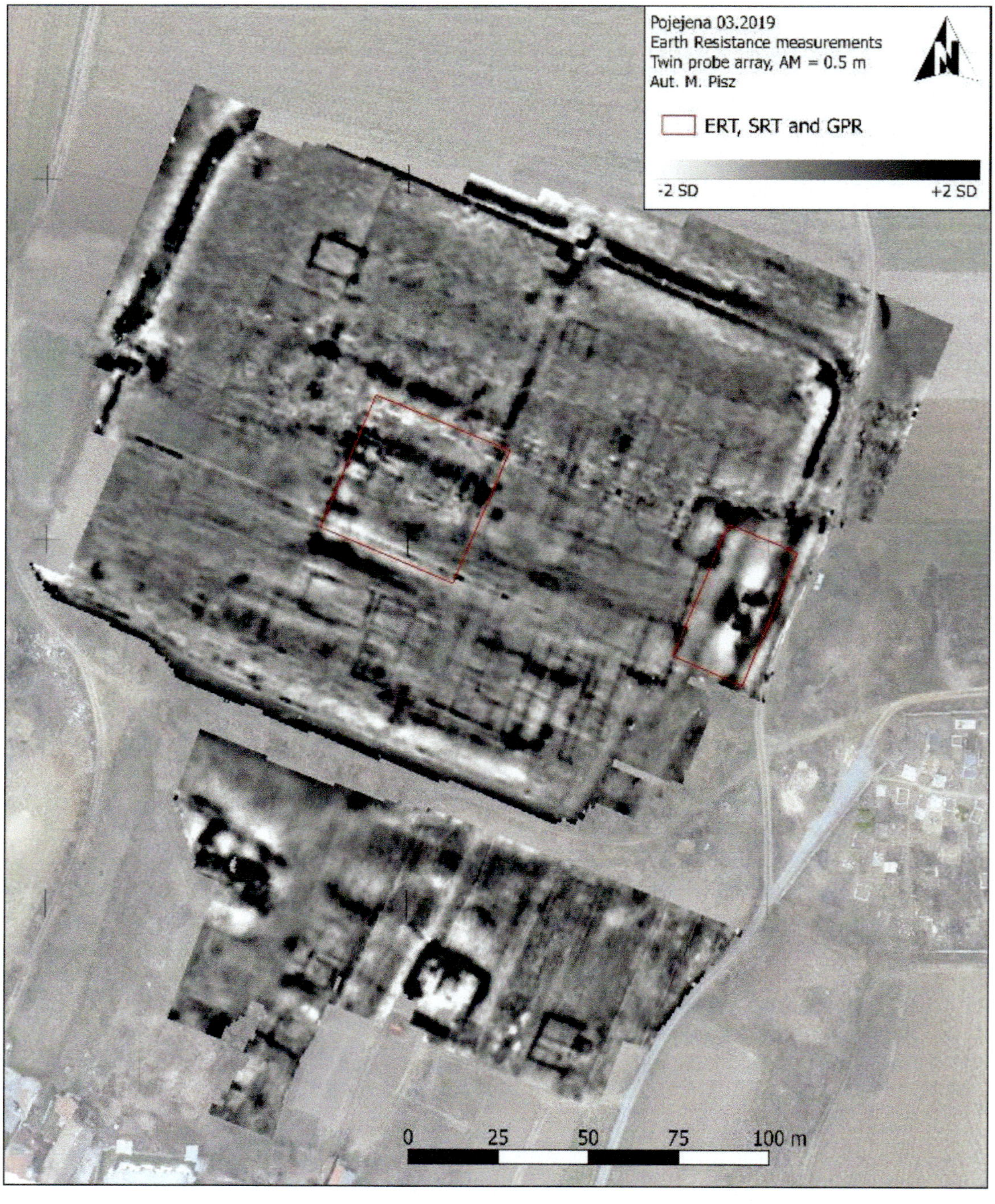

Fig. 1. Earth resistance results. Twin probe array, AM = 0.5m. Polygon 1 and Polygon 2 are outlined with red line.

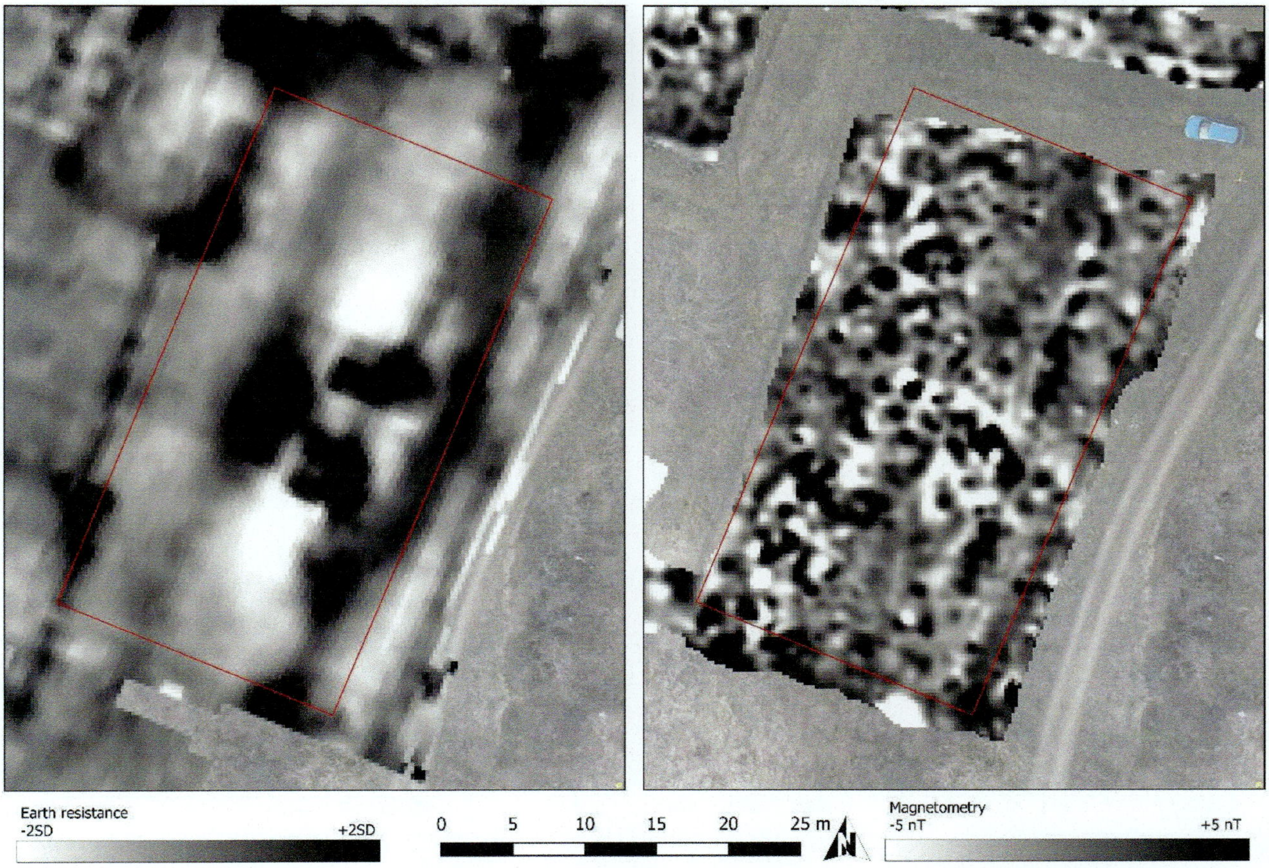

Fig. 2. Polygon 1 (marked with red line) over visualisations of the earth resistance (left) and magnetometry (right).

ERT measurements were performed with a gradient electrode array and provided good quality data (Fig. 3). The maximum depth of penetration reached up to 3m in Polygon 1 (21 profiles, each 40m long with 81 electrodes and 0.5m separation distance between them) and up to 5m in Polygon 2 (28 profiles, each 40m long with 61 electrodes and 0.75m separation distance between them). A 1m interval between profiles occurred in both polygons. SRT measurements registered higher velocities of seismic waves in an area of archaeological features, but no particular structures could be distinguished in the data.

Conclusions and Perspectives

At this stage of the research a few conclusions could be made. Almost all of the applied methods provided clear, comprehensive results. We have obtained information about both the horizontal and vertical distribution of anomalies. Structures made of schists are the most clearly detectable with earth resistance methods. Magnetometry has also provided good imaging of the structures, with negative linear features corresponding with stone walls and positive anomalies which could be interpreted as ditches and pits. GPR survey registered clear anomalies, but the depth of penetration was not big. SRT results indicated occurrence of stone remains, however the resolution of the results was not good enough to distinguish any objects.

Geophysical prospection delivered a lot of new information about the landscape of the fort and its hinterland. Six other forts have been selected for further studies. Geological prospection will be undertaken in the first instance to determine the natural environment of the sites. Subsequently, geophysical methods will be applied from the most extensive to the most intensive ones.

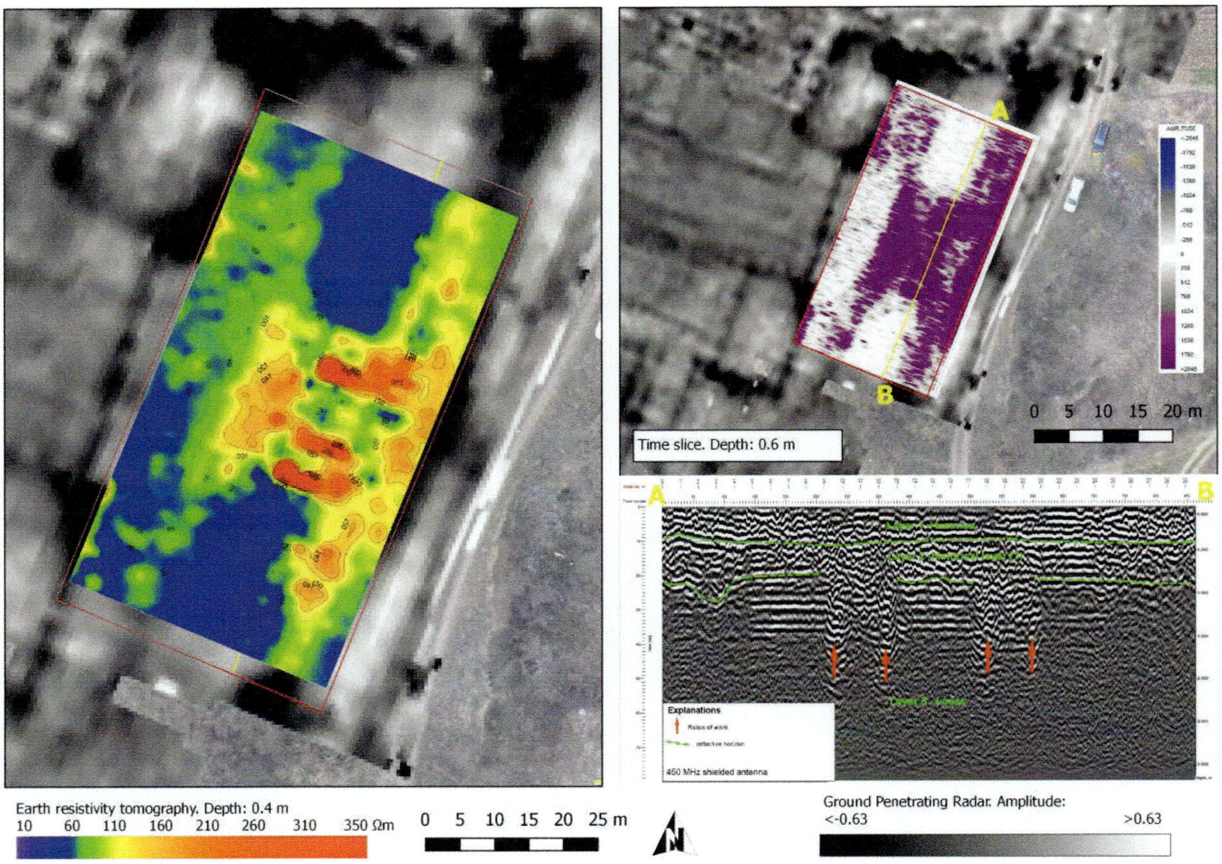

Fig. 3. Left: ERT results – a slice at 0.4m depth. Right: GPR time-slice and a profile from Polygon 1 (marked A-B).

Acknowledgements

The UTA Project research was supported by a STSM Grant and an ITC Conference Grant from COST Action SAGA: The Soil Science & Archaeo-Geophysics Alliance - CA17131 (www.saga-cost.eu), supported by COST (European Cooperation in Science and Technology).

Archaeological investigations in Pojejena were conducted within the framework of a Preludium 12 grant (number 2016/21/N/HS3/02933) awarded to Emil Jęczmienowski by the National Science Centre, Poland.

References

Meyer, C. 2016. The Enigmatic Vanishing of the Geophysicist in Archaeological Prospection at the Beginning of the 3rd Millennium. *Conference NSGG: Recent Work in Archaeological Geophysics*, Burlington House, Piccadilly, London, 6th December 2016.

Gudea, N. 1975. Date noi despre castrul roman de la Pojejena, *Banatica*. 3. 332-343.

Geophysical and archaeological research of the baroque church of Saint Nicolas in Kovarce, Slovakia

Mário Bielich[1], Ján Tirpák[2]
[1]Archeological Institute SAV, Nitra, Slovakia, [2]Faculty of Natural Science, UKF, Nitra, Slovakia

The archaeological research of the baroque church of St. Nicholas in Kovarce took place in 2016-18. The village of Kovarce is situated on the floodplain of the Nitra river in south-west Slovakia, along the alluvial zones of its left-hand tributaries. The flat and highland areas of the region are formed by young-eruption deposits, covered with river deposits and loess. The north-western slopes of the Tríbeč mountain range are crystalline rocks and quartzites from the older Mesozoic period (Vlastivedný slovník obcí 1978: 83). The site is located on a plain near the road connecting Kovarce with Ludanice. The area around the church is populated by village houses and the nearby manor house.

The excavation itself was preceded by ground-penetrating radar measurements (by RAMAX) in the interior and also exterior of the church in 2016. Based on the study of written sources, we found that the Baroque church was preceded by an older Gothic church of St. Nicholas. In the vicinity of this church was a cemetery. The first goal was to identify these older architectures. The second goal was to identify the interior crypts that came from the second half of the 18th century and the second half of the 19th century.

Gradually, we measured 9 areas (7 inside the church and 3 areas outside). Area 1 was passing through the centre of the church in a east-west direction. The width of the survey was 2m and the length was 16m, we started the survey in the north-east corner. Area 2 was located in the west of the church, surveyed in a north-south direction, the width of the area was 2m and the length was 8m. Area 3 was located in the middle of the church, surveyed in a north-south direction. Areas 4 and 5 were located in the eastern part of the church in front of the Baroque presbyterium, where measurements ran in a north-south direction, over a width of 5m and length of 8m (Area 4), and 2.5m width and length of 8m (Area 5). The purpose of the probes in the church interior (Areas 1-5) was to identify the stone foundations of the older church, which was mentioned in the written sources. Area 6 was located in front of the altar in the eastern part of the church, over a 5m × 3m area. Area 7 was located in the sacristy in the northern part of the church, where measurements were made in a west-east direction, over a 2m × 5.5m area. The aim of the probes in the eastern part of the church (Areas 6-7) was to identify the crypts mentioned in the written sources. Area 8 was located in the northern part of the church, measurements were made in a west-east direction, over a 40m × 4m area. The aim of the measurement was to capture the line and course of the cemetery wall. Area 9 was located in the eastern part of the church, where measurements were made in a north-south direction, over a 10m × 20m area. The aim of the measurement was again to capture the line and course of the cemetery wall. Area 10 was located south of the church. Measurements were made in a west-east direction over a 40m × 6m area. The aim of the measurement was to capture the line and course of the cemetery wall. An important issue for Area 10 was that it was situated under an asphalt road and was disturbed by engineering networks. The interpretation of the geophysical results is based on the horizontal and vertical GPR data at a of depth 80-120cm (Fig. 1).

Despite its small extent, the archaeological and geophysical research at the Church of St. Nicholas in Kovarce, created a lot of new knowledge about the development of this building and its surroundings in the Middle Ages and Modern Times. The direct cause of the research was the excavation of a hot water connection between the parish and the church. Several archaeological deposits were documented during the earthworks. Research in the interior of the church identified the original brick floor in the room north of the tower (dated to the 2nd half of the 18th century). In the main nave, under the new floor, we have discovered the original stone floor next to the northern wall.

There are 2 crypts in the church. The younger crypt is under the northern sacristy (dated to the 2nd half of the 19th century). The older crypt is located below the altar and enters it through the stairs under the triumphal arch. This crypt is in superposition with the older crypt of the Gothic church and their relationship is still unclear. One crypt is part of an older church and the other belongs to a donor of the baroque church, Judith Berényi (Klčo and Illášová 1993: 23). Under the sacristy, we identified - by means of geophysical research – a crypt, which was later investigated archaeologically (Fig. 2).

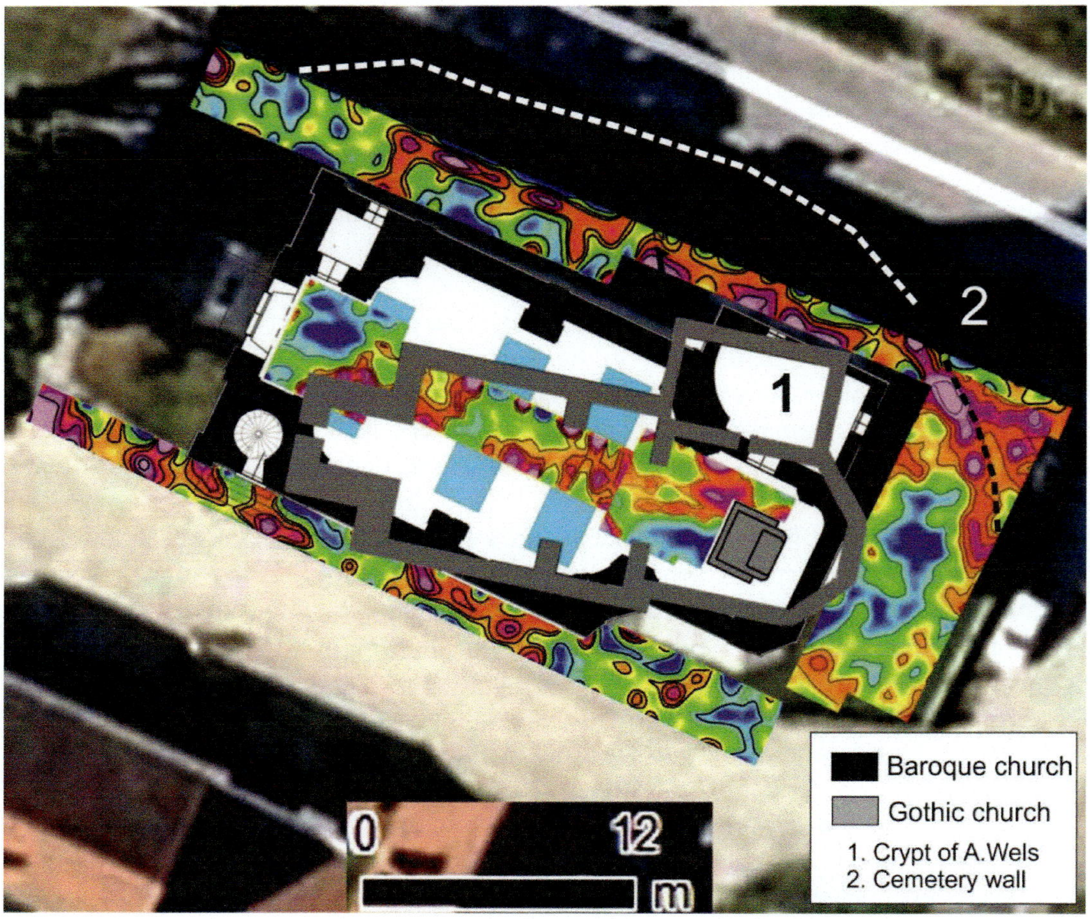

Fig. 1. Kovarce, St. Nicolas Church. The results of geophysical measurement - horizontal slides, depth 80-120cm. The Baroque church (black) and the foundations of the Gothic church (grey).

In the western part of the church, below the tower we caught the western wall of an older Gothic church. This wall is very poorly preserved. In the exterior of the church we examined 3 skeletal graves (E-W orientation), without any findings. These are the remains of the church cemetery, which can only be dated on the basis of the existence of the Gothic church, between the 14th-17th centuries. Since the 18th century the only burial place is the cemetery in the village, outside the sacral building. This cemetery is also marked on military mapping. At the eastern wall of the church we documented the polygonal presbytery of the original Gothic church of St. Nicholas. At the southern wall of the baroque church's main nave we managed to capture the supporting pillar of the southern wall of the Gothic church. Based on these results, we can estimate the size of the original Gothic church of St. Nicholas. At the northern wall of the baroque church, we explored the staircase to the crypt beneath the northern sacristy, probably built in the second half of the 19th century. In the crypt under the northern sacristy, we discovered the remains of A. Wels' – the great-grandfather of British actress Audrey Hepburn. The crypt was opened twice, once before the Second World War, when the original exterior entrance to the crypt ceased to be used, and at a second time during the war, when robbers got into the crypt from the interior of the northern sacristy. When we conserved the remains, we found cufflinks with an Egyptian motif (Fig. 3).

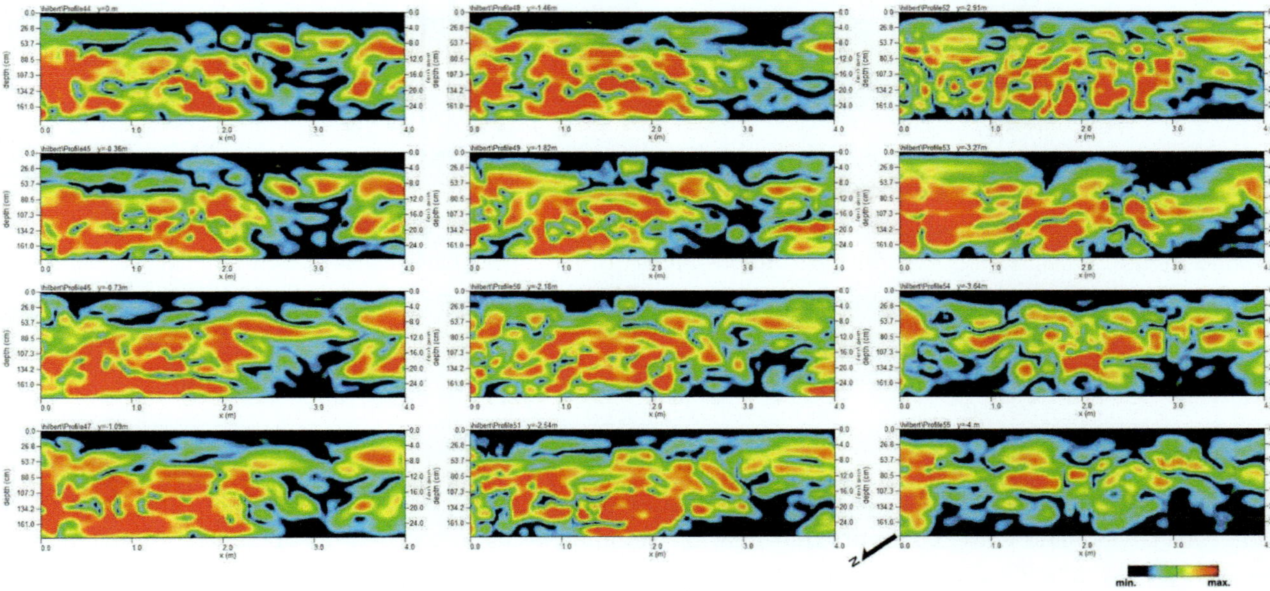

Fig. 2. Kovarce, St. Nicolas Church. Vertical slides from GPR measurement of crypt.

Fig. 3. Kovarce, St. Nicolas Church. Cufflink with Egyptian motif (19th-century).

The oldest sacral district in Kovarce was enclosed in the Renaissance by a fence wall, which was investigated by archaeological research in 2018. We only caught the wall east and west of the church. The northern and southern parts of the wall were destroyed by road construction. The findings of Renaissance ceramics from the 16th-17th centuries were uncovered here, including fragments of pot-shaped containers and tiles. Fragments from the 18th century appear among the findings of the tiles. These fragments certainly come from noble houses. Based on recent archaeological research, we can conclude that Kovarce played an important role in Central Ponitrie region during the late Middle Ages and the Modern Age.

References

Klčo, M. and Illášová, Ľ. 1993. *Historia Kovariec*. Nitra.

Vlastivedný slovník obcí 1978. *Vlastivedný slovník obcí*, 2 časť, heslo: Kovarce, 83.

Outlining the karst: ERT and GPR surveys to detect karstic morphologies in the Sierra de Atapuerca sites (Burgos, Spain)

Lucía Bermejo Albarrán[1, 2], Ana Isabel Ortega[1, 3, 4], Roger Guérin[5], Lawrence Conyers[6], Josep M. Parés[1], Isidoro Campaña[7], José María Bermúdez de Castro[1], Eudald Carbonell[8, 9]

[1]*CENIEH (Centro Nacional de Investigación sobre la Evolución Humana), Burgos, Spain,* [2]*Doctorado en Evolución Humana, Burgos, Spain,* [3]*Fundación Atapuerca, Burgos, Spain,* [4]*Grupo Espeleológico Edelweiss, Excma, Burgos, Spain,* [5]*Sorbonne universités, Paris, France,* [6]*Department of Anthropology, University of Denver, Denver, United States,* [7]*Independent researcher, Burgos, Spain,* [8]*IPHES, Institut Català de Paleoecología Humana i Evolució Social, Tarragona, Spain,* [9]*Universitat Rovira i Virgili (URV), Tarragona, Spain*

Background

The karstic system of the Sierra de Atapuerca is located 14km from the city of Burgos (Spain) and consists of numerous caves filled with sediments that bear outstanding archaeo-palaeoanthropological remains from the Early and Middle Pleistocene (Bermúdez de Castro *et al.* 1997; Carbonell *et al.* 2008; Ortega *et al.* 2014). 4.7km of this karstic system are accessible and have been explored for centuries, while another part was discovered by a 19th century railway trench that cut the southwestern flank of the hill, unveiling various caves silted up with sediments.

Identifying the development and morphology of this karst is therefore crucial for understanding the sites' formation processes as well as for planning excavation strategies. To that end, the Sierra de Atapuerca sites have been the target of many ERT (Electrical Resistivity Tomography) (Ortega *et al.* 2010; Bermejo *et al.* 2017) and GPR (Ground-Penetrating Radar) surveys, which have revealed the location, direction and dimension of different karstic morphologies. Many of these interpretations have been verified by test pits.

In this work, we present the more relevant karstic morphologies identified with these geophysical techniques, namely, sediment-filled caves, air-filled caves, silted up cave entrances and ancient valleys. We also illustrate how, in most of the cases, combining both the interpretation of the ERT and GPR results was necessary in order to detect the different characteristics of the targeted karstic morphologies.

Methods

The ERT acquisition was performed using a Syscal Pro resistivity meter (IRIS Instruments), which was connected to a linear array of 72 electrodes. The spacing of the electrodes changed in relation to the lateral and/or vertical resolution needed for each case. The inversions of these profiles were carried out with Res2Dinv software (version 4.04.01, Geotomo Software).

Some of the GPR profiles were collected with VIY®3 (Transient Technologies) 300MHz antennas. The data were processed with Synchro 3 (version 3.10.1.6, Transient Technologies) and GPR-SLICE v7.0 (c) software to produce reflection profiles. We also collected parallel GPR profiles with 270MHz antennas using a GSSI SIR-3000 system. With this data we were able to create horizontal time slices using the GPR Process software (Conyers 2010). These slices allowed us to detect the lateral changes of some karstic morphologies.

Results

The high contrast existing between the resistivity values of the sediments (very conductive) and the limestone host rock (very resistive) have allowed us to locate sediment-filled caves and silted up cave entrances (Fig. 1). However, the ERT results tend to overestimate the dimension of these conductive anomalies, as the electricity flows preferentially through these materials. In this sense, the GPR results were more accurate in terms of determining the right depth at which these sediments appear as well as delimiting its lateral contacts.

As for ancient valleys, ERT was the only suitable method for its characterization, given that these are filled with clayey sediments that rapidly attenuate the radar wave.

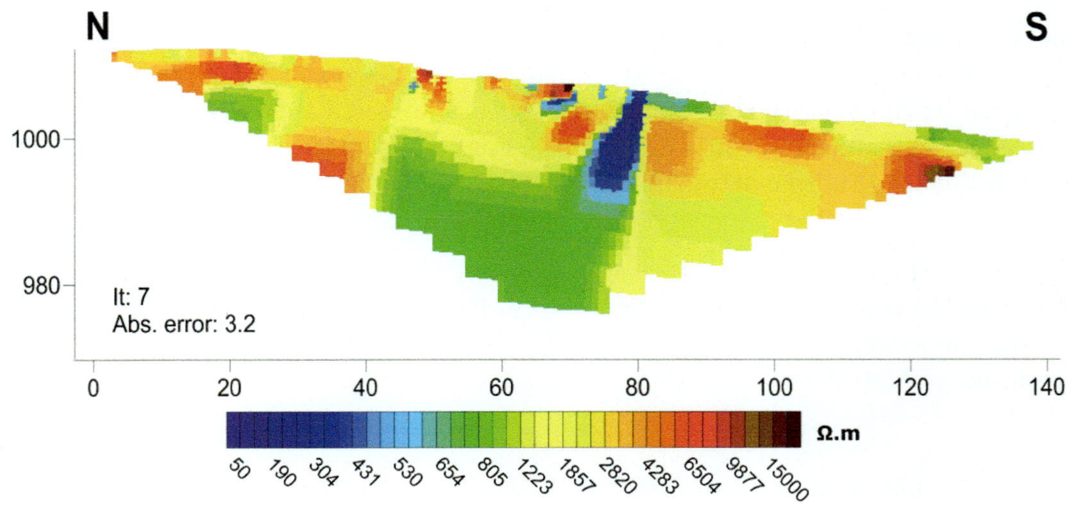

Fig. 1. ERT profile showing a silted-up entrance and a small sediment-filled cave. Note that the values for sediments are estimated to be < 400 Ω.m.

GPR survey was needed for locating air-filled caves, which are disguised in the ERT results because of the high resistive values of the Sierra de Atapuerca limestones. GPR is very useful for this, as the radar energy increases in velocity when entering an air-filled cave, creating high amplitude reflections (Fig. 2). In addition, the void can be confirmed by checking for reversed polarities in the radar wave (Conyers 2012).

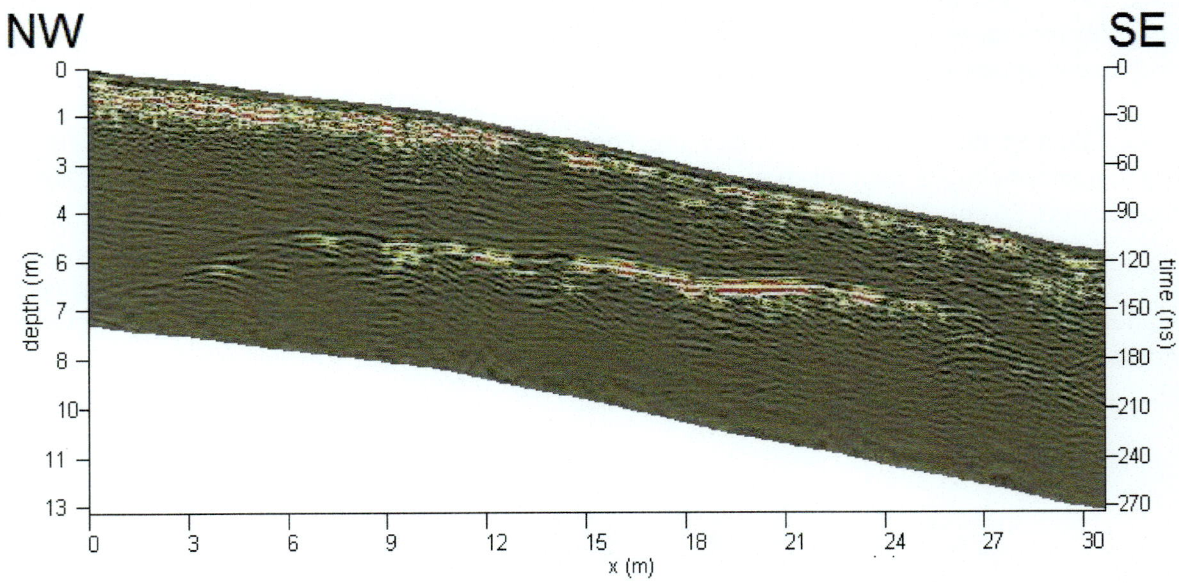

Fig. 2. GPR profile showing a high amplitude reflection that corresponds to the ceiling of an air-filled cave.

Discussion and Conclusion

Both ERT and GPR are suitable geophysical methods for studying karstic environments and combining the interpretation of their results can be very useful, especially when trying to understand the source of high resistive anomalies or high amplitude reflections. Two or more geophysical methods should be employed in any survey, but particularly in complex environments such as karsts. In the case of the Sierra de Atapuerca, combining both interpretations, together with the results from the test pits, have provided relevant information to improve the interpretations of further surveys made in the same area.

Acknowledgements

The first author is a beneficiary of a predoctoral grant of the Junta de Castilla y León (EDU/310/2015) financed with European funds. The authors wish to thank the multidisciplinary Research Team of the Sierra de Atapuerca (EIA) for its constant scientific and logistic support. Research at the Sierra de Atapuerca sites is supported by the MICINN project PGC2018-093925-B-C31 and archaeological fieldwork is also supported by the Consejería de Cultura y Turismo of the Junta de Castilla y León.

References

Bermejo, L., Ortega, A.I, Guérin, R., Benito-Calvo, A., Pérez-González, A., Parés, J.M., Aracil, E., Bermúdez de Castro, J.M. and Carbonell, E. 2017. 2D and 3D ERT imaging for identifying karst morphologies in the archaeological sites of Gran Dolina and Galería Complex (Sierra de Atapuerca, Burgos, Spain). *Quaternary International*. 433. 393-401.

Bermúdez de Castro, J.M., Arsuaga, J.L., Carbonell, E., Rosas, A., Martínez, I. and Mosquera, M. 1997. A hominid from the Lower Pleistocene of Atapuerca, Spain: possible ancestor to Neandertals and modern human. *Science.* 276. 1392-1395.

Carbonell, E., Bermúndez de Castro, J.M., Parés, J.M., Pérez-González, A., Cuenca-Bescós, G., Ollé, A., Mosquera, M., Huguet, R., Made van der, J., Rosas, A., Sala, R., Vallverdú, J., García, N., Granger, D.E., Martinón-Torres, M., Rodríguez, X.P., Stock, G.M., Vergès, J.M., Allue, E., Burjachs, F., Cáceres, I., Canals, A., Benito, A., Díez, C., Lozano, M., Mateos, A., Navazo, M., Rodríguez, J., Rosell, J. and Asuaga J.L. 2008. The first hominin of Europe. *Nature.* 425. 465-470.

Ortega, A.I. 2009. *La Evolución Geomorfológica del Karst de la Sierra de Atapuerca (Burgos) y su relación con los pacimientos pleistocenos que contiene*. Ph.D. Thesis. Universidad de Burgos, Facultad de Humanidades y Educación, Departamento de Ciencias Históricas y Geografía, Burgos, Spain.

Ortega, A.I., Benito-Calvo, A., Pérez-González, A., Porres, A. and Martín, M.A. 2010. Applying electrical resistivity tomography to the identification of endokarst geometries in the Pleistocene sites of the Sierra de Atapuerca (Burgos, Spain). *Archaeological Prospection.* 17. 233-245.

Ortega, A.I., Benito-Calvo, A., Pérez-González, A., Carbonell, E., Bermúdez de Castro, J.M. and Arsuaga, J.L. 2014. Atapuerca Karst and its palaeoanthropological sites. In Gutiérrez, F., Gutiérrez, M. (eds.), *Landscapes and Landforms of Spain*, World Geomorphological Landscapes. Springer Science+Business Media Dordrecht, 101-110.

Archaeological feedback of a GPR survey at Labeagako Santa Maria (Navarre): confirmation of survey interpretation and few more surprises

Ekhine Garcia-Garcia[1], Roger Sala[1], Alexandre Duró[2], Helena Ortiz-Quintana[1], Rosario Mateo[2], Javier Nuin[2]
[1]*SOT Archaeological Prospection, Barcelona, Catalonia,* [2]*OLCAIRUM Estudios Arqueológicos, Iruñea-Pamplona, Basque Country*

Background

The Monastery of Santa Gema is located in Labeaga, a rural village located *c.* 40km south-west of the capital, Iruñea-Pamplona. It has been abandoned long time ago and is nowadays partially in ruins.

One of the singularities of the site is that the church of the Monastery contains a crypt whose construction typology does not match with the walls of the building. The specific characteristics of this structure generated doubts about its chronology, because it could be earlier than the Monastery. Moreover, the crypt is connected with a 9m long corridor that nowadays appears to be collapsed and, because of that, the global length of the corridor and the location of the access was uncertain.

The Heritage Service of the Government of Navarre promoted a geophysical intervention prior the excavations. The objective was to trace the corridor and to locate its entrance, as well as to detect other structures that could host the courtyard-cloister of the Monastery.

The excavation occured after the geophysical survey, validated the results, but unexpected remains were also detected. This paper compares the results of the GPR survey to the excavation results, in order to understand why those remains were not detected.

Methods

Geophysical Survey

The survey was performed using an IDS Hi-Mod instrument equipped with two multi-frequency antennae (200MHz and 600MHz). This device provides two datasets, one for each centre-frequency, avoiding a choice between spatial resolution and penetration. Selected frequencies are suitable to detect archaeological remains larger than 12-15cm, depending on the electrical permitivity of the subsoil (Schmidt *et al.* 2015).

Only a small flat area of 235m^2 was suitable to be explored. Measurements were taken every 2.5cm along parallel profiles and a separation of 20cm was applied between the parallel profiles, all taken in the same direction. The time window was set at 90ns for the 200MHz antenna and at 60ns for the 600MHz antenna. Marked measuring tapes were used for positioning and the corners of the grid were georeferenced by GPS.

The data were processed using GPR-SLICE software. Standard data processing was applied to the profiles by means of time-zero adjustment, gain application and band-pass filter (Goodman and Piro 2013). Afterwards the time slice technique was used to generate the horizontal maps (Goodman *et al.* 1995). The average wave velocity was obtained from the analysis of the diffraction hyperbolae in the radar sections and estimated at 0.09m/ns. The selected slices were imported into a QGIS project for interpretation.

Archaeological Excavations

Layers corresponding to modern fills or geological sediments of no archaeological value were removed by mechanical excavation to the top of the first level of archaeological significance. From this point onwards, excavation was carried out using hand tools and following the stratigraphic system based on that postulated by E.C. Harris.

The excavation was documented by digital terrestrial photogrammetry with topographic support. Planimetries were made over the orthophotography obtained from the photogrammetry, and then imported into the QGIS project for comparison.

Results

Geophysical Results

A comparison between the two datasets established that the 200MHz dataset could not reach a large depth of penetration, probably because the conductivity of the terrain (Conyers 2013). Because of that, that data has been discarded due its lower resolution.

GPR results show a complex stratigraphy where most of the strongest reflections were located at the centre of the investigated area (Fig. 1). After some superficial reflections attributed to possible levelling or debris layers, the time slice sequence shows a reflective elongated area, indicating a clear limit at the east and dipping slowly to the west. Based on its non-homogeneous reflectivity and its morphology, it has been attributed to a geological unit.

At the southeast of the explored area the corridor has been located successfully. Results suggested that it ends abruptly, without any indication of superficial access. No other structures were identified in the GPR results.

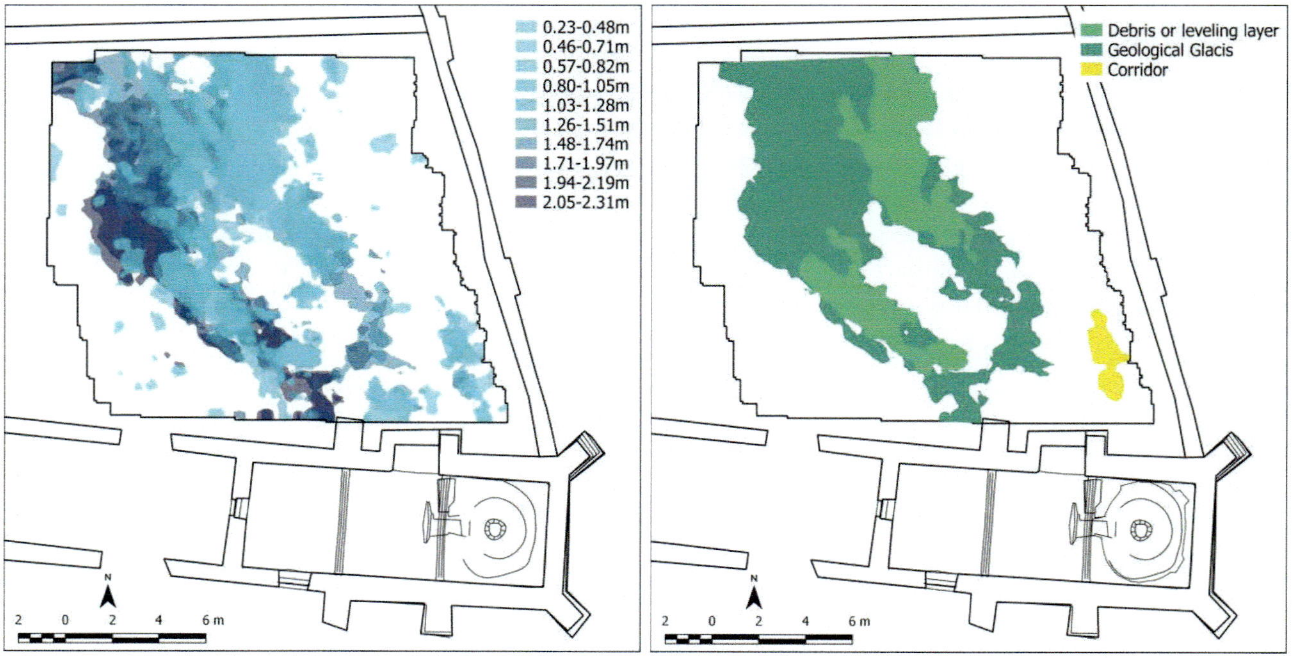

Fig. 1. Outline of the detected reflections organized by depth and interpretation scheme.

Archaeological Feedback

Excavation confirmed the position and the end of the corridor predicted by the geophysical survey. However, it also revealed some constructed walls that were not identified in the geophysical results (Fig. 2). They are mainly located at the end of the corridor, conforming to a rectangular lounge (a 3D model can be seen following the link: https://bit.ly/2G3MsyA). The geological glacis was cut and walls were constructed with irregular stones without any mortar between them. Several debris layers, with different amounts of stone material, were found in the surface layers.

Conclusions

While the void of the corridor has been easily detected by the GPR, the walls did not produce enough contrast to be isolated from the complex stratigraphy of demolition and geological units. It can be expected that the amount of stones within the demolition layer generates a low contrast on electromagnetic properties between the walls and the surrounding environment. The cut of the geological glacis was well resolved by the GPR, but it has not been interpreted as being of interest for the survey objective.

Fig. 2. Pictures of different phases of excavation.

A comparison between archaeological results show that some of the walls, the most superficial ones, had enough reflected radar energy to be detected and are slightly visible on the amplitude maps. However, they were not identified for the interpretation drawings (Fig. 3). One of the reasons is the low continuity of them, making them easily attributed to local heterogeneities. In fact, other similar anomalies did not correspond to any excavated wall. Also, as the objective of the geophysical survey focused on the corridor, no other structures were expected.

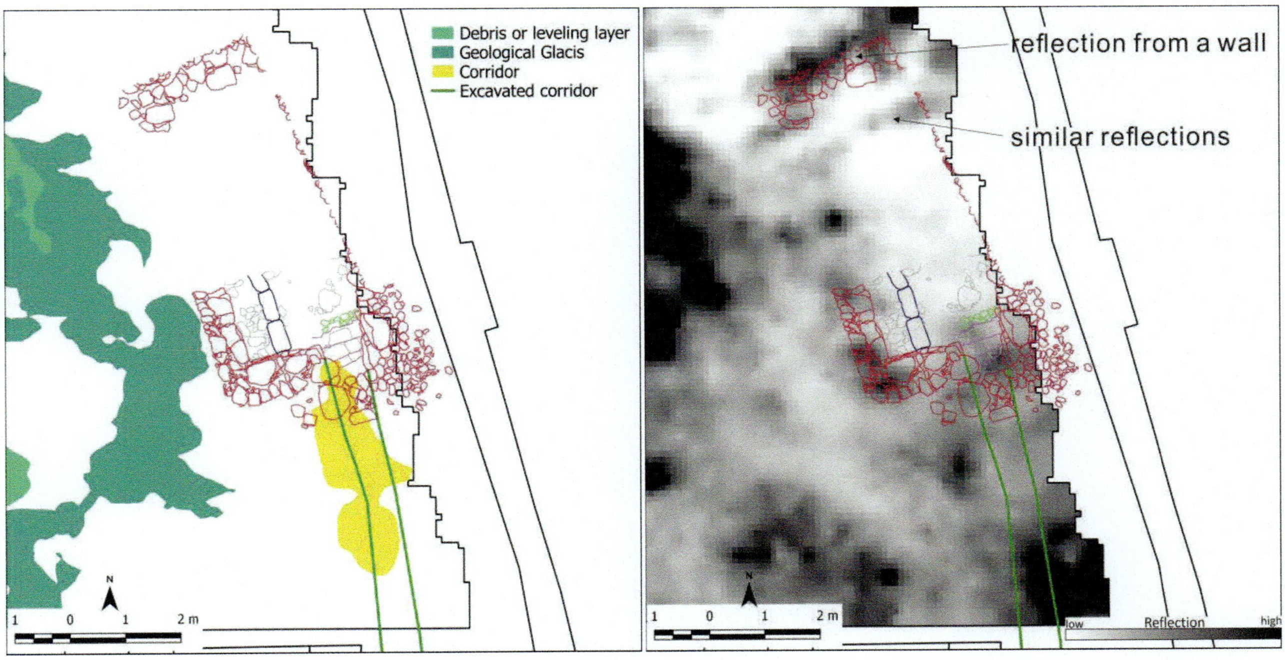

Fig. 3. Excavation results over the interpretation scheme based on GPR survey, and over the amplitude map corresponding to a depth between 0.80-1.05m.

Acknowledgements

The authors would like to thank Jesús Sesma and Jesús García, archaeologists of the Heritage Service of the Government of Navarre, for their advice and coordination of the project. Also, this project would not have been possible without the financial support of this institution.

References

Conyers, L. 2013. *Ground-Penetrating Radar for Archaeology*. Third Edition. Plymouth.

Goodman, D., Nishimura, Y. and Rogers, J.D. 1995. GPR time slices in archaeological prospection. *Archaeological Prospection.* 2. 85-89.

Goodman, D. and Piro, S. 2013. *GPR Remote Sensing in Archaeology*. New York.

Schmidt, A., Linford, P., Linford, N., David, A., Gaffney, C., Sarris, A. and Fassbinder, J.W.E. 2015. *EAC Guidelines for the Use of Geophysics in Archaeology: Questions to Ask and Points to Consider*. Belgium, Europae Archaeologia Consilium.

Multi method investigation of submerged features at Semblister, Shetland

Claire Christie[1], Michael Stratiagos[2], Benjamin Jennings[3]
[1]University of Aberdeen, Aberdeen, United Kingdom, [2]Scottish Universities Environmental Research Centre, United Kingdom, [3]School of Archaeological Sciences, University of Bradford, Bradford, United Kingdom

Summary

The Shetland Islands are well known for their archaeology, with many features from later prehistory surviving to an exceptional degree, with brochs being an excellent example. Less is known of submerged prehistoric sites, though the offshore profile of the islands means that ancient occupation of the current submerged zone was likely, and also that the sites are likely preserved.

One such archaeological site, previously identified, occurs near Semblister. The site has been known locally, and is recorded as a broch in the local Historic Environment Records, however, its dimensions and location do not strongly suggest that it is a broch. Combining satellite imagery – the site is in relatively shallow water and visible – with targeted side-scan sonar survey and data processing, will aid the development of a process approach to the re-classification / confirmation of historically recorded sites, and archaeological prospection in the shallow water zone.

Introduction

The Shetland Islands have a wide range of prehistoric archaeology sites, with settlement forms including enclosures, brochs, and crannogs. Many of these sites have been extensively researched, but others are known only through historic reference and limited historic environment records. One such site is known in the area of Semblister, on a sheltered bay on the western central portion of mainland Shetland (Fig. 1), which is identified as a Broch in the Historic Environment Record.

Fig. 1. Location of Semblister on Shetland Mainland. (Image source: Google Maps).

Satellite Imagery

Aerial imagery of the site clearly highlights a submerged feature on the sea bed, submerged by between 1m and 6m of water, dependent on tide conditions and profile (Fig. 2). The aerial imagery does not depict the submerged structure itself, but shows kelp growth; the kelp is anchored to stone, and does not grow on the surrounding sandy gravel material of the sea bed. The kelp growth is in a sub circular arrangement, with a diameter of *c.* 30m to 40m, with no growth in the centre. While the kelp will appear as a larger dimension than the underlying stone material, due to the spreading of kelp fronds, the dimensions are clearly substantial – much larger than the modern church on the centre left of the image, and larger than known brochs, which typically extend up to 15m internal diameter, with 3m thick walls (21m total diameter).

Fig. 2. Satellite imagery of the site, showing a distinct submerged feature picked out by kelp growth. This is in stark contrast to the predominant surrounding sea bed material, which supports little organic growth. (Image source: Bing Maps).

Sonar

With the aerial imagery indicating a site of interest, it was decided to conduct a survey of the site. Manual survey via diving would be possible, but obtaining a clear view of the submerged features through the kelp growth is problematic, and complete removal of the kelp is not possible (Fig. 3). Therefore, it was decided to conduct a sidescan sonar survey of the site, using a Tritech StarFish 990F, using StarFish Scanline to record data. With the shallow nature of the site, it was not possible to tow the sonar unit behind the survey vehicle, and a custom rig was constructed to attach the sonar to the stem of the boat, allowing variation in submersion depth of the sonar as required (Fig. 4). Mounting on the stem of the vessel necessitated a low speed to reduce the influence of bow waves on the sonar scan data.

Fig. 3. Edge-set stones within the feature after kelp removal. (Image source: the authors).

Fig. 4. Mounting of the Tritech StarFish (orange submerged block) to the bow of the survey vessel (20ft RIB). The pole can be moved within the wooden housing to allow the sonar unit to be raised or lowered depending on survey conditions and needs.

Scanning of the site was conducted in a grid pattern, running across the site in a series of N-S, E-W, and NE-SW directions, allowing the comparison of a variety of viewpoints when considering features. It is evident that the site contains a wide spread of stone, which is in distinct contrast to the natural exposed seabed of the surrounding area – compare the highlighted stones on the sonar plot to the black background of the natural sediment (Fig. 5). Within the sonar data it is clear that a considerable spread of stone material occurs, which is sharply defined at its edge. This edge may represent the remnants of an enclosure wall which has collapsed and gradually been eroded in to a stone mound rather than a sharp wall. It is also possible to see a sub-division attached to the perimeter edge of the stone spread, this could represent a structure or sub-enclosure attached to the main enclosure wall. Again, this has been eroded to a low mound of stones.

Fig. 5. Sonar data of a section across the site. The white / grey areas represent features standing proud of the seabed, in this case stone material, with the natural (flat profile) seabed being dark coloured / black. This demonstrates the distinct unusual aspect of the material, and shows a clear delimitation of the site. The perimeter of the sites may represent a collapsed wall, with a small structure attached to the enclosure wall at the top centre of the scan. The central white band of the image highlights where no data has been captured as this is directly below the sonar unit. (Image source: authors).

From the sonar data it is clear that there are a range of stone sizes included in the material, the larger ones may be used to pick out the main line of walls; this is an ongoing process. It is also possible to observe what appear to be edge set stones in the sonar data (not imaged here); an initial diver survey identified what may appear to be edge set stones underwater, but further confirmation is required, including the clearing of small sections of kelp.

Surrounding environment

The wider environment of the site highlights a number of other features, both anthropogenic and natural. Within 1km of the site is a circular stone enclosure and prehistoric house adjacent to the Loch of Semblister. This type of site is common to the West Mainland of Shetland and are typically thought of as Neolithic or Bronze Age in date. Superficially the remains around the Loch of Semblister resemble the submerged Semblister site investigated as part of this study. In terms of natural features, extant peat surfaces continuing to the sea edge (Fig. 6), and preserved peat surfaces are known to the east of the submerged site. The presence of surrounding preserved peat surfaces indicates that the bay has typically been a relatively sheltered and low energy environment, increasing potential for preservation of standing archaeological features in the shallow water zone. The preserved peat also offers the potential to obtain dating information – a strategy of coring to identify the extent of the submerged peat surface, and obtain a series of dating information, may considerably help the understanding of Holocene marine transgression and its relationship to the submerged site at Semblister.

Fig. 6: Extent of peat surface to the south of the submerged site. The peat continues to the very edge of the sea, being actively eroded at extreme high tides; recent high tide line is visible as debris in the foreground. Submerged peat surfaces are also recorded below the high tide line. (Photo source: authors).

Conclusion

The investigation of a submerged site at Semblister, Shetland, using multi-scalar methods – ranging from satellite and aerial imagery to site sonar survey and ground truthing – has highlighted a potential prehistoric activity site. The structure does not, under current interpretation, appear to tally with its former designation as a broch; rather it appears to be a small enclosure with two possible, small, circular structures. Further investigation, involving small targeted excavation to confirm the structures, and core sampling to identify the date and extent of the submerged peat surface, will help clarify the identification of the site type, and provide possible dating evidence for its construction.

Acknowledgements

The authors express their sincerest thanks to: Stephen Anderson for assistance on site, the use of his boat, and piloting skills; Archaeology Shetland for use of GPS equipment; the Shetland Amenity Trust for their support in surveying the site; the University of Bradford for providing the financial backing to conduct the survey under the Research Development Fund grant 003JEN.

The Archaeology of 20th Century Sports and Leisure: topophilia, interiography and texture

Chris Gaffney[1], Tom Sparrow[1], Alex Corkum[1], Helen McCreary[1], Chrys Harris[2], Jason Wood[3]
[1]University of Bradford, Bradford, United Kingdom, [2]Magnitude Surveys Ltd, Bradford, United Kingdom, [3]Heritage Consultancy Services, United Kingdom

The archaeology of modern sites has become a significant aspect of the work carried out by some research and commercial groups. The work is often directed at site types that have equivalent sites or features that can be recognised from earlier periods. For example, burials are common throughout human history and occasionally surveys over 20th century remains are reported (Davis *et al.* 2000, Hansen *et al.* 2014, Gaffney *et al.* 2015). There are also discussions of unique events, such as those linked to warfare, that are also within the literature (Gaffney *et al.* 2004, Capps Tunwell *et al.* 2016, Rees-Hughes *et al.* 2016, Saey *et al.* 2016, Stichelbaut *et al.* 2017).

Alongside these 'conventional' areas of study have grown an interest in sport and leisure as a discrete part of the archaeological record. While some of this is focused toward understanding sporting motifs or events on, say, Greek vases (Turner 2012), others have reached into the modern record to create a past that is imbued with local relevancy. It is in the latter context that the projects reported here are positioned.

We have been exploring the often unique environments in which sport and leisure is undertaken in the 20th century. This can incorporate regular but low 'impact' sporting activities (e.g. football) or large scale but short-lived events (such as festivals or exhibitions). These include survey data derived from:

- The 1904 Bradford Industrial Exhibition
- An open air swimming pool (Bradford Lido)
- A former football stadium (Bradford Park Avenue)

They are linked by local folk memory, despite the fact that they are now invisible or inaccessible to the general public. We must consider the language that is used to describe our data with reference to (locally) well known events as these will provide better interpretations of our data. In doing so the usefulness, or otherwise, of texture (Gaffney and Gater 2003) and 'interiography' (Sunseri and Byram 2017) should be considered.

In comparing and contrasting different sport and leisure events we also consider topophilia (a sense of identity and regard for a place) in explaining the loss or reinforcement of the local understanding of space. This analysis is strengthened by the narrative of place and heritage that is now part of the increasing value of sporting space that can be understood from both a cultural historical and a geographical perspective (e.g. Ramshaw and Gammon 2010; Bale 1988).

There is an inherent value for the investigation of recent sites that are outside the usual repertoire of the archaeological geophysicist. To fully understand a landscape, we must appreciate the value of engaging with the recent past, one that contains significant elements of folk memory.

References

Bale, J. 1988. The place of "place" in cultural studies of sports. *Progress in Human Geography*. 12. 507–24.

Capps Tunwell, D., Passmore, D.G. and Harrison, S. 2016. Second World War bomb craters and the archaeology of Allied air attacks in the forests of the Normandie-Maine National Park, NW France. *Journal of Field Archaeology*. 41. 312-330.

Davis, J.L., Heginbottom, J.A., Annan, A.P., Daniels, R.S., Berdal, B.P., Bergan, T., Duncan, K.E., Lewin, P.K., Oxford, J.S., Roberts, N., Skehel, J.J. and Smith, C.R. 2000. Ground penetrating radar surveys to locate 1918 Spanish flu victims in permafrost. *Journal of Forensic Sciences*. 45. 68–76.

Gaffney, C.F. and Gater, J. 2003. *Revealing the Buried Past: Geophysics for Archaeologists*. Tempus.

Gaffney, C., Gater, J. Saunders, T. and Adcock, J. 2004. D-Day: geophysical investigation of a World War II German site in Normandy, France. *Archaeological Prospection*. 11. 121-128.

Gaffney, C., Harris, C., Pope-Carter, F. Bonsall, J., Fry, R. and Parkyn, A. 2015. Still Searching for graves: an analytical strategy for interpreting geophysical data used in the search for "unmarked" graves. *Near Surface Geophysics*. 13. 557-569.

Hansen, J.D., Pringle, J.K. and Goodwin, J. 2014. GPR and bulk ground resistivity surveys in graveyards: locating unmarked burials in contrasting soil types. *Forensic Science International*. 237. B245-B255.

Peterson, R. and Robinson, D. 2012. Excavations and the afterlife of a professional football stadium, Peel Park, Accrington, Lancashire: towards an archaeology of football. *World Archaeology*. 44. 263-279.

Ramshaw, G. and Gammon, S. 2010. On home ground? Twickenham Stadium Tours and the construction of sport heritage. *Journal of Heritage Tourism*. 5. 1743-873.

Rees-Hughes, L., Pringle, J.K., Russill, N., Wisniewski K. D. and Doyle P. 2016. Multi-disciplinary investigations at PoW Camp 198, Bridgend, S. Wales: site of a mass escape in March 1945. *Journal of Conflict Archaeology*. 11. 166-191.

Saey, T., Gheyle, W., Stichelbaut, B., Bourgeois, J., Verplaetse, S., Eetvelde, V., Note, N. and Meirvenne, M. 2016. The Characterization of a Former World War I Battlefield by Integrating Multiple Signals from a Multireceiver EMI Soil Sensor. *Geoarchaeology*. 31. 267-281.

Stichelbaut, B., Gheyle, W., Van Eetvelde, V., Van Meirvenne, M., Saey, T., Note, N., Van den Bergh, H. and Bourgeois, J. 2017. The Ypres Salient 1914–1918: historical aerial photography and the landscape of war. *Antiquity*. 91. 235-249.

Sunseri, J.U. and Byram, S. 2017. Site Interiography and Geophysical Scanning: Interpreting the Texture and Form of Archaeological Deposits with Ground-Penetrating Radar. *Journal of Archaeological Method and Theory*. 24. 1400–1424.

Turner, S. 2012. In cold blood: dead athletes in Classical Athens. *World Archaeology*. 44. 217-233.

A New Look at Old County Number Records: Geophysical Reassessment of Scheduled Roman Villas

Neil Linford, Paul Linford, Andrew Payne
Historic England, Portsmouth, United Kingdom

Background

Since 1913 significant archaeological sites in the United Kingdom have been protected as Scheduled Monuments, a designation system intended to protect and manage a representative sample of the most important types of archaeological remains. However, sites scheduled in the early to mid-20th century, often relied on scant evidence such as surface finds or poorly documented antiquarian excavations to determine what archaeological remains might be present. These are referred to as "old county number" (OCN) records after the original monument numbering system. It is, perhaps, unsurprising that some of these original designations would benefit from reassessment with more recent methodologies developed since the monument was first protected. Surveys of three OCN monuments, all designated as Roman villas, demonstrate the potential for modern geophysics to improve understanding of what has been scheduled, in some cases changing the interpretation of the monument.

Method

Given the expectation from the OCN records, aerial photography and previous historic excavation for the presence of Roman structural remains, a combination of wide area magnetic survey together with more targeted Ground-Penetrating Radar (GPR) coverage was envisaged for all of the sites.

Magnetometer data were collected using an array of six high sensitivity Geometrics G862 caesium vapour magnetometer sensors mounted on a non-magnetic sledge, with a central gradient sensor mounted 1m above the array, towed behind a low impact All-Terrain Vehicle (ATV) (Linford *et al.* 2018a). A sampling density of ~0.15m by 0.5m was obtained along successive swaths with positional control achieved using a Trimble R8 Global Navigation Satellite System (GNSS) receiver mounted on the sensor platform.

For the GPR survey a 3d-Radar MkIV GeoScope Continuous Wave Step Frequency GPR system was used, collecting data with a multi-element DGX1820 vehicle towed, ground coupled antenna array and Trimble R8 GNSS receiver for positional control (Linford *et al.* 2010, Eide *et al.* 2018) . Data were acquired at a 0.075m x 0.075m sample interval to allow a series of amplitude time slices to be created to aid visualisation of the results.

Results

Low Ham, Somerset

Excavations in the 1940s uncovered the south-western range of buildings, now partially colonised by a badger sett. However, geophysical survey (Fig. 1) reveals potentially four ranges of buildings arranged around a rectangular central courtyard with a roadway running north-east towards the nearby river and surrounded by an extensive field system (Linford *et al.* 2018c). Intriguingly, one building uncovered by an extension to the original excavation proved to be post-medieval suggesting some of the remains may relate to a later 17th century mansion known to have been built somewhere in the vicinity but until now thought to be beneath the modern farm buildings. Further small-scale excavations conducted in the autumn of 2018 have provided significant additional insights within the immediate vicinity of the scheduled Roman site, including a precursor Iron Age enclosure revealed by the magnetic survey.

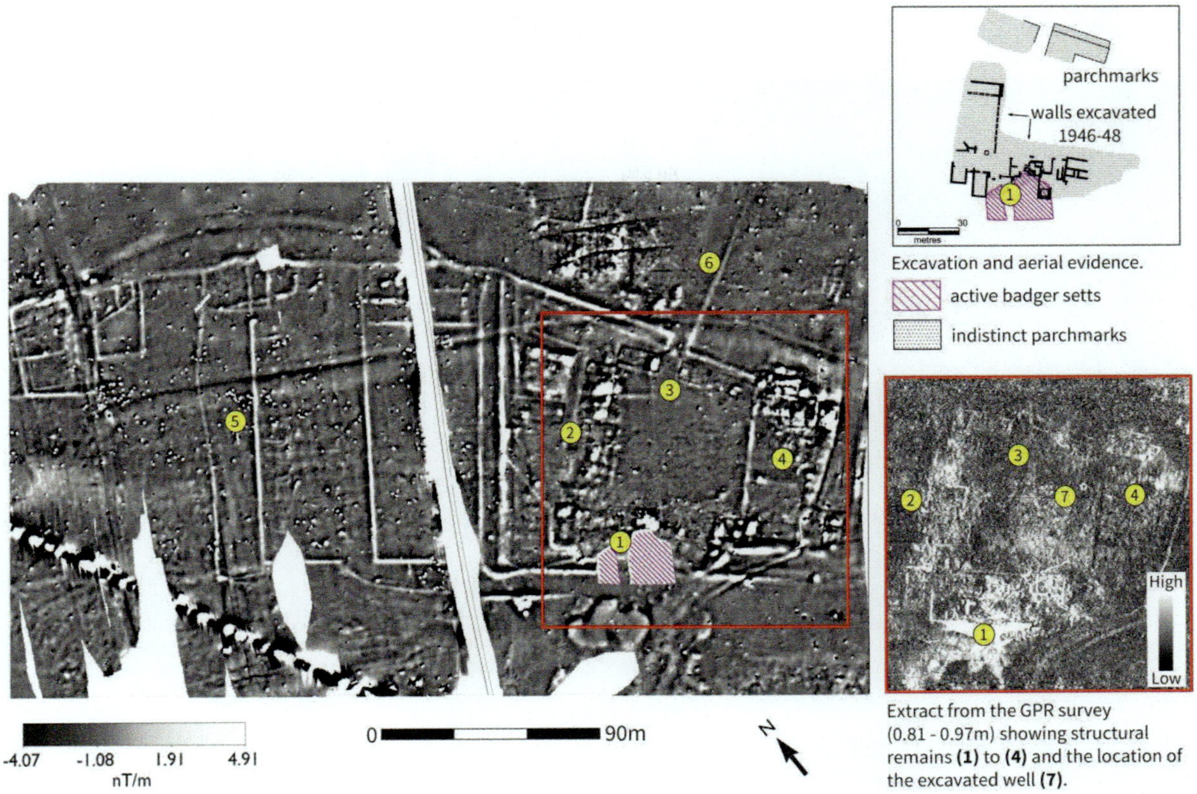

Fig.1. Extract from the magnetic survey showing the previously excavated west wing (1) of the scheduled villa and damage from an active badger sett threatening the survival of the remains. Additional building ranges (2), (3) and (4), only partially recognised from parchmarks, have been better defined in both the magnetometer survey and GPR data (inset bottom right). A much wider landscape of ditched enclosures (5) and an avenue (6) heading towards the river from the villa compound are also evident.

Nuthills Roman villa, Wiltshire

The Nuthills villa is known from some Roman finds and crop mark remains of a fragmented coaxial ditch-defined field system mapped by aerial photography with a rectangular double-ditched enclosure nearby. However, magnetometer survey (Fig. 2) reveals the ditch system to be an elaboration of two Iron Age banjo enclosures while GPR reveals the wall footings of six buildings on different alignments (Linford *et al.* 2018b). The site plan is suggestive of a Roman temple or shrine complex like those at Uley and Lyndey in Gloucestershire, rather than a Roman villa.

Bradford Abbas, Dorset

Roman remains were discovered at Bradford Abbas in the 19th century by a local professor of geology and further limited excavations were carried out by amateur archaeologist Charles Bean in the 1950s, uncovering the footings of a Roman building assumed to be part of a villa. After interpretation of new aerial photographic evidence in 2013 revealed a Roman marching camp on the other side of the same farm, this interpretation was called into question (Winton and Grady 2013) and it was decided to re-evaluate the existing scheduled remains. Geophysical survey results (Fig. 3) reveal a complex picture more suggestive of a manufacturing settlement than a villa (Linford *et al.* 2019).

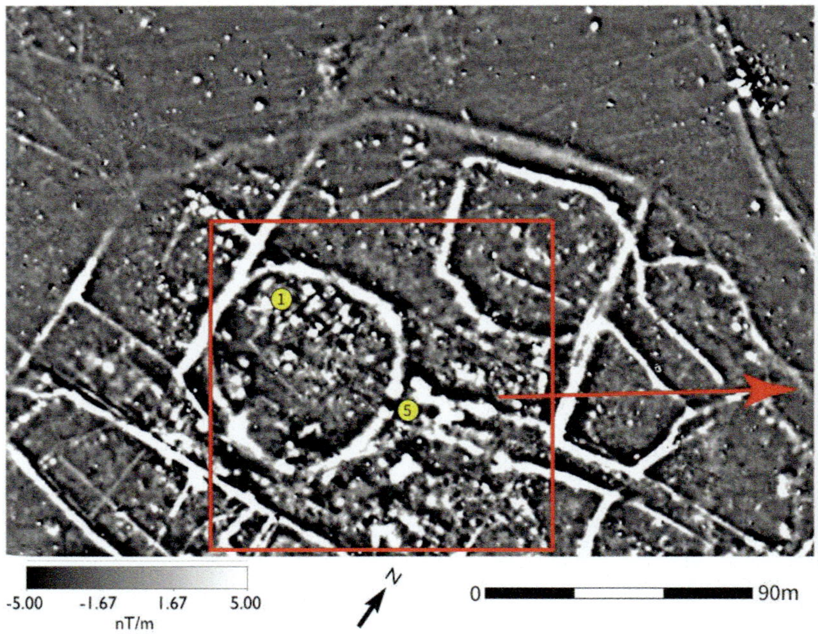

GPR survey in progress during difficult field conditions, particularly the maize stubble creating striping artefacts in the GPR data.

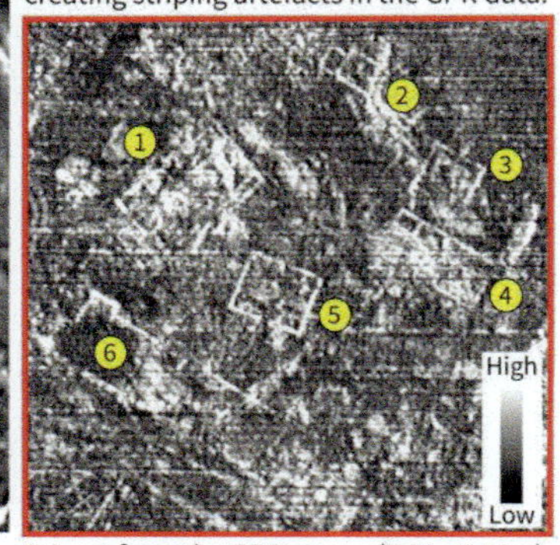

Extract from the GPR survey (0.81 - 0.98m) showing a complex of 5 buildings surrounding the main villa (1).

Fig. 2. Extract from the magnetic survey showing a complex multi-phase landscape including two IA banjo enclosures. The scheduled villa (1) is located within the larger banjo enclosure and one building (5) appears to be built over the entrance to the enclosure on the alignment of the approach ditches. Additional building remains were revealed by the targeted GPR survey (inset bottom right).

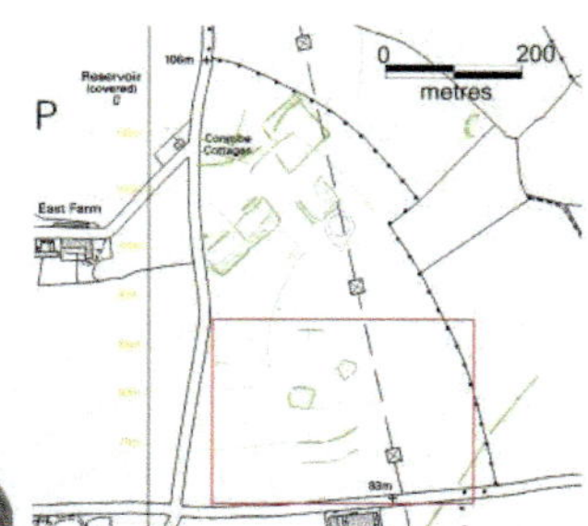

Aerial photograph transcription showing crop marks in green and extract of the survey area (red rectangle)

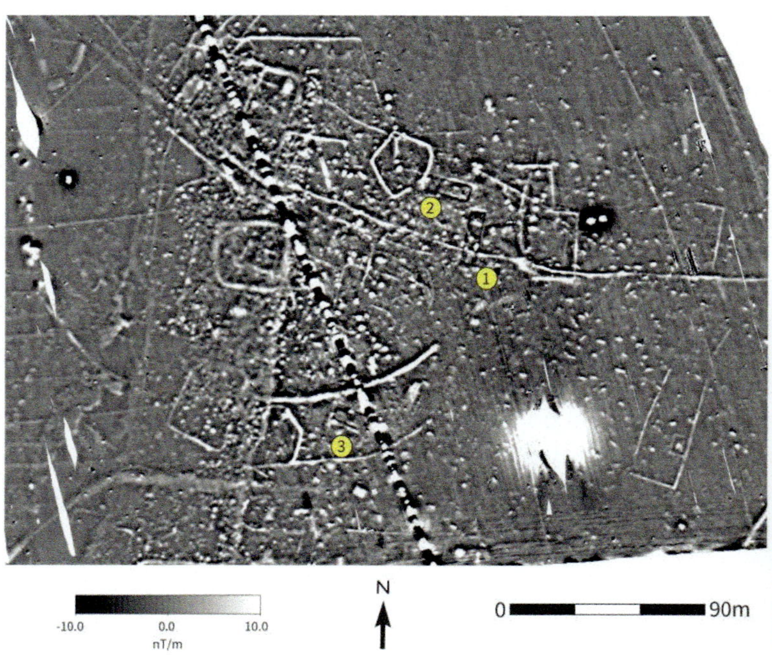

Photograph of the excavated walls

Fig. 3. Extract from the magnetic survey over the scheduled area which reveals a complex of pits and intercutting enclosures. Rectilinear negative anomalies at (1), (2) and (3) suggest buried wall footings. From its position the building anomaly at (1) is likely to be that excavated by Charles Bean in the 1950s (inset bottom right).

Conclusions

Geophysics is a rapid, non-invasive investigative method and these surveys demonstrate its potential to improve knowledge even for a class of monument considered relatively well understood. In two of the three cases a completely new interpretation for the monument is suggested, showing rich potential for further research. Geophysical reassessment of a wider range of OCN scheduled sites, including prehistoric sites with extensive excavation, has also revealed significant evidence to assist with the interpretation and management of the remains. It is hoped that improved protection of the sites could be achieved through revised arable regimes guided by the georeferenced data, with long term monitoring potentially provided through repeat geophysical survey to assess the condition and survival of the remains.

References

Eide, E., Linford, N., Persico, R. and Sala, J. 2018. Advanced SFCW GPR systems. In Persico, R., Piro, S. and Linford, N. (editors) *Innovation in Near-Surface Geophysics Instrumentation, Application, and Data Processing Methods*. 1st edition. Amsterdam. Elsevier. 253-285.

Linford, N., Linford, P., Martin, L. and Payne, A. 2010. Stepped-frequency GPR survey with a multi-element array antenna: Results from field application on archaeological sites. *Archaeological Prospection*. 17 (3). 187-198.

Linford, N., Linford, P. and Payne, A. 2018a. Advanced magnetic prospecting for archaeology with a vehicle-towed array of cesium magnetometers. In Persico, R., Piro, S. and Linford, N. (editors) *Innovation in Near-Surface Geophysics Instrumentation, Application, and Data Processing Methods*. 1st edition. Amsterdam. Elsevier. 121-149.

Linford, N., Linford, P. and Payne, A. 2018b. *Verlucio And Environs Project, Bowood Estate, Calne Without, Wiltshire, Report on geophysical surveys, February 2018*. [Research Reports Series]. London. Historic England. 41/2018. http://research.historicengland.org.uk/Report.aspx?i=15414

Linford, N., Linford, P., Payne, A. and Rees, L. 2018c. *Low Ham Roman Villa, High Ham, Somerset: Report on Geophysical Survey, July 2018*. [Research Reports Series]. London. Historic England. 72/2018. https://research.historicengland.org.uk/Report.aspx?i=16183

Linford, N., Linford, P., Payne, A. and Salkeld, E. 2019. *East Farm, Bradford Abbas, Dorset. Report on Geophysical Survey, September 2018*. [Research Reports Series]. London. Historic England. 12/2019. https://research.historicengland.org.uk/Report.aspx?i=16229

Winton, H. and Grady, D. 2013. *A Roman Camp at Bradford Abbas, Dorset, Aerial Investigation and Mapping, Archaeological Report*. [Research Report Series]. English Heritage. 37-2013.

The influence of buried archaeology on equine locomotion: results from the Burghley Horse trials cross country course

Neil Linford[1], Russell Mackechnie-Guire[2,3]

[1]*Historic England, Portsmouth, United Kingdom,* [2]*Centaur Biomechanics, Moreton Morrell, United Kingdom,* [3]*Royal Veterinary College, Hatfield, United Kingdom*

The aim of this research is to examine the potential influence that buried archaeological remains may have on the ridden horse and whether this has a measurable impact on performance or animal welfare when historic assets are found under eventing courses. Geophysical survey was used to investigate two areas of the world renowned Burghley Horse Trials site, near Stamford, Peterborough, U.K., to suggest a location over archaeological remains for subsequent controlled kinematic equine motion analysis trials, as part of the "Going over Old Ground" research project, coordinated by University College London.

The characteristics of the riding surface can increase the likelihood of lameness in the dressage horse, particularly over patchy or uneven surfaces, when the firmness of the surface may change compared to uniform conditions (Murray *et al.* 2010a; 2010b). The sub-surface also has an impact on the way a horse propels its mass forward, in supporting its mass the surface properties are crucial to optimise the locomotor system, and to give energy back to the horse to aid its locomotion. Some (artificial) surfaces have been designed to be firm and, although there may be some mechanical benefits, there is concern that there is an increase in ground reaction forces and an increased breaking force which over time could lead to injury (Hobbs *et al.* 2014). This study examines the effect that turf surfaces with underlying archaeology may have on the locomotion of the horse, related to the depth and extent of the buried remains.

Method

The presence of archaeological remains was initially assessed from known sources recorded in the local Historic Environment Record (HER) and the National Heritage List for England (NHLE). A geophysical survey, using a Ground-Penetrating Radar (GPR) array was used in advance to accurately locate and characterise significant archaeological remains to determine the best site to conduct the motion analysis study. The vehicle towed GPR survey was conducted with a 3d-Radar MkIV GeoScope Continuous Wave Step-Frequency GPR system collecting data with a multi-element DXG1820 vehicle towed, ground coupled antenna array (Linford *et al.* 2010, Eide *et al.* 2018). Data were acquired at a 0.075m x 0.075m sample interval across a continuous wave stepped frequency range from 40MHz to 2.99GHz in 4MHz increments using a dwell time of 3ms.

Once a suitable test site had been identified subsequent equine motion analysis trials were conducted (Fig. 1) with a convenience sample of five event type horses (8 years) all ridden by their associated rider. Five inertial measurement units collecting data at 60Hz per individual sensor channel where attached to each horse to measure variations in movement during the test, together with nine reflective markers attached with double sided tape over anatomical landmarks identifying joint centres and segment ends. A high speed (300 fps) video camera system was then used to record the gait of the horse using biomechanical software to capture the relative position of the reflective markers from each frame. Data were collected from both the left and right rein with three repeats per direction, with a field of view capturing two complete strides in trot and canter (Willmott and Dapena 2012).

Fig. 1. Equine motion analysis in progress - the horse has been instrumented with 5 triple axis inertial measurement units to monitor any changes in gait when passing over the location of the Ermine Street Roman road.

Results

The GPR survey over the site of the main event arena (2.2ha), close to Burghley House itself, revealed a plethora of anomalies most likely associated with the infrastructure for the horse trials spectator stands, including a large number of service runs (Orr 2017, Linford 2018). More significant results were found over the course of the Ermine Street Roman road, in the vicinity of the Cottesmore Leap on the cross-country eventing course. The survey here (1.6ha) revealed a well-preserved section of Roman Road, although the survival seems compromised by ploughing to the east of Queen Anne's Avenue. The GPR data suggested the location for two test tracks, one situated over the remains of the Roman road and one immediately north of the road with little or no apparent archaeology present (Fig. 2).

Results from the pilot equine motion analysis data, show that when horses were travelling over ground which had underlying archaeology, this was associated with an alteration in gait when compared to ground where the archaeology was absent. A variation in gait was also found when the horses were travelling from the right of Track 1, commencing over the best-preserved portion of the Roman road agger, possibly suggesting a greater degree of firmness and more uniform surface for the horses to propel their mass. No invasive means of determining the hardness of the ground was possible in this pilot study, but a repeat GPR survey was conducted immediately after the equine motion analysis trial which provided useful information regarding very near-surface ground deformation due to the horses over the experimental tracks.

Conclusions

Whilst the results from this study are limited in its sample size, a correlation between equine locomotion and presence of buried archaeological remains has been demonstrated and warrant further investigation. Some direct parallels are suggested between the relative survival of the archaeological remains and observations made on the condition of artificial equine sports surfaces. In essence the horse is able to stabilise and propel their mass more efficiently when the ground has significant archaeology present. Also partially reflected is the greater degree of track wear evident where the underlying Roman road created a firmer riding surface.

Fig. 2. Extract from the GPR survey showing an amplitude time slice between 5.0ns and 7.5ns (0.27m to 0.4m) of Ermine Street Roman road, superimposed over an aerial photograph of the Burghley House equestrian cross country course. The location of the two equine motion analysis tracks are shown by the green boxes, positioned over the centre and immediately to the north of the underlying Roman road.

References

Eide, E., Linford, N., Persico, R. and Sala, J. 2018. Advanced SFCW GPR systems In Persico, R, Piro, S and Linford, N. (eds.), *Innovation in Near-Surface Geophysics Instrumentation, Application, and Data Processing Methods*. Amsterdam. Elsevier, 253-285.

Hobbs, S.J., Northrop, A. J., Mahaffey, C., Martin, J. H., Murray, R., Roepstorff, L. and Peterson, M. 2014. *Equine Surfaces White Paper*. Fédération Equestre Internationale.

Linford, N. 2018. *Burghley Park, St. Martin's Without, City of Peterborough, Report on Geophysical Survey, April 2018*. Historic England Research Reports Series 76/2018.

Linford, N., Linford, P., Martin, L. and Payne, A. 2010. Stepped-frequency GPR survey with a multi-element array antenna: Results from field application on archaeological sites. *Archaeological Prospection*. 17(3). 187-198.

Murray, R., Walters, J., Snart, H., Dyson, S. and Parkin, T. 2010a. How do features of dressage arenas influence training surface properties which are potentially associated with lameness? *The Veterinary Journal*. 186(2). 172-9.

Murray, R., Walters, J., Snart, H., Dyson, S. and Parkin, T. 2010b. Identification of risk factors for lameness in dressage horses. *The Veterinary Journal*. 184(1). 27-36.

Orr, S.A. 2017. *Remains of St Michael's Priory, Burghley Park, Ground penetrating radar report, November 2017*. School of Geography and the Environment, University of Oxford.

Willmott, A. and Dapena, J. 2012. The planarity of the stickface motion in the field hockey hit. *Journal of Sports Sciences*. 30(4). 369-77.

When finding nothing is interesting

Kris Lockyear
UCL/WAS/CAGG, Wheathampstead, United Kingdom

Introduction

The Roman city of Verulamium, which lies some twenty miles north of London, was the third largest city in the province of Britannia, if judged by the area enclosed within its walls. The city was preceded by a so-called Iron Age "territorial *oppidum*". The Roman town seems to have become quite important quite quickly judging by the recent find of a writing tablet at the Bloomberg site in London (Tomlin 2017, Tablet 45) dated to AD 62. It was famously one of the three cities sacked by Boudicca in AD 60/61 but recovered to become a *municipium*. Around AD 80, an area of 52ha was enclosed by an earthwork now known as the "1955 ditch". An area to the NW of the town was also enclosed by a large ditch and bank, now known as the Fosse, but the dating and purpose of this feature is unclear. In the third century, new town walls were built with token bastions, and these enclosed 81ha. In the medieval period the focus of settlement moved up hill to the area next to the medieval abbey and the shrine of St Alban, Britain's first saint. As a result, the Roman town now largely lies under agricultural pasture and a municipal park making it the largest town from Roman Britain not built over by a modern city. A useful overview of the town is given by Niblett (2001), and a more detailed assessment was provided by Niblett and Thompson (2005).

The Community Archaeology Geophysics Group (CAGG)

CAGG grew out of an AHRC-funded project entitled *Sensing the Iron Age and Roman Past: geophysics and the landscape of Hertfordshire* that started in 2013. One aim of the project was to train and support a team of volunteers from local archaeological societies in geophysical survey techniques. The project purchased a Foerster Ferex cart system and work by the group initially focused on gradiometry survey (Fig. 1, middle). Although the team surveyed a number of sites, the largest was the half of Verulamium that lies under the municipal park (Lockyear and Shlasko 2017). Although the project funding ceased in 2014, the group has continued to undertake surveys and has now worked on over 31 sites. Thanks to the Collaborative Doctoral Training school *Science and Engineering in Architecture, Heritage and Archaeology* (SEAHA), the group has been, since 2015, able to use a Mala GX GPR with a 450MHz antenna (Fig. 1, top), and from 2016 the Institute of Archaeology, UCL has loaned us a Geoscan RM85 Earth Resistance Meter (Fig. 1, bottom). Additionally, one of our volunteers, Peter Alley, has been undertaking detailed topographic surveys using a UAV and Structure from Motion. The project maintains a blog (hertsgeosurvey.wordpress.com) which makes the initial results of surveys available quickly, often on the same day as the survey.

Continuing Surveys at Verulamium

Since 2015, the team has been able to undertake surveys on the northern half of the town which lies under pasture and forms part of the Gorhambury Estate. The magnetometry survey of the area inside the town walls was completed by the end of the 2016 season, and in 2018 we were able to expand the survey to the fields to the NW which includes the area enclosed by the Fosse (Fig. 2). The GPR survey has completed 19ha within the town walls at a 0.5m transect spacing (Fig. 3). Since 2016, we have also been able to undertake Earth Resistance survey using the Institute's Geoscan RM85 at a 0.5m by 0.5m data density. Although dry spells have slowed progress, we have completed almost 7ha. In August 2019 we hope to continue the magnetometry survey outside the walls, complete the GPR survey of the NW quadrant of the town and extend the Earth Resistance survey.

Fig. 1. The CAGG team in action. Top: GPR, middle: magnetometry, bottom: Earth Resistance.

Fig. 2. The Magnetometer survey at Verulamium as of the end of August 2018.

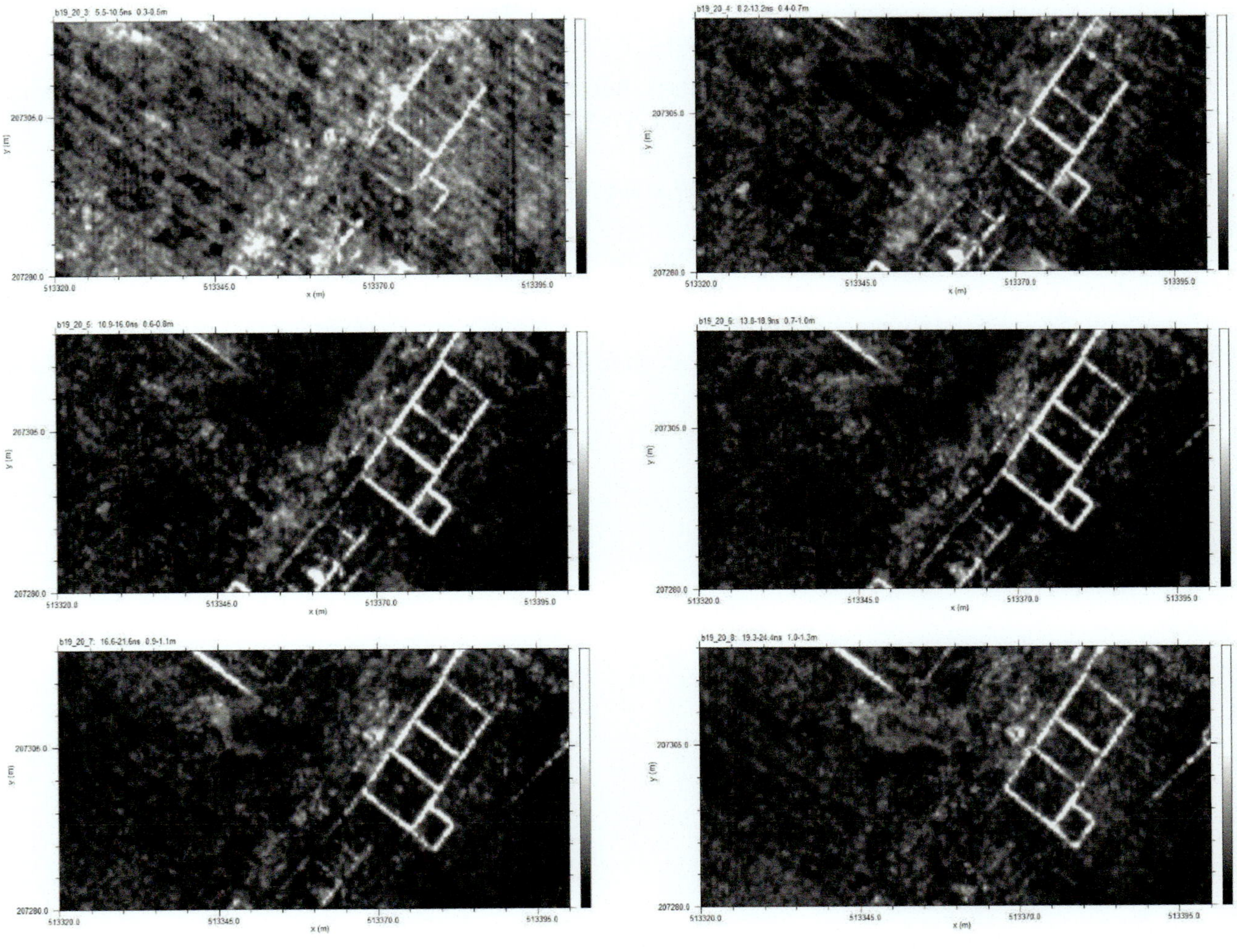

Fig. 3. Six time slices across the buildings to the east of the Insula XVI temple.

Empty Spaces

In addition to the features one would expect within a Roman city: town houses, roads, an aqueduct, temples and so on, the plan of Verulamium also includes a number of areas with minimal features detected by the surveys. There are two main zones. The first is the area within "the Fosse" but outside the third century walls. The magnetometry survey has revealed almost no features here at all. The curious nature of this area is magnified when the topography of the area is considered, especially when the line of the formally massive earthworks is considered. The interpretation of this area has been much debated, and the complete lack of finds adds an extra twist.

The second zone consists of the majority of the upslope areas within the town walls. These are also very "quiet", and may well have never been densely settled, if at all. This is in contrast to other towns in Roman Britain which have had almost-complete surveys, such as Silchester and Caistor-by-Norwich. For the northern half of the town, that within the Gorhambury Estate, we have the additional evidence of a large-scale programme of test-pitting undertaken in 2000 by Oxford Archaeology. The distribution of finds and soil depths provides useful additional evidence. Why did the city walls enclose so much seemingly empty space? Seen in the context of Roman cities on the continent, Romano-British cities are an oddity, but Verulamium appears to be even odder!

Acknowledgements

We would like to thank the AHRC for the original grant (ref: AH/K007602/1) which purchased the Foerster Ferex and enabled the formation of the group; UCL for the loan of the dGPS and Geoscan RM85; and SEAHA for the loan of the Malå GX GPR. We would also like to thank Lord Verulam for allowing and facilitating access to the Gorhambury Estate. I, personally, would like to thank all the members of the group for all their hard work collecting the data over the last six years.

References

Lockyear, K. and Shlasko, E. 2017. Under the Park. Recent Geophysical Surveys at Verulamium (St Albans, Hertfordshire, UK). *Archaeological Prospection.* 24. 17–36.

Niblett, R. 2001. *Verulamium. The Roman City of St Albans*. Stroud.

Niblett, R. and Thompson, I. 2005. *Alban's Buried Towns. An assessment of St Albans' archaeology up to AD 1600*. Oxford.

Tomlin, R. 2017. *Roman London's First Voices: Writing Tablets from the Bloomberg Excavations, 2010–14*. London.

Co-creation and archaeological prospection: LoCATE – The Local Community Archaeological Training and Equipment Project

Kate Welham[1], James Brown[2], Paul Cheetham[1], Mike Gill[3], Lawrence Shaw[2]
[1]Bournemouth University, Poole, United Kingdom, [2]New Forest National Park Authority, United Kingdom, [3]Avon Valley Archaeological Society, United Kingdom

This paper is based on the co-creation of research through an innovative partnership focused around archaeological prospection techniques. LoCATE (Local Community Archaeological Training and Equipment) is a project that brings together archaeologists at Bournemouth University and the New Forest National Park Authority with archaeological societies and community groups from across Dorset and Hampshire. LoCATE provides access, training, and support for the use of advanced survey equipment that can otherwise be hard to get hold of. It supports the work that all partners already do by extending the range of techniques and skills they can use and expanding their capacity to undertake research.

The idea for LoCATE was first instigated in 2015 when members of the Avon Valley Archaeological Society approached the University and asked them to consider providing access to older, but serviceable geophysical equipment that was not being used regularly for teaching and other activities. Working with the New Forest National Park Authority, LoCATE was developed, and the first instrument made available was a Geoscan Research FM36 followed a year later by a Geoscan Research RM15. Most recently a total station has been added to the equipment pool, funded through Heritage Lottery Funding (Our Past and Our Future, Landscape Partnership Scheme) and the Hampshire Field Club and Archaeological Society. Access to the equipment is managed through the New Forest's volunteer equipment loan system. LoCATE members are given access to free training on these techniques using a variety of expertise situated across the partnership (Fig. 1), and LoCATE members sign up to a code of responsible survey and data sharing (Fig. 2). Open data is a core value, and LoCATE also encourages members to use open access materials and software, for example Snuffler freeware geophysics software (Staveley 2018). Inter-partner support is fostered through a variety of means including shared prospection activities, and project social media channels.

Fig. 1. LoCATE training day (Photo: L. Shaw).

Fig. 2. LoCATE member report on the FM36 gradiometry results from Bustard Farm, East Martin (Hampshire).

Now in its fourth year, LoCATE has become well established, enabling relationships that support the research agendas of all partners. It has developed both capacity and expertise in the use of archaeological prospection activities in the local region. Example of the success of the project can be seen through the diversity of the outcomes from the work of LoCATE members from prehistoric monuments, including previously understudied Neolithic long and oval barrows (Fig. 3) and Bronze Age double ring ditches, to extensive Romano-British sites along the Avon Valley and on Cranborne Chase (Hampshire) (Gill 2019a; 2019b). Through these surveys LoCATE members have achieved their own research aims, but also contributed to the collective goal of the project in enabling an improved understanding of the rich archaeological heritage of our region.

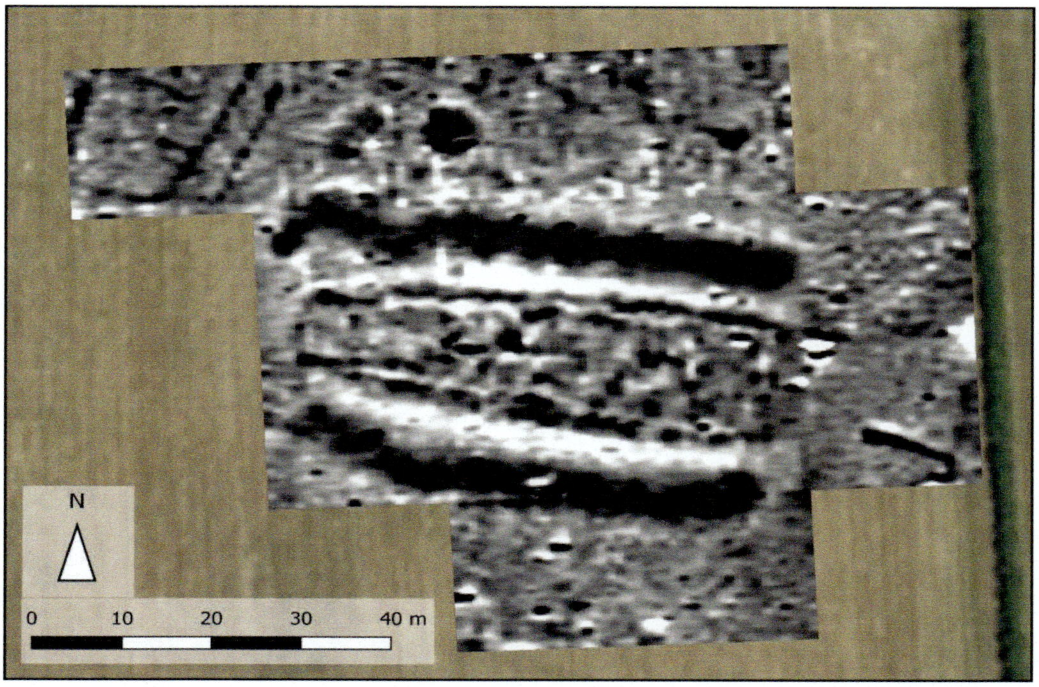

Fig. 3. The results of an initial LoCATE member survey using the FM36 gradiometer to identify a ploughed out long barrow near Fordingbridge (Hampshire).

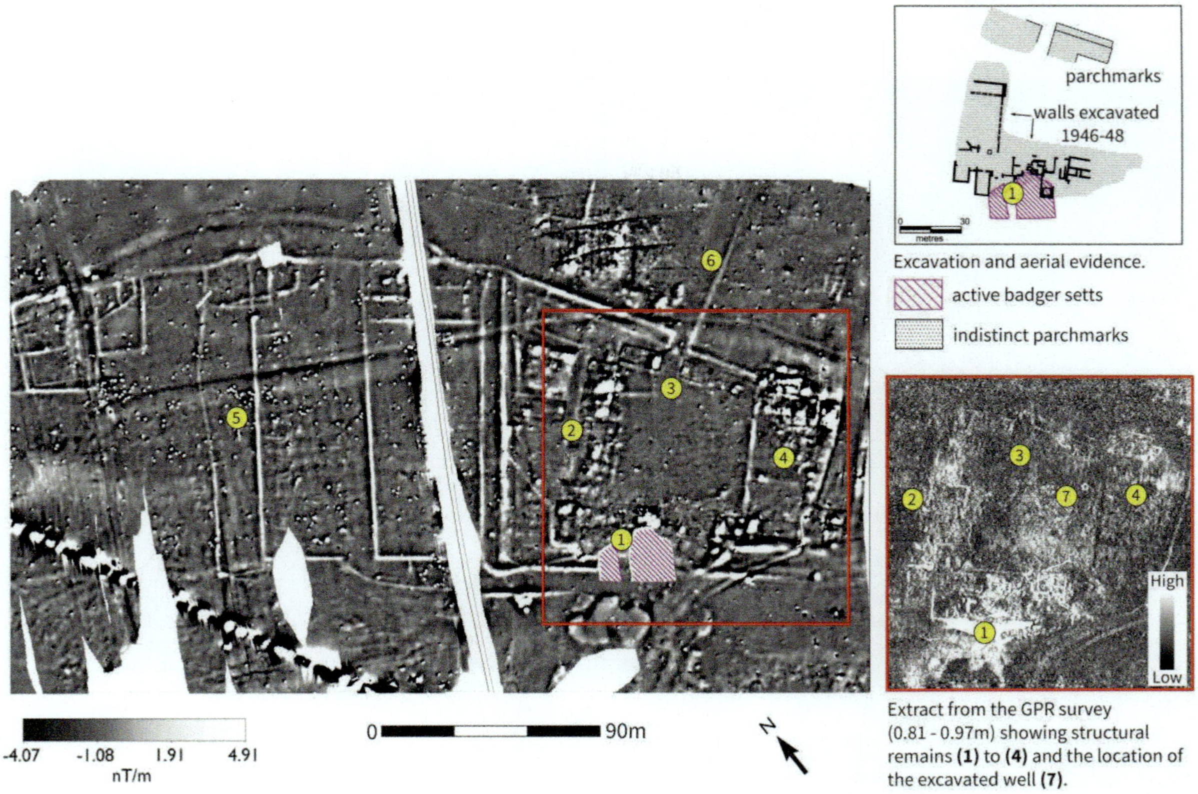

Fig.1. Extract from the magnetic survey showing the previously excavated west wing (1) of the scheduled villa and damage from an active badger sett threatening the survival of the remains. Additional building ranges (2), (3) and (4), only partially recognised from parchmarks, have been better defined in both the magnetometer survey and GPR data (inset bottom right). A much wider landscape of ditched enclosures (5) and an avenue (6) heading towards the river from the villa compound are also evident.

Nuthills Roman villa, Wiltshire

The Nuthills villa is known from some Roman finds and crop mark remains of a fragmented coaxial ditch-defined field system mapped by aerial photography with a rectangular double-ditched enclosure nearby. However, magnetometer survey (Fig. 2) reveals the ditch system to be an elaboration of two Iron Age banjo enclosures while GPR reveals the wall footings of six buildings on different alignments (Linford *et al.* 2018b). The site plan is suggestive of a Roman temple or shrine complex like those at Uley and Lyndey in Gloucestershire, rather than a Roman villa.

Bradford Abbas, Dorset

Roman remains were discovered at Bradford Abbas in the 19th century by a local professor of geology and further limited excavations were carried out by amateur archaeologist Charles Bean in the 1950s, uncovering the footings of a Roman building assumed to be part of a villa. After interpretation of new aerial photographic evidence in 2013 revealed a Roman marching camp on the other side of the same farm, this interpretation was called into question (Winton and Grady 2013) and it was decided to re-evaluate the existing scheduled remains. Geophysical survey results (Fig. 3) reveal a complex picture more suggestive of a manufacturing settlement than a villa (Linford *et al.* 2019).

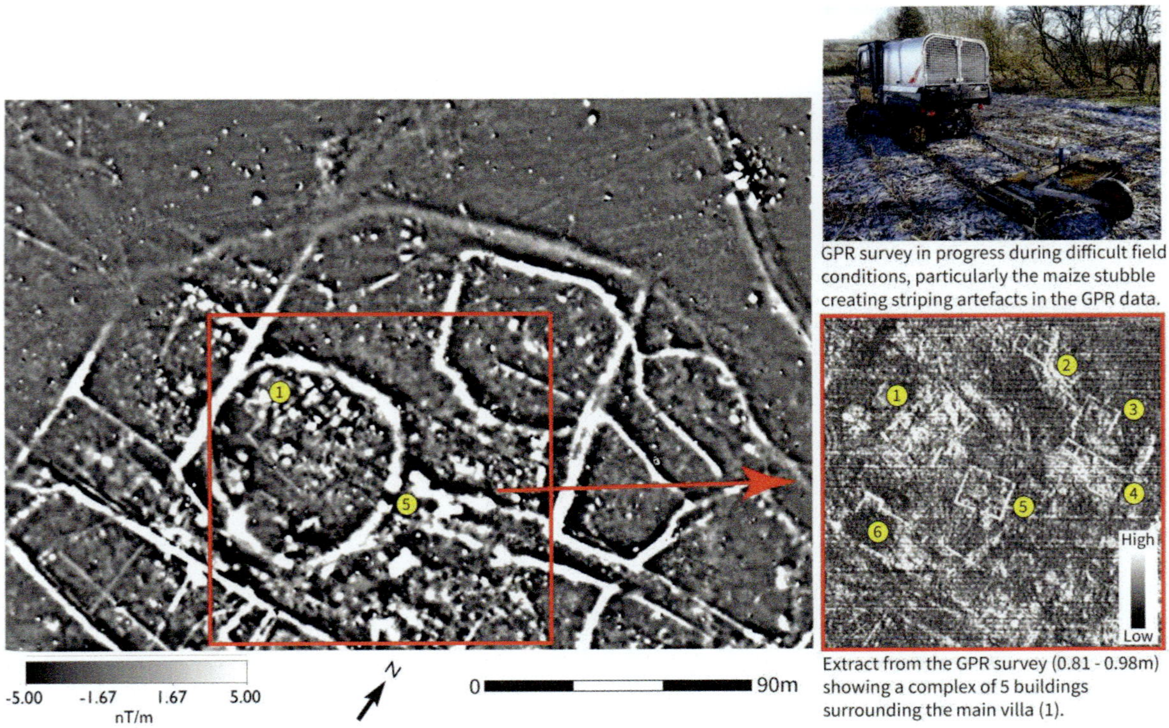

Fig. 2. Extract from the magnetic survey showing a complex multi-phase landscape including two IA banjo enclosures. The scheduled villa (1) is located within the larger banjo enclosure and one building (5) appears to be built over the entrance to the enclosure on the alignment of the approach ditches. Additional building remains were revealed by the targeted GPR survey (inset bottom right).

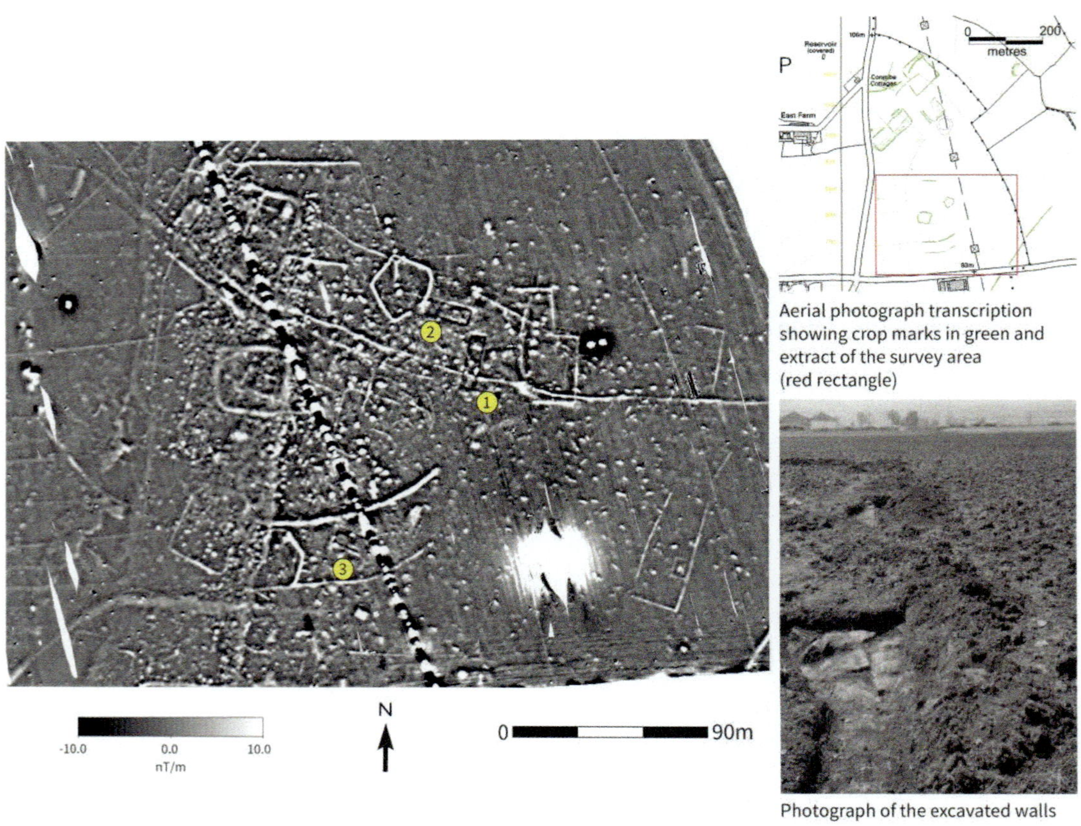

Fig. 3. Extract from the magnetic survey over the scheduled area which reveals a complex of pits and intercutting enclosures. Rectilinear negative anomalies at (1), (2) and (3) suggest buried wall footings. From its position the building anomaly at (1) is likely to be that excavated by Charles Bean in the 1950s (inset bottom right).

In conclusion, LoCATE provides a new model for community engagement in archaeological prospection projects. In an era where the integration of techniques and data are central themes, it is perhaps timely to also consider the integration of people, and how we best work with a variety of different communities to create a shared understanding of our collective past.

References

Gill, M. 2019a. New long barrow discoveries in the vicinity of the middle Avon Valley and Cranborne Chase. *PAST*. 91, 5-7.

Gill, M. 2019b. Putting old kit to good use, the LoCATE geophysical project. *British Archaeology*. March/April 2019. 30(5).

Staveley, D. 2018. *Snuffler - Freeware Geophysics Software*. Available from: http://www.sussexarch.org.uk/geophys/snuffler.html

Part Two – Archaeological Prospection in Africa

Old Plans versus Geophysical Surveys: the case study of Gurob (Fayum, Egypt)

Lionel Darras[1], Tomasz Herbich[2], Marine Yoyotte[3]

[1]Archeorient, Lyon, France, [2]Institute of Archaeology and Ethnology, Warsaw, Poland, [3]Institut français d'archéologie orientale, Cairo, Egypt

Introduction

The site of Gurob extends beyond the remains of the ancient town of Mi-wer, on the southern side of the Fayum oases entrance (Fig. 1). The town was built by Thutmose III (BC 1479-1425) in order to shelter a palatial complex. This was served by a harbour and was occupied at least until the reign of Ramesses V (BC 1147-1143). The site preserves the only palace from the Pharaonic period identified as a royal harem. The remains of the palace, the adjoining necropolis as well as written evidence found *in situ*, provide a better understanding of the organisation and administration of a harem during the New Kingdom. Therefore, this site proves to be of major interest. The harbour was connected with the Bahr Yussef, which is a waterway between Fayum and the Nile.

In the framework of the Gurob Harem Palace Project led by Dr Ian Shaw (University of Liverpool), investigations of the site, mainly around the harem (Shaw 2012, Yoyotte 2012), were conducted between 2005 and 2015. In 2016, the Institut français d'archéologie orientale (Ifao) gathered an interdisciplinary team in order to complete the results of the previous work and to initiate a new research project. Archaeological, geoarchaeological and geophysical studies, which have been carried out since then, have been focused on two major structures: the harem and the harbour. The objective is to: (i) gain more detailed knowledge of the spatial organisation of the area, (ii) identify possible settlements around the palace and (iii) locate the channel connecting the harbour with the plateau. Furthermore, two neighbouring areas, previously identified as a "fort" and kilns, are being re-investigated in order to establish a better chronology of the site.

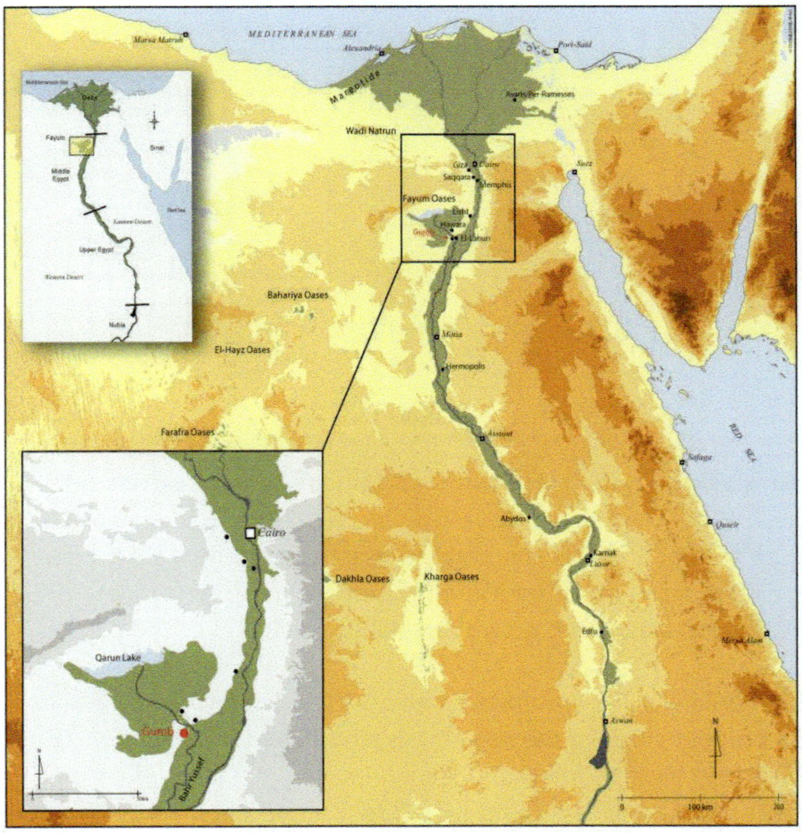

Fig. 1. Geographical location of Gurob. © Ifao.

Old excavations and plans

Flinders Petrie conducted two archaeological seasons on the site between 1888 and 1890. Following these works, he incorrectly identified the main building on the site as a temple of large dimensions, of which he drew a plan (Petrie 1891). In 1903-1904, Charles Currely and Leonard Loat spent three weeks in Gurob, although they were interested in the animal and predynastic cemeteries (Loat 1905).

In 1920, Guy Brunton and Reginald Engelbach (Brunton and Engelbach 1927) drew the first general map of the town and its nine cemeteries (Fig. 2). Northeast of the outer enclosure of the harem, they discovered a structure which they identified as a "fort" because of the thickness of its walls. It was dated to either the First (BC 2160-2055) or the Second Intermediate Period (BC 1650-1550). To the west of the fort, there was a small square enclosure, on top of which glass workshops and lime-kilns were located.

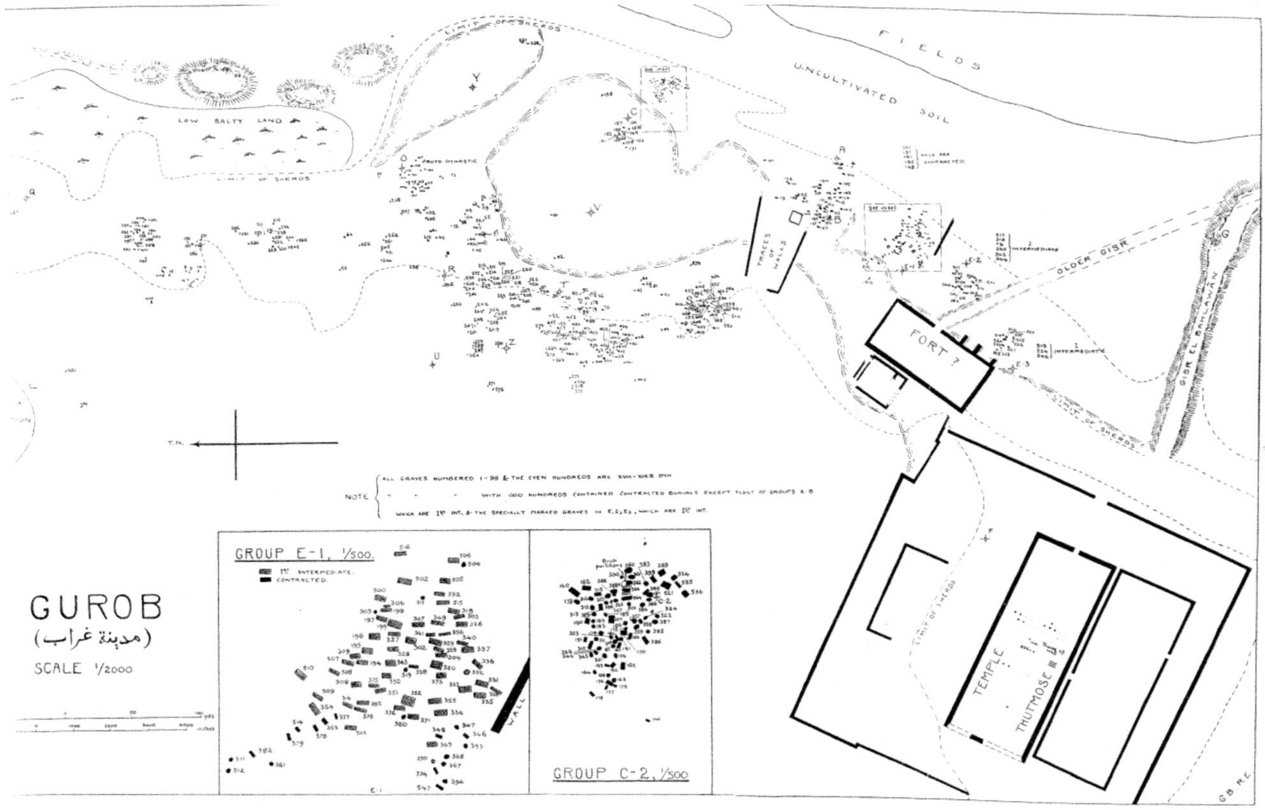

Fig. 2. Brunton and Engelbach's plan (1927).

Geophysical surveys

To verify old plans and excavations, several geophysical campaigns were carried out:

- In 2006 and 2007, an area of 5ha was investigated using a FM36 fluxgate gradiometer (Geoscan) in the framework of the field programme of the Gurob Harem Palace Project (Shaw 2008). The main objective was to trace the internal structure of the two rectangular buildings of the harem.

- In 2017 and 2018, an area of 13ha was investigated using a G858 caesium gradiometer (Geometrics) in the framework of a broader study of the site and its landscape (Yoyotte *et al.* 2019). The main purposes of the geophysical surveys were as follows:
    - confirm the location of an access ramp mentioned on an old plan by Brunton and Engelbach and locate the ancient harbour;
    - complete the geophysical survey and understand the link between the harem palace and the fort;
    - acquire more details about the occupation area shown on several old plans (Brunton and Engelbach 1927, Kemp 1978).

The results of these magnetic surveys were processed with the open-source software WuMapPy (Marty *et al.* 2015, see also Vitale *et al.*, this volume), and combined with the archaeological structures mentioned in Brunton and Engelbach's plan using GIS software (Fig. 3).

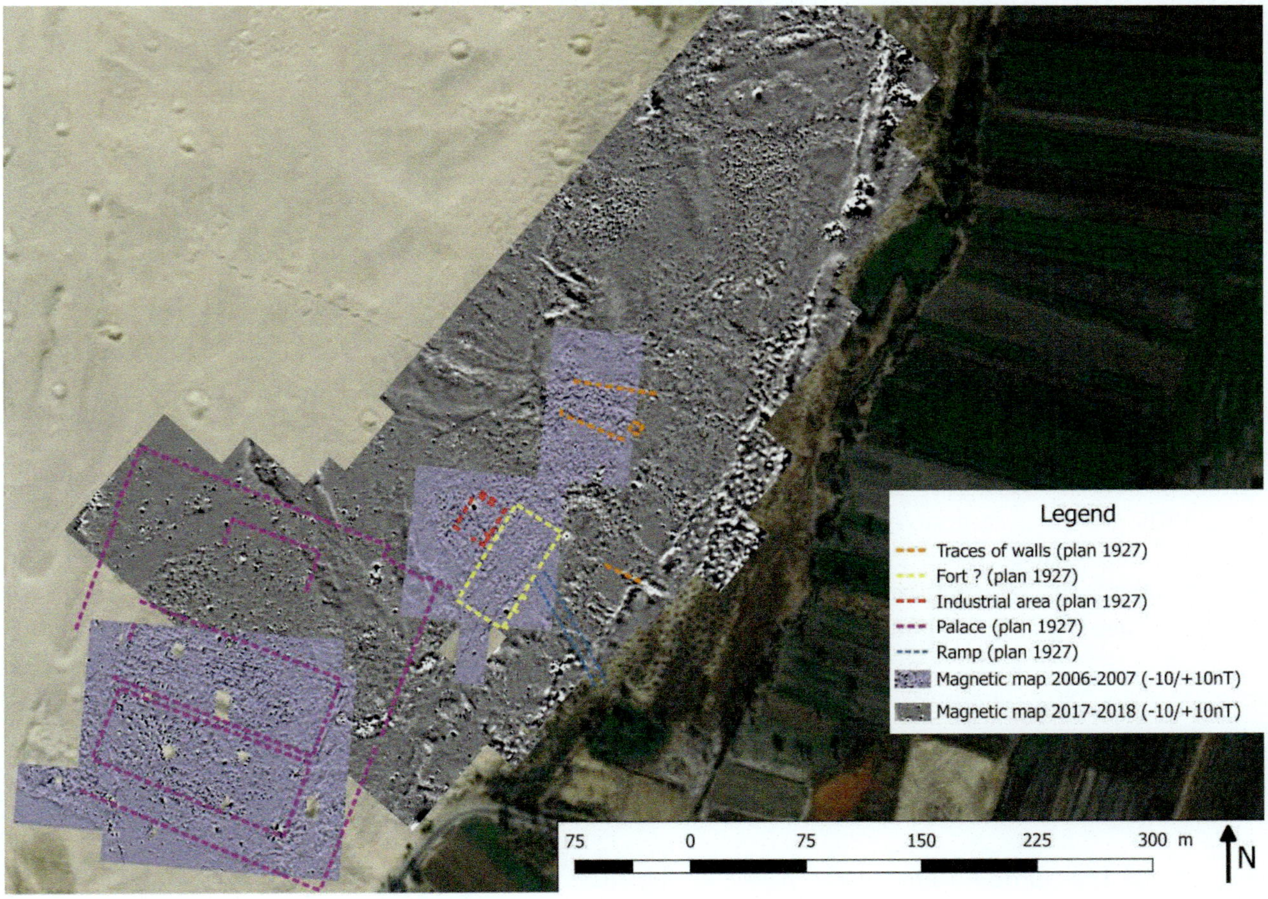

Fig. 3. Magnetic map (-10nT, white / +10nT, black) of the site of Gurob.

On the large-scale magnetic map, a number of homogeneous and heterogeneous areas are visible. An estimation of the anomaly source depth helped to better characterize the previously inhabited areas and deep structures of the north town (Kemp 1978).

On the eastern edge of the magnetic map, numerous anomalies seemed to match with modern pits visible on the surface. However, a small-scale excavation in the south-eastern part revealed archaeological structures which might correspond to a New Kingdom jetty or quay. A closer analysis of these anomalies revealed other alignments that could be evidence of archaeological structures.

In the centre of the map, near the fort mentioned by Brunton and Engelbach, a lot of isolated circular positive anomalies are visible. Some alignments seem to correspond to the fort (Fig. 3, yellow lines), while other positive discrete anomalies are probably more related to earlier buildings (Fig. 3, red lines). Some of them are strong positive discrete anomalies corresponding to kilns: the kiln area was verified (and kilns confirmed) by excavations in 2009 and 2012 (Hodgkinson 2017: 348-360). Furthermore, other, less strong, positive discrete anomalies could correspond to ancient pits or buried boulders.

In the case of the harem palace (Fig. 3, pink lines), several plans have been drawn based on the results of the field surveys (Petrie 1891; Brunton and Engelbach 1927) or by combining older plans (Kemp 1978). The results of the magnetic surveys were also used to draw the final plan.

Conclusion

The four geophysical campaigns carried out between 2006 and 2018 made it possible to verify and complete old archaeological plans. Furthermore, the purpose of the surveys was to improve our knowledge of the north town and to find the link between the fort and the harbour. Finally, the prospection allowed to precisely locate the inner walls of the two main structures of the harem palace. Future excavation will help to better understand the connection between the geophysical image and archaeological structures hidden below the surface.

References

Brunton, G. and Engelbach, R. 1927. *Gurob*. London.

Hodgkinson, A.H. 2017. *Technology and Urbanism in Late Bronze Age*, Oxford Studies in Egyptian Archaeology. Oxford.

Kemp, B.J. 1978. The Harim-Palace at Medinet el-Ghurab, *Zeitschrift für Ägyptische Sprache und Altertumskunde*. 105. 122-133.

Loat, L. 1905. Gurob, accompanied by Murray M.A., *Saqqara Mastabas Part I*, BSAE 10. London.

Marty, P., Darras, L., Tabbagh, J., Benech, C., Simon F-X. and Thiesson, J. 2015. WuMapPy – an open-source software for geophysical prospection data processing, *Archaeologia Polona*. 53. 563-566.

Petrie, W.M.F.I. 1891. *Illahun, Kahun and Gurob*. London.

Shaw, I. 2008. Report to the Supreme Council of Antiquities of work conducted at Medinet el-Gurob, 25 March-3 April 2008 (Gurob Harem Palace Project). Online report, 1-11.

Shaw, I. 2012. The Gurob Harem Palace Project, spring 2012. *Journal of Egyptian Archaeology*. 98. 43-54.

Yoyotte, M. 2012. *Le "harem" royal dans l'Égypte ancienne. Enquête philologique, archéologique et prosopographique*. Unpublished PhD thesis, Paris.

Yoyotte, M., Pillon, A., Darras, L., *et al.* Forthcoming 2019. Nouvelles recherches sur le site de Gourob. Des fouilles du xixe siècle à la campagne 2017. *Bulletin de l'Institut français d'archéologie orientale*. 118. 1-66.

Geophysical surveying in Egypt and Sudan: periodical report for 2017–2018

Tomasz Herbich, Robert Ryndziewicz
Institute of Archaeology and Ethnology, Polish Academy of Sciences, Warsaw, Poland

This paper follows the format of Herbich (2017), reporting on the latest geophysical surveying projects implemented in both Egypt and Sudan by the Institute of Archaeology and Ethnology at the Polish Academy of Sciences, Warsaw. The reduced number of projects in Egypt compared to previous years is due to administrative restrictions as well as an unstable political situation in some areas of the country resulting in limited security clearance for foreign teams. At the same time there is a growing interest in research on sites in ancient Nubia and northern Sudan, which is reflected in the rise of the number of geophysical projects carried out in this region (see also Herbich and Ryndziewicz's second paper in this volume, and Ryndziewicz and Drzewiecki, also in this volume).

In Egypt, geophysical surveys were conducted at Hierakonpolis, Heliopolis, Dahshur, Kom el-Hisn and Kom Ombo (the latter two are new projects, the others a continuation of earlier research, see Herbich 2019; Alexanian and Herbich 2015). In Sudan, the prospection was carried out in Sedinga, Soleb, Kawa and Old Dongola. The first two were new projects, the others took up on earlier work (Welsby 2017, Godlewski 2013).

The magnetic method was predominant, using Geoscan Research FM256 fluxgate gradiometers with a sampling density of 0.5m by 0.25m. The electrical resistivity method (with a use of ADA05 apparatus by Elmes) was applied at one site, that is, Heliopolis. In the wake of the GPR survey in Dongola, the GPR methodology will certainly be included as a standard for prospecting sites with dried mud-brick architecture. The Dongola experiment was undertaken in view of the positive results of the German prospection of a Meroitic settlement at Hamadab (Ullrich and Wolf 2015). The radar used was a MALA GroundExplorer with 450MHz antennae. The spacing of the transects was 0.5m. The fieldwork was carried out by the authors and Krzysztof Kiersnowski (in Dahshur and Sedinga).

Egypt

Hierakonpolis *(Predynastic, Middle/New Kingdom period, Expedition to Hierakonpolis, project director Renee Friedman).* The objective was to identify more industrial features related to pottery production, brewing beer and food processing at the Predynastic site HK11; features of the like were discovered and tested archaeologically in previous years (Herbich 2019). However, the nature of the anomalies indicates a settlement rather than production area. Prospection at a Pan-Grave culture cemetery (HK21A, Middle/New Kingdom) registered a number of anomalies tentatively corresponding to graves.

Dahshur *(Old Kingdom period, German Archaeological Institute in Cairo, project director Daniela Rosenow).* The prospection supplemented a survey carried out by H. Becker in 2007–2009 on an extensive plateau with a necropolis situated west of the Red Pyramid. No new structures were discovered.

Heliopolis/El-Matariya (*Old Kingdom and Late period; Leipzig University, project director Dietrich Raue*). Prospection in the central and eastern part of the Amon-Ra temple enclosure (Fig. 1), in a spot already partly excavated, detected stone blocks from the late temple below the present ground water level. Concentrations of blocks, mostly of basalt, were located with the magnetic method, as well as electrical resisitivity. Archaeological verification of the survey results was carried out immediately following the prospection.

Kom el-Hisn *(Old Kingdom period, Kom el-Hisn Provincialism Project, project director Leslie Anne Warden).* Digging for *sebakh* (soil from cultural deposits used as field fertilizer) has dramatically changed the topography of this settlement site. Nonetheless, a few structures interpreted as the remains of early occupation were identified as a result of the geophysical prospection carried out on the site.

Kom Ombo *(Old Kingdom – Ptolemaic periods, Austrian Archaeological Institute in Cairo, project director Irene Forstner-Mueller)*. The survey on the summit and slopes, as well as at the base of the artificial mound (Arab. *kom*) formed of cultural deposits led to the discovery of a presumed production complex (pottery making?) and architecture on the hilltop, apparently below the remains of a modern fort. A detailed mapping of the fort was another outcome of the work.

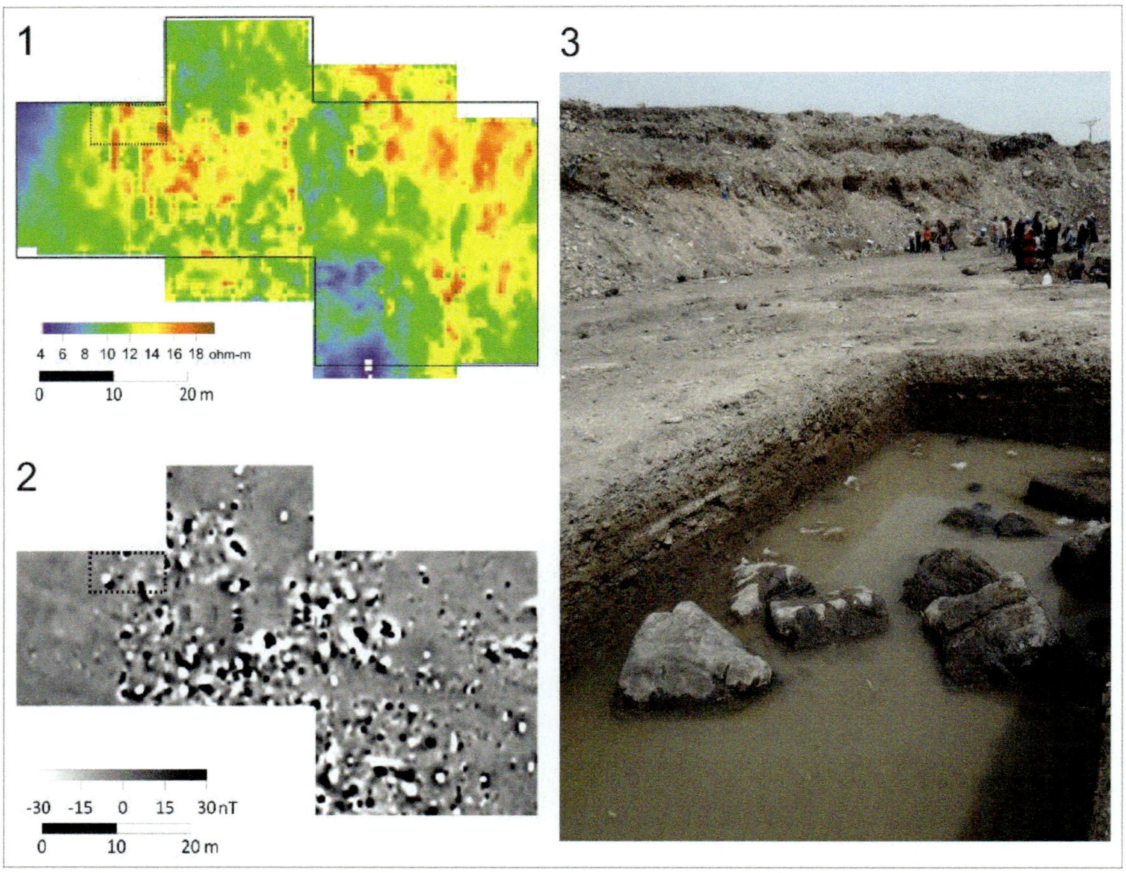

Fig. 1. Heliopolis / Matariya, temple of Nektanebo I. 1: resistivity map (Schlumberger asymmetrical array, AM=3.5m; MN=1m); trench in dotted outline, solid line marks the extent of the magnetic survey; 2: magnetic map; 3: stone blocks found in the trench.

Sudan

Kawa *(Kushite period, Sudan Archaeological Research Society, project director Derek Welsby)*. The survey was carried out in the cemetery and in an area south of the temple of Amon. The search for a temple enclosure was unsuccessful, but the prospection of the cemetery mapped a few hundred graves of similar orientation and construction (for more on this project, see Herbich and Ryndziewicz, Surveying Kushite sites in Sudan: town and cemetery in Kawa, in this volume).

Old Dongola *(Medieval period, Polish Center of Mediterranean Archaeology, University of Warsaw, project director Artur Obluski)*. Prospection of an urban area within the citadel of Dongola was carried out under the European Research Council program "UMMA - Urban Metamorphosis of the community of a Medieval African capital city". The phase under investigation represents the last occupation of the city from a period of transformation of the community from Christian to Islamic. The two methods used were GPR and magnetic survey. GPR neatly mapped the architecture including streets and the layout of particular buildings (Fig. 2). The anomalies observed on the magnetic map (Fig. 3) were due mainly to the use of red brick beside the dried mud brick that was the staple building product in the period under investigation. The radar mapping was much more precise than the magnetic image in reconstructing architecture in this case. A very clear image of urban architecture observed on the magnetic map of the eastern part of the prospected area concerns late buildings (still inhabited in the early 20th century) built entirely of mud brick.

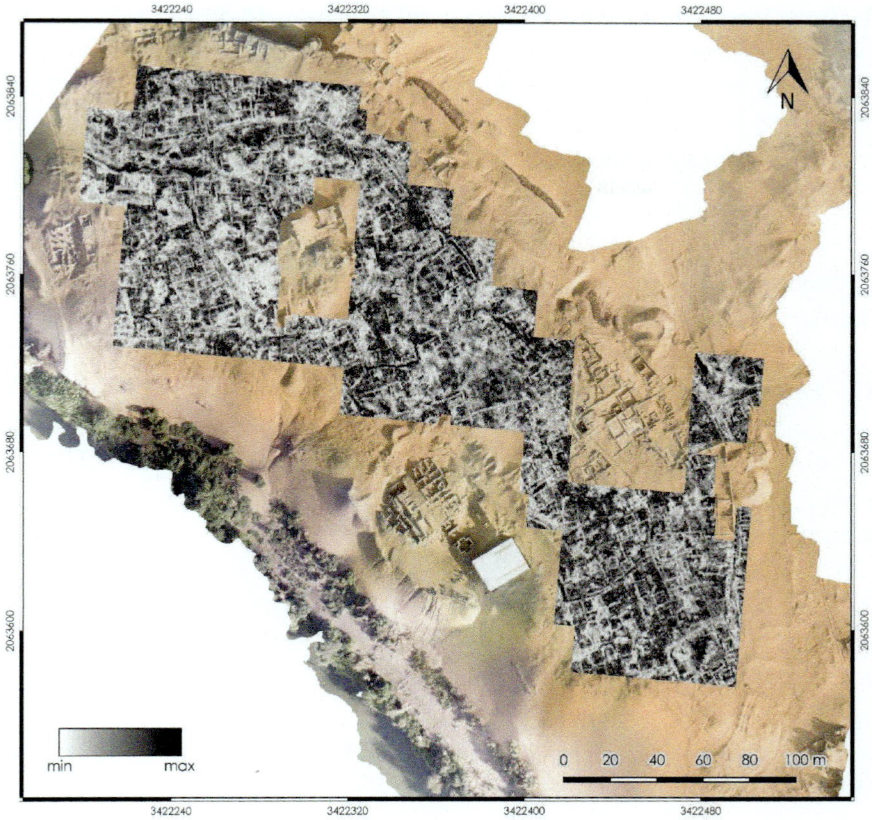

Fig. 2. Old Dongola, citadel. GPR survey. Timeslice 12sn to 14ns (approximate depth 0.6m–0.7m) on a vertical aerial photograph of the citadel

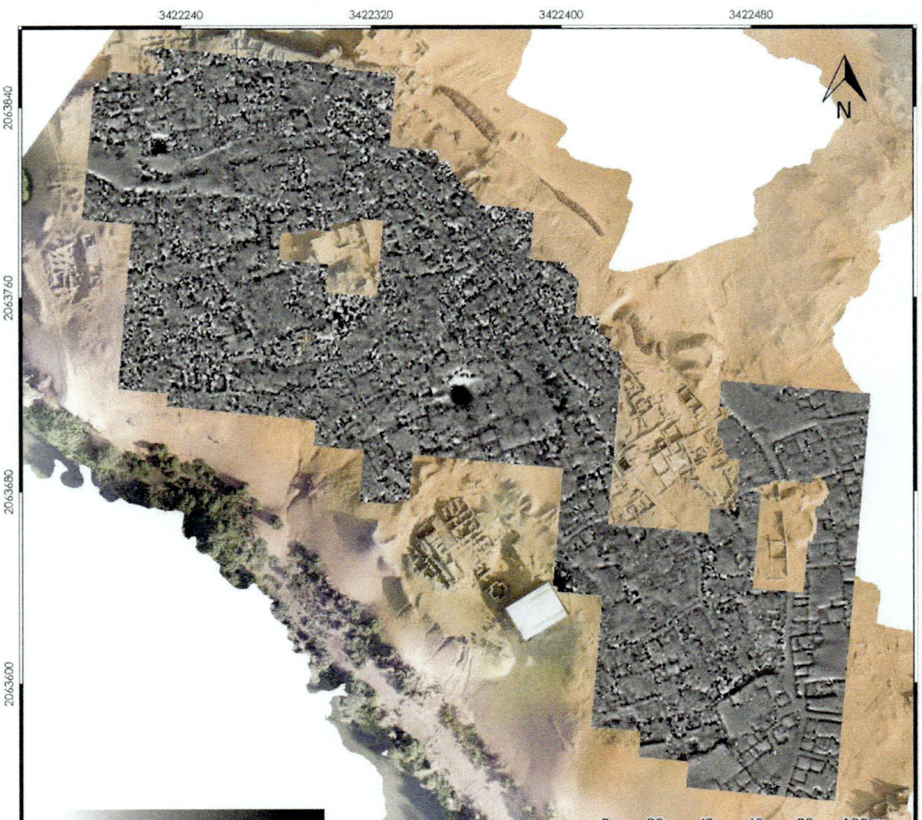

Fig. 3. Old Dongola, citadel. Magnetic map on a vertical aerial photograph of the citadel.

Sedinga and Soleb *(New Kingdom, Kushite period, French Mission to Sedinga and Soleb, project director Claude Rilly)*. The two sites represent Pharaonic Egyptian culture from a period when the southern border of Egypt reached the territory between the Third and Fourth Nile cataracts. The area around the temples of Amenhotep III in Soleb and his royal consort Tyiy in Sedinga were explored, recording stray fragments of urban architecture and numerous graves (the latter at Sedinga).

References

Alexanian, N. and Herbich T. 2015. The workmen's barracks south of the Red Pyramid at Dahshur. *Mitteilungen des Deutschen Archaeologischen Instituts Abteilung Kairo*. 70/71. 13-23.

Godlewski, W. 2013. *Dongola – ancient Tungul. Archaeological guide*. Warsaw.

Herbich, T. 2017. Geophysical surveying in Egypt and Sudan: periodical report for 2015-2016. In Jennings, B., Gaffney, C., Sparrow, T. and Gaffney, S. (eds.) *AP2017. 12th International Conference of Archaeological Prospection*. Archaeopress. Oxford. 107-109.

Herbich, T. 2019. Efficiency of the magnetic method in surveying desert sites in Egypt and Sudan: case studies, In R. Persico, N. Linford and S. Piro (eds.), *Innovation in Near-Surface Geophysics: Instrumentation, Application, and Data Processing Methods*, 195–251. Amsterdam.

Ullrich B., Wolf P. 2015. Hamadab near Meroe (Sudan): results of multi-technique geophysical surveys. *Arcaheologia Polona*. 53. 392-395.

Welsby, D.A. 2017. *The Pharaonic and Kushite town of Gematon. History and archaeology of the site*. London.

3-D Electrical Resistivity Tomography in an Urban Environment: the case of Shallalat Gardens, Alexandria, Egypt

Nikos Papadopoulos[1], Dimitris Oikonomou[1], Nasos Argyriou[1], George Kritikakis[1], Calliope Limneos Papakosta[2]
[1]GeoSat ReSeArch Lab, IMS-FORTH, Rethymno, Greece, [2]Helenic Research Institute of the Alexandrian Civilization, Alexandria, Egypt

Introduction

Shallalat Gardens are located at the center of Alexandria in Egypt (Fig. 1). The coverage area (~0.1km^2) replaced the northern bastions created during the years of Muhammad Ali, who modernized the medieval city walls of Alexandria. In antiquity, this area was extending between the Royal Palaces at the seaside and the Canopic street, the main longitudinal street of Hellenistic Alexandria. From the archaeological point of view, the site was chosen for further investigation due to its position in the topography of Ptolemaic Alexandria; it belonged to the Basilea, that is the Royal Quarter, according to ancient sources and particularly Strabo (Book 17.8).

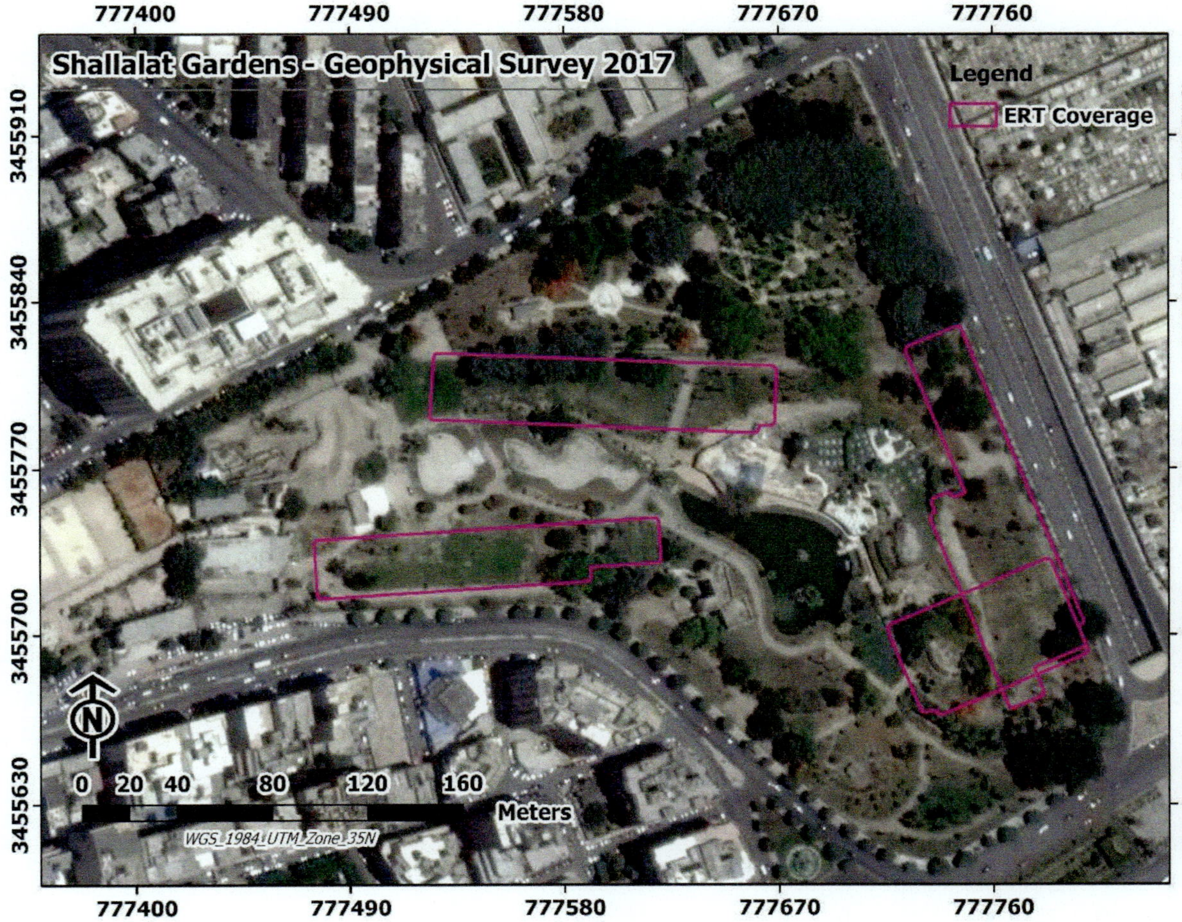

Fig. 1. Area coverage with ERT in Shallalat Gardens during the 2017 field campaign.

The Hellenic Research Institute of Alexandrian Civilization (H.R.I.A.C.) initiated an archaeological project in 2007, which is still active and up-to-date with significant archaeological results (see http://www.hriac.com/projects/). One of the landmarks of this diachronic archaeological research, is the unique early Hellenistic statue of Alexander the Great that was found in the NW part of the park which is currently exhibited in the National Museum of the Alexandria.

The excavated archaeological remains are composed of bulky and well-preserved foundation or wall stones and lie at a depth of about 7-9m below the ground surface. The general stratigraphic units are characterized as mud within the first 3m, mixed limestone and sandstone pieces up to 7.5m, solid sandstone as the bedrock, and the water table is found just a few metres below the ground surface.

Methodology

Based on the significance of the archaeological site, a geophysical survey was conducted during October of 2017 employing 3-D Electrical Resistivity Tomography in the east, south and north part of the garden (Fig. 1), using a dense network of 2-D lines and subsequent 3-D resistivity inversion, as described by Papadopoulos *et al.* (2007). The purpose of 3-D resistivity imaging was to map the subsurface properties in different parts of Shallalat Gardens in order to indicate geophysical signatures that should be further investigated, with either drilling or excavation, to reveal their nature (archaeological or geological origin). A total area of almost 12,000 square metres was covered in different parts of the park along 41 ERT lines having a cumulative length of 4.6km (Fig. 1). The dipole-dipole electrode configuration was used to complete the ERT survey with a basic electrode distance along the lines of 3m. Ultimately, the ERT survey was completed along multiple parallel transects in an effort to map the resistivity properties below the surface within a 3-D context (Papadopoulos *et al.*, 2007).

A systematic method was followed for the pre-processing of the ERT sections. Initially, data having potential values less than 10μVolt were removed from the datasets. Additional despiking filters removed data with high geometric factors and low injection currents. The topography variations along the lines were captured with a differential GPS unit and this information was included in the inversion procedure. The network of parallel lines in each different area were combined and a 3-D inversion algorithm (Papadopoulos *et al.* 2011) was used to reconstruct the 3-D subsurface resistivity variation. The final inverted models were sliced every 3m and the Google earth satellite image was used as background to overlay the respective resistivity depth slices.

Results

The depth slice of 6.75m in the **northern area** (Fig. 2) outlines the resistive anomaly N5 with dimensions 9m by 7m. The anomaly N1 occupies a section of 9.5m x 6.5m in the western part of the area within the depth range of 6.75-9.25m. The shape, the burial depth and the resistive signature of the specific anomalies render them as candidate areas for further archaeological investigation. The anomalies N3 and N4 share comparable horizontal dimensions (~4.5m x 7m). The elongated resistive anomalies N5a and N5b have the strongest response in the depth slice of 6.75m. They enclose a conductive linear anomaly with SW-NE direction that is correlated with the pathway leading to the north side of the park. Most probably N5a and N5b comprise "shadow effects" of the inversion procedure due to the reduced quality of the data in that specific region. Finally, anomalies N2 and N6 are of low confidence due to the known inherent limited resolution of the ERT in the edges of the grid.

The **eastern area** is also occupied by anomaly E1 separated in three segments. The anomaly E2 is further to the south having dimensions 3m x 5m and a thickness of about 2m. Anomalies E3a and E4a lie on a low resolution area resulted by the elimination of extreme noisy data and most probably this fact along with the existence of the east-west conductive zone amplified the signatures of these anomalies during the inversion procedure. The anomaly E4b (4.5m x 3.5m) is outlined in the depth slice of 5.25m and it is connected to E4a in deeper levels. On the contrary the square anomaly E3b seems isolated within the depth slices of 6.75-8.25m and only at a depth of 9.25m merges with E3a (Fig. 2)

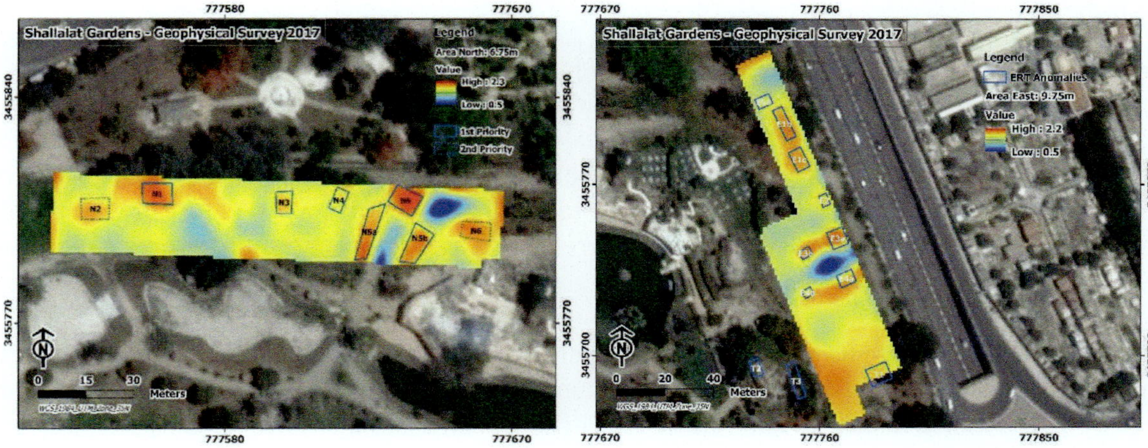

Fig. 2. Resistivity depth slice of 6.75m and 9.75m of the northern and eastern area respectively in Shallalat gardens. The axis units in metres and the resistivity is in Ohm.m.

Within the **southern area** (Fig. 3), the diagonal resistive anomaly S5 to the east has a continuous vertical continuation into the deeper levels with a moderate resistive signal, thus signifying its geological origin related to the local stratigraphy. Moving further to the west the elongated resistive anomaly S4 with horizontal dimensions 12.5m x 5m appears within the depth slices of 5.25-6.75m. A relatively orthogonal resistive region with dimensions 9m x 6.5m (S3) is fully formulated in the depth range of 6.75-9.75m having a thickness of about 3m. At a distance of about 12m west of S3, a smaller (7.5m x 5m), less thick (~2m) and less resistive anomaly (S2) is registered. The small square (~4.5m x 4.5m) anomaly S1 is also outlined on the NW edge of the grid, which seems to continue further to the north.

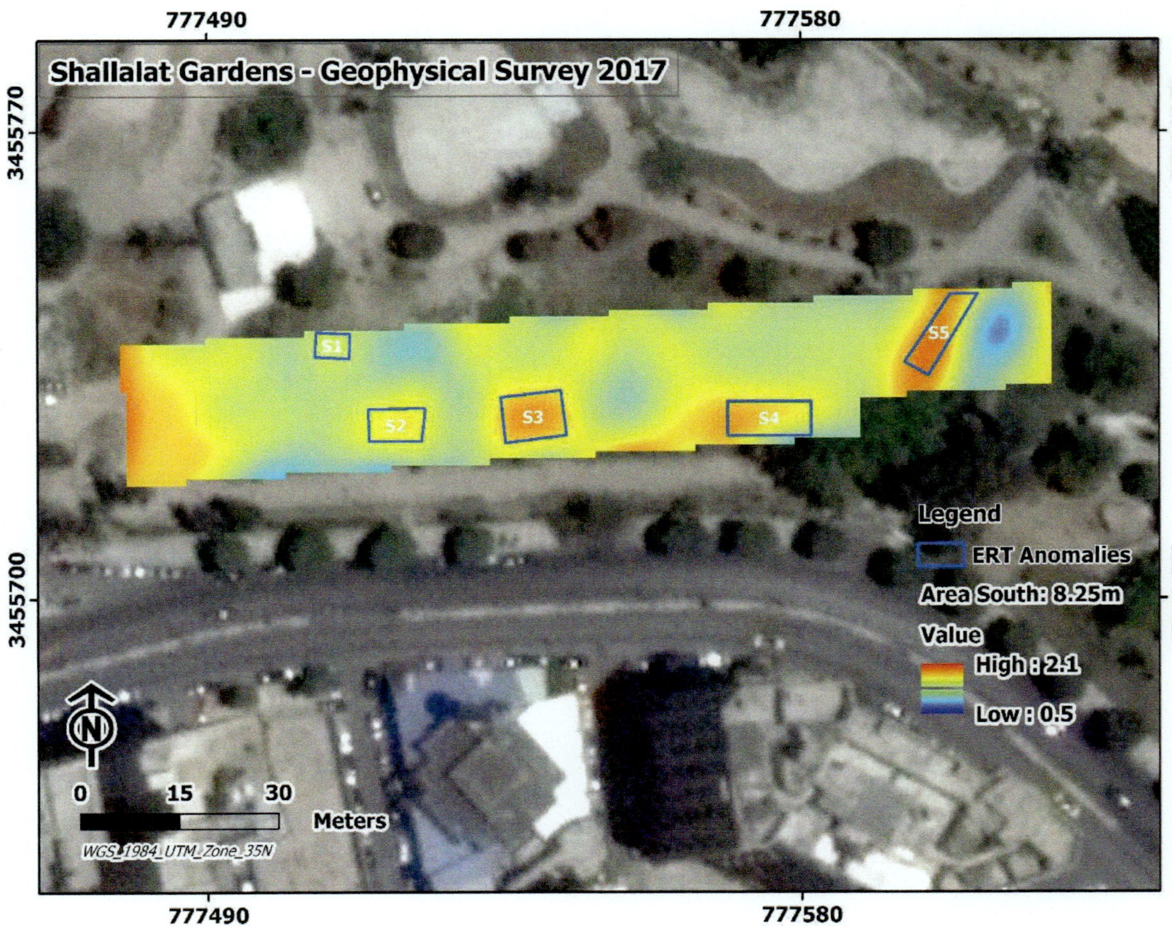

Fig. 3. Resistivity depth slice of 8.25m of the southern area in Shallalat gardens. The axis units in metres and the resistivity is in Ohm.m.

Conclusions

Ongoing large-scale urbanisation is a feature of developing countries and many urban areas have a long historical and archaeological background that is hidden in the subsurface. Geophysical methods can significantly contribute to understanding the complex changes in the physical environment in urbanized regions. In contrast to conventional geophysical investigation of archaeological sites, a geophysical survey in an urban area may face some objective difficulties due to the highly heterogeneous nature of the upper layer and the ambient noise in cities caused by electrical currents and electromagnetic radiation (Papadopoulos *et al.* 2009).

In Shallalat Gardens, the 3-D ERT method was considered the optimum technique to achieve the specific objectives of the project, taking into account the nature and dimensions of the targets in combination to their relatively large burial depth. Despite the *in situ* difficulties imposed by the demanding environment, as well the interventions from natural or modern obstacles, the ERT survey was able to indicate a number of candidate targets mainly based on their dimensions and their burial depth that can be further investigated through test excavations or targeted drillings. The results also enhance the contribution of the geophysical exploration in the development of the relatively new field of "urban geophysical prospection".

References

Papadopoulos, N.G, Tsourlos, P., Tsokas, G.N. and Sarris, A., 2007. Efficient ERT Measuring and Inversion Strategies for 3D Imaging of Buried Antiquities. *Near Surface Geophysics.* 5(6). 349-362.

Papadopoulos, N., Sarris, A., Yi, M. and Kim, J. 2009. Urban archaeological investigations using surface 3D Ground Penetrating Radar and Electrical Resistivity Tomography methods. *Exploration Geophysics*. 40. 56–68.

Papadopoulos, N.G, Tsourlos, P., Papazachos, C., Tsokas, G.N., Sarris, A. and Kim, J.H. 2011. An Algorithm for the Fast 3-D Resistivity Inversion of Surface Electrical Resistivity Data: Application on Imaging Buried Antiquities. *Geophysical Prospecting*. 59. 557-575.

Going back to Medamud: Excavation feedback on processing, interpretation and planning

Julien Thiesson[1], Félix Relats Montserrat[2]
[1]Umr 7619 Metis Sorbonne Université, Paris, France, [2]Institut français d'archéologie orientale, Cairo, Egypt

Introduction

This study follows the work previously done on the site of Medamud (Relats *et al.* 2016: 325-384, Thiesson *et al.* 2017: 245). One of the results was a map of archaeological features obtained in combining various apparent geophysical maps (mainly conductivity and susceptibility at various depths and magnetic anomalies). These maps where used to guide the excavation planning in an area that had never been excavated. This paper will address the comparison between the maps obtained in 2015 and the results of the excavation carried out from 2017 to 2019. It will show the issues that we encountered and discuss how the excavation helps enhance the integration of the geophysical data in the archaeological understanding of the site.

Excavation and archaeological results

Exploration of the Medamud site took place mainly between 1924 and 1940 (Relats *et al.* 2016). During these times, the first excavators focused on stone masonry and left the brick structures surrounding the sanctuary. The nature of the latter was investigated thanks to a pedestrian survey that showed a significant number of over-fired ceramics and firing waste. It strengthens the existence, in Medamud, during antiquity, of a ceramic workshop sector.

In addition to the archaeological contribution offered by geophysics in an unspoiled context, the following points guided our reflection:

- the probable presence of kilns presenting a strong magnetic signal
- the vicinity of the temple leading to a road network - constituted of mud brick walls - offering a clear contrast

As a result, it was clear that the geophysical response should be a source of reflection rather than a solution to compensate for the absence of excavations. Among the areas surveyed in 2015, we initially focused on the M4/K9 zone, which contained the largest surface area and was closest to the temple. The objective of the 2019 excavation was therefore to understand the organization of urban planning and crafting activity in the area near temenos, while verifying the geophysical results.

According to the data, which illustrates the existence of a large mud brick structure crossing the area from east to west, it was decided to open a 33m by 10m trench, oriented north-south (Fig. 1a). In this area, several archaeological structures were found, linked to an artisanal area, including several kilns, mudbrick walls and some ill-defined features linked with the kilns and building collapse (not shown).

Comparison with the 2015 synthetic map

A synthetic map (Fig. 1b) based on the geophysical results obtained by magnetometry and electromagnetic induction (EMI) demonstrates a cross positioning problem. It appears that the geographical coordinates obtained for the 2015 survey are not equal to the ones in current use for the excavation and site survey. As they were derived from points measured with a handheld GPS receiver with a 3m-10m absolute error, this was not totally unexpected. However, as the geophysical surveys are an important part of the Medamud archaeological study, corrections were required. We decided to use the excavation results to identify the features to their corresponding anomaly in order to correct the positioning of the 2015 maps. Furthermore, additional detailed surveys also helped to calibrate the synthetic interpretation based on the geophysics.

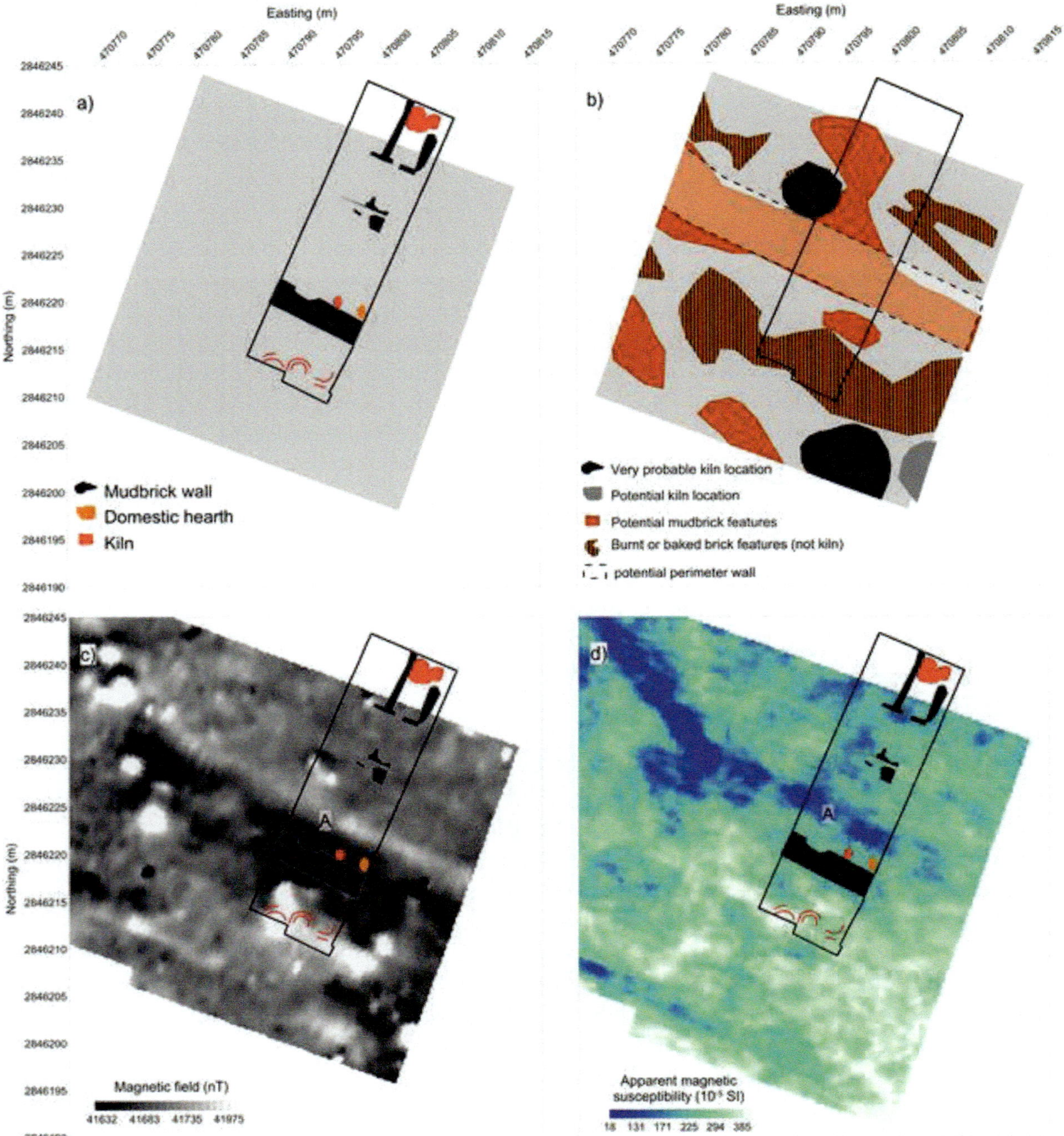

Fig. 1. a) the simplified archaeological results of the area excavated in 2019, b) the geophysical driven synthetic map proposed in 2016, c) the magnetic field map obtained in 2015, d) the apparent susceptibility map obtained in 2015.

Identifying excavated features to calibrate the positioning

As several features on the synthetic map could correspond to the ones excavated (and vice versa), we need to evaluate the values of magnetic susceptibility for the features excavated in 2019. It appears that the expected mud bricks wall corresponds well in susceptibility to the feature identified as A (Fig. 1c, 1d). Based on the hypothesis that this feature did not move between 2015 and 2019, we computed the X and Y offsets needed to correct the coordinates of the 2015 surveys.

Detail mapping of an area

As the archaeological features are covered by pottery sherds up to 1m thick, we conducted another experiment. We repeated twice the measurement over the small area south of the southern mudbrick wall. The first time, the first 0.15m was excavated. Then, after the measurement, another 0.15m was excavated. We obtained two maps of the same area giving some clues on the magnetic susceptibility of the features.

Testing TDR measurements

In addition, some TDR measurements were carried out over some of the features in order to evaluate the interest of using dielectrical permittivity for further surveying and interpreting based on geophysical surveys. We used a Hydrosense 2 from Campbell Scientific with a 0.12cm probe. The devices give a time in µs which is linked to the permittivity. The device was quite hard to stick in the ground mainly due to the large amount of pottery sherds. Table 1, below, shows qualitative results other some well identified features. The results suggest that it could be useful for some finer discrimination.

Archaeological feature	Mudbricks (well-shaped)	Yellow mudbricks (ill-shaped)	Black mudbricks (ill-shaped)	Upper filling (with shards)
Time in µs (linked to electrical permittivity)	1.831 to 1.942	1.337 to 1.659	1.780 to 1.822	1.182 to 1.681

Table 1. Ranges of times (linked to apparent dielectrical permittivity) for some identified remains.

Conclusion

This paper aims to show some feedback on the synthetic map obtained over the Medamud site. The measurements exhibit some issues concerning the positioning of the 2015 geophysical map in the actual coordinate system of the site. The measurements over some excavated features permit to correction and ensure that further planning based on the geophysical survey should be accurate.

It also means that the threshold values taken for the synthetic map should be reassessed (in 2015, they were based on a statistical assumption only). An important step consists in measuring the properties directly on the features excavated. In addition to magnetic susceptibility, the opportunity was taken to test dielectrical permittivity as another discriminating parameter for future survey, but it gives very patchy results mainly due to the difficulty of sticking the probe in to the surficial pottery sherds. The choice of relevant parameters, as it is site dependent, must be tested and validated during the whole archaeological campaign. Unfortunately, we could only use magnetic susceptibility and dielectrical permittivity which are a sparse sample of the geophysical parameters that could be measured over the excavated features.

References

Montserrat, F.R., Thiesson, J., Mendieta, Z.B., Sanchez, C., Rejiba, F. and Guérin, R. 2017. Une première campagne de prospection à Médamou: méthodologie et résultats préliminaires (Mission Ifao/Paris-Sorbonne/Labex Resmed de Médamoud). *Bulletin de l'Institut Français d'Archéologie Orientale du Caire*. 116. 325-384.

Thiesson, J., Montserrat, F.R., Sanchez, C., Guérin, R. and Réjiba, F. 2017. Prospection at the Medamud (Egypt) site: building archaeological meaning from the geophysical in situ measurements. In Jennings, B., Gaffney, C., Sparrow, T. and Gaffney, S. (eds.) *AP2017. 12th International Conference of Archaeological Prospection*. Archaeopress. Oxford. 245.

Magnetic prospection close to the magnetic equator: Case studies in the Tigray plateau of Aksum and Yeha, Ethiopia

Sandra Ostner[1], Jörg W.E. Fassbinder[1], Mandana Parsi[1], Iris Gerlach[2], Sarah Japp[2]
[1]*Ludwig-Maximilians-Universität, München, Germany,* [2]*Deutsches Archäologisches Institut, Berlin, Germany*

Introduction and historical background

Yeha and the UNESCO world heritage site Aksum, both situated in the Ethiopian highlands, was the centre of the ancient kingdoms of Di'amat and Aksumite. The ruins of these kingdoms spread all over the Tigray Plateau, show the wealth and influence of these civilisations. Both kingdoms achieved economic and political importance since there were focal points for the trade routes of ancient Egypt, South-Arabia and the Roman Empire to Africa and the Indian Ocean (D'Andrea *et al.* 2008: 152-154). Yeha is deemed to be the capital of Di'amat formed in the early first millennium BC (Gerlach 2014: 5-7). It shows a strong influence of Sabean culture, which might be connected to their dominance of the Red Sea region. However, the Ethio-Sabaean culture disappeared towards the mid-first millennium BC by unknown cause and was succeeded by the Aksumite kingdom. The area of Yeha has remained inhabited through the millenniums to present day and is today known for its ruins of the monumental structure Grat Be'al Gebri as well as the Great Temple (Fig. 1). The motivation of the geophysical survey was to obtain more information about the existence of surrounding settlement structures and their organisation. The selection of the different prospecting areas was predetermined by the accessibility and the requirement for areal magnetometer prospection. The total Earth's magnetic field measurements and the subsequent interpretations are complicated by the shallow magnetic inclination of 15°.

Fig. 1. The Great Temple of Yeha (right). Located in the highlands of northern Ethiopia, Yeha was the capitol of the ancient kingdom Di'amat. Today the Ethiopian Orthodox Tewahedo Church is situated on the same site.

Measurements

The survey was conducted with a Scintrex Smartmag SM4G-special magnetometer and a Geometrics G-858 magnetometer. Both were applied as total field magnetometers in duo-sensor configuration. Since they are optically pumped caesium magnetometers, their output frequency and consequently the output signal depend on the orientation of their sensors to the ambient magnetic field (Scintrex Limited 1993: 2-2-4-7).

For highest precision, it is suggested to align the sensors at 45° to the local field direction. Nevertheless, the construction of modern magnetometers enables the measurement at angles of around 10° to 85°, depending on the instrument, without considering the so-called heading error. If the sensor however is orientated almost parallel or perpendicular to the ambient magnetic field, either the optical pumping or the light modulations fail to work. Consequently, the instrument cannot be operated in these dead zones. Also the measurement noise starts to rise rapidly outside of the active areas and close to the dead zones. Thus, for working at a magnetic inclination of around 15° in Yeha, we tilted the sensors of the Scintrex accordingly from the usual set-up, perpendicular to the soil. However, we didn't re-orientate the sensors for the Geometrics magnetometer. Therefore it was operating very close to its dead zone of 15° and lower (Geometrics Inc. 2001: 20). While measuring in Ethiopia, we noticed on occasions that the signal of the sensors was lost for couple of seconds. An explanation could be that with movement of the instrument, for example by changing survey lines, the sensors tilted into the dead zone. Therefore we suggest orientating the sensors of optically pumped magnetometers into the centre active zone and neglect the different azimuth of the sensors which can occur during measurement, since the heading error is more or less insignificant in that case.

Latitude dependence of the signals

One of the first reports on archaeological magnetometer surveys at the equator was conducted by Mike Tite (1966: 24-31). In his work he also discusses the inclination dependence on the measured anomaly for a uniformly magnetised sphere with the dipole moment orientated along the Earth's magnetic field. Fig. 2 shows the signal of the anomaly for different inclinations, calculated with his derived formula. At European latitudes, the signal has a minimum in the North and a maximum in the South. The signal intensity of the maximum is almost three times higher. With shallower inclination, the maximum drops, while the minimum increases and moves towards the origin. For a signal at 15°, we expect a sharp negative anomaly and a wider positive anomaly one with just around 60% of the signal strength compared to the negative. The influence of the inclination on the signal, especially in the near magnetic equator regions, was already shown in the work of Fassbinder and Gorka (2011: 45-48).

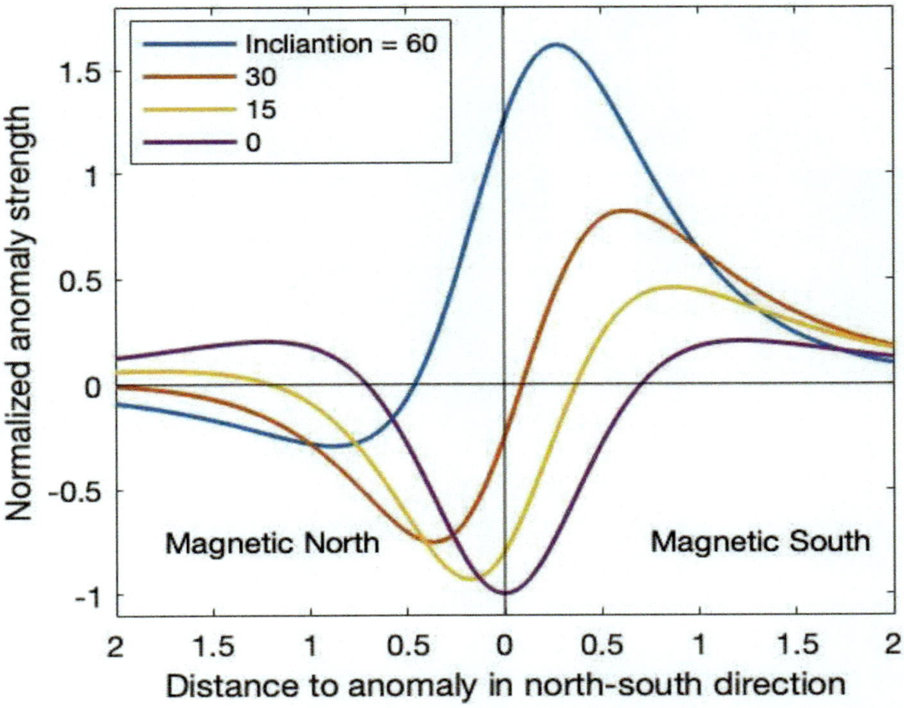

Fig. 2. Anomaly strength of the total field intensity as north-south traverse through the anomaly's centre for different inclinations. Horizontal and vertical component in units of burial depth. After Mike Tite (1966: 26).

Results

In Yeha we prospected five different sites whose extents were severely limited by modern buildings or installations and insurmountable field boundaries. Unfortunately the modern use of these areas is also visible in the magnetogram, including subsurface pipes or disturbance by iron constructions (Fig. 3a). Moreover, recent field boundaries are detected as unusually prominent anomalies in the magnetograms and often with an inverted orientation of the negative and positive anomaly (Fig. 3).

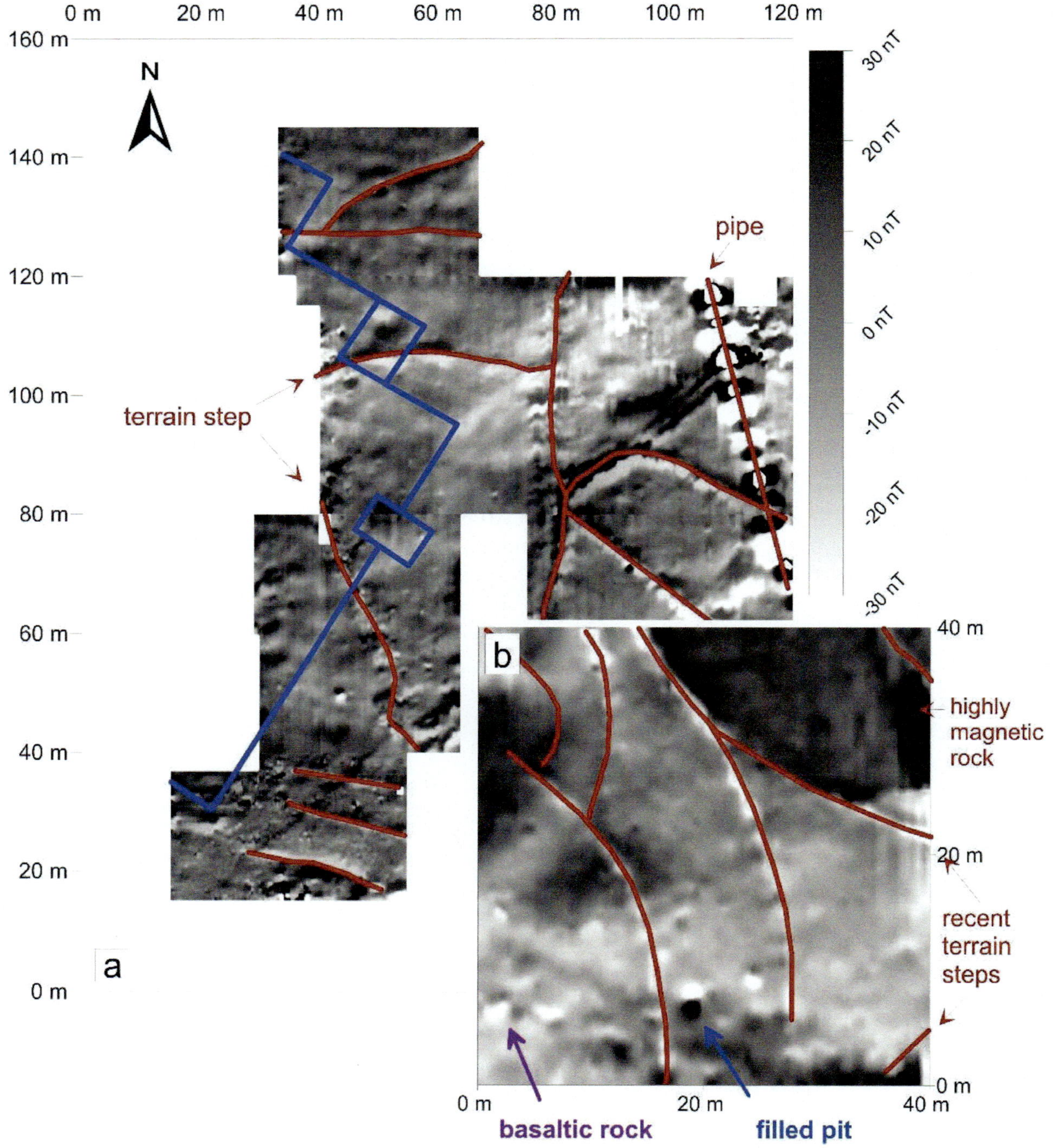

Fig. 3. a) Magnetogram of the survey area in the south of the Grat Be'al Gebri. Caesium Magnetometer, Scintrex SM4G-Special and Geometrics 858 in duo-sensor configuration, sensitivity +/-0.01nT, sampling interval 25cm x 50cm, total field in Yeha 11/2019 *c.* 37,500 +/- 100nT. Magnetic inclination of around 15°. Geological structures and recent field boundaries are marked red. In blue, the detected archaeological feature with similarities to the monumental buildings of Yeha.

b) Magnetogram of the survey area north of the Temple (survey parameters are the same as (a)). Geological features marked in red. The pit-filling, marked in blue, shows the typical anomaly expected for an inclination of 15°. The basaltic rock, marked purple, shows a different anomaly probably due to its high thermal remanence.

On-site measurements with the SM-30 magnetic susceptibility meter (Zh-Instruments) and geological inspection reveal that field boundaries and terrain steps are fixed with highly magnetic basalts. A reasonable explanation for the inversion of their anomaly might be that their thermal remanence outweighs their induced magnetisation. Since recent boundary lines are so plainly visible in the magnetogram, we excluded every curved feature from further interpretation as they are most likely of modern origin. Focusing on linear features, in Fig. 3a we see rectangular lines (marked blue) with risalit-like corners similar to the layout of the already partly excavated Grat Be'al Gebri. Rescue excavation performed at the beginning of this year revealed wall structures of archaeological origin in these areas. These indicate another monumental building. Fig. 3b shows the above discussed appearance of induced anomalies at the magnetic latitude of 15°. The filled pit (blue arrow) clearly shows an almost equally pronounced negative and positive anomaly. The negative (white) anomaly is sharper while the positive (black) anomaly is broader towards the south and has only 70% of the signal strength. In contrast, a basaltic rock (purple arrow) causes an inverted anomaly which is and less distinct due to the above discussed inherent thermal remanence.

Conclusion

We were able to detect archaeological features presumably of Ethio-Sebaean origin in the sub-surface. Furthermore, we observed the necessity of orientating the magnetometer sensors into their active zones to ensure uninterrupted signal recording. Additionally an inspection of the survey area, considering the latitude dependency of the signal, helps to interpret the magnetogram.

Acknowledgements

The work is partly funded by the DFG-Project: "Achaemenidische Residenzen und ihre Paradiese". Special thanks to Iris Gerlach of the Orientabteilung Deutsches Archäologisches Institut DAI Berlin.

References

D'Andrea, A.C., Manzo, A., Harrower, M.J. and Hawkins, A.L. 2008. The Pre-Aksumite and Aksumite settlement of NE Tigrai, Ethiopia. *Journal of Field Archaeology*. 33(2). 151-176.

Fassbinder, J.W.E. and Gorka, T. 2011. Magnetometry near to the geomagnetic Equator. *Archaeological Prospection: 9th International Conference on Archaeological Prospection*. 45-48.

Geometrics, Inc. 2001. *G-858 MAGMAPPER Operation Manual*. San Jose.

Gerlach, I. 2014. *Yeha - A Cultural Tourist Guide*. Berlin.

Scintrex Limited. 1993. *The CS-2 Magnetometer Operating Manual*. Concord.

Tite, M. S. 1966. Magnetic prospecting near to the geomagnetic equator. *Archaeometry*. 9(1). 24-31.

Surveying Kushite sites in Sudan: town and cemetery in Kawa

Tomasz Herbich, Robert Ryndziewicz
Institute of Archaeology and Ethnology, Polish Academy of Sciences, Warsaw, Poland

Location and geology

Kawa lies in Nubia (northern Sudan) on the alluvial plain in the Nile valley, approximately 80km upstream of the Third Cataract. The site lies directly on the east river bank. Over the millennia, the area has been subject to strong wind erosion, the surface is covered by sand, which in places form low dunes.

Early research, history and architecture

The first known excavation of the site took place in 1885 when Col. J. Colthorne, a correspondent of the Daily News with the Gordon Relief Expedition, undertook the clearance of the temple. Large scale excavations were begun by F.L. Griffith for the Oxford Excavation Committee in 1929-1931. After Griffith's death in 1934, the excavations were conducted by L. Kirwan in 1935-1936. These activities confirmed that the town was probably founded by Egyptian pharaoh Akhenaton (BC 1352-1336) and bore the name of Gematon ("The Aton is perceived"). The earliest building discovered so far was constructed during the reign of Tutankhamun (BC 1336-1327). After the collapse of Egyptian control over Nubia around BC 1070, Kawa became an important town of the Kingdom of Kush, which seems to have developed during the 9th century BC. During the period of its greatest prosperity (7th century BC), the Kushite rulers controlled a vast empire extending form Palestine in the north to central Sudan in the south. Kawa's temple of Amun, built between BC 684-680 by Taharqo, was one of those visited annually by the Kushite king during the reaffirmation of his right to rule (Welsby 2010). By the late Kushite period, Kawa had dramatically declined. Archaeological evidence suggests that the town was not occupied after 3rd century AD.

At its peak, the town covered an area of about 40ha, and stretched over 1km along the river bank. The ruins, covered by the wind-blown sand, form a prominent mound reaching up to a maximum height of 11.5m above the plain. The majority of the town is covered only by a layer of dust or sand, and there are only a few standing remains, including Amon's temple built by Taharqo, and a temple erected by Tutankhamun. The basic building material used was the sun-dried Nile mud brick; stone was only used in temple buildings (Welsby 2004).

SARS project

Following an almost 60 year period without archaeological activity on the site, in 1993 research was undertaken by Derek Welsby, on behalf of the Sudan Archaeological Research Society (SARS) and The British Museum. The research covered the city area and its immediate environs, as well as a cemetery located at a distance of about 0.5km to the east of the city.

The research of the city provided a unique opportunity to investigate the plan of the early Kushite settlement, largely unencumbered by later remains. Two strategies were used in the archaeological research of the city: regular excavations aimed to fully uncover the structures, at the same time providing knowledge of the stratigraphy of the site, and the survey aimed to produce a detailed topographical map, combined with cleaning of the tops of the walls. This allowed drawing plans of buildings of considerable size (including one over 85m long and 56m wide), registering areas with dense agglomerations of well-built mud brick dwellings, sacral buildings, and an industrial district with kilns. All investigated buildings date to the early Kushite period (8th-6th centuries BC).

While doing research of the city's main cemetery, both a surface survey and excavations were carried out. This defined the plan and range of the cemetery, determined the approximate number of burials (estimated at about 1000 graves) and determined the types of tombs. The most common type of grave consisted of a long stepped or sloping passage cut into the alluvium. At the end of the passage an arch-shaped doorway led into the tomb (Fig. 1). After placing the deceased in the chamber, the doorway would be blocked by a wall of mud-brick and the descendary was refilled with earth (Welsby 2004: 150; 2017: 27-30). The graves were covered by tomb monuments - tumuli, mastabas or mud-brick or stone pyramids. As a result of erosive processes, only the remains of stone pyramids, in the area of elite burials, have been preserved to our times.

Fig. 1. Kawa, eastern cemetery. Grave 215 seen from the east, and cross-section of a typical tomb. Photograph and drawing from the SARS Project.

Magnetic survey

The magnetic survey was undertaken in 2008. Measurements were carried out using a fluxgate gradiometer Geoscan Research FM256. Sampling grid was 8 points per 1 square metre (0.25m x 0.50m). During two seasons (the second in 2009) measurements were taken in the central and southern areas (as well as a small area in the northern part) of the city, covering a total area of 10ha. In 2018, measurements were carried out in the city and in the cemetery, covering a joint area of over 8ha (5.5ha in the cemetery and 2.7ha in the city).

Measurements taken in the city captured the western edge of the town (indicating that during the city's functioning, the river bank was located eastward, and determined the detailed plans of a number of buildings, which were only registered following the cleaning of the wall tops (Fig. 2b-c). In the area located at a higher level due to the overlaying of several building phases, the research allowed in some places to capture the two last building phases - of different orientations (Fig. 2a). However, the measurements did not allow to register the remains of the temenos which surrounded the Temple of Amun.

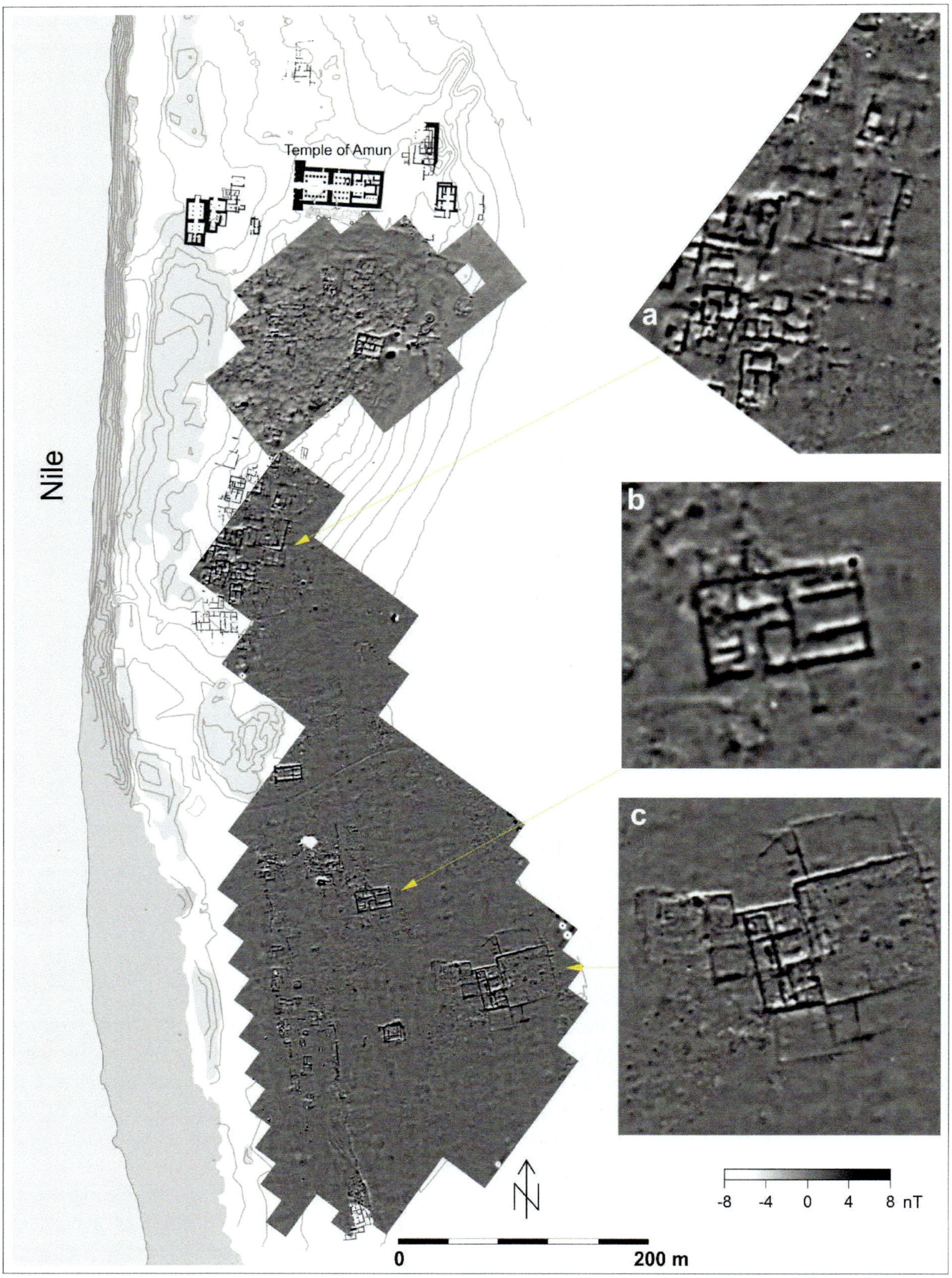

Fig. 2. Kawa, town. Magnetic map superimposed on the map of the site.

The research carried out in the cemetery allowed for an accurate registration of several hundred burials. The anomalies that correspond to tombs have a reduced magnetic field value. They consist of two elements: one has a longitudinal shape, and connects to the other: an oval anomaly with lower values than the longitudinal anomaly (Fig. 3). The longitudinal anomaly corresponds to a dromos fill, and the oval anomaly corresponds either to a shaft dug by robbers who were trying to get into the chamber, or a chamber opened from above in result of a collapsed vault caused by erosion. Anomalies corresponding to graves have similar orientations, and one can see that in many areas of the cemetery the graves were dug in rows.

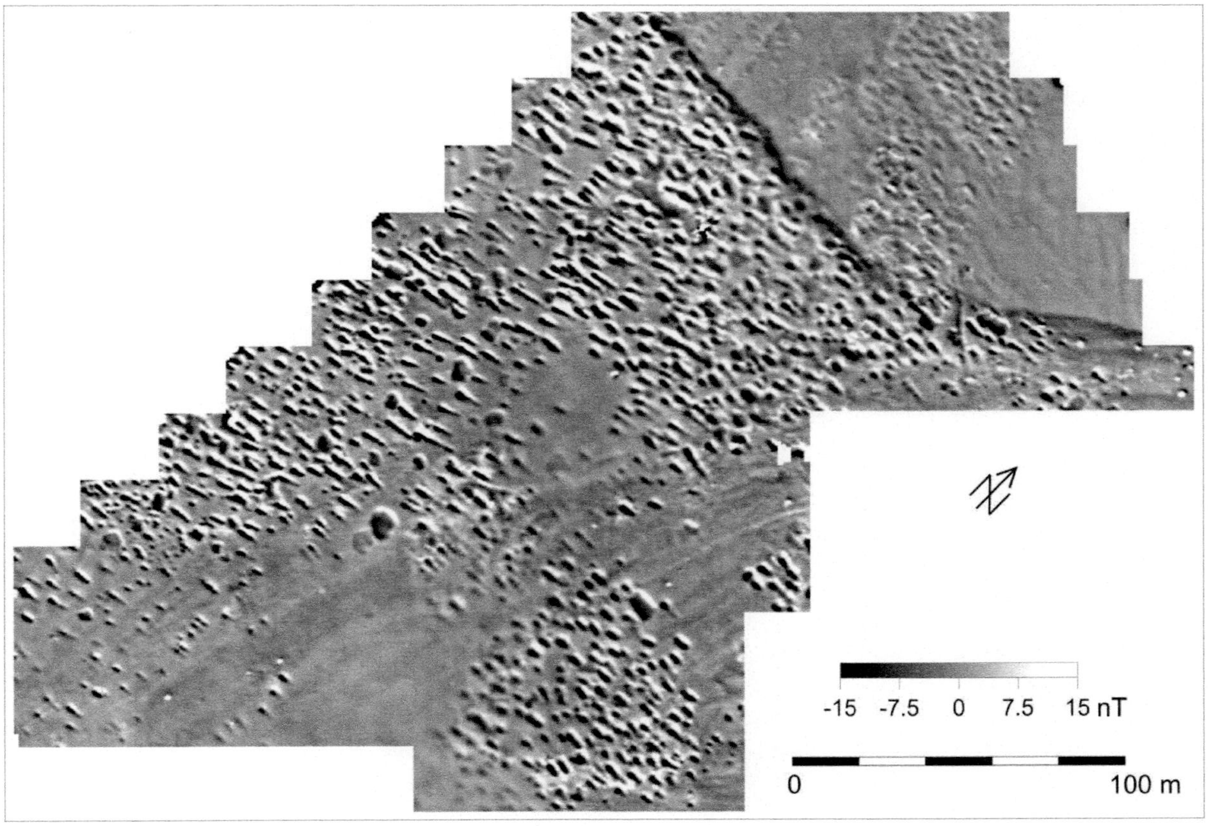

Fig. 3. Kawa, eastern cemetery. Magnetic map.

Conclusion

The measurements taken in the city provided information about the density of the building development in individual areas, as well as the dynamics of changes of the riverbed. In a number of cases, they showed earlier unknown elements of building plans, which had been unrecognized during shallow excavations. In the cemetery area, the measurements allowed for a precise reconstruction of its plan - with the exact location of individual tombs, as well as an indication of areas with a varied intensity of graves. These observations have been crucial to the reconstruction of funeral rituals among the peoples of the Kingdom of Kush.

References

Welsby, D.A. 2004. Coffee. In D.A. Welsby, J. Anderson (eds.). *Sudan. Ancient treasures*, 148-157. London.

Welsby, D.A. 2010. The Pharaonic and Kushite town of Gematon (Kawa, Northern Sudan). In F. Raffaele, M. Nizzolo and I. Incordino (eds.), *Recent discoveries and latest researches in Egyptology*, 349-358. Wiesbaden.

Welsby, D.A. 2017. *The Pharaonic and Kushite town of Gematon. History and archaeology of the site*. London.

Archaeo–geophysical prospection of forts in the North Omdurman (Sudan)

Robert Ryndziewicz[1], Mariusz Drzewiecki[2]
[1]*Institute of Archaeology and Ethnology, Polish Academy of Sciences, Warsaw, Poland,* [2]*Polish Centre of Mediterranean Archaeology, University of Warsaw, Warsaw, Poland*

Numerous remains of defensive architecture are highly visible parts of the archaeological landscape in Upper Nubia (present-day northern and central Sudan). They are mainly located along the banks of the Nile on both the alluvial soils of the river valley and rocky ridges and hilltops. This study focuses on three southernmost medieval forts in the Nile Valley: Hosh el-Kab, Abu Nafisa and Umm Marrahi. They are located relatively close to each other. Umm Marrahi is approximately 3.5km south from Hosh el-Kab and Abu Nafisa is approximately 0.5km northeast from Hosh el-Kab (Fig. 1).

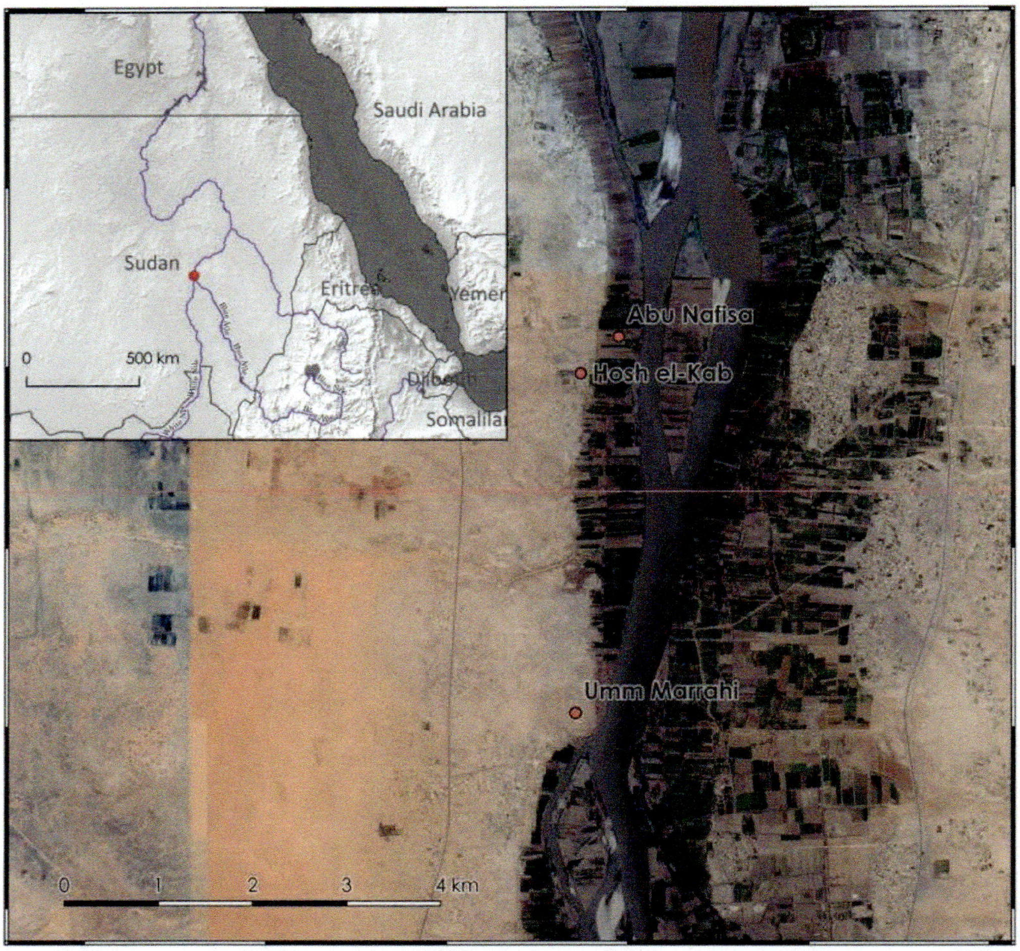

Fig. 1. Map showing location of the study sites (image source: www.naturalearthdata.com and www.maps.google.com).

The presence of these forts has been interpreted to be related to the first rulers of the Kingdom of Alwa from the 6th century AD onwards (e.g. Welsby 2014). However, some researchers (e.g. Crawford 1953) suggest that at least some of the forts might have been built during the Meroitic (*c.* 4th century BC – 4th century AD) or Late-Meroitic period (*c.* 2nd – 4th centuries AD). The unusual concentration of defensive structures in the area of North Omdurman raises question about their cultural origins and history (Drzewiecki *et al.* 2018). A non-destructive geophysical survey along with small, targeted excavations, aerial survey, pottery research, and field walking survey was chosen as a component of the scientific programme developed to investigate the forts, which were carried out in January/February and November/December 2018.

Geophysical investigation of the fortified sites in Nubia began in the late 1960s when French geophysicist Albert Hesse carried out a survey on the remains of the Middle Kingdom fortress in Mirgissa with a proton magnetometer (Hesse 1970). This research proved the usefulness of the magnetic method in tracing archaeological remains in the Nile Valley. Hesse observed the magnetic contrast between mudbrick structures and surrounding aeolian sand and noticed that anomalies he recorded were caused by a high magnetic susceptibility of the Nile silt – the main building material in Nubia. The result of the survey was a hand-drawn magnetic map, helpful in reconstructing the internal layout of the fortress (Hesse 2015: 124). Furthermore, it is noteworthy that the research at Mirgissa was the first implementation of the magnetic method on an archaeological survey in the entire valley of the Nile (Herbich 2015: 32).

Methods

The main aim of the geophysical survey was to identify and document unknown structures inside the forts, as well as to obtain new information on the construction of the defences. Equally important was to evaluate the state of preservation of the sites. A Geoscan Research FM256 fluxgate gradiometer was used for the magnetometry survey. Measurements were collected with a sampling interval of 0.25m along transects spaced 0.5m apart, within 20m x 20m grids. The data were processed in Geoplot software to produce grayscale plot magnetic maps. Simultaneously, a series of aerial photos with ground control points measured by total station was taken to create 3D models, orthomosaics, and digital elevation models (DEMs) of each site. The results were superimposed in QGIS software. This enabled detailed analysis and comparisons between magnetic anomalies, remains of the walls visible on the surface, and visible changes in composition and consistency of the soil.

Survey and results

The largest of the forts, Hosh el-Kab, is located on the edge of the flood plain (see Fig. 1) on an alluvial substrate mixed with aeolian sand. The area surveyed comprised the whole area of the fort, except the part destroyed by a modern trench dug for a large irrigation channel (Fig. 2). The magnetic anomalies fit into the spatial arrangement of the site as previously determined by topographical studies in 2011 (Drzewiecki and Polkowski 2016: 87) and the orthophotomosaic created for this study. Remains of the walls, still visible on the surface forming a rectangle *c.* 96m x 89m, are reflected on the magnetic map. Concentrations of small, high amplitude anomalies correspond to stones used in the construction. Some anomalies can be interpreted as a reflection of mud-brick structures adjacent to the inner face of the enclosure.

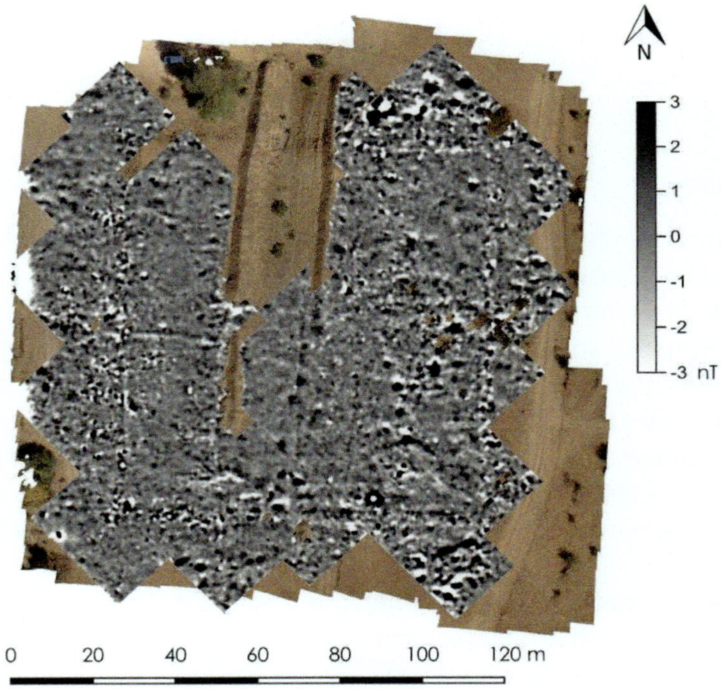

Fig. 2. Hosh el-Kab magnetometry results. Data collected with a Geoscan Research FM256. Sampling grid 0.25m x 0.5m. (Data processing by Robert Ryndziewicz; orthophotomosaic by Mariusz Drzewiecki).

Umm Marrahi fort was built on the top of a hill, in a location convenient for defense and to observe the surrounding area (see Fig. 1). The survey in this location provided information about the geological background of the site but very little about anthropogenic structures.

Abu Nafisa, the third fort, was built close to the river bank, on very homogenous Nile alluvial soil (see Fig. 1). The site is covered by vegetation (trees and bushes) and stone debris are scattered on the surface. However, some remains of the fort enclosure wall are still clearly visible on the surface. The magnetic survey provided a clear image of the fort's curtain sunken in the alluvial soil in the northern part of the site as well as some indicators of buildings inside the fort (Fig. 3).

Fig. 3. Abu Nafisa magnetometry results. Data collected with a Geoscan Research FM256. Sampling grid 0.25m x 0.5m. (Data processing by Robert Ryndziewicz; orthophotomosaic by Mariusz Drzewiecki).

Discussion and Conclusions

Magnetometry of the North Omdurman forts added valuable information to the knowledge of the sites. Some potential structures were located inside the forts. In the case of Hosh el-Kab, the survey suggested the presence of buildings in the interior of the fort. In addition to excavations, magnetic measurements demonstrated also the segmental structure of the curtain walls. Comparison of images obtained in Hosh el-Kab and in Abu Nafisa allows us to draw general conclusions about the geological substrate. Numerous irregularly distributed anomalies recorded in Hosh el-Kab reflect a minorchanges in the soil caused by natural factors. More homogeneous soil at Abu Nafisa makes the magnetometry result clearer. It indicates that Abu Nafisa might have been affected by Nile floods. Results obtained in all three forts contains numerous disturbances caused by modern destruction of the sites which confirms the poor state of preservation.

This study shows that even in relatively adverse conditions magnetometry surveying is an applicable tool in archaeological investigation in northern Sudan, and one of the most optimal ways to obtain useful and reliable archaeological data in a short period of time.

Acknowledgements

Fieldwork in the forts of North Omdurman was carried out as part of the project 'Did Meroitic rulers build fortifications? Fortified sites and politics in Upper Nubia during the fall of Meroe and rise of the Kingdom of Alwa', funded by the National Science Centre, Poland based on agreement no UMO-2016/21/D/HS3/02972, and directed by Dr Mariusz Drzewiecki. The authors are grateful to Tomasz Herbich for his kind support in the geophysical part of the project and Dr Abdelrahman Ali Mohammed (The General Director of the National Corporation for Antiquities and Museums of Sudan) for support.

References

Crawford, O.G.S. 1953. *Castles and Churches in the Middle Nile Region*. Sudan Antiquities Service Occasional Papers No. 2. Khartoum.

Drzewiecki, M., Cedro, A., Ryndziewicz, R., Elmontaser, D.E. and Rączkowski, W. 2018. Forts of North Omdurman: First season of fieldwork in Hosh el-Kab and Abu Nafisa. *Sudan & Nubia.* 22. 127–141.

Drzewiecki, M. and Polkowski, P. 2016. *Fortified Sites in the Area of the Fifth and Sixth Cataract in Context: Preliminary Reflections*. Gdańsk Archaeological Museum African Reports 8. 79–92.

Herbich, T. 2015. Magnetic prospecting in archaeological research: a historical outline. *Archaeologia Polona.* 53. 21-68.

Hesse, A. 1970. Introduction géophysiques et notes techniques. In J. Vercouter (ed.), *Mirgissa* I. 51–121. Paris.

Hesse, A. 2015. Early experiments with the use of new surveying methods in the archaeology of the Nile Valley. *Archaeologia Polona.* 53. 123-128.

Welsby, D.A. 2014. The Kingdom of Alwa. In J. R. Anderson and D.A. Welsby (eds.), *The Fourth Cataract and Beyond. Proceedings of the 12th International Conference for Nubian Studies* (British Museum Publication on Egypt and Sudan 1). Leuven-Paris-Walpole, MA: 183–200.

Part Three – Archaeological Prospection in Asia

Magnetic signal prospecting in a former Achaemenid 'palace': the example of Gumbati (Georgia)

Julien Thiesson[1], Sébastien Gondet[2], Jörg W.E. Fassbinder[3], Florian Becker[3], Marion Scheiblecker[3], Sandra Ostner[3], Mandana Parsi[3], Solène Bourgeois Espèron[4], Kai Kaniuth[5]

[1]Umr 7619 Metis Sorbonne Université, Paris, France, [2]UMR 5133 Archéorient (Univ Lyon, CNRS/Lyon 2 University), Lyon, France, [3]Geophysics Department Ludwig-Maximilians-University München, München, Germany, [4]Polytech Sorbonne, Sorbonne Université, Paris, France, [5]Institut für Vorderasiatische Archäologie Ludwig-Maximilians-Universität München, München, Germany

The Gumbati site

The archaeological site of Gumbati is located on the right bank of the Alazani River in the Kakheti region, Georgia. Excavations in the 1990s (Knauß 2000: 119-130) revealed parts of a rectangular building made with mudbrick walls of 2m width on average. Its southern extension is almost 40m in length. Stone column bases, typical of an Achaemenid royal style well-known from ancient Persia, were found around the building. Because of its architectural characteristics, it was interpreted as the remains of one of the administrative complexes controlling the Transcaucasus, the northernmost province of the Achaemenid Empire (mid-6th/mid-4th century BC).

Project framework and purposes

In July 2018, archaeological investigations were resumed at Gumbati within the frame of a German-French joint project named "Paradise" and funded from 2017 to 2020 by the *Deutsche Forschungsgemeinschaft* (DFG) and the *Agence National de la Recherche* (ANR). Its main goal is to produce comparable datasets regarding the spatial organization of several Achaemenid centers of power located in central Iran and in the Caucasus. As these sites were created within planned landscapes, covering dozens of hectares and encompassing spacious gardens, residences and administrative buildings, an improved understanding of their layout depends on complementary surveys. Consequently, one of the main tasks implemented is to develop suitable geophysical approaches for these sites. Since they are of a complex archaeological nature, combining mudbrick architectural remains and lighter garden infrastructures such as flow channels, we focus on combined magnetic and electromagnetic methods. Also, since all sites are located in intensively farmed regions, we aim to evaluate whether a deeper study of the soil magnetic properties would enable us to map the different sectors of the site, e.g. gardened and inhabited areas, despite surveying within landscapes of destruction.

Implementing geophysics at Gumbati

Devices used

In order to comprehensively evaluate the magnetic signal of soils of the surroundings, we used magnetometry and EMI prospecting.

The magnetic anomaly map was obtained by using and processing data from three devices made available by the Munich University Geophysics Department:

- Förster Ferex Gradiometer in the so-called "quadro-sensor" configuration (Foerster Ferex 4.032 Datalogger with four CON650 probes). The probes were mounted on a frame and carried in zigzag mode, 40cm above the ground.
- Scintrex Smartmag Sm4G-special Cesium-magnetometer applied as total field magnetometer, in a so-called "duo-sensor" configuration.
- Geometrics G-858 MagMapper Cesium-magnetometer applied as total field magnetometer, in a so-called "duo-sensor" configuration.

The EMI maps were produced using a GF Instruments CMD-Mini Explorer made available by the University Lyon 2/CNRS Archéorient lab. Measurements were recorded in both horizontal coplanar/vertical dipole (HCP/VD) and vertical coplanar mode/horizontal dipole (VCP/HD) configurations. It provides in-phase and apparent electrical conductivity maps for three different transmitter-receiver (Tx-Rx) distances of 0.32m, 0.71m and 1.18 m.

Survey strategy

The main target of the first 2018 geophysical prospection campaign at Gumbati was to obtain more information on archaeological remains in the vicinity of the partially excavated building. The trenches that brought to light architectural features were located in cultivated fields southwest of a large farmstead built in Soviet times. Consequently we mainly focused the magnetic survey on that area. A grid of 40m by 40m squares was staked out on the field over an area of about 5.5ha. The southern part covered the refilled trenches of the former excavation. The prospection was accomplished using all three magnetometer instruments for different parts in order to evaluate their efficiency in that particular context. The Ferex gradiometer was used for the two eastern and southern 40m rows. The two cesium magnetometers were alternatively used on the northwestern part.

Once the magnetograms were edited, we selected an area where we implemented the EMI survey. This area shows remarkably discrete features combined with more continuous soil magnetic property variations. By surveying it with the EMI device, we aimed to test the capability of this method to describe accurately both types of anomalies (geometries and/or magnetic properties). The whole area was surveyed twice with the EMI, at first in HCP mode, then in VCP mode, collecting 12 datasets in all. The probe was held as close as possible to the surface. Regular measurements were taken on a reference point in order to correct the drift.

Results

Here, for brevity, we focus principally on the magnetic signal of the site, rather than a large treatise on the entire set of 12 EMI apparent properties maps.

Magnetic map

The magnetogram (Fig. 1) shows that the site is poorly preserved. The excavators had already stressed the impact of repeated ploughings on the architectural remains. In the northern part, a number of roundish dark anomalies can be interpreted as signatures of pits. The linear features, pointing to the existence of mudbrick walls, are only just visible in the center of the image. These faint lines, oriented approximately at a 45° angle from north, may belong to building remains such as the ones excavated. However this direction is also parallel to the actual ploughing and field limits. Further analysis is therefore needed to ascertain the interpretation. In the southern part of the surveyed area, the magnetogram shows blurry features possibly caused by the filling of the old trenches.

EMI and transformed magnetic anomaly map

The magnetic signal can be described as the whole set of parameters which could be measured in prospecting. It includes the magnetic field anomaly (total or gradient), the magnetic susceptibility (measured by EMI devices) and ideally the magnetic viscosity. Presently, only the first two parameters were measured. The magnetic map (Fig. 2a), and its conversion as a susceptibility layer (Desvignes and Tabbagh 1995: 122-132) (Fig. 2b), may be compared to apparent magnetic properties using EMI (Fig. 2c). The general trends are the same. However, since the EMI is characterized by a less spatially resolved sampling interval than the magnetometry, some discrepancies appear. We cannot evaluate if these discrepancies are linked to the spatial sampling, measurement errors or the difference between the total and the induced magnetization. The next step will be to use the processing described by Benech *et al.* (2016: 103-112) for inverting the data to a magnetic susceptibility model to be transformed in to a magnetic anomaly. If the discrepancies are confirmed, some sampling and trench testing could be useful to understand its origins.

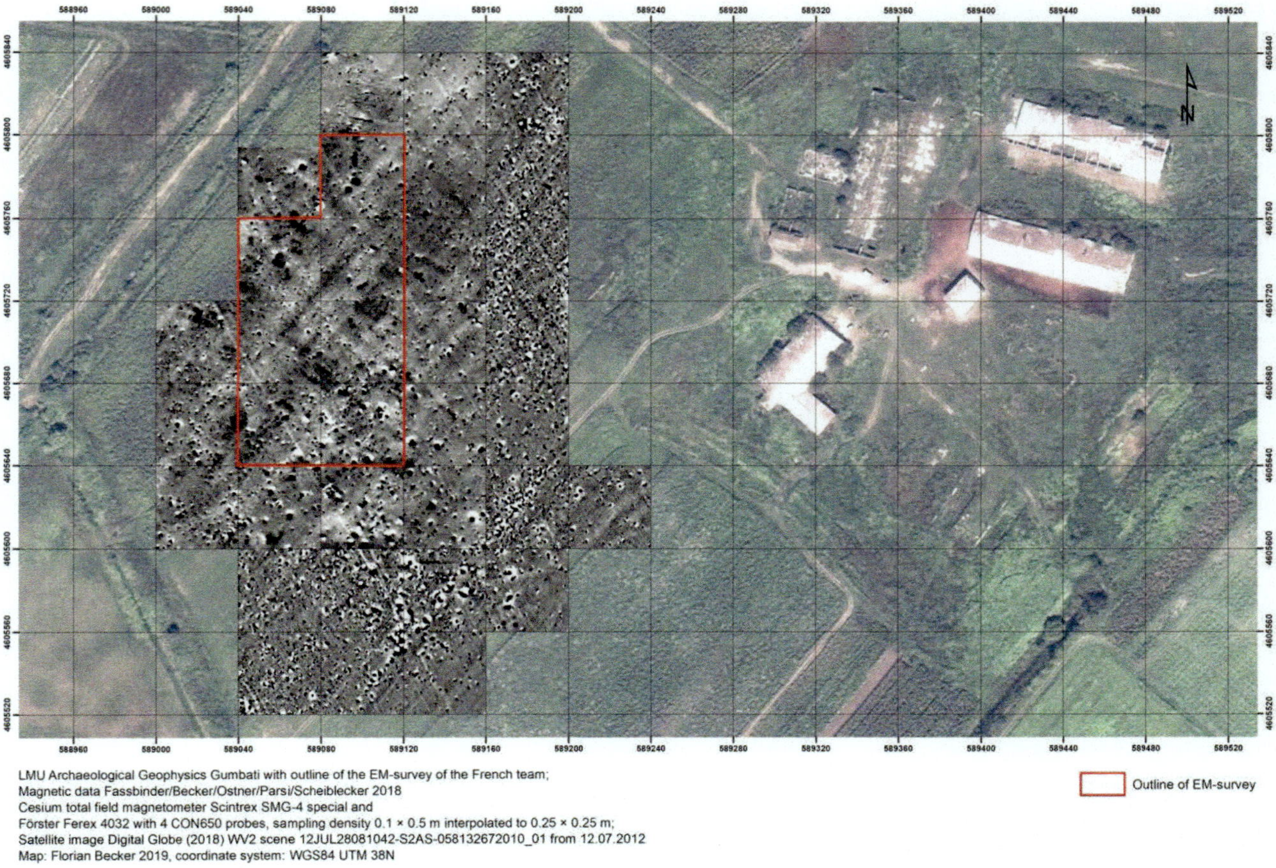

Fig. 1. Results of the magnetic prospection on Gumbati (scale: white/-10nT to black/+10nT).

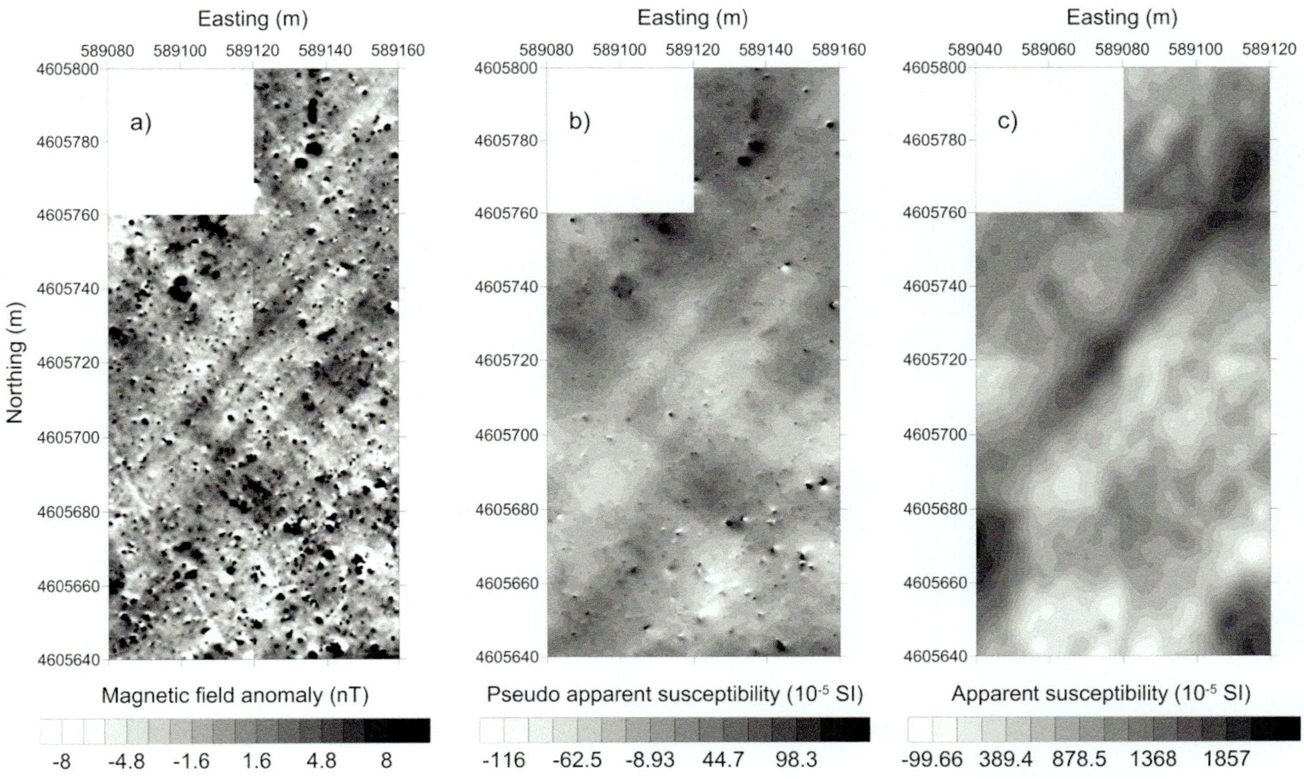

Fig. 2. (a) Extract of the magnetic anomaly map, (b) map of the pseudo-apparent magnetic susceptibility map computed from the magnetic anomaly map, (c) the apparent magnetic susceptibility map obtained from the in-phase measurement of the VCP Tx-Rx =1.18m channel.

Conclusion

The magnetic signal of the archaeological remains at the site of Gumbati was investigated. Magnetic and EMI survey results are generally in good accordance. The discrepancies between the pseudo-susceptibility map and the apparent susceptibility map could evidence the presence of material bearing remnant magnetization. However, the uncertainties linked to the EMI map lead us to process the data further and we will compute an inverted susceptibility model. If differences are confirmed, then it might be worthwhile to combine the survey with a complete magnetic description of soil samples that would help us to clarify their origins.

References

Benech, C., Dabas, M., Simon, F-X., Tabbagh, A. and Thiesson, J. 2016. Interpretation of shallow electromagnetic instruments resistivity and magnetic susceptibility measurements using rapid 1D/3D inversion. *Geophysics*. 81(2). 103–112.

Desvignes, G. and Tabbagh, A. 1995. Simultaneous interpretation of magnetic and electromagnetic prospecting for characterization of magnetic features. *Archaeological Prospection*. 2. 129–139.

Knauß, F.S. 2000. Der "Palast" von Gumbati und die Rolle der Achaimeniden im transkaukasischen Iberien. *Archäologische Mitteilungen aus Iran und Turan*. 32. 119–130.

Venice in the desert: Archaeological geophysics on the world's oldest metropolis Uruk-Warka, the city of King Gilgamesh (Iraq)

Jörg Fassbinder[1], Sandra Ostner[1], Marion Scheiblecker[1], Mandana Parsi[1], Margarete van Ess[2]
[1]*Geophysics, Department of Earth and Environmental Sciences, Ludwig-Maximilians-University Munich, Munchen, Germany,* [2]*Deutsches Archäologisches Institut, Orient-Abteilung, Berlin, Germany*

Historical background

Uruk-Warka, UNESCO-world heritage site together with Ur and Eridu, can be claimed to be the world's oldest megacity. Here the invention of handwriting and the scene of action of the oldest epic of humankind, the famous "Epic of Gilgamesh", took place. The inner city covers an area of *c.* 555ha and was populated by *c.* 40,000 people already in BC 3000. The diameter of the enclosed city is 4-5km; the city wall has a length of *c.* 9km and is up to 8-25m wide. Uruk was inhabited for nearly 5000 years till the 3rd century AD. Its occupation ended when the Euphrates River changed its way towards west and since this time the site remained untouched as a huge heap of adobe mudbricks with a Ziggurat on top (Fig. 1). Magnetometer measurements revealed a sophisticated water canal system, which provided access to the different city quarters, but also protected the inhabitants from the danger of annual flooding.

Fig. 1. Uruk. Panorama view of the city centre with the Ziggurat.

Prospecting Methods

Magnetometer prospecting was initiated in 2001 and continued in 2002, resumed in 2016 and carried out for a larger area in 2018 and 2019 (Fig. 2). For the survey, we applied three different types of magnetometers: a caesium Scintrex Smartmag SM4G-special, the caesium Geometrics G-858G magnetometer (both applied as total field magnetometers in a so called duo-sensor configuration) and a Foerster Ferex 4.032 fluxgate gradiometer in a so-called "quadro-sensor" configuration. Ground conditions are of soft and muddy, or dusty and salty soil. To get further information about the depth of the canals and the adobe city wall, we applied an ERT (4point light 10W) system with 60 active electrodes (ActEle), which allows spacing of 0.5m to 5m.

Results

Our magnetometer prospection now covers an area of *c.* 70ha and revealed a network of waterways, ship canals, harbours and moles, water gates and landing places that gave access to different city quarters. The water network crosses the city from north to south, provides water for the irrigation of gardens inside the enclosed city, and protects the inner city from floodwaters (Fig. 2).

Fig. 2. Magnetogram of the survey areas 2001-2019. Caesium-magnetometers Scintrex SM4G-Special and Geometrics G-858, both in duo-sensor configuration, dynamics ±25nT in 256 greyscales, spatial resolution 25cm x 50cm. The average total Earth's magnetic field intensity in Uruk increased from 2001 to 2019 from 45,180 ±20nT to 46,000 ±20nT.

The magnetometer surveys focus on the south-western part of the city inside and close to the city wall. The magnetometer results reveal one of the main canals, coming from the north gate and leading to a large harbour in the western part of the city, passing settlement areas in the east of the "Sinkashid" palace and a settlement area southwest of the palace (Becker *et al.* 2013). In the south, this harbour is limited by a mole and a water gate, obviously to regulate a junction to another branch of the canal. This branch reaches towards the south-east direction, connecting another settlement area, which due to archaeological field survey results, can be dated into Obeid and early Uruk-period (Finkbeiner 1991). Further to the south the main canal leads to a spacious agricultural field system with a complex network of irrigation canals.

A second large area was measured at the southern city, bringing to light the detailed structures of the city wall with bastions that are described in the Gilgamesh Epic. In the south, the city wall and a canal crossing it can be seen. Here, the course of the city wall and, at regular intervals, its bastions, known from previous excavations and documentation elsewhere in the city, are clearly visible. The data moreover seems to indicate that parts of the wall on its inner and outer faces are made of fired bricks; a detail that was not known before (Fassbinder *et al.* 2005).

Excavations undertaken in late March 2019, however, furnished the proof that these adobe bricks were tempered by pottery debris and thus behaved magnetically like burned bricks. It is also apparent that the wall was made out of several separate layers that were previously unknown and that the canal that circled the city ran just outside it. The entire wall complex was nearly 20-40m wide. The wall itself, with its inner and outer shells of tempered bricks, is some 9m thick, an observation that corresponds with the excavation results. Further details about Uruk's structure are provided by the magnetogram of the southwest gate, which is nearly 15m wide and can be interpreted as a floodgate, where the inner city's large east, west and central canals flowed out through the wall. On the outside, the gate was flanked by towers and strengthened very probably with fired bricks. In front of the floodgate at a distance of 240m, a small side canal branches off to the southeast, expanding roughly midway in front of a large building of fired bricks into a small harbour-like structure.

ERT measurements

In spring 2019, for the first time, we applied complementary ERT measurements on a range of selected profiles in order to get some additional information on the construction details of the city wall and the construction of the canal. One profile crosses the western canal (Fig. 3). The results clearly provide us with the evidence that this canal was an anthropogenic construction and not a natural riverbed. The magnetogram already indicates an edging of the canal with burned mudbrick. Now the ERT measurements clearly confirm this assumption.

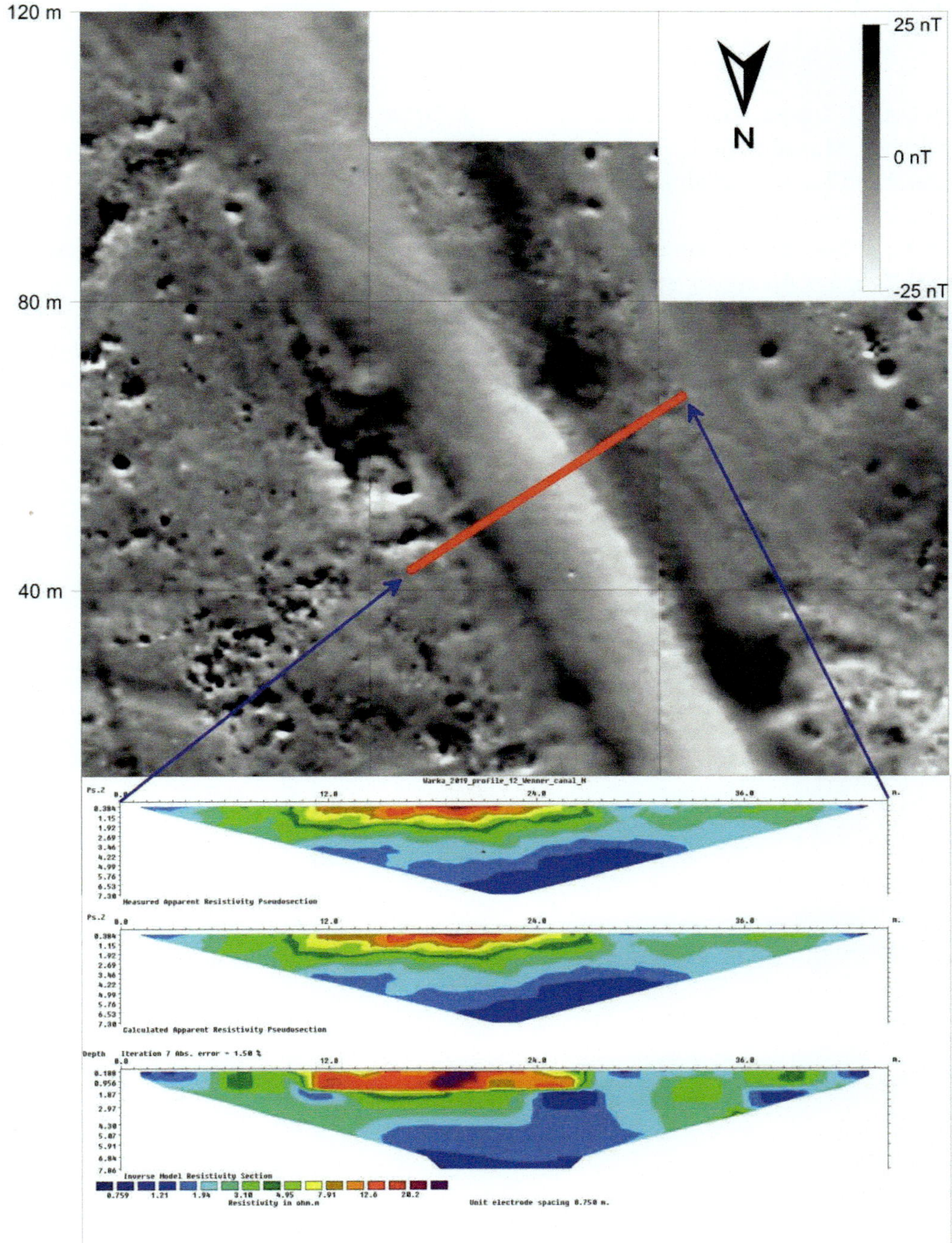

Fig. 3. At the top: Magnetogram extract of the main canal near the "Sinkashid" palace, the location of the 45m ERT-profile is marked in red. Below: ERT profile across the canal (0.75m electrode spacing, Wenner configuration).

Conclusions

All in all the magnetometer measurements gave insight into settlement areas of different occupations, gardens and fields inside the city wall as well as the network of canals that obviously served as the main arteries of Uruk. The main (western) canal, traced throughout the western part of the city, has a length of *c.* 1500m. It is up to 15m wide and *c.* 3m-4m deep. At several points slightly smaller canals branch off for the irrigation of fields. Left and right of the bordered canal, we find little landing stages, settlement areas, harbours and moles.

Detailed analysis of the magnetograms, complemented by rock magnetic analysis and further ERT measurements, the topographical information, as well as the available archaeological data, will possibly allow closer insights into the development, structure and function of the city and will support future excavations. The magnetometer survey hopefully will be continued soon and will offer a comprehensive picture of the structure of Uruk through time. Magnetometer prospecting is supposed to be carried out in all accessible areas of the city. Excluded from large scale prospections are the central district of Uruk and the Ziggurat, where already extensive excavations took place during the last 100 years.

References

Becker, H., van Ess, M. and Fassbinder, J.W.E. 2013. Uruk: Urbane Strukturen im Magnet- und Satellitenbild. In Vorderasiatisches Museum Berlin and Reiss Engelhorn Museum Mannheim (ed.). *Uruk: 5000 Jahre Megacity. Begleitband zur Ausstellung "Uruk - 5000 Jahre Megacity".* Petersberg. 355-361.

Fassbinder, J.W.E., Becker, H. and van Ess, M. 2005. Prospections magnétiques à Uruk (Warka). La cité, du roi Gilgamesh (Iraq). *Dossiers Archéologie.* 308. 20-25.

Finkbeiner, U. 1991. Uruk Kampagne 35-37, 1982-1984. *Die archäologische Oberflächenuntersuchung (Survey).* Mainz.

Ancient Charax Spasinou (Iraq) – Interpreting a multi-phase city based on magnetometer survey data

Lena Lambers[1], Jörg Faßbinder[2], Stuart Campbell[3], Stefan Hauser[1]
[1]Archaeology of Ancient Mediterranean and Near Eastern Cultures, University of Konstanz, Konstanz, Germany, [2]Geophysics Department, Ludwig-Maximilians University, Munich, Germany, [3]Near Eastern Archaeology, University of Manchester, Manchester, United Kingdom

Historical background and project aims

The ancient city of Charax Spasinou was situated in southern Iraq near Basra, between the rivers Tigris and Eulaios, at the modern location Jebel Khayaber. It offers the opportunity to study the layout and functionality of a major urban city dating from the Seleucid to the Sasanian period. The city was originally founded by Alexander the Great and given the name Alexandria (Campbell *et al.* 2019: 215). After its destruction by flooding, it was re-founded in BC 166/165 by the Seleucid king Antiochos IV and re-named Antiochia. This settlement was again destroyed by flooding. It was rebuilt under Hyspaosines and named Charax Spasinou (ancient Greek for 'palisade of *[Hy]spa[o]sines'*). Due to its favourable location Charax became a very important harbour in the Persian Gulf area and a major trading point between India and Babylonia, supplying goods further up to the Mediterranean (Campbell *et al.* 2019).

Charax was first identified with Jebel Khayaber in 1965, when distinctive ramparts with an average height of 4m to 6m were documented (Hansman 1967: 39). In 2016 Jane Moon, Robert Killick and Stuart Campbell (University of Manchester), together with Stefan Hauser (University of Konstanz) and the Iraqi State Board for Antiquities & Heritage, started a project to document and protect the ancient city of Charax Spasinou. The aim is to investigate the site through an integration of remote sensing technologies and surface survey as well as limited excavations in order to reconstruct the city layout, its chronology and to document its state of preservation for purposes of conservation and site management.

Methods

The mapping of *c.* 12km^2 was carried out using a UAV, covering the probable extent of the city and part of the wider landscape. Using Agisoft Photoscan, the images were compiled into a 3D model of the surface of the site. Together with over 60 ground control points, this allows the creation of digital elevation models (DEM) and orthomosaics with a resolution of *c.* 4cm per pixel. The upper levels of the soil are heavily salinated. Different rates of evaporation and precipitation above and between buried walls lead to visible, variable salt deposits on the surface after rain. This allows the outlines of buildings to be observed over some portions of the site (Fig. 1).

For the magnetometer survey, in March 2016 Jörg Fassbinder used a Scintrex SM4G-Special Caesium magnetometer in a duo-sensor and total field configuration. The device was carried *c.* 30cm above the ground at a sampling rate of 25cm x 50cm. The total Earth's magnetic field at Charax in October 2016 was *c.* 45700,00 ±20nT, sensitivity ±0.01nT. An area of almost 10ha was surveyed.

During the following three field seasons in 2017-18, Stuart Campbell used two Bartington Grad601 dual channel fluxgate gradiometers with a sampling rate of 12.5cm x 50cm. Area A was measured with a range of ±1000nT and Areas A1 to F with a range of ±100nT, sensitivity ±0.3nT (Fig. 2). These further surveys have covered a total of nearly 90ha.

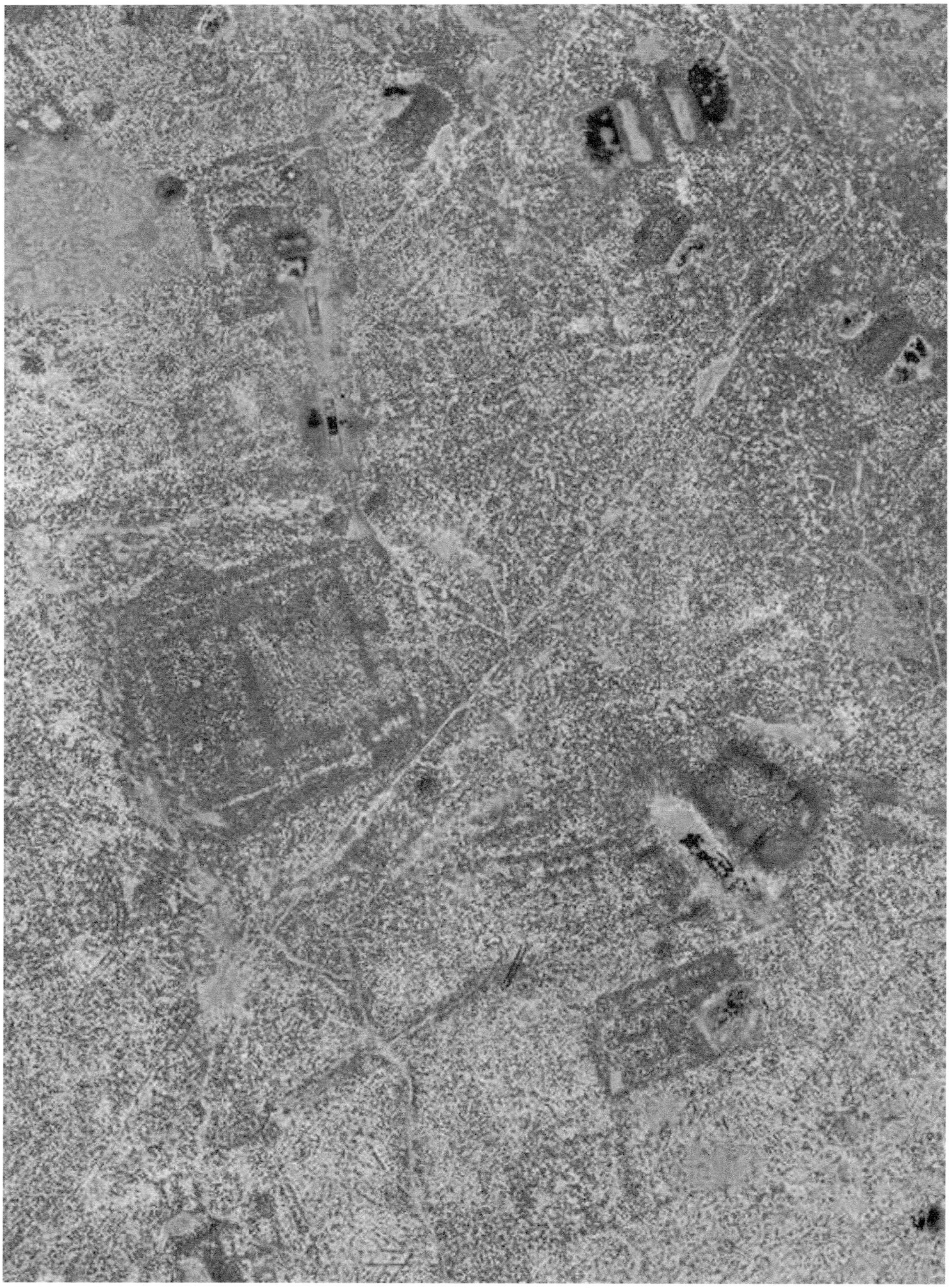

Fig. 1. Drone photo with salt lines.

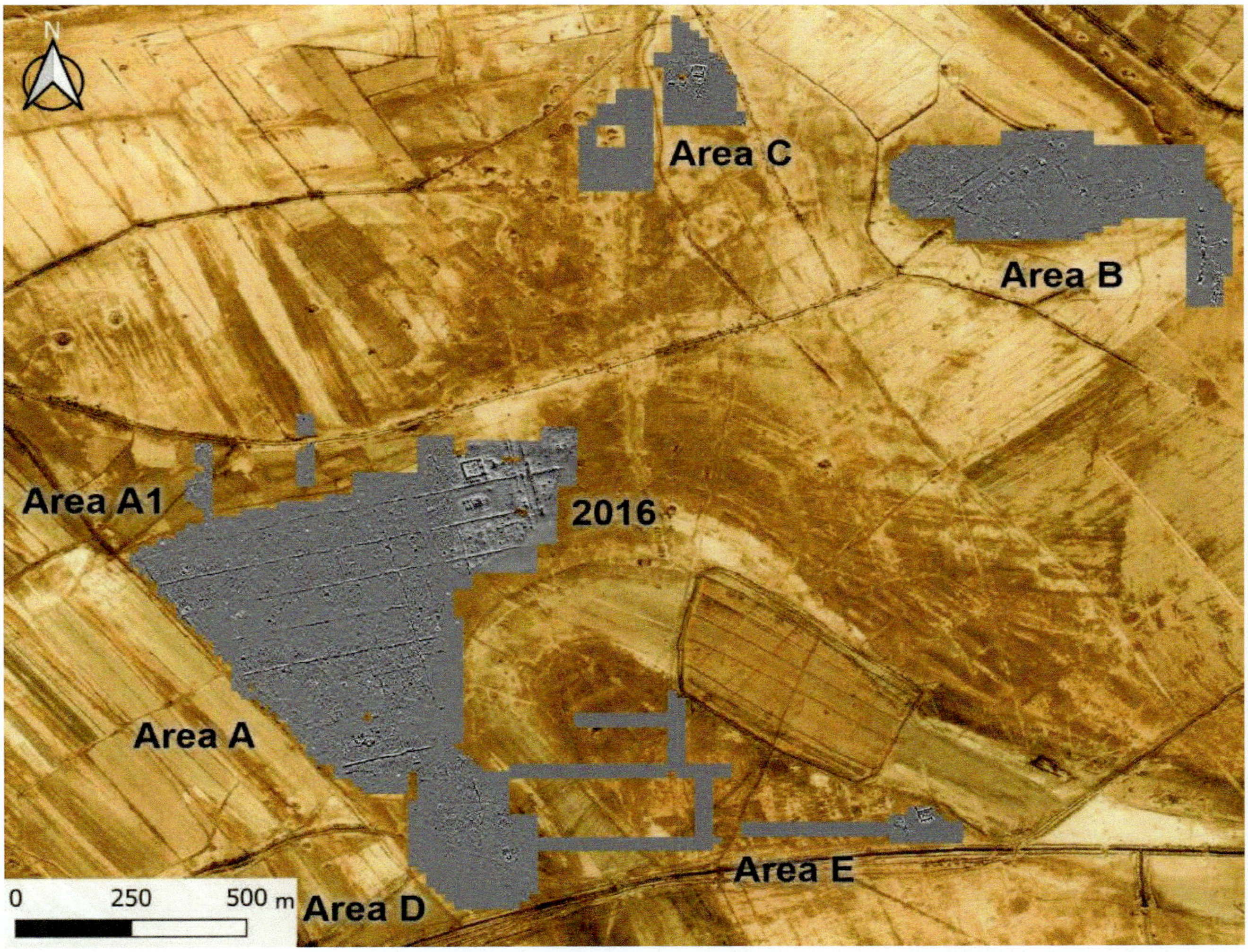

Fig. 2. Orthophoto from drone pictures with overlay of the magnetic prospection data.

Results and interpretation

The northern and eastern city limits are well preserved and marked by sections of the huge rampart over 2.4km long in the northern part and 1.1km in the north-eastern part. Magnetic prospection (Fig. 2. Area B) shows that the rampart continued in the east of the site, changing its direction to run to directly south. The former river course of the Karun has heavily eroded the southern part of the city, beyond the riverbed now visible in satellite images, but the geophysical data confirms the evidence from the surface survey that some settlement still survives to the south of the riverbed. Based on the survey results so far, we might estimate a city area of 700ha. However, the city was not only subject to repeated flooding in antiquity but was damaged by flooding into the 1970s. To date we can only verify good preservation of archaeology over an area of *c.* 150ha.

The 2016 geophysical survey suggested that the streets of the city centre follow the typical Hippodamian grid system with a grid size of around 161m x 88m (550 x 300 Attic Ionic feet), which is one of the largest we know of from the ancient world (Campbell *et al.* 2019: 220). However, the surveys in Areas A and A1, which extend the coverage of the street plan, suggest a more complex picture. The long east-west streets retained a regular spacing and orientation but the north-south streets that defined the city blocks become less regular to the west of the area. While this may be due to long-term subversion of an original urban plan by subsequent construction, it may also suggest that the apparent uniform plan always had some elements that were less tightly planned.

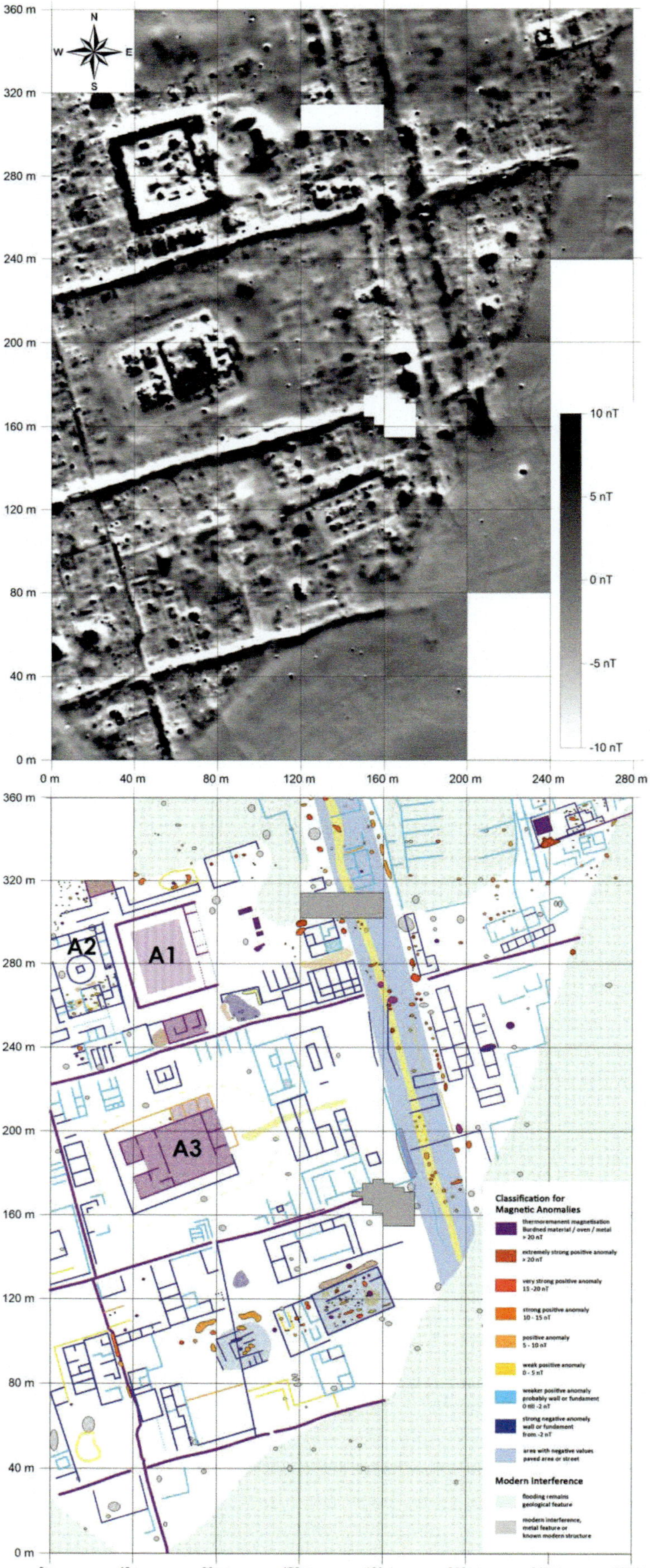

Fig. 3. Magnetic map of the 2016 survey area (sampling rate 25cm x 50cm, interpolated to 25cm x 25cm). Classification and interpretation of the magnetogram based on magnetic value.

The measurements also revealed several monumental buildings, such as temples and royal palaces, but also a wide range of residential buildings. In some places we can easily identify two building phases, mainly through their different orientations (Fig. 3). Around Building A3 we can also recognize removed and levelled ground with later building activities upon it. Building A1 intersects the walls of underlying buildings and the place in front of that building seems levelled, too. Along a central street or canal, which runs North-South and is up to 25m wide, earlier or later building phases are visible, too. In some instances the building sequence can be easily discerned. Studying the properties of the anomalies of these cases is expected to assist in the stratigraphic interpretation of other, less clear examples.

Some buildings have a higher density of high magnetic values, such as Buildings A2 and A4, which might be related to their function and use. Building A2 shows a noticeable combination of a round inner structure with surrounding small rooms. Such a ground plan suggests a *macellum*, which would be quite particular for this region.

While the magnetogram of most portions of the site show a clear picture, many areas show traces of flooding, which eroded the main inner city structures. Nevertheless, our investigations give us a coherent impression of the urban layout and planning. A number of interesting contexts revealed through magnetometry have been or will be further investigated through small-scale excavations, providing an additional data source along with archaeological, airborne and geophysical prospection for the interpretation of the site.

Two evaluation trenches showed that the east-west streets at least appear to have been placed over a drainage sub-structure of re-used storage jars (Campbell *et al.* 2019: 221-222). This arrangement is marked by a thermoremanent signal in the magnetogram. Such a complex drainage system both indicates a high level of investment in urban planning and the importance of resilience in the urban structure in the face of repeated flooding.

References

Campbell, S., Hauser, S., Moon, J., Killick, R., Shepperson, M. and Doležálková, V. 2019. Charax Spasinou: New Investigations at the Capital of Mesene. *Zeitschrift für Orient-archäologie.* 11. 212-239.

Hansman, J. 1967. Charax and the Karkheh. *Iranica Antiqua.* 7. 21.

Revealing the Hidden Structure of the Ancient City Ur (Iraq) with Electrical Resistivity Tomography

Mandana Parsi[1], Jörg Fassbinder[1], Nikos Papadopoulos[2], Marion Scheiblecker[1], Sandra Ostner[1]
[1]*Geophysics, Department of Earth and Environmental Sciences, Ludwig-Maximilians-University, Munich, Germany,*
[2]*Institute for Mediterranean Studies-FORTH, Rethymno, Greece*

Historical background

Ur, the city of moon god and "Home of Abraham" was founded by settlers in the 4th millennium BC (Fig. 1). The remains of the site are located 345km south of Baghdad and 257km away from the Persian Gulf. It is one of the most prominent cities in Mesopotamia (Wooley 1934-1976). There is evidence that the occupation was ended by a flood, formerly attributed to the flood described in Genesis. Although the city is much smaller than Uruk, in the next (Early Dynastic) period Ur became the capital of southern Mesopotamia under the Sumerian kings of the 1st dynasty of Ur (25th century BC). The last king, who left his traces both at Ur and Uruk, was the Achaemenian Cyrus the Great, whose inscription on bricks was found in recent excavations. The cities survived until the reign of Artaxerxes II. It was perhaps at this time that the Euphrates changed its course. With the breakdown of the whole irrigation system of Ur, its fields reduced to a desert and were finally abandoned.

Fig. 1. Ziggurat, the temple for the God of the moon in Ur, (Iraq).

Field survey

In spring 2017 and 2019, we applied large scale magnetometer prospection to uncover the city plan of Ur (Fassbinder *et al.* 2019). Due to the salty and clayey soils, radar prospecting is non-applicable to receive information on the depth of adobe mudbrick walls, canals or the harbour. One of most suitable methods turned out to be Electrical Resistivity Tomography (ERT). Here we show for the first time that ERT is a robust geophysical method to detect and image shallow and deep underground structures made from sundried mudbricks in the adjacent mud in detail and considerable resolution, by determining the underground resistivity distribution (Schmidt 2013). In recent years, the application of ERT became more and more important to bridge the gap between magnetometer and ground-penetrating radar prospection, in enhancing, completing and integrating the information context retrieved by these methods. Sophisticated computer programs to trigger the multichannel electrodes, combined with inversion and three-dimensional (3-D) analysis software, allow for the tracing of apparent electrical resistivity in substantial accuracy, illustrating the stratigraphic composition of underground layers (Tsokas *et al.* 2012). For the ERT measurements, we applied the Earth resistivity meter 4point light 10W (Lippmann Geophysikalische Messgeräte, Germany). The design of the active electrodes (ActEle) allows spacing of 0.5 to 5m, together with short, thin and easy to repair cables and a lightweight readout unit of only 750g, the instrument is easy to transport and suitable for archaeological prospection.

Results

The first objective of this work was to compile a 3-D representation of the subsurface resistivity related to an area hosting a buried house (new Babylonian period) that is traditionally made from adobe bricks. In total 25 parallel ERT profiles were acquired with a profile and electrode spacing of 0.5m in a specific area covering almost half of the house. Fig. 2 shows the horizontal depth slices extracted by the 3-D resistivity inversion model over the area of the house. The light blue lines have been drawn to visualize the outer walls of the house. The map shows two full rooms on the right side and a bigger room on the left side. According to the resistivity of the walls, we can deduce that the walls have been made from mudbricks. The analysis of the magnetometer measurement reveals a "negative" anomaly for the walls of the house, which is also an indicator for mudbrick. The top of the house's walls are at 1m depth and they reach approximately 2m downward. This is one of the first studies that detected mud-brick walls by ERT method.

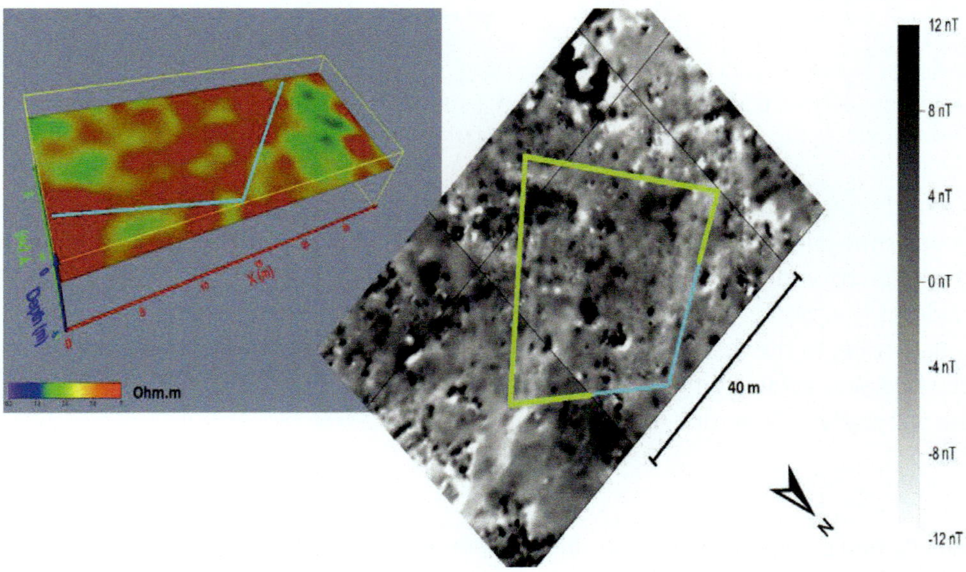

Fig.2. The left picture shows a 3-D model of the house. The light blue lines indicate the outer walls of the house (ERT instrument 4point light 10W, electrode spacing 0.5m x 0.5m). On the right, a magnetogram of the same area is shown. The light blue lines indicate the walls of the house that are shown in the 3-D model and the green lines are continuations of the walls of the house (magnetogram: Caesium magnetometer, Geometrics G-858 in Duo-sensor configuration spatial resolution 25cm x 50cm).

One other objective of the geophysical survey in Ur was, to map the horizontal extent of the city wall, its vertical dimensions and to acquire information related to the stratigraphic layering between the two main wall structures. An ERT profile was laid out perpendicular to the direction of the wall, which had already been revealed through magnetometer prospection conducted prior to the ERT survey.

The 2-D vertical resistivity inversion section shows the location of the city wall. It has a height of around 1m, which corresponds to the archaeological information retrieved by Woolley (1934-1976). The width of the inner and outer wall is around 4m and 2m respectively (see Fig. 3a; the black circles show the position of these walls). Moreover, the right part of the Fig. 3a outlines the refilled archaeological trench from the 19th century mentioned by Woolley. It has comparable depth extent to the respective wall and it is marked with a black rectangle. The middle walls between the main walls are displayed with pink circles.

The water content in the soil is a limiting factor that potentially could affect the interpretation of the results. The moisture content was monitored with repeated ERT measurements over the same profile for a whole day and the tomographic data were also supported by the collection of direct soil moisture, of temperature and conductivity measurements over the city wall with a Time Domain Reflectometry (TDR) instrument. The preliminary results show a maximum moisture content change of about 10 vol%.

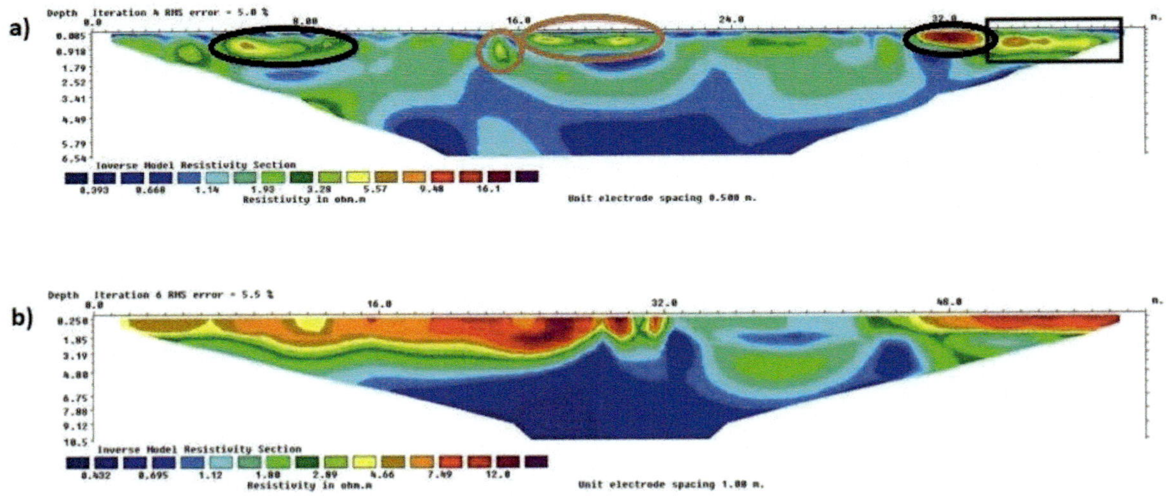

Fig. 3. (a) ERT profile over the city wall in Ur (electrode spacing 0.5m, dipole-dipole configuration). The black circles indicate inner and outer wall, the pink circles show the walls in between and the black rectangle indicates Woolley's excavation, which was refilled afterwards. (b) 2-D map of the harbour (1m electrode spacing, Wenner configuration).

An ERT survey, consisting of several profiles with different electrode spacings, was completed over the location of the harbour to verify its existence, which was originally mapped through the magnetometer survey. The left side of the map in Fig. 3b shows the wall. The width of the harbour is around 15m and its depth about 8m. According to the result of the ERT, the wall is made out of baked bricks and this information matches the archaeological evidence that exists for Ur. Further interpretations will be added to this study, after receiving the soil analysis from our American colleagues.

Conclusion

ERT measurements turned out to provide a suitable complementary prospection method. It delivers not only reliable information on the depths of archaeological features that are situated in clayey, salty and waterlogged soils, but also depicts a prospecting method that can detect adobe bricks in the adjacent clay and mud.

Acknowledgements

I greatly appreciate the funding of my short term scientific mission (STSM) in the Institute for Mediterranean Studies-FORTH in Greece by the European Cooperation in Science and Technology (COST) under the number CA17131 and thank the IMS-FORTH lab for being a great host. The work is partly funded by the DFG-Project: "Achaemenidische Residenzen und ihre Paradiese". Special thanks to Adelheid Otto (LMU), Elisabeth Stone and the University of Pennsylvania group.

References

Fassbinder J.W.E., Ostner, S., Scheiblecker, M. and Parsi, M. 2019. *Geophysical prospection campaign 2019: Magnetometry and Earth Resistance Tomography (ERT) at the archaeological site of Ur, Iraq* (Unpublished report Directorate of Antiquities, Iraq).

Schmidt A. 2013. *Earth resistance for archaeologists*. Lanham.

Tsokas, G.N., Van de Moortel, A., Tsourlos, P.I., Stampolidis, A., Vargemezis, G. and Zahou, E. 2012. Geophysical survey as an aid to excavation at Mitrou: A preliminary report. *Hesperia.* 81(3). 383-432.

Woolley, C.L. 1934-1976. *Ur Excavations, vol. 2-10.* Oxford University Press.

Geophysical survey of single phase archaeological sites: Magnetometry in Wadi Shamlu, Kurdistan, Northern Iraq

Marion Scheiblecker[1,2], Simone Mühl[2], Jörg Faßbinder[1]
[1]*LMU München, Geophysics, Department of Earth and Environmental Sciences, München, Germany,* [2]*LMU München, Near Eastern Archaeology, München, Germany*

Introduction

Starting in 2009, the international team of the Shahrizor Survey Project (SSP) investigated archaeological sites in the Shahrizor plain, from tell sites with several settlement layers to almost invisible little elevations indicating single phase settlements. According to the chronological evidence provided by superficial pottery and other finds, the archaeological sites located in this area date from the Neolithic to the late historic period. Further research carried out in the area combined archaeobotany, geology, chemical analyses, micromorphology and geophysics as well as excavations and historical sources (Altaweel *et al.* 2012).

In the southern center of the Shahrizor plain, many water streams fed the fertile region with enough water to create ideal conditions for settlements. A meandering stream in the middle – Wadi Shamlu – is running north-south entering the artificial lake Darband-i Khan, formerly flowing into the Tanjero river (Mühl *et al.* 2018). Along Wadi Shamlu, four tell sites are known, accompanied by 31 smaller sites, probably single phase, which were archaeologically surveyed in 2012/2013 and partly prospected using geophysical means in 2017/2018 (Fig. 1). The project aims to classify the architecture of different periods and reconstructing the ancient landscape and settlements. This paper focuses on the main mound of Gird-i Shamlu and its neighbouring mound.

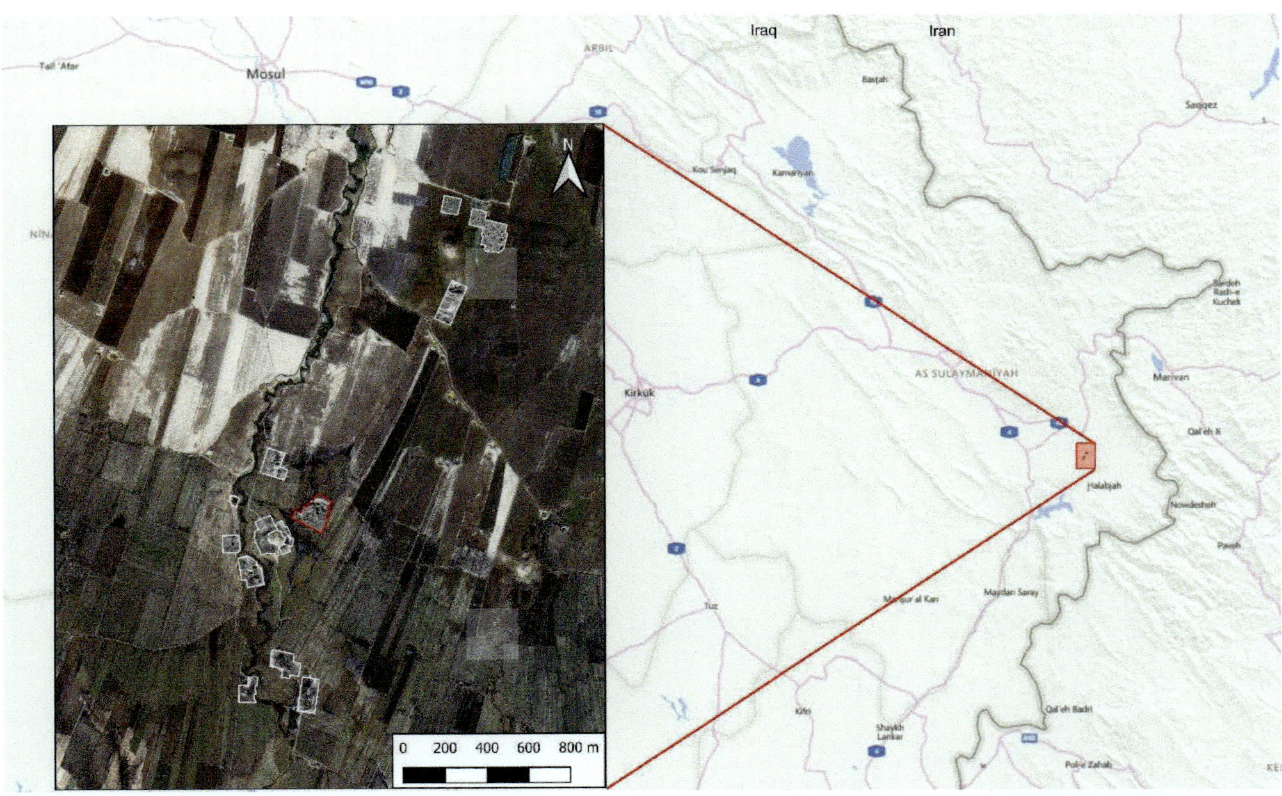

Fig. 1. Overview of the survey area of Wadi Shamlu (Northern Iraq) with overlaid magnetograms (background: Bing maps).

The main mound of Gird-i Shamlu (SSP-33) is characterized by strong disturbances due to the remains of military constructions and their iron waste, making it impossible to trace weak magnetic signals derived from subsurface mudbrick remains with magnetometry. Therefore, the first magnetic survey in 2014 focused on the neighboring lower mound rising 3m-4m in the northeast (SSP-34), indicating an extension of the Gird-i Shamlu site (SSP-33; Mühl and Faßbinder 2015: 483-484). Recent excavations and new investigations in the surrounding area led to a revision and new interpretation of the 2014 measurements.

Methods

For the Shahrizor plain with its alluvial fans and clayey rich soil, the most suitable and fastest method is magnetometry (Mühl and Fassbinder 2016: 230-231), precisely because of the limited accessibility of the southern part of Wadi Shamlu due to the Darband-i Khan lake and intensive agriculture.

In autumn 2014, we undertook the first geophysical survey in this region. We used a total field magnetometer (Scintrex) Smartmag SM4G-Special in duo-sensor configuration (spatial resolution 25cm x 50cm) to investigate three mounds (Mühl and Faßbinder 2015: 483). It offers high sensitivity to image archaeological features easily, including mudbrick architecture, which is typical for this region, as well as geological structures (Mühl and Faßbinder 2016: 231). In 2017 and 2018, the total field magnetometer Geometrics G-858 was introduced (Fig. 2) to investigate flat sites along Wadi Shamlu in duo-sensor configuration (spatial resolution 12.5cm x 50cm).

Fig. 2. Measuring SSP-40 with the handheld Geometrics G-858, in the back Gird-i Shamlu (SSP-33; Photo: J. Faßbinder, 09/2017).

New results and interpretation

Northeast of Gird-i Shamlu, an approximate area of 160m x 160m was magnetically surveyed, covering the site SSP-34 completely. Modern ploughing lines are visible in the magnetogram (Fig. 3), almost in a north-south direction. The reservoir is surrounding/flooding the mound annually, causing the fine features visible in the northern and southern part of the magnetogram (Fig. 3, thin lines in cyan). Erosion and accumulation of magnetic minerals are forming these shorelines (Faßbinder 2015: 88-89). A former Wadi or streamlet is visible, limiting the mound to the north (Fig. 3, marked in cyan). Corresponding to the architecture, it is probably ancient and fits to the environmental investigations indicating irrigation for the late 3rd millennium BC in a more humid environment (Marsh *et al.* 2018: 966).

Linear structures of different magnetic properties show several orientations indicating at least two different ancient settlements visible in the magnetogram (Fig. 3). In the north-eastern part of the measured area, close to the streamlet in the north, negative magnetic anomalies are dominating, orientated north-west to south-east (Fig. 3, in yellow). They are interrupted by small-scale positive magnetic anomalies of rounded or rectangular shape (Fig. 3, in orange), showing a different orientation compared to the negative anomalies as well as other directions, possibly an indicator for different periods. The negative signal, caused by limestone or mudbrick enriched with less magnetizable materials than the surrounding soil, could be used for creating these walls. The smaller features in the northeastern part (Fig. 3, in orange and red) are possibly constructed from mudbrick, indicated by weaker and higher positive magnetic anomalies (Faßbinder 2017: 505-507).

In the remaining area (Fig. 3), the orientation of the archaeological features indicating architecture is rotated by 25° in comparison to the previously mentioned architectural structures. The positive magnetic signal (Fig. 3, in orange and red) is also suggesting mudbrick walls in this area. Kilns/fireplaces can be traced in the central and southern part of the magnetogram (Fig. 3, red circles), showing high magnetic values and a white (negative) shadow to the north, indicating thermoremanent magnetization (Faßbinder 2017, 504).

Conclusion

Magnetometry in Wadi Shamlu revealed at least the presence of two different settlements at SSP-34. Recent archaeological investigations show the use of mudbrick differing in their composition, which is responsible for the differences in magnetic signal. In combination with the mentioned methods and with functional and spatial analysis, it allows us to establish an archaeological typology and chronology for the area of Wadi Shamlu and to compare it with neighboring regions in the Shahrizor plain and beyond.

Acknowledgements

The new investigations at Wadi Shamlu were financially supported by the Gerda Henkel Stiftung for the dissertation: "Geophysikalische Prospektion in der Shahrizor-Ebene. Eine Analyse zur Raumnutzung in ländlichen Siedlungen und urbanen Einzugsbereichen altorientalischer Siedlungssysteme", in the framework of the DFG project "Flucht – Migration – Interaktion. Artefaktbezogene Diversität in altorientalischen Kontexten des 3. und 2. Jahrtausends v. Chr." and with the support of the Directorate of Antiquities Sulaymaniyah.

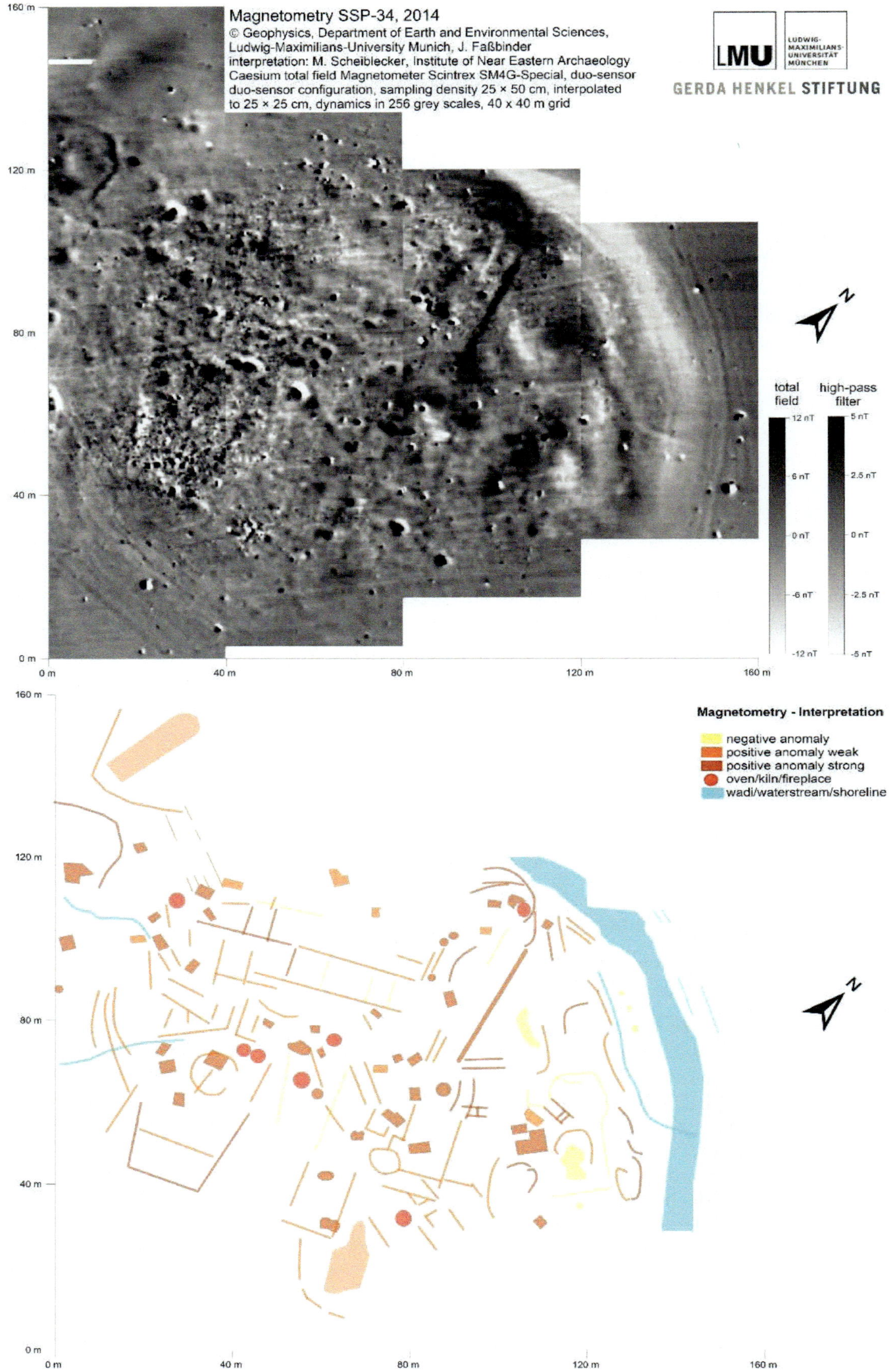

Fig. 3. Magnetogram and interpretation of SSP-34.

References

Altaweel, M., Marsh, A., Mühl, S., Nieuwenhuyse, O., Radner, K., Rasheed, K. and Ahmed Saber, A. 2012. New Investigations in the Environment, History, and Archaeology of the Iraqi Hilly Flanks: Shahrizor Survey Project 2009-2011. *Iraq*. 74. 1–35.

Faßbinder, J. 2015. Seeing beneath the farmland, steppe and desert soil: Magnetic prospecting and soil magnetism. *Journal of Archaeological Science*. 56. 85-95.

Faßbinder, J. 2017. Magnetometry for Archaeology. In A.S. Gilbert, P. Goldberg, V.T. Holliday, R.D. Mandel and R.S. Sternberg (eds.), *Encyclopedia of Geoarchaeology*. Encyclopedia of earth sciences series, 499–514. Dordrecht.

Marsh, A., Fleitmann, D., Mohammed al-Manmi, D.A., Altaweel, M., Wengrow, D. and Carter, R. 2018. Mid- to late-Holocene archaeology, environment and climate in the northeast Kurdistan region of Iraq. *The Holocene*. 28. 955–967.

Mühl, S. and Faßbinder, J.W.E. 2015. Archaeological Geophysics in the Shahrizor Plain (Iraqi Kurdistan). *Archaeologia Polona*. 53. 481–485.

Mühl, S. and Faßbinder, J. 2016. Magnetic Investigations in the Shahrizor Plain: Revealing the Unseen in Survey Prospection. In K. Kopanias and J. MacGinnis, J. (eds.), *The Archaeology of the Kurdistan Region of Iraq and Adjacent Regions*, 229–236. Oxford.

Mühl, S., Rösch, M., Mohammed Almanmi, D.A., Kadereit, A. and Qader Aziz, B. 2018. Irrigation in the Shahrizor Plain: The potential of archaeological and geoarchaeological archives to reconstruct ancient water management. In H. Kühne (ed.), *Water for Assyria. Studia Chaburensia* 7, 117–136. Wiesbaden.

Conclusions from Twenty Years of Electrical Resistivity Tomography (ERT) Surveys in Israel for Archaeological Prospection

Paul Bauman[1], Alastair McClymont[1], Eric Johnson[1], Richard Freund[2], Randy Shinduke[1]
[1]Advisian Worleyparsons, Calgary, Canada, [2]University of Hartford, Hartford, United States

Background

Over the last 20 years, electrical resistivity tomography (ERT) has evolved as an important and often essential tool for archaeological investigations in Israel. In the late 1990s, there was broad reticence in Israel to use ERT as an archaeological prospection tool due to its relatively poor spatial resolution, increasingly poor vertical resolution with depth, the generally inconvenient nature of utilizing cross-sectional images versus map view images, logistical challenges of field implementation, and the relatively slow acquisition rate as compared to more broadly used geophysical prospection methods. Nevertheless, ERT investigations at numerous sites since the year 2000 have shown that the technique is surprisingly quite versatile in terms of field implementation, the archaeological periods to which it can be applied, and the types of subsurface targets for which surveys can be planned.

Typically, geophysical surveys in support of archaeological prospection are designed to provide map view images. This is particularly true where a single, dominant layer of cultural activity is present. However, where multiple layers of human activity spanning thousands of years are present, and where these layers may extend over vertical extents of tens of metres, cross-sectional imaging with ERT can be an effective method of providing information at depth. This is particularly true in the Levant which has been a causeway for many civilizations, and tells (a hill or mound created from human debris) that may contain remains from 20 or more civilizations, extending to depths of 20m or more below the top of the mounds.

Surveyed Sites and Findings

Based largely on pottery finds at the surface, Tel Yavne in Israel has more than 25 layers representing various time periods from AD 1948 extending back through to the Bronze Age. Tel Yavne is the largest unexcavated tell in Israel. 21 ERT sections were collected over three investigation seasons, exploring to a depth of 30m below ground surface (bgs). Features successfully imaged included a likely water system, Roman period pottery works, the palaeo-relief of the site as defined by the Kurkar sandstone, the full vertical extent of archaeological debris across the entire site, and Roman period architecture in the unexcavated flood plain sediments surrounding the tel.

Tunnels and caves are features that usually have a very small footprint. Furthermore, many voids may be simply impossible to detect from surface geophysical methods. However, tunnels and caves can often be imaged as outstanding ERT resistivity anomalies as they are frequently infinitely resistive air-filled voids present in a moderately resistive host material. Perhaps the most important archaeological site in the Levant is Qumran where the Dead Sea Scrolls were excavated. All of the known Dead Sea Scrolls were recovered from 11 caves excavated between 1947 and 1956. ERT has been useful in discovering new, yet unexcavated caves at Qumran, as well as other important site features.

In the past, the use of ERT has often been dismissed over terrain where soil or contact resistance conditions create challenges of injecting electrical current into the subsurface, or measuring voltages in the subsurface. Improvements in field equipment and field acquisition methods, however, have greatly broadened the versatility of using resistivity methods at archaeological sites. In the summers of 2000 and 2001, in perhaps the first use of ERT in an archaeological prospection program in Israel, ERT successfully imaged a constructed, earthen floor within a refuge cave from the year AD 135. This was particularly challenging due to the lack of soil, the lack of moisture, and the large blocks of roof collapse rubble lying on the present-day cave floor. Additionally, ERT has successfully imaged through a marble floor into the hypocaust of a Byzantine period bath house in Nazareth, and through a stone patio of the Church of the Annunciation, into the foundation of what is likely the original Greek Orthodox church in Nazareth. Ongoing excavations at this site have already uncovered a 4th century mosaic identified as a resistive, slab like anomaly in the ERT survey.

Future Developments

While ERT instrumentation continues to improve with more rapid acquisition speed, increasing number of channels, the ability to simultaneously collect induced polarization measurements, and the development of distributed systems, it is likely in processing and visualization where the most significant changes will occur in the next few years. Greatly improved facility with 3D visualization, and the ease in which ERT datasets can be merged with a multitude of other datasets is already vastly improving our understanding of complex sites. Joint inversion with induced polarization, seismic refraction, and other geophysical datasets may improve interpretations particularly in complex sites. Cloud based systems are allowing geophysicists and archaeologists to integrate their findings and thoughts of any given site in real time, while they are situated continents apart. It is these improvements in global and cross-disciplinary communication that may provide the greatest step changes in the understanding of various sites, and prevent geophysical datasets from simply being collected and then filed away, never to be fully utilized.

Restoring burial mounds damaged by disasters — Contribution of archaeological prospection to collect information at the *Idera* burial mound, Japan

Akihiro Kaneda, Hiroshi Yamaguchi
Nara National Research Institute for Cultural Properties, Nara-shi, 日本

Introduction

This paper introduces the application of archaeological prospection on buried mounds that have been damaged by disaster. Earthquakes frequently occur in Japan. In recent years, a wide area of terrestrial deformation has been caused by such huge earthquakes as the Tohoku Earthquake (2011) and the Kumamoto Earthquake (2016). As a result, many cultural properties were seriously damaged.

With regard to the Kumamoto earthquake, many of the archaeological burial mounds (called *Kofun* in Japanese) have been affected by the disaster. These burial mounds were built in the Kofun period (3rd-7th century AD) on the Japanese archipelago. Many of them in the Kumamoto area were damaged; in particular, many stone chambers in the mounds were severely damaged. The Nara National Institute for Cultural Properties has carried out cooperative restoration activities with concerned organizations. These activities include archaeological prospection, 3D measurement, environment monitoring and archaeological survey.

The *Idera* burial mound was built in the late 5th century AD. It is located in Kumamoto prefecture in the southwestern part of Japan. It has a stone chamber decorated by drawn patterns, lines and arcs. These geometric patterns are represented by pictures and reliefs (Hamada 1917).

It has been confirmed that the Kumamoto Earthquake caused a large-scale deformation and stone collapse in the stone chamber, and surface cracks have appeared on the mound. These damages require immediate restoration work. However, even if the dismantling of the stone chamber could be carried out, excavating without knowing the status of the underground stone chamber or the construction condition of the mound might further impair the stability of the stone chamber (Hashiguchi 2016, Ohmi 2017). Therefore, non-destructive methods were selected to collect underground information.

The Nara National Institute for Cultural Properties carried out the archaeological prospection of the site with the support of the local government, the Board of Education of Kumamoto Prefecture and the town of Kashima.

Methods

The GPR method was chosen in this study. The GPR equipment used included a GSSI SIR-3000 and 400MHz antenna with a trigger and a Mala geoscience X3M and 250/500MHz antenna with hip-chain and RTK-GPS. The survey lines were determined along the stone chamber. The distance between each survey line was measured at 25cm intervals for the GPR survey. The analytic software employed was GPR-SliceV7.0 (produced by the Geophysical Archaeometry Laboratory Inc.).

Due to a concern about the collapse of the stone chamber, sandbags and scaffolds were installed for safer exploration. The GPR survey was carried out above the stone chamber and the ground cracks using a rope and sled for the antenna mobile (Fig.1).

The acquired data were processed with a Bandpass filter and Background filter. Time slice methods were utilized to make the plan maps at every even depth. Finally, a 3D volume isosurface render model was generated from the GPR reflection.

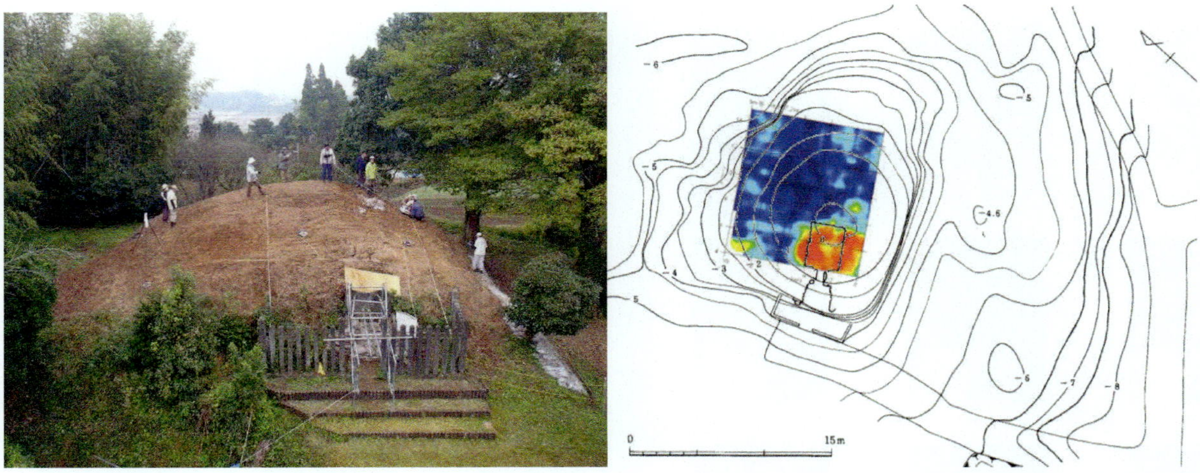

Fig. 1. Survey photo and area at the *Idera* burial mound (Contour map: Kumamoto Prefecture 1984).

Results

The cross-sectional shape of the anomaly formed a triangle in the upper half and a square in the lower half of the GPR survey. Its shape is characteristic of the Kumamoto area. It is referred to as a *Higo* (a historical name in the Kumamoto area) type stone chamber. This anomaly starts in a shallow area (6.4ns) from the ground surface (Fig. 2).

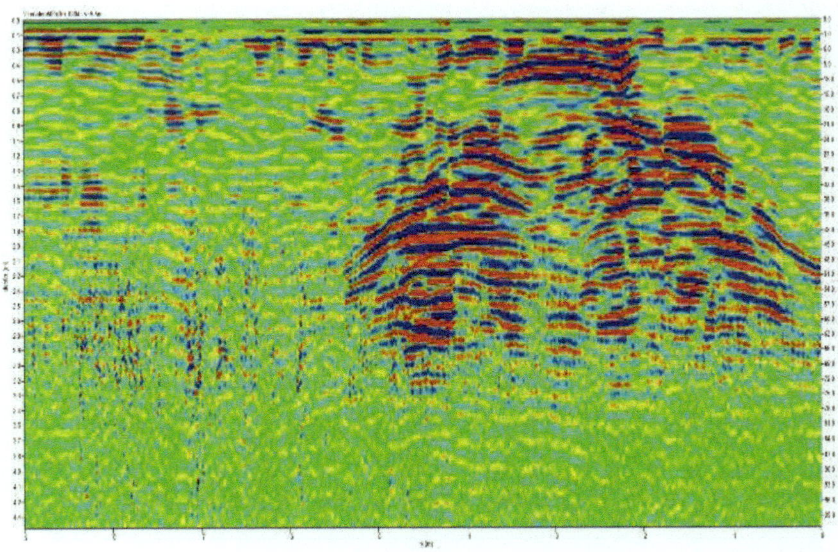

Fig. 2. GPR section profile of the stone chamber (6.5m line from the chamber entrance).

Although the shape of the ground plan for the stone chamber is a long rectangle toward the back wall, the GPR anomaly shows a wide rectangle in the direction of the sidewall. This anomaly contains surface stones and backfilled stones. It suggests there are many backfilled stones behind the side surface stones. In contrast, there are few stones behind the back wall (Fig. 3).

In the isosurface volume rendering based on the GPR results, it was possible to three-dimensionally grasp the strong range of reflection (Fig. 4).

The origin of the *Higo* type stone chamber is thought to be a combined stone coffin. It is also known to pack backfilling stones thickly on the sidewalls. By acquiring information on the back of the stone chamber, it is possible to obtain information about the technical genealogy of the origin of the *Higo* type stone chamber.

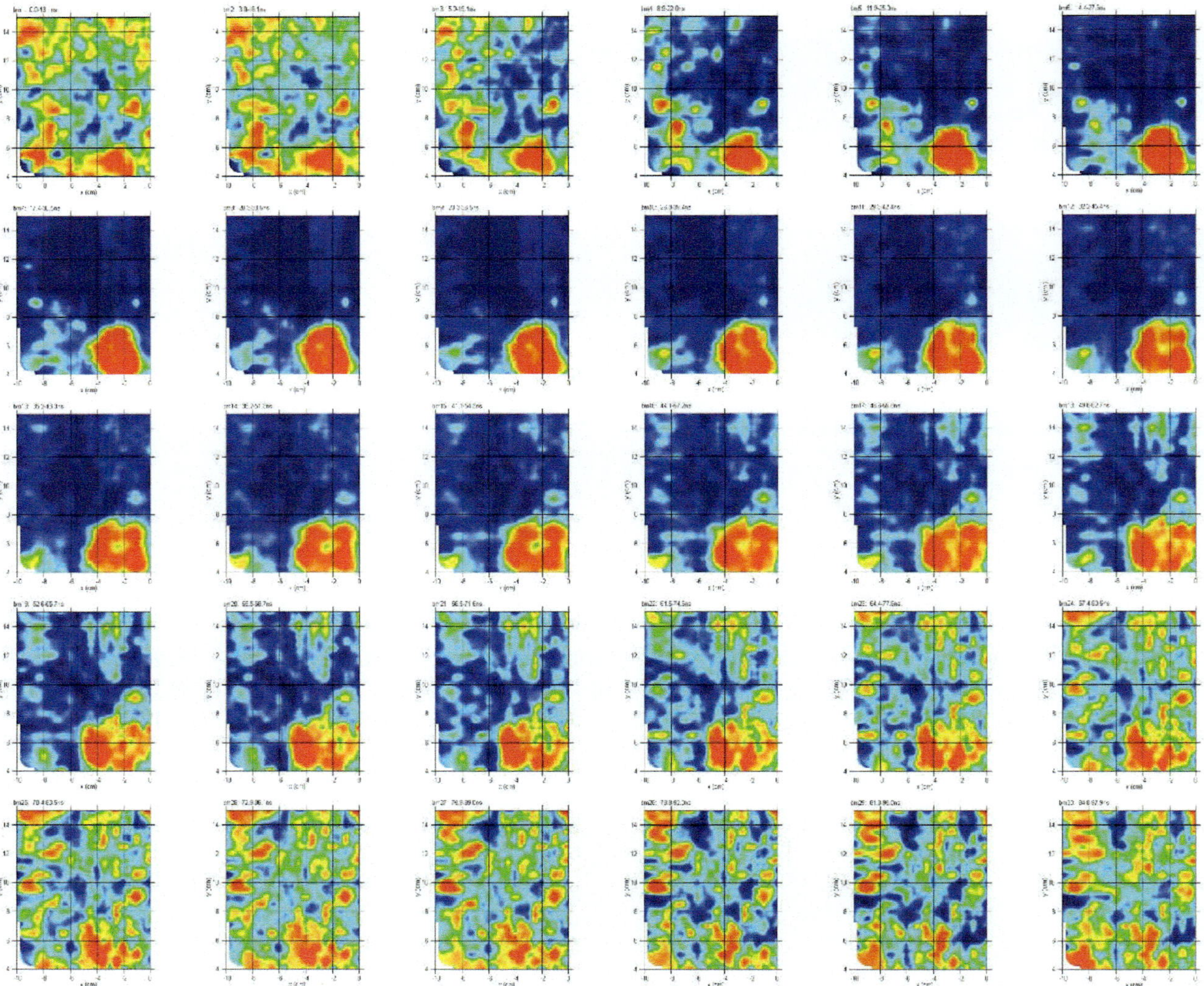

Fig. 3. GPR Time Slices.

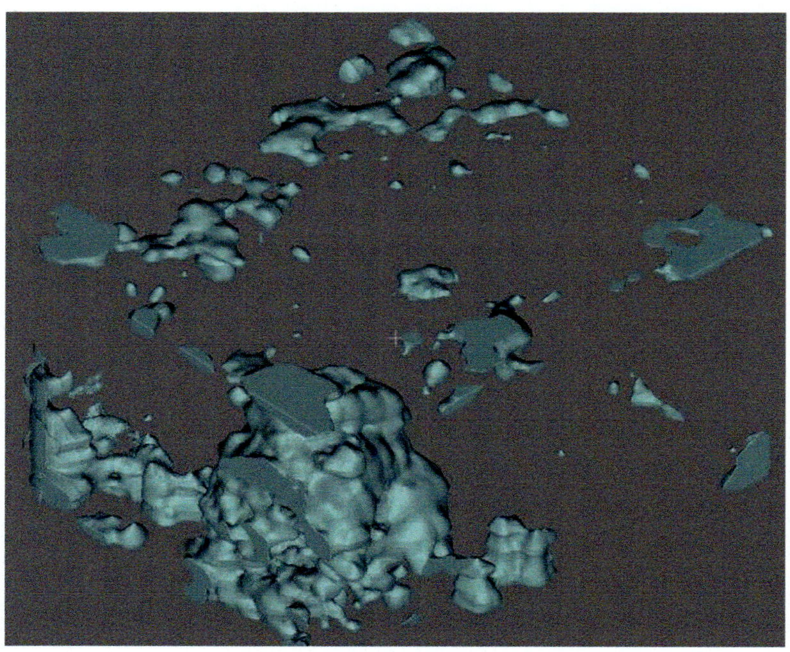

Fig. 4. Volume rendering by the underground reflection at the *Idera* burial mound.

In addition, there is a weak reflection under the stone chamber, especially in the south-east. It is unknown whether it was caused at the time of the stone chamber's construction or acquired earlier. However, it is necessary to consider the observation from the stone chamber, the falling of the stones in the stone chamber, and the ground crack of the mound.

In the vicinity of the stone chamber, there is a linear reflection in the area corresponding to the crack on the ground surface, on the south side of the stone chamber, and it is assumed that a void or uneven soil is included here.

Conclusion

This result shows that it is possible to collect information on burial mounds at risk of collapse in a non-destructive way. Even now, the collapse of the stone chamber of *Idera* Kofun is in progress, and measures are urgently needed to mitigate this. Geophysical exploration is very useful to determine the unseen part of the ruins beneath the surface. This survey demonstrates the effectiveness of archaeological exploration when planning restoration and conservation. This information, combined with UAV and/or Pole surface observations such as LiDAR and SfM on the ground, proved useful when considering protection and restoration planning.

References

Hamada, K. 1917. *Ornamented tombs in the province of Higo*. Kyoto University.

Kumamoto prefecture, Board of Education. 1984. *Kumamoto decoration old burial mounds synthesis research report*. (Printed in Japanese).

Hashiguchi, T. 2016. *The present conditions and problems of the tombs damaged by the 2016 Kumamoto earthquake*. http://hdl.handle.net/2298/37525 (Printed in Japanese).

Ohmi, T. 2017. *Towards Recovery from the Kumamoto Earthquakes*.
http://archaeology.jp/remains/towards-recovery-from-the-kumamoto-earthquakes/ (Printed in Japanese).

UAV-based Airborne Laser Scanning in densely vegetated areas: Detecting Sue pottery kilns in Nakadake Sanroku, Japan

Michael Doneus[1], Maria Shinoto[2], Irmela Herzog[4], Naoko Nakamura[6], Hideyuki Haijima[7], Tomokazu Ōnishi[3], Shōhei Kita'ichi[7], Baoquan Song[5]

[1]*University of Vienna, Austria,* [2]*Universität Heidelberg, Germany,* [3]*Kagoshima International University, Japan,* [4]*LVR-Amt für Bodendenkmalpflege im Rheinland, Germany,* [5]*Ruhr-Universität Bochum, Germany,* [6]*Kagoshima University, Japan,* [7]*Nakanihon Air Service, Japan*

Since 2013, surveys by field walking and geophysical prospection as well as excavations have documented the remains of several ancient Sue pottery kilns at Nakadake Sanroku, Minami-Satsuma City, Kagoshima Prefecture, Japan. Sue kilns are dug into the slopes of hills, often as tunnels. A kiln being excavated in Nakadake since 2014 reached 8m in length and nearly 2m in width; it was found about 1.5m below the ridge of a steep slope (Nakamura and Shinoto 2015: 10-16). Recent survey results suggest that about 30 to more than 60 of such kiln sites dating from the beginning of the 9th century onwards are located in this region, concentrated in an area of nearly 6km^2, as can be seen in a detailed description of sub-clusters in the area by Matsusaki (2018). Like most other Sue kilns, these are situated in a mountainous area. Intensive land use during the Middle Ages and later times changed the relief of the study area. Nowadays, next to the remains of rice fields, dense forest cover (mainly Japanese cedar planted after WWII, clusters of bamboo and a mixture of native trees and fern) impedes the survey approaches mentioned above.

Airborne laser scanning (ALS) has proved to be very effective to document archaeological sites in vegetated areas. However, it was clear that in this kind of dense and mostly evergreen vegetation (see background in Fig. 1), data acquisition would be a challenge and that a high point density from a scanning device with small laser footprints would be desirable in order to be able to derive a detailed digital terrain model (DTM). Combined with the fact that the area is characterized by steep slopes with large height differences (heights range between 15m and 286m above sea level within the scanned area of roughly 0.5km^2), this seemed to be only achievable using a drone-based scanning platform.

Therefore, in 2018, a small scaled project was launched from the Japan Society for the Promotion of Science (JSPS) project ("Production and Distribution of Sue ware at the Southern border - Complementary Research in Archaeology and Natural Sciences", JSPS A-15H01902, leader: N. Nakamura) and Nakanihon Air Service, which aimed to document kiln remains and other archaeological features in part of the Nakadake region using airborne laser scanning.

Data acquisition was undertaken in May 2018 by Nakanihon Air Service utilizing a Riegl VUX sensor mounted on a multicopter (Fig. 1). Due to the slow flying capacity of the multicopter combined with a low flying height and a small scanner-footprint, the system seemed to be a good choice for the difficult site conditions. Altogether, an area of roughly half a square kilometre was scanned with a high point density of more than 100 points per square metre.

While data acquisition was a fast process, post-processing the data recorded in this area with abundant vegetation was a challenge. This is mainly due to the fact that the UAV-based laser scan resulted in an extremely high point density. The high density of the UAV-recorded points allowed the capturing of an adequate number of ground points even in densely overgrown areas. However, standard filtering did not identify these ground points reliably; misclassifications could mainly be found in difficult terrain with dense vegetation cover (Fig. 2A).

Fig. 1. Multicopter and Riegl VUX during data acquisition from 16th of May 2018. The background shows the dense, evergreen vegetation cover of the project area (Photo: M. Shinoto).

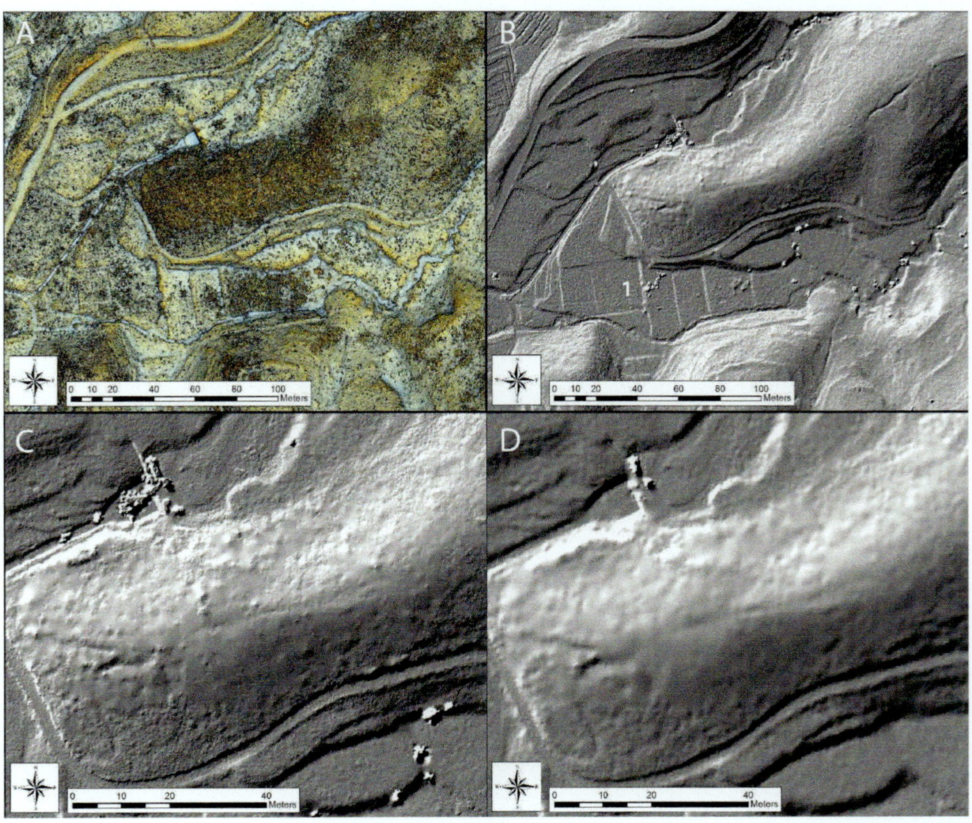

Fig. 2. (A) Filtering of point cloud using standard parameters. While open areas and less vegetated areas show the topography in detail, a large number of non-ground points are still clearly visible and hinder archaeological interpretation. (B) Filter settings tweaked for high detail displaying difficulties in areas with extremely dense vegetation. (C) Ridge densely overgrown with fern. Here, the filter settings used in (B) create problematic results. (D) Filter settings adapted to extremely dense vegetation with a more generalized digital terrain cell size of 0.2m. There is less noise and the archaeologically relevant paths can be identified more clearly.

Fig. 3. Height-coded combination of slope and hillshade displaying a small valley with at least 5 potential Sue pottery kiln sites (1-5).

A more sophisticated filtering that resulted in a more satisfying digital terrain model for archaeological interpretation (Fig. 2B) was achieved by using specialized software (SCOP++). The software supports applying a robust interpolation with an eccentric and asymmetric weight function within a hierarchical framework (more details can be found in Doneus and Briese 2006). Although the filtering process is highly sophisticated and could be adapted to the specific requirements of the combination of topography, vegetation and archaeological structures, due to the complexity of the terrain and the strongly varying vegetation, it was impossible to achieve a qualitatively uniform filtering for the total area.

Therefore, two different parameter sets were used, which resulted in two terrain models: (1) A model that shows fine details, as regular rows of a former sugi plantation in today's rice-fields (Fig. 2B), but produced a noisy surface in very densely overgrown areas (Fig. 2C); and (2) A smoothed terrain surface, which is adequate in densely overgrown areas but shows less details (Fig. 2D). The examples given illustrate the need for a more "intelligent" filter software supporting local filter parameter adaptation to vegetation density.

Both versions of the filtered DTM were visualized using various techniques (hillshade, local relief model, Sky-view factor, positive and negative openness) and archaeologically interpreted. As a first result possible kilns, path remains, and other archaeological features could be located and verified during on-site visits (Fig. 3). Still, given the fact that airborne laser scanning has only rarely been applied in Japan for archaeological prospection (Akashi 2010, Fujii *et al.* 2015), and never in such difficult terrain, the next steps to be undertaken include the creation of interpretation keys. This means that we will have to: (1) specifically search for and investigate areas that have been interpreted as archaeologically interesting in the ALS-based terrain model, in order to gain experience for the interpretation of future scans; and (2) compare the area with the terrain model and document how they appear in the visualized datasets and why they may not be recognizable in the scan.

References

Akashi, Y. 2010. Views of ancient mountain castles: The scenery from a digital map with elevation data. *Chizu Chūshin.* 453. 7-15.

Doneus, M. and Briese, C. 2006. Digital terrain modelling for archaeological interpretation within forested areas using full-waveform laserscanning. In M. Ioannides, D. Arnold, F. Niccolucci and K. Mania (eds.), *The 7th International Symposium on Virtual Reality, Archaeology and Cultural Heritage VAST* (2006), 155-162. Goslar.

Fujii, N., Saito, K. and Chiba, T. 2015. High-Resolution Aerial Survey: Evolution of Measurement and Display Techniques, from LiDAR to Red Relief Image Map. *Archaeology and Natural Science.* 69. 77-94.

Matsusaki, H. 2018. Surveys on distribution in the Nakadake Kiln Site Cluster, *Kyūshū Kōkogaku.* 93. 133-142.

Nakamura, N. and Shinoto, M. (eds.) 2015. *Studies on the Nakadake Kiln Site Cluster*. Research Center for Archaeology, Kagoshima University (http://hdl.handle.net/10232/23138).

Interpreting GPR data from Jaffna Fort, Northern Sri Lanka, using historic maps and new excavations

Armin Schmidt[1,2], Prishanta Gunawardhana[3], Christopher Davis[1], Duncan Hale[4], Robin Coningham[1], P. Pushparatnam[3,5], Richie Villis[4], Mark Woolston-Houshold[4], Mark Manuel[1]

[1]Department of Archaeology, Durham University, Durham, UK, [2]Geodatawiz, Remagen, Germany, [3]Central Cultural Fund, Government of Sri Lanka, Colombo, Sri Lanka, [4]Archaeological Services, Durham University, Durham, UK, [5]Department of History, University of Jaffna, Jaffna, Sri Lanka

The first recorded fort at Jaffna, in northern Sri Lanka, was established by the Portuguese in the early 17th century, to protect trade and the surrounding settlement. Quadrangular in design, it was captured by the Dutch in 1658, remodelled according to the latest pentagonal design and completed in 1792. Three years later the Dutch garrison surrendered to British forces, who then held it until the end of colonial governance in 1948. One of the best-preserved colonial forts in Asia, the site was badly damaged during hostilities between Sri Lankan government forces and the Liberation Tigers of Tamil Eelam.

Fortifications and internal buildings were changed, destroyed and newly erected in all these phases and the monument is now a palimpsest of different features above and below ground. Given its prominence for the northern part of Sri Lanka, a project was established to investigate the fort's history through subsurface investigations and to make the results available for the local population and tourists.

Detailed GPR surveys of the interior (500MHz, 0.25m line spacing) were undertaken in 2017 and, after removal of several very large stone piles, in 2018, lead to a detailed dataset of the area (Fig. 1). However, interpretation of the data was difficult due to a weak contrast between built structures and surrounding subsurface strata, and the juxtaposition of seemingly unrelated features. It was therefore necessary to complement the GPR data with historic maps of the fort's different building phases, place excavation trenches in particularly relevant areas and create a detailed UAV orthophoto (0.025m ground resolution) of the whole area. This allowed for attribution of the major GPR anomalies to structures from different historical phases.

Of considerable interest is a Portuguese church, shown in the centre of the fort on several seventeenth century maps, with a courtyard garden or cloister. To establish its location it was necessary to relate these maps of the quadrangular Portuguese fort to the pentagonal Dutch design. Damage to Dutch-era ramparts revealed that some of the original semi-circular Portuguese bastions were incorporated within them and good correlation of the maps was possible. In the approximate location of the thus mapped church a set of rectilinear GPR anomalies (*c.* 1.3m wide, over an estimated depth range from 0.7m-1.0m) is visible (Fig. 2), delineating several structures around a large open space (*c.* 28m square).

To characterise this structure, an excavation trench was positioned over its north-eastern corner. The removal of turf and then topsoil revealed the old parade ground surface. This overlaid a brown soft sandy material, which was cut by a linear north-east to south-west feature, *c.* 2m wide, filled with concentrations of broken coral pieces, brickbats and tile fragments. This material then sat on large irregular limestone blocks, which were encountered at a depth of 0.80m from the top of the trench (Fig. 3). Further excavation revealed that these limestone blocks were laid directly on top of the natural limestone bedrock and measured *c.* 0.20m-0.25m in thickness, with the feature measuring *c.* 1.20m in width at its base.

It is hypothesised that during the Dutch occupation of Jaffna Fort, existing structures were robbed out of ready-cut coral and limestone blocks for reuse in the re-modelling of the fort. Whilst the footing of this structure survived, the upper stone blocks were removed and replaced with rubble to level voids. This explains the low geophysical contrast of the feature. The location of the footings and robber cuts in relation to historical maps, as well as identified features cutting through earlier cultural phases containing pre-colonial contact material, suggest that the structure was a Portuguese-era church.

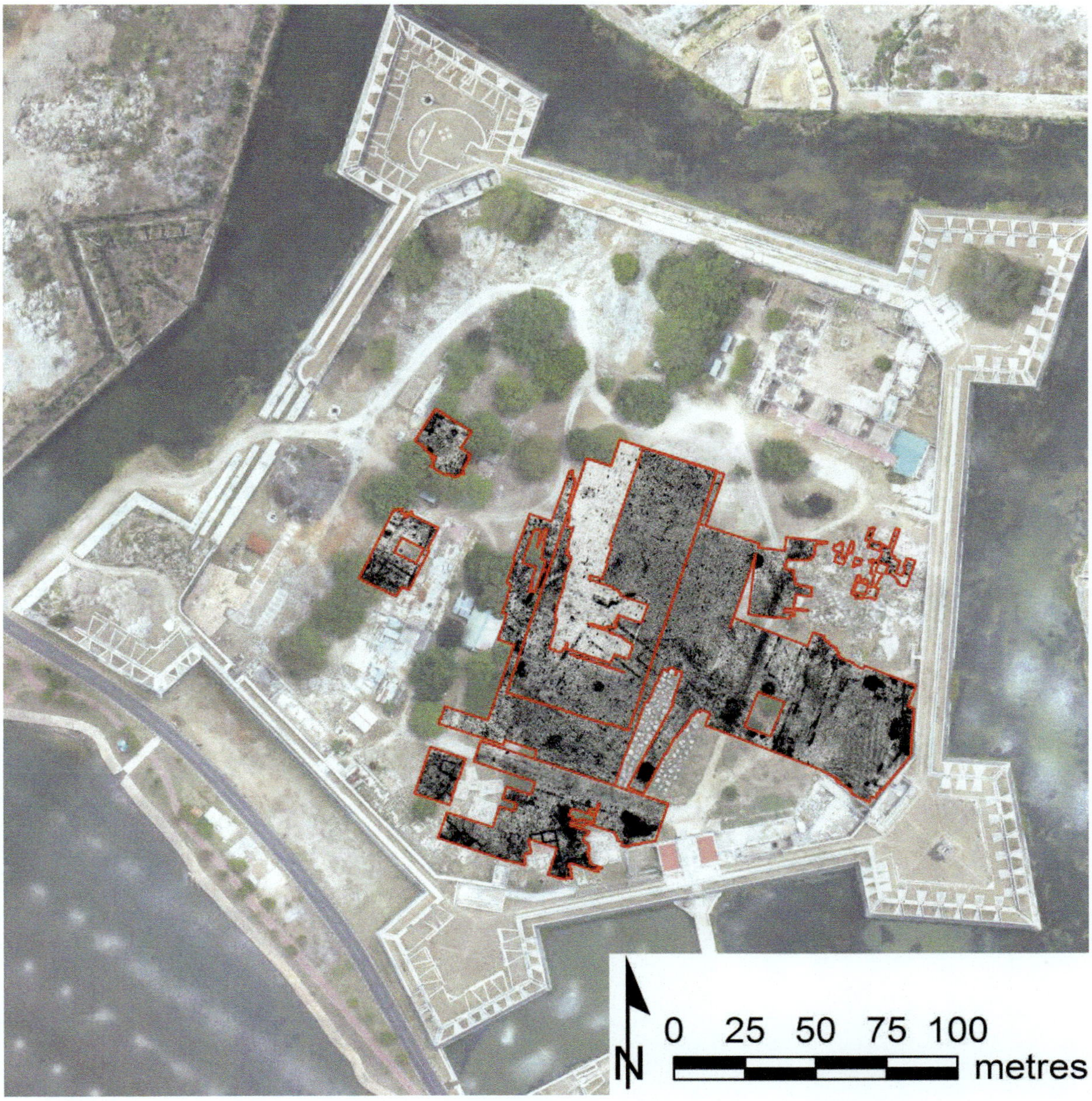

Fig. 1. UAV orthophoto of Jaffna Fort with GPR data outlined in red.

Further features are apparent in the GPR data to the north of this structure, including parts of an octagonal anomaly that can be seen in an early Dutch map which shows the whole of such a building, 20m wide. However, it is depicted in the same way as a south-eastern block of barracks that was probably never built and it was therefore assumed that the octagonal feature had also not been constructed. This has now been refuted by the GPR data, but since no internal structures could be detected, the building's function remains unclear. This Dutch map represents the current ramparts and adjoining installations very accurately and it can be assumed that its depiction of barracks and buildings was equally precise. However, GPR data, current surface markings mapped in the orthophotos and excavation results show a different layout of walls and rooms for these buildings. It is hence possible that the now recorded remains reflect a remodelling of the barracks by the British military.

The large Dutch church (Kruys Kerk) constructed as a Greek cross in 1706 in the north-eastern part of the fort was destroyed during recent hostilities within Jaffna. Only parts of the foundations are now visible on the surface and a GPR survey was undertaken between the scattered building remains. It showed several localised rectangular anomalies that are interpreted as burials below the church's floor. Despite the destruction of the church's superstructure it appears to still preserve important remains. To clear the site of the destroyed church for the subsequent GPR measurements a post-disaster excavation methodology was applied that had initially been developed in Nepal in the wake of the 2015 Gorkha earthquake. It proved effective in shifting large amounts of rubble onto preserved and *in situ* floor surfaces, while simultaneously examining and recording the area carefully. A detailed excavation of the church's northern wall provided insights into the methods and materials that were used for its construction.

Fig. 2. GPR timeslice of central area (estimated depth 0.8m) with major anomalies outlined in red. Southern anomalies relate to the Portuguese church, northern anomalies to the octagonal Dutch building.

None of the available maps show structures in the south-eastern part of the fort, possibly because it is the location of the original Portuguese moat and hence of poorer ground conditions. It was therefore surprising to find a very large rectangular structure (14m x 33m) in the GPR data of this area. It is delineated by long and very narrow (0.2m) anomalies of at least 1.7m depth that could be made of metal. The fill is only about 1.1m deep and appears to have a ramp on the southern side. It is possible that this structure was built during recent hostilities in Jaffna, potentially as a large storage pit.

Fig. 3. Excavation of parts of the Portuguese church showing the limestone footings that created the GPR anomalies.

Overall, the multi-method investigations of subsurface remains were able to identify several features from different phases of the fort's history. Through the combination of above-ground mapping from UAV photogrammetry, keyhole excavations and map regression, complementary data were collected that allowed the interpretation of weak GPR signals from complex and fragmented structures.

Acknowledgements

This project was supported by HEFCE-GCRF; the British Academy; the Institute of Medieval and Early Modern Studies, Durham University; the Central Cultural Fund, Government of Sri Lanka; the Department of Archaeology, Government of Sri Lanka; the University of Jaffna and the Postgraduate Institute of Archaeological Research, University of Kelaniya.

Common interpretation of induced polarization tomography (IPT) results with other geophysical methods in an archaeological site

Meriç Aziz Berge[1], Mahmut Göktuğ Drahor[1,2], Caner Öztürk[2], Atilla Ongar[1]
[1]*Dokuz Eylül University, İzmir, Turkey,* [2]*GEOIM Engineering LTD, İzmir, Turkey*

Introduction

In the last decade, applications of Electrical Resistivity Tomography (ERT) are frequently seen, while Induced Polarization (IP) is not preferred due to its comparatively low data acquisition speed. However, Induced Polarization Tomography (IPT) might produce good results especially in mud-brick based structures. The clay content of the mud-brick material is directly responsible for the IP response. On the other hand, we expect that the IPT method should respond to the buried conductive bodies such as slag heaps, ditches and trenches including anthropogenic materials. Firstly, Aspinall and Lynam (1968; 1970) presented examples of the IP method in archaeological applications, and after a long time De Domenico *et al.* (2006), Florsch *et al.* (2012) and Leucci *et al.* (2014) used this method in their studies. By using this method together with ERT, we might expect good results related to buried archaeological features. However, the origin of IP anomalies in relation to the archaeological structures is quite unclear. In this study, the results of ERT, GPR and magnetic gradiometry obtained from an archaeological site were presented as a common interpretation with IPT results. Thus, we tried to clarify the cases that form the IP anomaly in the archaeological environment.

Methods

The measuring area is located in one of the ancient cities of Hittite Empire, Šapinuwa (Çorum, Turkey). The test site (10m x 30m) was selected according to magnetic gradiometry results of the area, and ERT-IPT measurements were performed on 11 N-S oriented profiles using 1m electrode and traverse intervals. Time-domain IP data were collected using an AGI-SuperSting resistivity and IP system, which enables the measurement of both resistivity and chargeability data. Thus, we tested again the capability of the IPT method after the first results given by Berge *et al.* (2017) at the same archaeological site. This time, we preferred the dipole-dipole arrangement instead of Wenner-Schlumberger array in the data acquisition. In order to improve the contact resistance and data quality related to dry top soil, we watered the electrodes for a good current injection. During the data processing of the resistivity and chargeability data, bad datum points were removed from the datasets, and we used the robust inversion routine of Res2Dinv and Res3Dinv software. In addition, previously collected magnetic gradiometry and GPR data were reprocessed for this area.

Results

In order to compare the results of different geophysical methods in the study area, magnetic gradiometry, depth slices of ERT, IPT and GPR data are given (Fig. 1). As there is no archaeological excavation in this particular part of the area, we compared the overall geophysical results to obtain a reliable interpretation. In the magnetic image, there are high magnetic values in the middle of the area. According to the previous excavation in the Šapinuwa archaeological site, we think that this kind of high magnetization should be characterized by burned structures. Besides this magnetic variation in the gradiometry image, especially white (negative) coloured anomalies have a regular orientation that might represent buried architecture of the area (Fig. 1a). According to the ERT result of the depth slice between 0m and 0.6m, the high resistivity values correspond to the anomalies determined in the magnetic image (Fig. 1b). Thus, we defined these structures as buried stone walls at the site. Note that the ABS error values did not exceed 2.8% after six iterations in the inversion of ERT data. The GPR results from the same area show a complex of features on the depth slices between 0.5m and 0.9m. A direct relationship between the previous two methods and the GPR was not achieved in the context of subsurface archaeology (Fig. 1c). According to previous excavations and GPR results obtained from the other parts of the archaeological site, we decided that this case might be due to the undulated geometry of the buried structures, which were affected by ancient earthquake(s) in the region. High chargeability values in the IPT depth slice (Fig. 1d) coincide with the resistive structures defined in the ERT result. These particular chargeability values were generated between the wall structures.

However, we could not determine a regular orientation of the structures in the IPT slice as it is defined in magnetic and ERT methods.

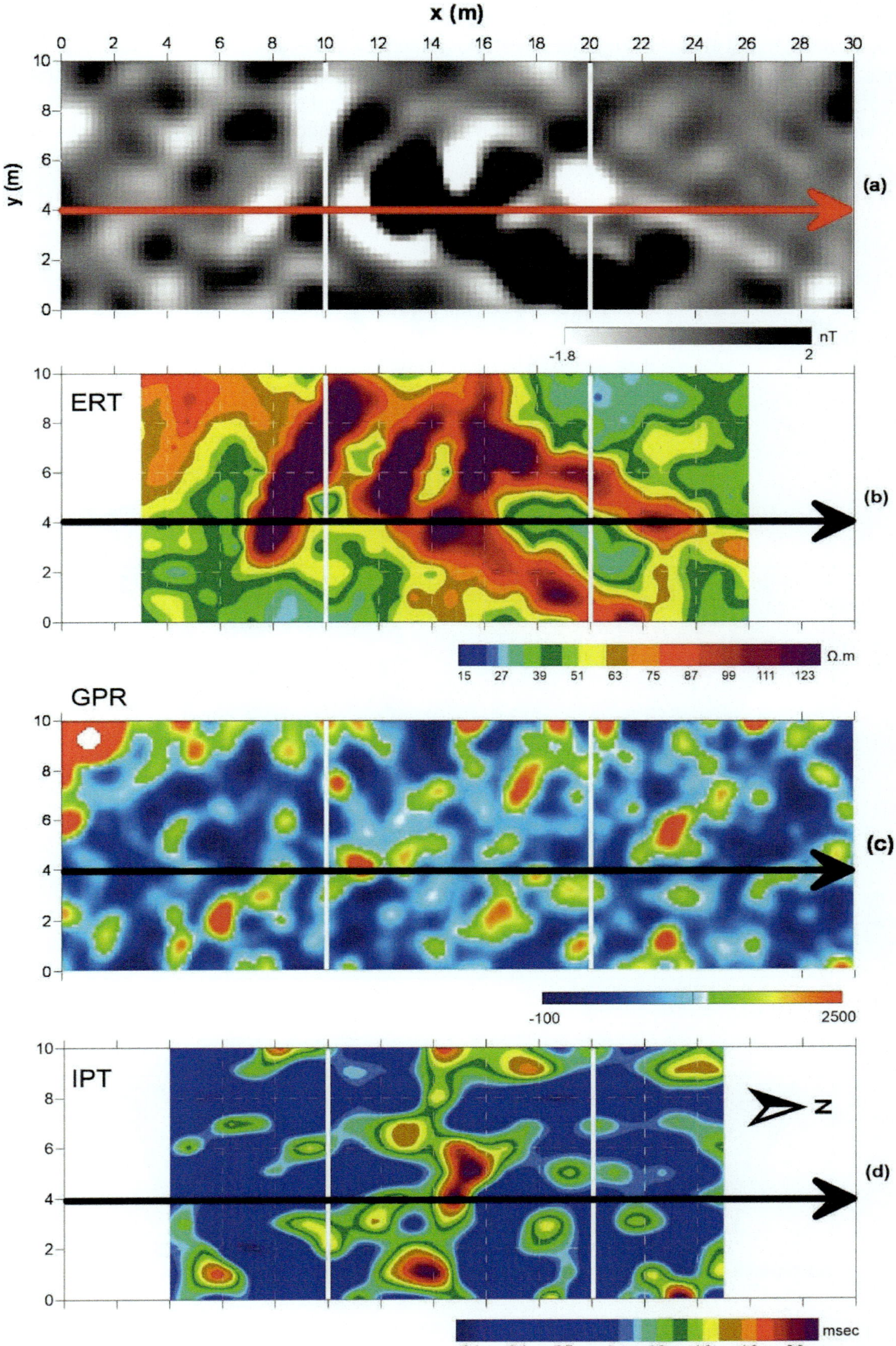

Fig. 1. Magnetic gradiometry image (a) and depth slices of ERT (b), GPR (c) and IPT (d) obtained from the test site. The arrows on these images indicate the example profile given in Fig. 2.

In order to correlate these geophysical methods, we also present the results of an example profile (Fig. 2). The ERT section and radargram of the GPR particularly demonstrate that the buried structures no longer exist beyond a depth of 1.9m (the deepest part is seen along the 13th metre of the profile). High resistive structures seen on the ERT section directly correspond to the magnetic anomalies in this particular profile (shown by arrows). It is also interesting that the IP chargeability values increased where the magnetic values reach their maximum values in this profile. We think that this case might be due to the burned structures at this part of the measuring site. However, to verify this hypothesis and the geophysical results obtained from this site, we require test excavation results, which will be performed by archaeologists in the 2019 field season.

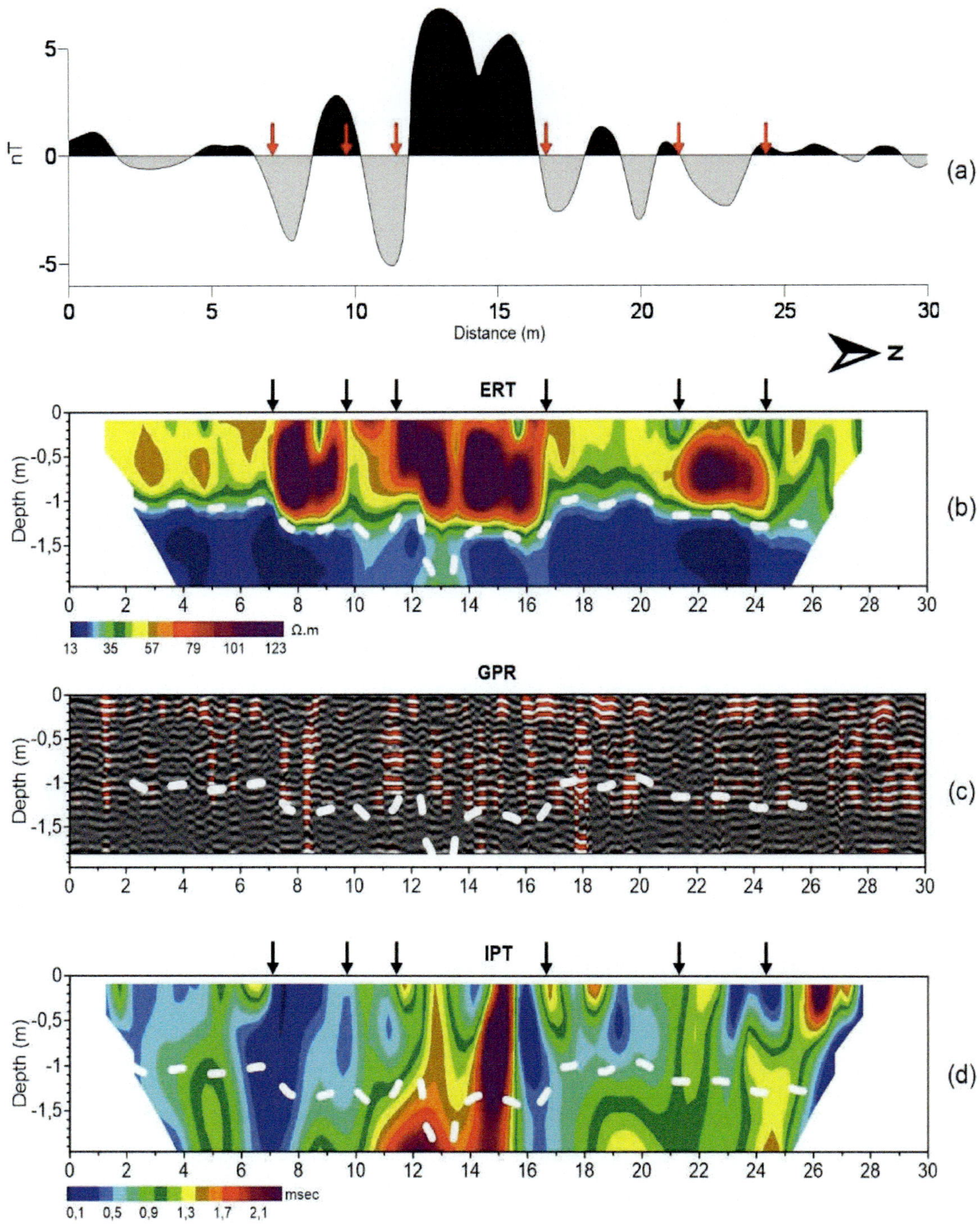

Fig. 2. Geophysical results obtained from the example profile shown in Fig. 1, (a) magnetic gradiometry, (b) ERT, (c) GPR and (d) IPT. Arrows indicate the boundaries of particular magnetic anomalies and resistive structures. White dashed lines show the transition between archaeological context and bedrock according to the ERT result.

References

Aspinall, A. and Lynam, J.T. 1968. Induced polarization as a technique for archaeological surveying. *Prospezioni Archeologiche.* 3. 91–93.

Aspinall, A. and Lynam, J.T. 1970. An induced polarization instrument for the detection of near-surface features. *Prospezioni Archeologiche.* 5. 67–75.

De Domenico, D., Giannino, F., Leucci, G. and Bottari, C. 2006. Integrated geophysical surveys at the archaeological site of Tindari (Sicily, Italy). *Journal of Archaeological Science*. 33. 961–970.

Drahor, M.G., Berge, M.A., Öztürk, C. and Ortan, B. 2015. Integrated geophysical investigations at a sacred Hittite Area in Central Anatolia, Turkey. *Near Surface Geophysics.* 13. 523–543.

Florsch, N., Llubes, M. and Téreygeol, F. 2012. Induced polarization 3D tomography of an archaeological direct reduction slag heap. *Near Surface Geophysics.* 10. 567–574.

Leucci, G., De Giorgi, L. and Scardozzi, G. 2014. Geophysical prospecting and remote sensing for the study of the San Rossore area in Pisa (Tuscany, Italy). *Journal of Archaeological Science.* 52. 256–276.

Comparison of excavation results with GPR and magnetic gradiometer surveys at a workshop area in Šapinuwa, Central Anatolia/Turkey

Mahmut Göktuğ Drahor[1], Caner Öztürk[2], Meriç Aziz Berge[1], Aygül Süel[3], Atilla Ongar[1], Sedef Ayyıldız[4], Semih Gerçek[5], Murat Aydın[6]

[1]Dokuz Eylül University, İzmir, Turkey, [2]GEOIM Engineering LTD, İzmir, Turkey, [3]Hitit University, Çorum, Turkey, [4]Bitlis Eren University, Bitlis, Turkey, [5]Gazi University, Ankara, Turkey, [6]Ankara University, Ankara, Turkey

Introduction

GPR and magnetic gradiometry are two of the most commonly used geophysical techniques for assessing archaeological sites. These two techniques, which generally give successful results in many areas, can also give negative results when there are unusual physical and chemical changes in the soil. While there are studies on the physical and chemical changes occurring in the soil affecting geophysical data, there is not a study investigating the changes due to active tectonic phenomena in the soil. Especially in the regions where strong active tectonic events are observed, significant soil deformations occur in the archaeological areas exposed to large earthquakes. Due to these deformations, the significantly mixed archaeological context could importantly affect the geophysical studies and possibly lead to negative results. For this reason, geophysical interpretations in such areas should include these effects.

Šapinuwa was one of the kingdom cities in the Hittite Empire. The archaeological site of Šapinuwa is located in the Çorum region of Central Anatolia, Turkey. The use of integrated geophysical surveys have continued since 2012 on this site (Drahor *et al.* 2015). In these studies, methods such as magnetic gradiometry, GPR, ERT, IPT, SRT and MASWT have been used together and architectural plans of the ancient city have been created using multiple parameters (Drahor *et al.* 2017). In this study, the results obtained from a workshop area, which was determined by geophysical results and confirmed by archaeological excavations, are presented. As we know from archaeoseismology and geoarchaeology studies carried out in the area, the Šapinuwa archaeological site has been significantly affected by earthquakes. Archaeoseismological data revealed that historical earthquakes should originate from known active faults to the north of the site. In the workshop area particularly, an important in-soil deformation caused by such earthquakes, has been identified. This deformation, resulting from an earthquake was documented with photographs, and detailed GPS studies were carried out on archaeological structures deformed by earthquakes. Thus, the deformations in the structures are visualized in three dimensions. When these results were compared with the geophysical results, it was found that the changes due to tectonic effects significantly affect the geophysical results. This study discusses the results obtained.

Data Acquisition and Processing

In this study, the geophysical data corresponding to the workshop area obtained from the large-scale magnetic gradiometry and GPR data of Šapinuwa was used. A GSSI-SIR3000 system equipped with a 400MHz shielded antenna was used during data collection of the GPR method while the Geoscan FM256 Fluxgate gradiometer was used to collect the magnetic data. The measurement profiles were aligned N-S at 0.5m intervals. Standard processing steps were applied to obtain interpretable geophysical results. The archaeological structures found after the excavations were measured using a sensitive GNSS system to reveal the exact location, depth and deformations of the recovered structures.

Results and Discussion

The magnetic gradiometer and GPR depth slice (0m to 2m) images of the workshop area with 20m x 40m dimension are given in Fig. 1. The magnetic gradiometer image displayed with greyscale shows the high magnetic properties as black and low magnetic properties as white colours. The successive black-and-white traces have NE-SW and NW-SE directions. These traces, which are observed intensively in a certain part of the area, are in the same direction as the general building extensions of the structures that were excavated in the area. Especially high (black) magnetic anomalies show that the structures in the area were affected by significant fire. The dark greyscale traces are also in a similar distribution, indicating that there is a building complex close to the surface. Negative traces indicate the walls, made of limestone and similar rocks, which have low magnetic properties (Fig. 1a). This causes a very low magnetic contrast between the limestone walls and their fill, which would make them magnetically disappear within the background noise. In order to compare magnetic gradiometer and GPR results, a broad GPR depth slice occured. The high reflection amplitudes seen in the GPR depth slices of the workshop area occur intensively in a certain part of the area. Although the high reflections (shown in red) do not give a descriptive image such as the magnetic data, it is interesting to note that the presence of a NW-SE and NE-SW trace distribution occur. The highly reflective traces seen especially in the south-eastern quarter suggest that this area may have been stone-filled. Reflections in the western part of the area are dramatically decreasing. The anomaly decreasing is also valid for the magnetic image. It is interesting, however, that there is not a clear reflection in the areas where there are high positive-negative traces in the magnetic field, and this indicates that the structures found in these areas may have been significantly damaged.

The remarkable difference seen in the east and west, especially starting from the middle of the area, reveals a significant change (Fig. 1b). An archaeological excavation was carried out between 2015-2018 in this area where the structure was determined from the geophysical studies, especially the magnetic results. As a result of the excavation, the building complex (Fig. 1c) has emerged. The excavated structure has an important deformation connected with the earthquake (Fig. 2). The sinusoidal displacement forming in the walls of the building, slope, collapse, shear and other vertical and horizontal displacements of the structures demonstrate that the subsurface has been damaged as a result of an earthquake. This result suggests that the geophysical studies on the structures that have undergone such important underground deformations can be negatively affected. While the structures are closer to the surface in the south, they are buried deeper in the north. It is believed that the complexity seen in the magnetic image is caused by this subsurface phenomenon. The radargrams and magnetic graphs obtained from the three lines shown in Fig. 1, are given in Fig. 3. In particular, the magnetic anomaly with a +40 and -10nT amplitude on line 39 indicates the presence of an important burning event. In this section, the amplitude of radar data has a significant loss. The general structure base is evident in the radar traces at 1.1m. The ringing characters seen in the traces that extend towards greater depth probably depend on the environmental characteristics. It is interpreted that the slopes seen in the traces near the surface may be related to the vertical displacements resulting from the earthquake. Extremely complex radargrams reveal the complicated character of the subsurface (Fig. 3).

Conclusions

This study revealed that the subsurface deformations caused by large earthquakes has complicated the archaeological environment, and therefore the interpretation of the geophysical studies in such areas was difficult. While magnetic images are not very clear, they reveal meaningful traces. However, the interpretation of GPR data in such environments is extremely difficult. Therefore, the reprocessing of GPR data according to the excavation results will provide an important contribution to the interpretation. In addition, the measurement of the excavated structures using GPS and the creation of 3D distributions of the structures will further enhance the interpretation of GPR data.

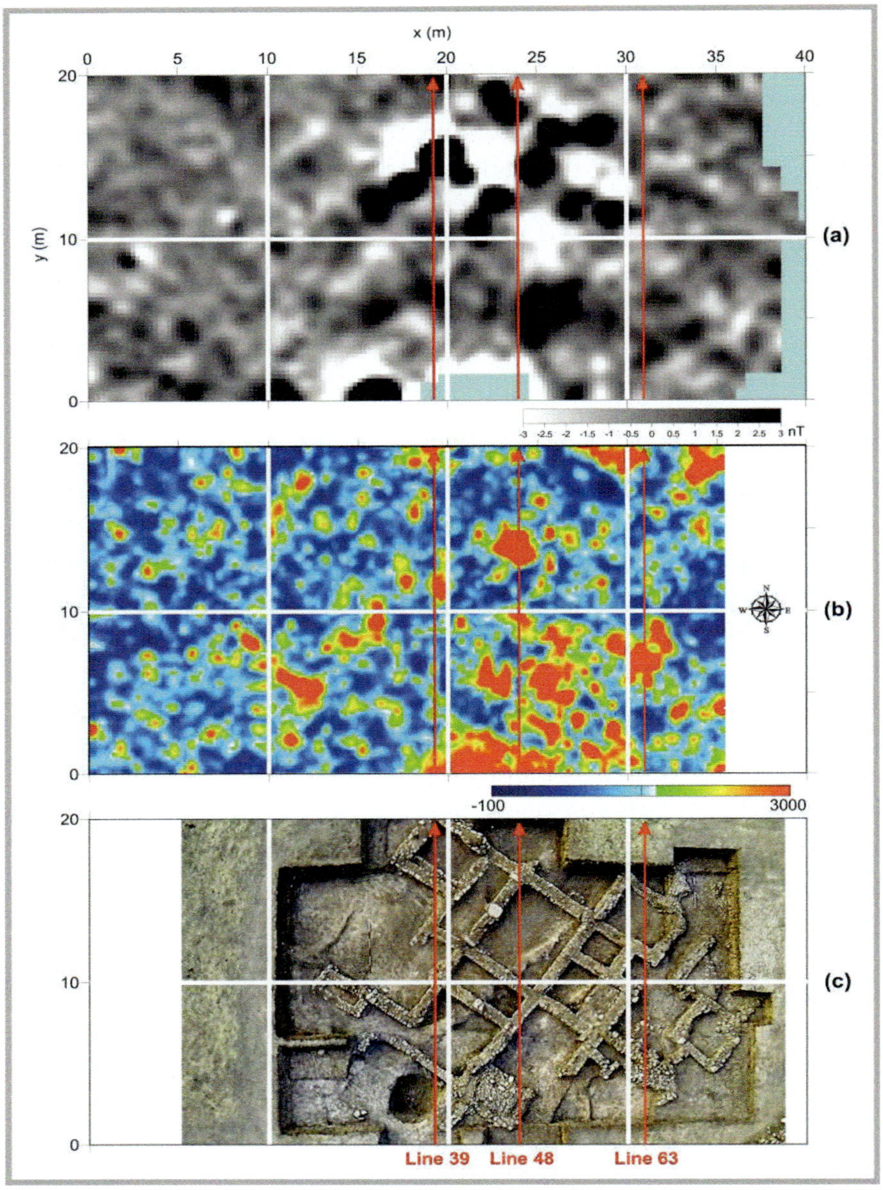

Fig. 1. (a) Magnetic gradiometry image, (b) GPR depth slice of 0m-2m and (c) aerial photo of excavated part of the workshop area.

Fig. 2. The photograph shows the result of the excavation at the workshop site.

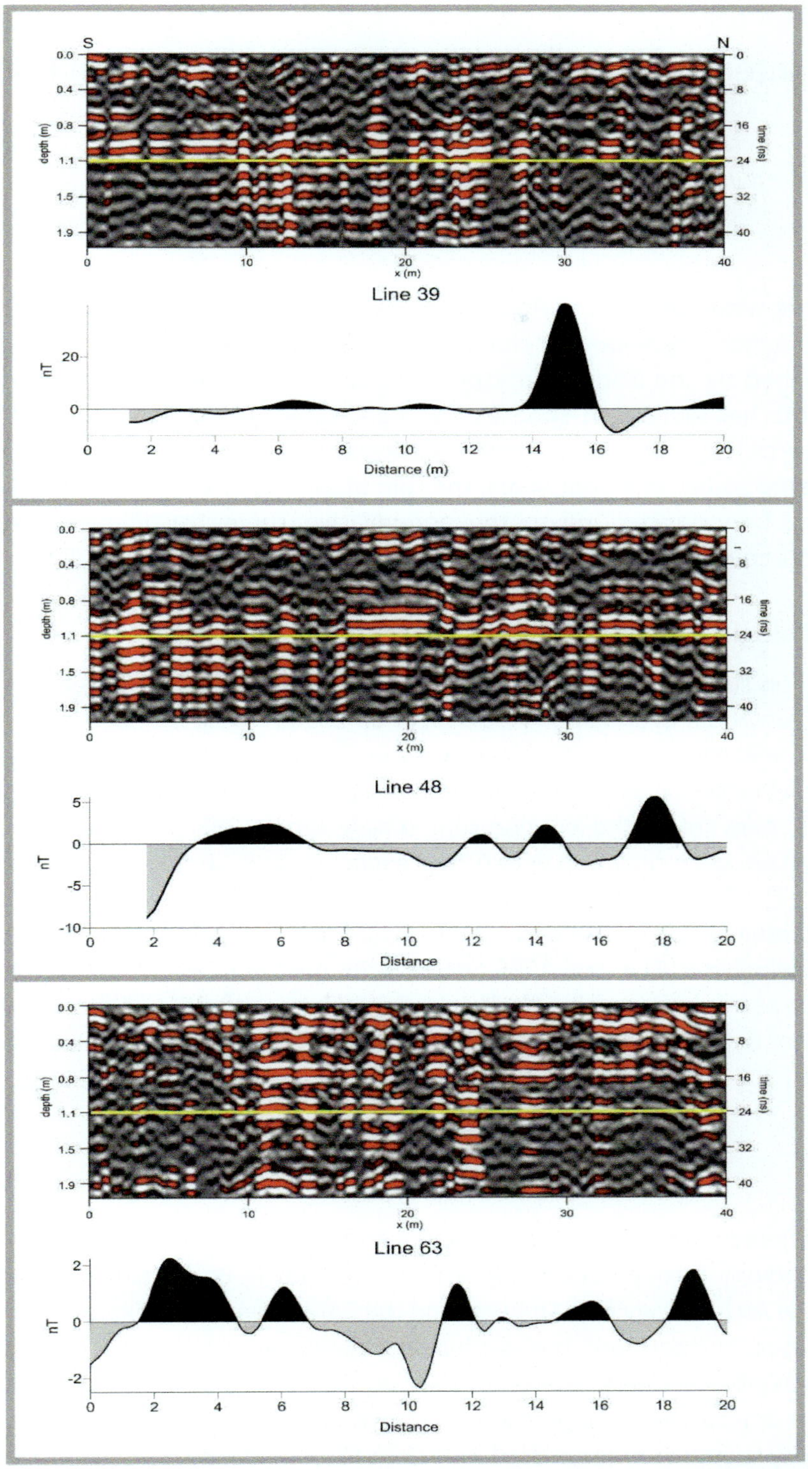

Fig. 3. Radargrams and magnetic gradiometry data of three selected lines. Yellow lines on radargrams show the general structure base

References

Drahor, M.G., Berge, M.A., Öztürk, C. and Ortan, B. 2015. Integrated geophysical investigations at a sacred Hittite area in Central Anatolia, Turkey. *Near Surface Geophysics.* 13. 523–543.

Drahor, M.G., Berge, M.A., Öztürk, C., Ortan, B., Ongar, A., Süel, A., Ayyıldız, A., Avsever, Ö.Ş. and İçke, F. 2017. Integrated geophysical prospection in a Hittite Empire city (Šapinuwa). In Jennings, B., Gaffney, C., Sparrow, T. and Gaffney, S. (eds.) *AP2017. 12th International Conference of Archaeological Prospection.* Archaeopress. Oxford. 73-76.

Archaeological interpretation of the prospection data from Ephesos

Jasmin Scheifinger
Austrian Archaeological Institute of the Austrian Academy of Sciences, Vienna, Austria

Introduction

Ephesos, located at the west coast of the Aegean Sea, is one of the most important archaeological sites in Turkey. The ancient city has a very long research history. The first archaeological excavations took place in the 1860s, commissioned by the British Museum and led by John Turtle Wood. Since 1895 the Austrian Archaeological Institute has worked in Ephesos (Wohlers-Scharf 1995, Wiplinger and Wlach 1995). Even though the main research focus has always been on excavations, only approximately 15% of the entire urban area has been excavated so far. In recent years, the role of geophysical archaeological prospection surveys has gained increasing importance. Due to the non-invasive survey techniques, information of buried archaeological features could be provided without exposing them.

Geophysical archaeological prospection in Ephesos

Geophysical prospection surveys have been conducted in Ephesos since 1995 (Seren, Neubauer and Eder-Hinterleitner 2001, Groh 2003, Groh 2006: 49, Seren *et al.* 2015). These included ground-penetrating radar (GPR), magnetometry, electrical resistivity, electromagnetic induction measurements, and seismology. The results of the GPR surveys have been particularly significant; thus this method and the results of the magnetic prospection survey are relevant for the work presented here. From 1995 until 2016 in total 175ha have been surveyed, while 65ha have been recorded with GPR prospection and 110ha by magnetometry.

The first survey was conducted by Geofyzika a.s. and ARGIS Archäologie und Geodaten Service in 1995, which continued in the following year. The year 1996 also saw the Austrian Central Institute for Meteorology and Geodynamics (ZAMG) perform research with a few exceptions until 2016. Other institutions involved in surveying Ephesos were S. Groh from the Austrian Archaeological Institute, who has carried out a magnetic prospection survey in 2006 and the University of Southampton led by S. Keay (ERC project 'Portus Limen') in 2015 and 2016.

Methods

All data from both GPR and magnetometry prospection methods, were re-processed according to the latest state of the art technology through cooperation with the Ludwig-Boltzmann Institute of Archaeological Prospection and Virtual Archaeology (LBI ArchPro) and the ZAMG. New data analysis and filtering algorithms were developed and used. Furthermore, the existing data classification was modified and complemented. By these means, an archaeological interpretation of the data was conducted. The evaluation of the extensive geophysical data represents a central aspect in the topographical analysis of Ephesos. Lastly, interpretative investigations of the city layout were conducted by S. Groh (Groh 2006: 47-116), whose work thus constitutes a very important basis for the research presented here. The gathered data will be recorded in a catalogue and also incorporated in a geographical information system (GIS). The application of GIS enables the structured capturing, analysis and presentation of geographic data. For additional remarks, excavation documentation, excavation diaries, aerial photographs as well as drone photographs, copperplate engravings, historical maps, and the results of research and secondary literature will be integrated.

Investigation of the harbour area

Although the entire prospection material shall be investigated, some areas will be analysed in greater detail. One of these areas is the harbour and the harbour channel (Fig. 1), which is largely unexcavated and has been the lifeline of the ancient city (Zabehlicky 1995, Steskal 2014, Ladstätter 2016). The first stratigraphically conducted excavations in the harbour area were carried out between 1987 and 1989 (Zabehlicky 1999). During these excavations, some late antique-early byzantine structures, a part of a quay wall and a jetty, were found. Further excavations were performed from 2008 until 2010 (Österreichisches Archäologisches

Institut 2008: 18; 2009: 14; 2010: 28-32) in the north and south of the harbour channel, which is regarded as the harbour necropolis. During these excavations, burial houses and graves cut into the local bedrock, were discovered (Steskal 2013: 245, 253; 2017: 177-179). Different surface surveys were conducted and brought to light on the one hand, further burial houses and graves (Österreichisches Archäologisches Institut 2010: 31; 2011: 12-14; 2012: 16-19, Steskal 2017: 179), while on the other hand, findings such as ceramics, fragments of wall coverings, glass, metal, wall plastering fragments, mosaics, bricks, and animal bones/shells were collected (Österreichisches Archäologisches Institut 2013: 26-28, Schörner 2016: 336). Thus, the results of both excavations and surveys can be included in this research; therefore questions may be asked about the extent of the harbour area and the layout of the cemetery. Additionally, questions about the different kinds of structures and their function can be asked.

Fig. 1. Overview of the Harbour area from east (© image: ÖAW-ÖAI/ Niki Gail).

Comparison between excavation and geophysical results

Another significant part of the research study is the comparison between the excavation- and the geophysical results. For this, the excavation documentation of the project 'late antique-medieval city quarter to the south of the church of Mary' (Schwaiger 2012: 192, 197-200; 2016: 87, Österreichisches Archäologisches Institut 2011: 23-25; 2012: 23-27; 2013: 21-24; 2014: 14-18; 2015: 5-11; 2016: 6-16) will be analysed. Based on the geophysical prospection survey (Fig. 2), about 2000m² area was excavated from 2011 until 2018 (Fig. 3).

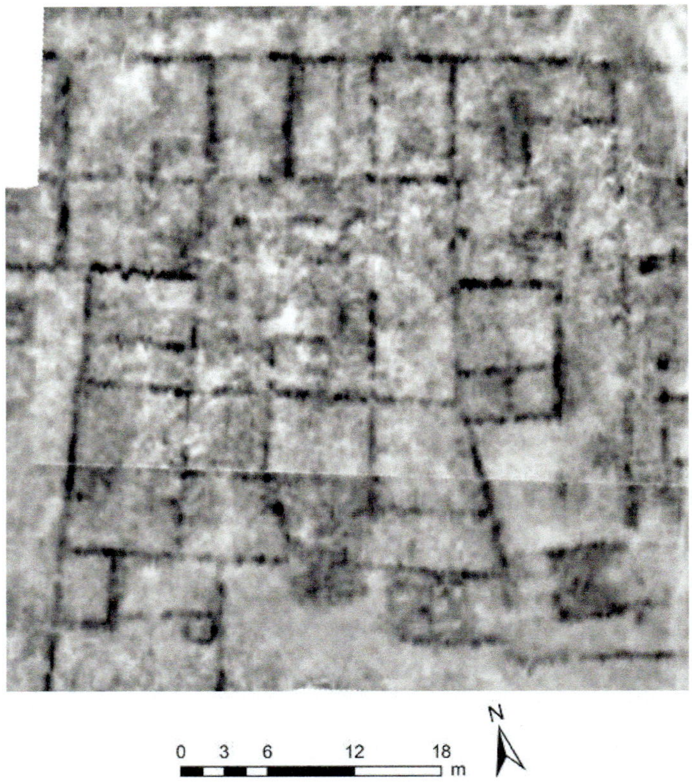

Fig. 2. Detail of the lower city quarter in Ephesos, GPR sum image 30cm-220cm; data collection and processing ArcheoProspections®, ZAMG A. Hinterleitner (© plan: ÖAW-ÖAI/Jasmin Scheifinger).

Fig. 3. Overview of the excavated city quarter (2011-2018) south of the church of Mary (© image: ÖAW-ÖAI/ Niki Gail)

During the entire project the excavations were carried out stratigraphically, and every single layer was measured by TachyCAD in combination with AutoCAD. For this reason it is possible to compare the GPR depth-slices with all the stratigraphic units. In the prospection data you can see anomalies starting at 60cm below the ground surface that could be interpreted as debris. These debris layers were also visible during the excavation. In addition, there are also reflections of darker areas deeper in the buildings, which probably provide hints about floors in the rooms. Throughout the excavation, floors made of *opus-sectile*, bricks, mosaic and stone plates have been found. The question is, whether every floor reflects the GPR wave in the same way or whether there exist differences to be discovered? How can the results of the geophysical survey be compared to those of the excavation? Where are the congruities, where are the differences?

Conclusion

Although a large amount of data is available, the archaeological interpretation of the data from Ephesos is at a very early stage. Thus, an abundance of survey results is often confronted with only rudimentary archaeological interpretations. The purpose of this research is, on the one hand, to record the city and its environs as completely as possible. On the other hand, the benefits of a culture-historical investigation will be carried out, allowing the geophysical survey results undertaken at Ephesos to be compared with the excavated material from a topographical as well as an architectonical perspective.

Acknowledgements

This research project is carried out as a doctoral thesis at the University of Vienna, funded by the Austrian Academy of Sciences within the DOC-Team scholarship, for which the author is very grateful. The author would like to thank the supervisors Sabine Ladstätter (Austrian Archaeological Institute of the Austrian Academy of Sciences) and Immo Trinks (LBI ArchPro) for the possibility to work with the data and their great support. Furthermore I am grateful to Alois Hinterleitner, Klaus Löcker (both LBI ArchPro/ZAMG) and Ralf Totschnig (ZAMG) for the data processing and their fruitful dialogue. I would also like to thank Helmut Schwaiger and Martin Steskal (both Austrian Archaeological Institute of the Austrian Academy of Sciences) for their advice and suggestions.

References

Groh, S. 2003. Integrated prospection in the Upper Town of Ephesus, Turkey – a case study. *Archaeologia Polona*. 41. 185.

Groh, S. 2006. Neue Forschungen zur Stadtplanung in Ephesos. *Jahreshefte des Österreichischen Archäologischen Institutes in Wien*. 75. 47-116.

Ladstätter, S. 2016. Hafen und Stadt von Ephesos in hellenistischer Zeit. *Jahreshefte des Österreichischen Archäologischen Institutes in Wien*. 85. 233-272.

Österreichisches Archäologisches Institut. 2008-2016. *Wissenschaftlicher Jahresbericht des Österreichischen Archäologischen Instituts*. Available at https://www.oeaw.ac.at/oeai/kommunikation/jahresberichte-des-oeai/ (accessed 4 March 2019).

Schörner, G. 2016. Survey in den *suburbia* von Ephesos und Metropolis: Fragestellungen – Methoden – erste Ergebnisse. In G. Grabherr and B. Kainrath (eds.), *Akten des 15. Österreichischen Archäologentages in Innsbruck 27. Februar – 1. März 2014, Ikarus. Innsbrucker Klassisch-Archäologische Universitätsschriften* 9, 331-346. Innsbruck.

Schwaiger, H. 2012. Comparing houses. Domestic Architecture in Ephesus from the Mid Imperial Period to Late Antiquity. In A. Anguissola (ed.), *Privata Luxuria – towards an Archaeology of Intimacy: Pompeii and beyond*. International Workshop Center for Advanced Studies, Ludwig-Maximilians-Universität München, 24-25 March 2011, Münchner Studien zur Alten Welt 8, 187-209. Munich.

Schwaiger, H. 2016. Domestic Architecture in Ephesus from the Hellenistic Period to Late Antiquity. In S. R. Huebner and G. Nathan (eds.), *Mediterranean Families in Antiquity: Households, Extended Families, and Domestic Space*, 79-91. Chichester.

Seren, S., Neubauer, W. and Eder-Hinterleitner, A. 2001. GPR surveys at Ephesos (Turkey). In M. Doneus, A. Eder-Hinterleitner and W. Neubauer (eds.), *Archaeological Prospection. Fourth International Conference on Archaeological Prospection Wien*, 19-23. September 2001, Graz.

Seren, S., Totschnig, R., Hinterleitner, A., Löcker, K. and Ladstätter, S. 2015. Archaeological prospection results of the Serapeion at Ephesus, Turkey. *Archaeologia Polona.* 53. 502-505.

Steskal, M. 2013. Wandering Cemetries. Roman and Late Roman burials in the capital of the province of Asia. In O. Henry (ed.), *Le mort dans le ville. Pratiques, contextes et impacts des inhumations* intra-muros *en Anatolie, du début de l'Âge du Bronze à l'époque romaine, 2èmes rencontres d'archéologie de l'IFÉA, Istanbul,* 14-15 Novembre, 2011, 243-257. Istanbul.

Steskal, M. 2014. Ephesos and its Harbors: A City in Search of its Place. In S. Ladstätter, F. Pirson and T. Schmidts (eds.), *Häfen und Hafenstädte im östlichen Mittelmeerraum von der Antike bis in byzantinische Zeit. Neue Entdeckungen und aktuelle Forschungsansätze, Istanbul, 30.05-01.06.2011,* Byzas 19, 1. Sonderschriften des Österreichischen Archäologischen Instituts in Wien 52, 1, 325-338. Istanbul.

Steskal, M. 2017. Reflections on the mortuary landscape of Ephesus: The archaeology of death in a Roman metropolis. In J. R. Brandt, E. Hagelberg, G. Bjørnstadt and S. Ahrens (eds.), *Life and death in Asia Minor in Hellenistic, Roman and Byzantine times. Studies in Archaeology and Bioarchaeology, Studies in Funerary Archaeology* 10, 176-187. Oxford and Philadelphia.

Wiplinger, G. and Wlach, G. 1995. Ephesos. *100 Jahre österreichische Forschungen*. Vienna, Cologne and Weimar.

Wohlers-Scharf, T. 1995. *Die Forschungsgeschichte von Ephesos*. Entdeckungen, Grabungen und Persönlichkeiten, Europäische Hochschulschriften 38, Archäologie 54. Frankfurt am Main.

Zabehlicky, H. 1995. Preliminary Views of the Ephesian Harbor. In H. Köstner (ed.), *Ephesos. Metropolis of Asia. An Interdisiciplinary Approach to its Archaeology, Religion, and Culture*. Harvard Theological Studies 41, 201-215. Valley Forge.

Zabehlicky, H. 1999. Die Grabungen im Hafen von Ephesos 1987-1989. In H. Friesinger and F. Krinzinger (eds.), *100 Jahre österreichische Forschungen in Ephesos. Akten des Symposions Wien 1995, Archäologische Forschungen 1.* Österreichische Akademie der Wissenschaften, Philosophisch-Historische Klasse. Denkschriften 260, 479-484. Vienna.

Part Four – Archaeological Prospection in Australasia

Multiple processing and interpretation methods of a complex 3-D GPR dataset: An example from northern Australia

Lawrence Conyers
University of Denver, Denver, United States

Introduction

The ground-penetrating radar (GPR) method has a unique ability to record reflections of electromagnetic waves in the near-surface within 3-D volumes of ground (Conyers 2013). All GPR processing software now commonly allows for large-scale grids of data to be analysed and many recent practitioners have streamlined their processing and interpretation procedures to move rapidly from field-collected data to slice-map images. This data processing and analysis approach is extremely popular, especially with many younger geophysicists, as it is much like what I think of as the "smart phone app" approach, where icons can be chosen and the internal workings of software then produce the desired result. The reasoning behind each of the processing steps is often not necessarily understood by the users, and the internal software algorithms that generate the final product are either absent or difficult to access. This "immediate gratification" method of GPR data visualization often neglects other important analyses such as understanding individual reflection traces, reflection profile analysis, and changing sampling and gridding procedures for specific site parameters (Conyers 2012).

The site and GPR data processing and analysis

GPR data were collected on two mounds in northern Australia (Fig. 1), which were interpreted using a variety of interpretation methods in non-standard ways in order to visualize certain internal mound features (Conyers *et al.* 2019). The mounds were found to contain a number of burials, recognized by reflection hyperbolae visible in profiles, which were visible at an appropriate depth for human burials, and which could be identified within at least three parallel GPR profiles spaced 50cm apart to provide the size and orientation of human remains (Conyers *et al.* 2018). The two mounds reported on here, and many hundreds of similar mounds nearby are between 15m–25m in diameter and average 2m–3m in height, with some reaching 4m.

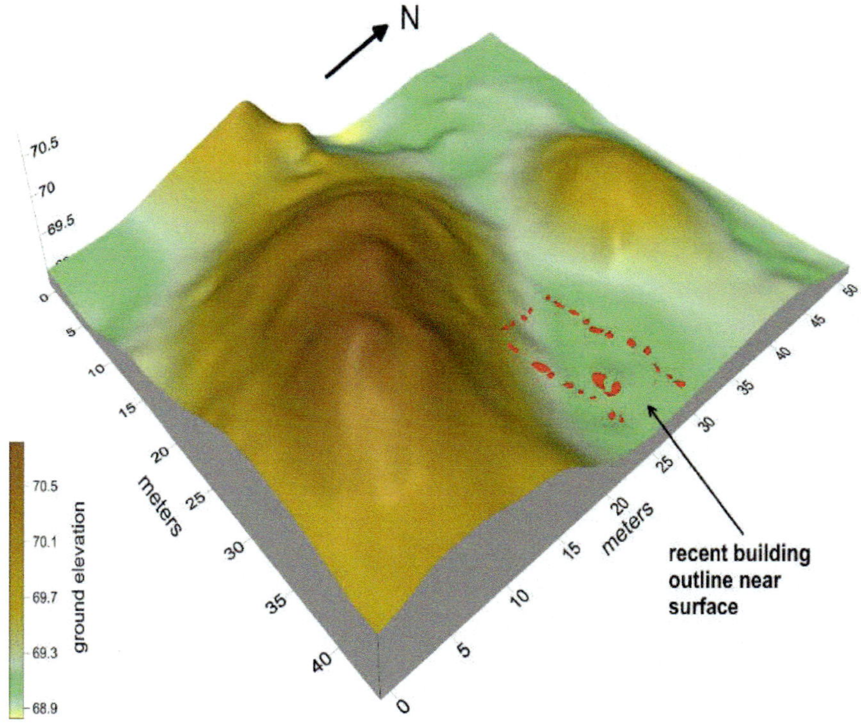

Fig. 1. Surface image of the mounds showing the location of the recent structure between the north and south mounds, which was discovered by standard amplitude slice-mapping.

The GPR data were collected with a GSSI SIR-3000 system using 400MHz antennas and a survey wheel for distance calibration. The reflection profiles were first sliced into 6ns slices (each about 44cm thick), constructed parallel to the ground surface. A first-pass reflection amplitude slice-map created parallel to the ground surface shows the outline of a rectangular structure in the top 44cm of the ground (Fig. 1). There is no surface evidence for this structure, and it is hypothesized to be a temporary enclosure used for ritual purposes.

When each of the reflection profiles were viewed and interpreted individually, a number of reflection hyperbolas are visible, which are distinct from those produced from shallow tree roots and were generated from burials as they are visible in three or more parallel profiles spaced 50cm apart and in no more than four profiles, they are the length and width of an adult human body (Fig. 2).

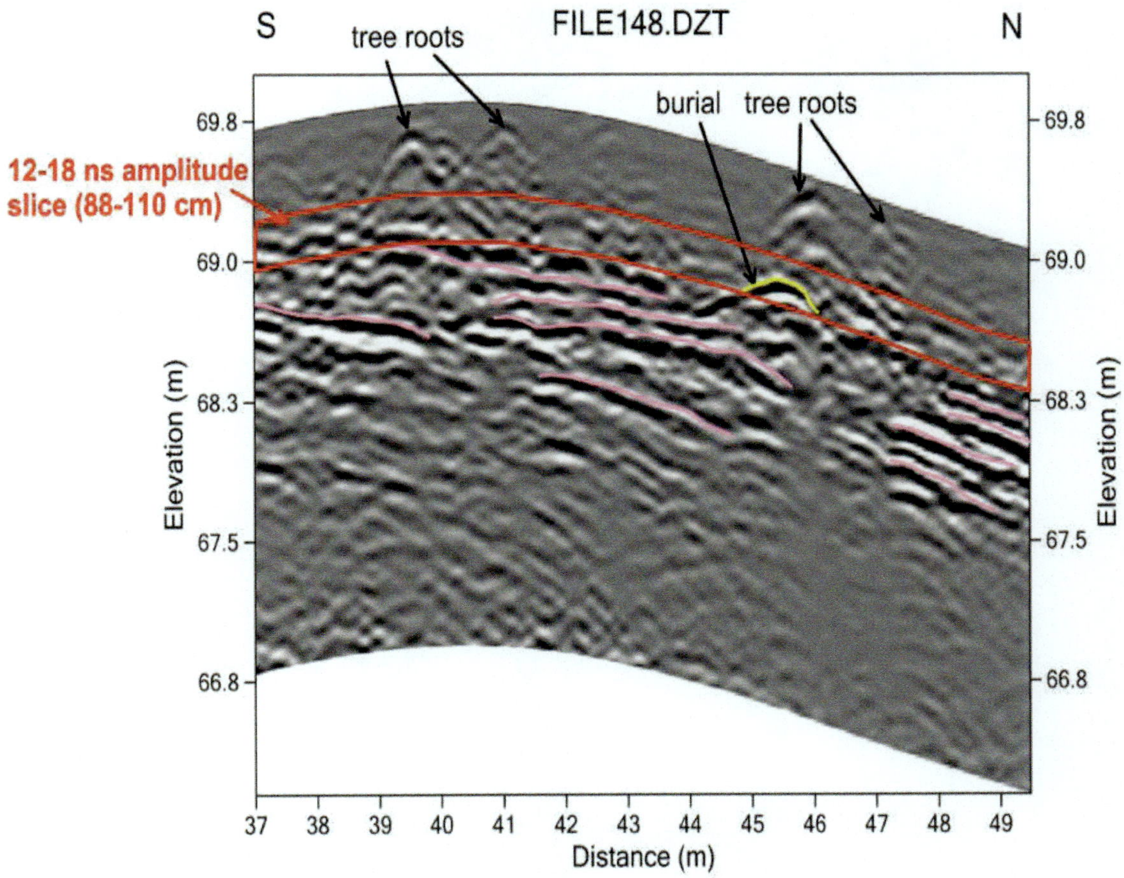

Fig. 2. Reflection profile corrected for topography, showing the reflection hyperbola produced by a human burial below the tree-root reflection. Mound fill units are shown in pink.

Standard amplitude slice-mapping gives no indication of the burials and they were only visible in detailed profile analysis after being corrected for topography. Once their presence was known, the reflection profiles were re-sampled in a smaller grid, resampling the amplitudes from the depth where the burial was identified in the profiles. Every reflection within the depth slice was sampled and given its own unique location in space, with a 1.2m search radius used in interpolation during the gridding procedure. The kriging gridding method was also used, which mathematically biases the interpolation values closer to the center of the search radius, producing a more detailed map of the burial feature. The burials were only visible after topographic adjustment of each reflection profile where the hyperbolic reflections were visible at the correct depth for human remains (below the depth of the tree roots). Once their location was approximately known from reflection profile analysis, the reflections within the profiles could be resampled and re-gridded and mapped, showing the correct spatial location of a burial. Fourteen other burials of this sort were visible in these mounds using the same methods.

In deeper slices within the mounds, four concentrations of high amplitude reflections were visible (Fig. 3), constructed after topography was adjusted so that slices were parallel to the pre-mound surface. These reflection features are located at or just above the original ground surface prior to mound construction. The amplitude map is interesting as it indicates something important at this depth under the mounds, but little else. The reflection profiles show that the high amplitudes were caused by layers of sediment that appear to have been piled up in layers (Fig. 3).

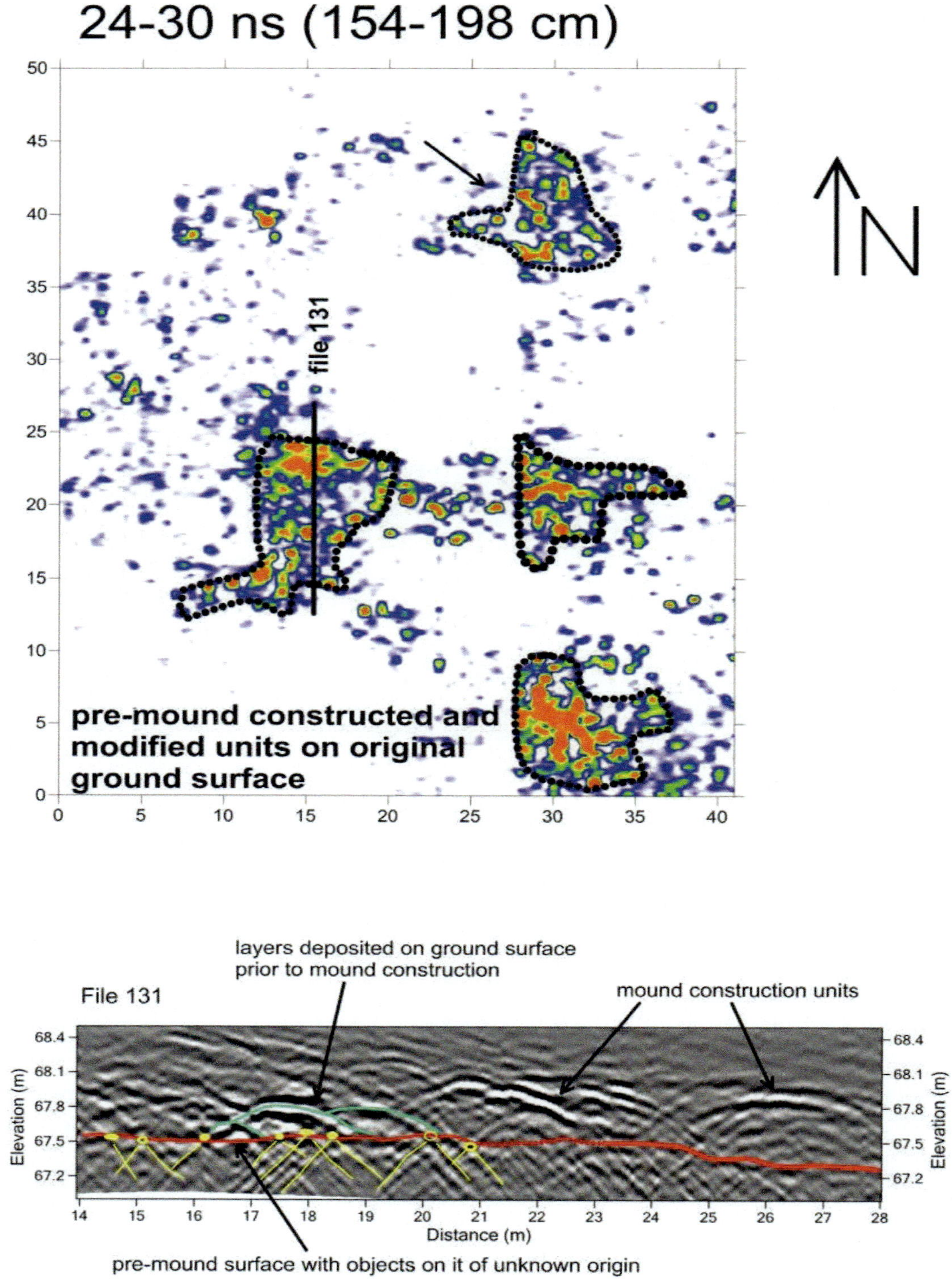

Fig. 3. Amplitude slice map of the deepest slice showing the concentrations of high amplitude reflections on the pre-mound surface that are concentrations of stones. A reflection profile produced after the reflections were frequency filtered to include only those between 400MHz and 600MHz shows the point-source reflection hyperbolae (axes of which are shown in yellow) from the stones on the original ground surface.

In order to obtain a more detailed view of the pre-mound ground surface and the initial layers of material constructed on it, a standard topographically-adjusted profile was frequency filtered so that only the 400MHz–600MHz reflections were used to display the smaller reflection features (Fig. 3). In this display of reflections the small objects (stones or coral pieces) generate distinct hyperbolic reflections consistent with standard point-source reflections. The pre-mound surface reflection is visible just below these point-source generating objects. The upper surfaces of three distinct piles of material that were placed on the pre-mound surface, covering the objects on it that can be seen, are consistent with piling sediment to create some kind of a living surface or at least elevated area of some sort. Only after frequency filtering the raw radar waves and then producing a detailed profile of only the stratigraphy of interest can the origins of the pre-mound units be determined. It is apparent that objects of some sort were first placed on the ground surface, and covered by intentionally mounded units to create elevated areas for some reason. Perhaps they were areas for people to keep dry during the rainy season or had some other function that cannot be determined without excavations. Later, this area was transformed into a mound, the remains of which we see today, which was used for the burial of human remains.

Conclusions

At these two mounds any one method of GPR analysis would have yielded only a partial interpretation of the cultural features below and within the mounds. Amplitude slice-mapping would have discovered the shallow rectangular enclosure and the general outline of the pre-mound construction layers but little else. The pre-mound layering units were only visible once profiles were adjusted for elevation and then re-sliced, and the profile used to interpret the specific origin of the pre-mound materials needed elevation adjustments and frequency filtering in order to see the individual stones.

Smaller features within the mound, such as the numerous human burials, were only visible after individual reflection profile analysis. These burials were effectively invisible using standard amplitude slice-mapping, but a re-sampling and gridding of a small area around one discovered burial showed its exact orientation and size.

Any one data display technique using the 3-D GPR datasets will only produce a limited picture of the ground from which to make interpretations about buried materials. Both amplitude slice-mapping and reflection profile analysis, used in unison and in an iterative way, can provide a more detailed view. Standard GPR displays must be modified and later reconstructed once something is known about the size and geometry of the features discovered.

Most important for this project is that the GPR analyses indicate these mounds were an important place on the landscape of northern Australia. It is apparent that people were using this area in an intensive way perhaps for feasting, ritual, and everyday activities. Those locations were later transformed into constructed mounds, which necessitated a good deal of coordinated labour, suggesting some authority and motivation by individuals within the societies. After the mounds were constructed, they were used for burials of some, but likely not all, members of society, also indicating social complexity and perhaps incipient social stratification. The mounds continued to be used for burials until recent times, and are remembered by elders as both burial and ritual locations. The shallow rectangular feature is likely the remains of a structure built in the last few centuries, related in some way to a continued ritual use of this important place on the landscape.

Acknowledgements

Thanks to the Traditional Owners of Mapoon, Australia. Also Mary Jean-Sutton and Emma St. Pierre for their collaboration in the field.

References

Conyers, L.B. 2012. *Interpreting Ground-Penetrating Radar for Archaeology*. Routledge, Taylor and Francis Group. New York, NY, USA, 2012.

Conyers, L.B. 2013. *Ground-Penetrating Radar for Archaeology*, 3rd edition. Rowman and Littlefield. Lantham, MD, USA.

Conyers, L.B. 2016. Ground-Penetrating radar mapping using multiple processing and interpretation methods. *Remote Sensing*. 8. 562.

Conyers, L.B., St. Pierre, E.J., Sutton, M-J. and Walker, C. 2018. Integration of GPR and magnetics to study the interior features and history of earthen mounds, Mapoon, Queensland, Australia. *Archaeological Prospection.* 26. 3-12.

Conyers, L.B., St. Pierre, E.J. and Sutton, M-J. 2019. Dissecting and Interpreting a Three-Dimensional Ground-Penetrating Radar Dataset: An Example from Northern Australia. *Sensors*. 19. 1239-1241.

Urupa - burial grounds - and remote sensing in Aotearoa (New Zealand)

Hans-Dieter Bader
Archaeology Solutions Ltd, Blockhouse Bay, New Zealand

Maori, the indigenous people of Aotearoa, have a strong connection to their local burial grounds. They are extended family (*whanau*) or sub-tribe (*hapu*) burial grounds. Some of them are still in use today. The burial grounds where the ancestors (*koiwi*) rest are one of the most important links of the people to the land. In many cases these links endured for centuries. Until the mid-20th century individual burial plots were rarely marked as each family knew the places their ancestors lay. With the displacement of many Maori to the cities this knowledge has waned. Furthermore the European surveying of the burial grounds at the beginning of the 20th century was very schematic and did not always encompass the entire extent of the burial grounds.

A resurgence within Maori communities connecting with their past is driving a desire to define the extent of the burials within the modern cadastral system, especially in those *urupa* which are still in use. Commonly in the burial grounds used since the mid-1800s, European practice of outstretched body burial with or without a coffin is followed. Before the spread of Christian beliefs within Maori communities shallow crouch burials were the most common burial practice amongst a number of others.

Intrusive investigations are out of the question due to cultural reasons, that is, to not disturb the peace of the ancestors. Remote sensing that is within the financial reach of the communities is the only option. From ground tested results of settlements with storage pits it has been assumed that the signature of burial pits will be quite similar to those of storage pits.

Archaeological geophysical surveys have been undertaken in New Zealand only sporadically in the past but within the last ten years a growing compendium of experience on various sites and soils has been built up. On the basis of this experience the following multi-method process was developed to indicate the presence of unmarked graves within a known burial ground. The target for all methods is the disturbance created by the burial pit and/or signature of the coffin furniture.

Using photogrammetry from low altitude aerial photos a point cloud of the surface is created. Small depressions show up. The high resolution orthophoto created from the same point cloud enables a survey of the marked graves and other elements of the burial ground such as fences, service huts, etc. A geomagnetic survey using a fluxgate gradiometer shows soil disturbances, buried metal fences around burials and sometimes grave furniture (Fig. 1A). A ground-penetrating radar survey along the same grid lines allows a different view onto soil disturbances and allows an interpretation where the geomagnetic survey is ambiguous, for example close to cast iron burial markers or basalt stone markers (Fig. 1B).

Overlaying all four datasets allows for the creation of a sufficiently accurate plan of the burial ground with some indications of unmarked graves (Fig. 1C, Fig. 1D). The location of individual marked graves is not within cadastral survey quality (cm accuracy in the survey grid) but the relationship between the marked graves is accurate enough to create a map of the cemetery and identify individual marked graves. Such plans are often missing or are so schematic that individual graves cannot be identified on them. The more accurate map of the burial ground allows the community to plan how to care for the burial ground, possibly to relocate fences, and give an indication for where future burials can be placed.

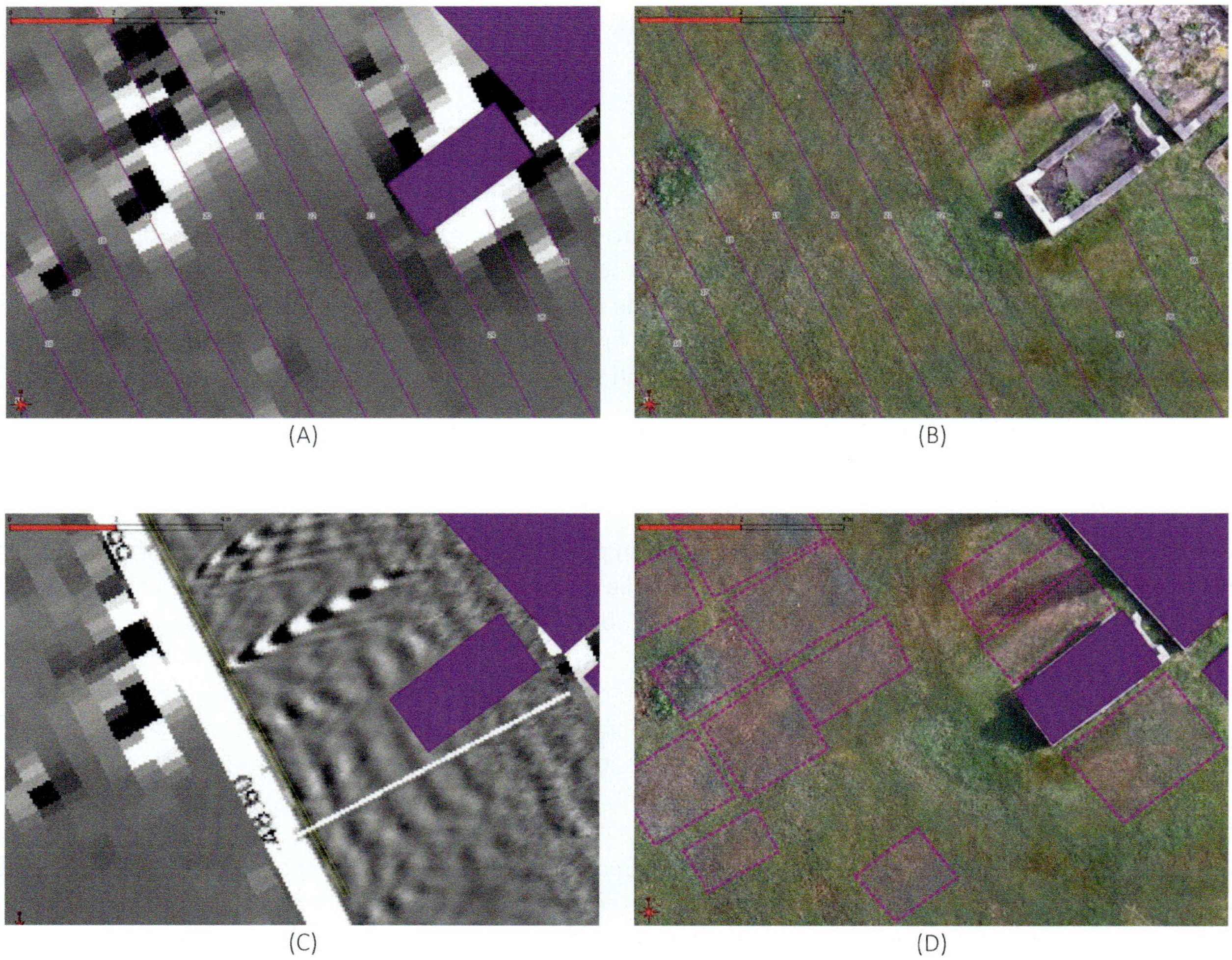

Fig. 1. (A) Geomagnetic survey around and over marked graves. (B) Ground-penetrating radar lines around the marked graves. (C) Marking "disturbed" areas on radar profile and comparing it with geomagnetic data. (D) Plan of marked graves and possible, unmarked graves.

Conclusions

This method is not perfect, but with limited resources it is a reasonable solution to a growing problem. Detecting burial pits is difficult and some soil disturbances from historic planting can be mistaken for burials. Some urials pits without a coffin do not leave enough of a signature to be identified. The nature of the soil also plays a part. A burial pit dug into clay is easier to detect than a burial pit dug into sand and backfilled with sand. And 3D block diagrams of the radar data in conjunction with closer spaced survey lines would improve the interpretation of the results.

As most burial grounds were already used in the 19th century or earlier, under New Zealand law they are classified as archaeological sites. It is not unreasonable to employ archaeological methods to help with a modern community problem.

Part Five – Archaeological Prospection in North America

Village Life in the Middle Ohio Valley, USA: Geophysical Survey and Anomaly Testing

Jarrod Burks
Ohio Valley Archaeology, Inc., Columbus, United States

Background

The advent of maize agricultural regimes swept through the Middle Ohio Valley in the eastern United States at about AD 950-1000. It transformed the way people used the landscape, from communities of dispersed horticulturalists to densely packed agricultural villages focused on the river floodplains. While locally domesticated, "weedy" plants such as *chenopodium* (a plant similar to quinoa) continued to be grown and harvested, maize came to dominate the diet. This cultural complex is referred to as the Fort Ancient, and their villages are quite common along the Ohio River in the states of Ohio, Indiana, Kentucky and West Virginia. However, determining village size and layout has yet to be measured at most sites.

Over the last decade, magnetometry and magnetic susceptibility have been employed to examine the structure of Fort Ancient villages and develop excavation strategies specific to the research interests of two Fort Ancient specialists: Robert Genheimer (Cincinnati Museum Center) and Robert Cook (Ohio State University).

Hahn Village

Genheimer and I focused our efforts on one very large site—the Hahn village site. There, a low-density (10m x 20m) magnetic susceptibility survey with GF Instrument's Multi Kappa system identified the approximate extent of the primary village midden, as well as a possible central plaza. Our magnetic gradiometer survey covered 5.2ha with a Geoscan Research FM256 fluxgate gradiometer, in the primary village area, and about 3.8ha of the area surrounding the village with a Sensys 5-probe fluxgate gradiometer system with RTK GNSS.

The magnetic gradiometer results in Fig. 1 show the Sensys data overlapping with the FM256 data, though it is hard to see where the two datasets meet. Hundreds of probable pit features were detected and their distribution appears to define the approximate boundaries of the Fort Ancient occupation. The small ring-shaped feature is an older ditch (*c.* AD 100-300) that was previously unknown until this magnetic survey. Note the apparent gap in the occurrence of pit features near the middle of the site. This likely is a plaza within a circular village. Based on excavations at other Fort Ancient sites, plazas tend to be ringed by small square-rectangular houses. However, no obvious house features were detected in the magnetic data, though houses are present. Since 2009, Genheimer has run a one-month-long summer field excavation at the Hahn site.

The challenge at this site is the sheer number of pit features, and the fact that each pit produces tens of thousands of artifacts. Thus, to generate a representative sample of pit feature data, since about 2010, each season the team has focused their excavations on just two or three 20m x 20m blocks. In each block a small selection of pits from three to four amplitude classes (weak, moderate, and strongly magnetic) was excavated (Fig. 2). Correlations between magnetic amplitude and pit size and contents exist. Deep, refuse-filled pits, as well as earth ovens (with baked edges and a large volume of burned igneous rock), have the strongest magnetic signatures. Though portions of an outer palisade line (moderately large post-holes) has been found in excavations, it is not evident in the magnetic data.

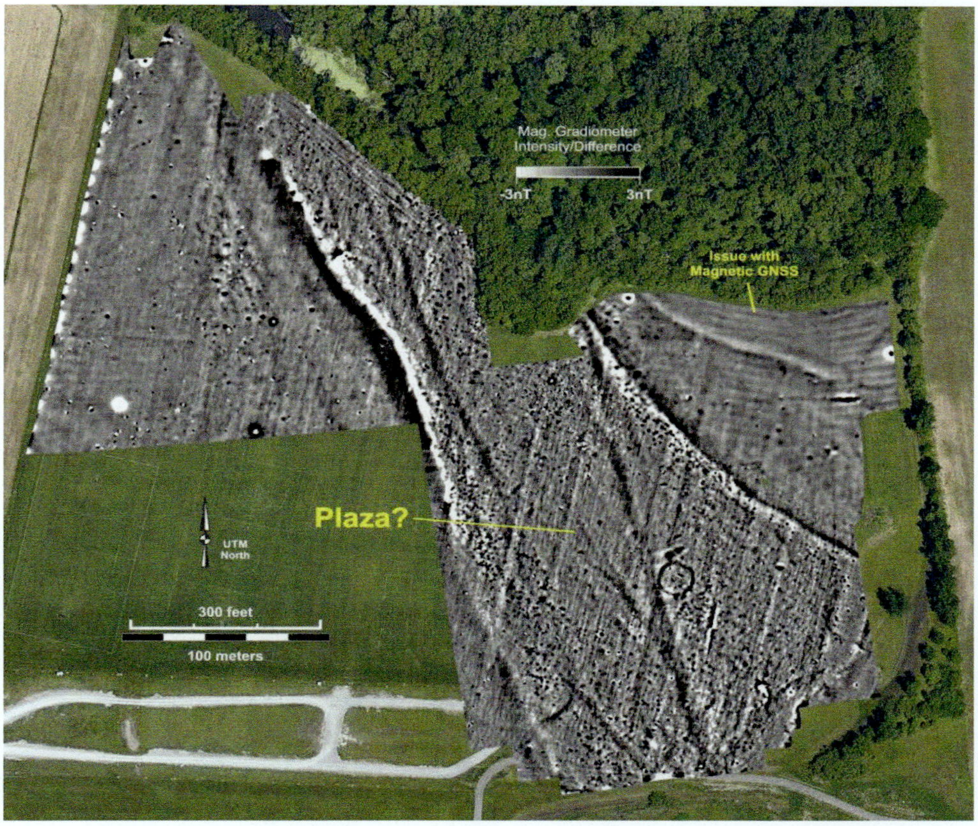

Fig. 1. Magnetic gradiometer data from the Hahn Village site, near Cincinnati, Ohio.

Fig. 2. Excavating large storage pits filled with refuse at the Hahn Village site.

Fort Ancient "villages"

Robert Cook and I have approached the body of Fort Ancient sites in the region in a slightly different way. We began some years ago by running gradiometer and susceptibility surveys at a wide range of sites, from small upland sites to large floodplain villages. In this way we discovered that not all "villages" are large and circular as expected (see Cook and Burks 2011). We also found a range in feature detectability, though houses eluded us until we surveyed our most recent site—the Guard site in south-western Indiana. The magnetic gradiometry survey at Guard began with a Geoscan Research FM256 magnetometer in 2011. This survey was slowly enlarged over the ensuing years during the very brief winter/spring survey period—floods often make the site inaccessible and farming begins once the floods recede. Despite the accessibility challenges, the Guard site has proven itself one of the most revealing (Fig. 3). Nearly all of the site's houses were detected (in the areas surveyed), and they are arranged in a distinctive circular pattern. Numerous pit features are present along the inside edge of the ring, between the houses and a probable central plaza (iron objects from a 19th century farmstead have disguised the probably empty plaza).

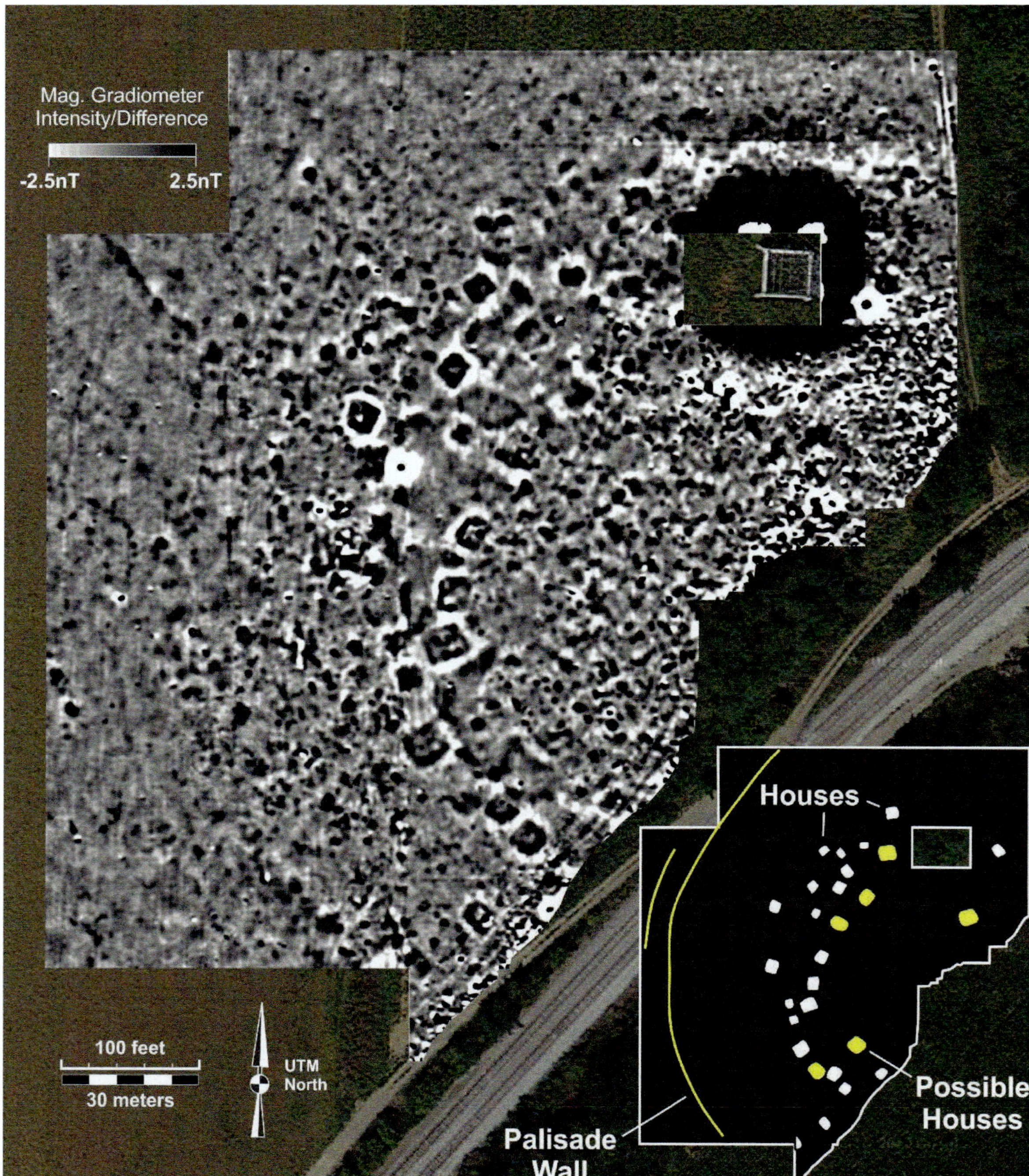

Fig. 3. Magnetic gradiometer data from the Guard site, in southeast Indiana near Lawrenceburg.

Excavations to date have focused on (1) small tests within each house, (2) more extensive investigations in a smaller number of houses, and (3) sampling a range of pit features (see Cook *et al.* 2015). These excavations have revealed the site to be one of the earliest Fort Ancient villages in the region. The excavations also answered the question of why the houses were so detectable at Guard: each was burned down and has a semi-subterranean floor that has been covered over by alluvium and remains untouched by modern agriculture. A recent addition to the map in 2019 was provided with data from a Foerster Ferex 4-probe survey along the west edge of the site. While attempting to define the outer boundary of the village, one or two faint but perceptible lines of anomalies were found running parallel to the arc of houses. These likely are the faint magnetic signatures of palisade lines.

Conclusion

When combined with shovel testing, wide-area magnetic susceptibility survey, and targeted excavation, magnetic gradiometry can quickly reveal the structure of most Fort Ancient villages. With a thoughtful sampling strategy, these limited excavations based on the magnetometry results can also provide samples across multiple households and reveal the age and structure of Fort Ancient villages.

References

Cook, R.A. and Burks, J. 2011. Determining Site Size and Structure in Cases of Low Surface Visibility: A Fort Ancient Example. *American Antiquity.* 76. 145-162.

Cook, R.A., Comstock, A.R., Martin, K.R., Burks, J., Church, W. and French, M. 2015. Early Village Life in Southeast Indiana: Preliminary Report of Recent Field Investigations at the Guard Site (12D29). *Southeastern Archaeology.* 34. 95-115.

Geophysical Evidence for the Timing, Pace, and Complexity of Construction at the Poverty Point World Heritage Site, Louisiana, USA

Michael Hargrave[1], R. Berle Clay[2], Diana Greenlee[3], Rinita Dalan[4]

[1]*CEERD-CN-C, Champaign, United States,* [2]*Retired, Lexington, United States,* [3]*School of Sciences, University of Louisiana-Monroe, Monroe, United States,* [4]*Minnesota State University Moorhead, Moorhead, United States*

Background

The Poverty Point World Heritage Site in north-eastern Louisiana (see also Dalan *et al.*, this volume) is a Late Archaic complex of monumental earthworks that includes six concentric earthen ridges partially surrounding a 17.4ha plaza, two other ridges, and at least six mounds, one of them (A), is the second largest earthen mound north of Mexico (DOI 2013).

Radiocarbon dates suggest that substantial occupation and some earthwork construction (Mounds B, E, and possibly C) occurred from BC 1650 to BC 1450. The ridges, Mound A, and much of the plaza date to a construction boom between BC 1450 to BC 1250. Some believe that earthwork construction was very rapid, with the concentric ridges built in as little as one generation and the massive Mound A built over the course of a few months (Gibson 2001, Ortmann 2010, Ortmann and Kidder 2013) (Fig. 1).

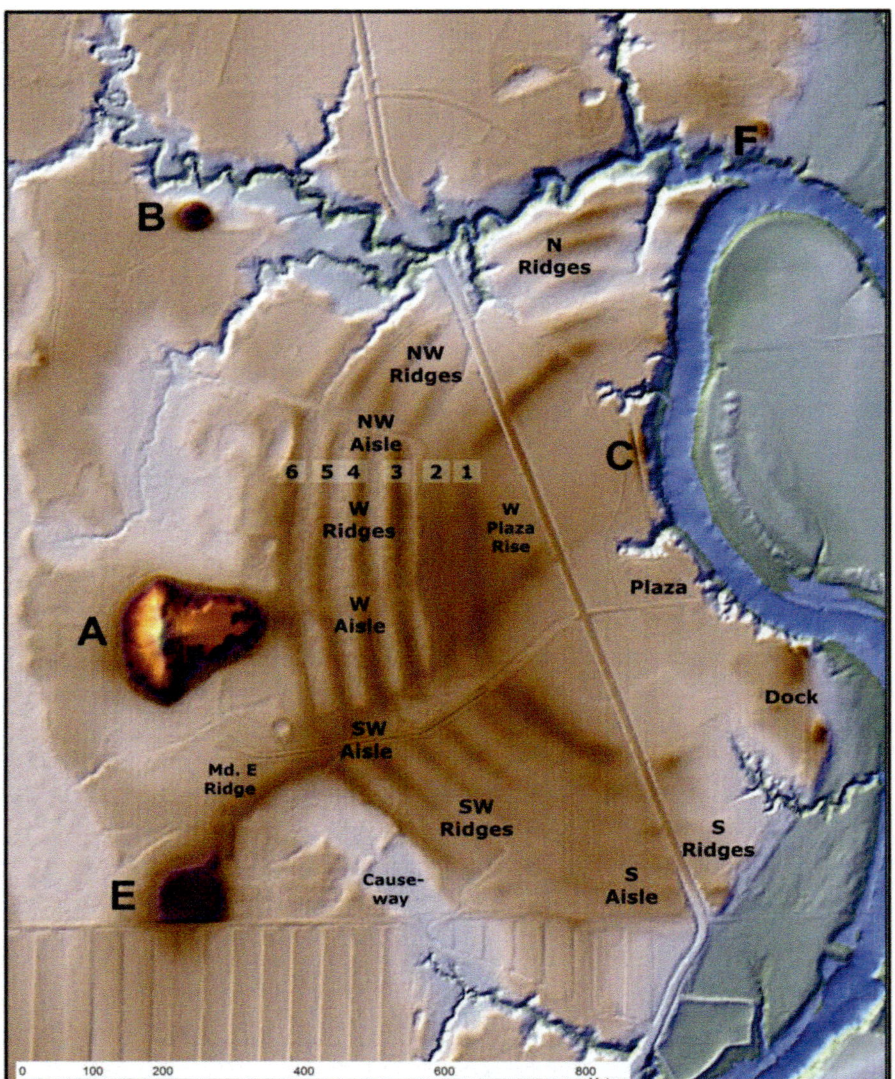

Fig. 1. LiDAR digital elevation model of the Poverty Point World Heritage Site showing the location of earthen ridges, plaza, mounds, and aisles. Ridges are designated using site area and numbered from the plaza outward, e.g. Ridge W-1 is the innermost west ridge.

Methods

Archaeologists conducted small area surveys at Poverty Point in 2001 to assess the usefulness of magnetic gradiometry, electrical resistance, electromagnetic induction, and ground-penetrating radar (Britt *et al.* 2002, Hargrave *et al.* 2007). Magnetic gradient data collected using a single Geoscan Research fluxgate FM36 revealed that subsurface portions of the ridges were well preserved and delimited from the adjacent swales and plaza by both positive and negative linear anomalies.

Hargrave and Clay returned to Poverty Point in 2006, this time using a duplexed pair of Geoscan Research FM256 fluxgate gradiometers, the first such system then available in the US. Results were so informative that we continued the survey as an intermittent, self-funded effort. By 2011, the magnetic gradient survey covered an area of 34.67ha, including most of the accessible portions of the ridges and plaza. Data were collected at a relatively high density (8 data values per metre along traverses spaced at 0.5m intervals) in hopes of detecting a wide range of feature types (Fig. 2).

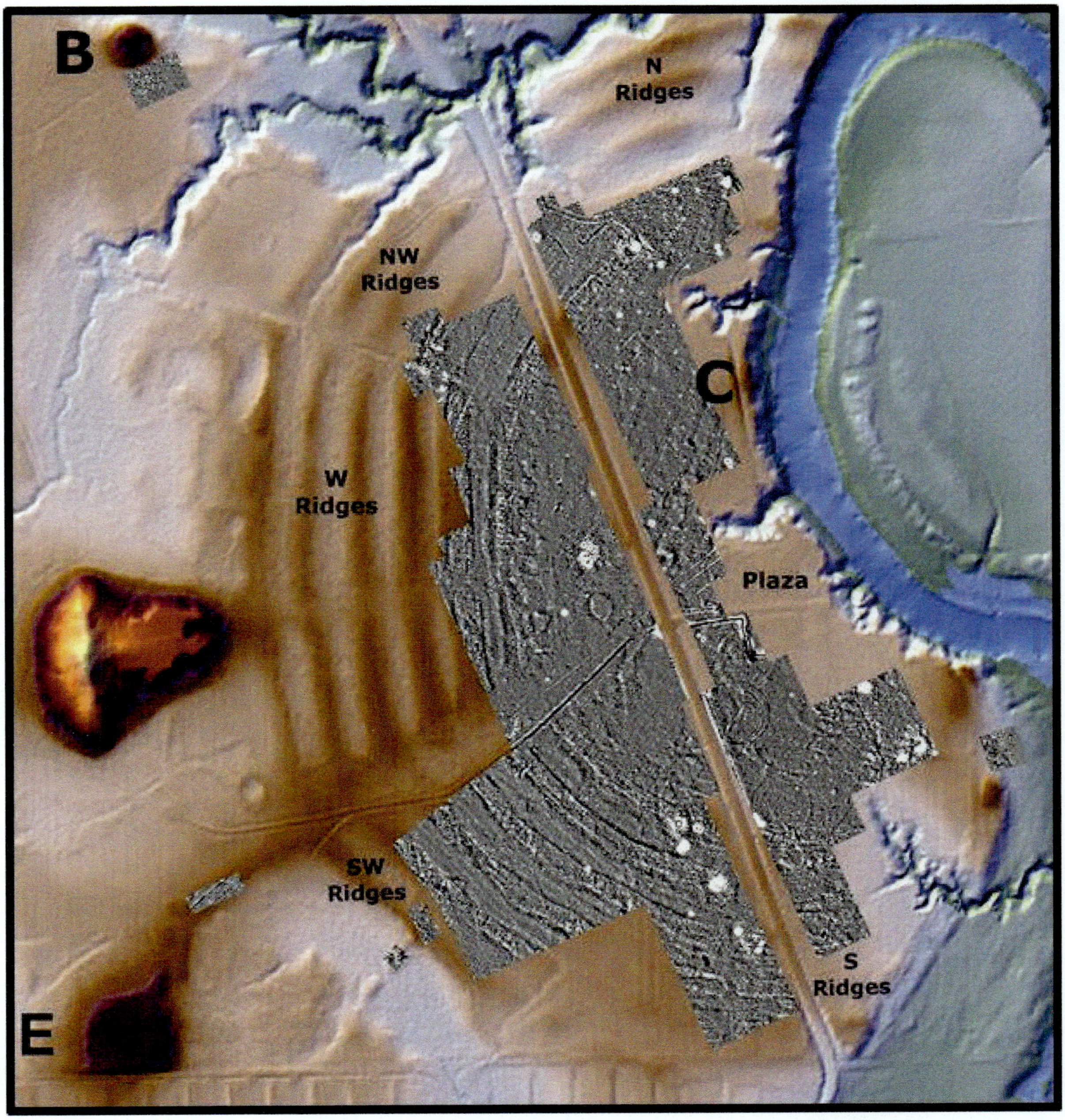

Fig. 2. Results of the magnetic gradient survey at Poverty Point. Note circles in and near the plaza and negative and positive linear anomalies associated with each concentric ridge.

Dalan and Greenlee also established a research program at Poverty Point in 2006, focusing on formation processes and stratigraphy of the ridges, swales, and plaza based on horizontal and vertical variation in magnetic susceptibility using a Bartington MS2H down-hole sensor, lab analyses of soil cores extracted with manual and truck-mounted systems, and use of a hand-held Bartington MS2K surface sensor in small excavation units targeted on features detected in the magnetic gradient data (Dalan *et al.* 2010, Dalan and Sharp 2012, Greenlee 2009).

Our integrated research design focused on several issues: 1) the site's construction history including the relative chronology and "pace" of ridge and plaza construction; 2) formation processes, particularly the origins of the negative and positive linear perimeter anomalies; and 3) the nature of massive circles detected in and near the plaza.

Results

The large-area magnetic gradient survey yielded much new data on the configuration and apparent integrity of the ridges, as well as unexpected features in the plaza. Magnetic susceptibility investigations show the negative perimeter anomalies appear to be associated with deposits of lower susceptibility materials that were emplaced during ridge construction episodes. Outer Ridges 3, 4, 5, and 6 in the SW sector each exhibit only one negative perimeter anomaly, indicating a single construction component. Innermost Ridges 1 and 2 include multiple negative linear anomalies, suggesting a more complex construction history (Fig. 2). Ridge W-1, with 8 components, is the most complex, and W-2 is the next most complex, with 5 components. The presence of midden, hearths, pits, and other features suggests that the concentric ridges were raised habitation platforms. The ridges are made primarily of loess, and intensive occupation of the ridge tops might contribute to erosion and deflation, requiring ridges to be repaired and periodically replaced. Older ridges that were ultimately used longer would have been rebuilt more extensively than later ridges that had not been used as long (Hargrave and Clay 2016).

Gradiometry survey detected, in and near the plaza, at least 38 large circles composed of post pits and possibly other pit types, the earliest such circle features known in North America (Hargrave and Clay 2016). Small excavations revealed closely spaced and overlapping postholes, indicating that the circles had been rebuilt by inserting new posts between earlier ones that had been removed (Greenlee 2011). Rebuilt circles were eventually replaced by new circles constructed in the same area, and most of the circles comprise three nested clusters (Fig. 3). Radiocarbon dates from four tested circles suggest that the earliest circles were constructed *c.* BC 1600-1500, and the latest, *c.* BC 1400-1100. The circles have diameters ranging from 9.45m to 66.3m (with a mean of 35.86m), too large to have been roofed structures.

Conclusions

Negative perimeter anomalies indicating the presence of multiple components suggests that the innermost two concentric ridges had complex construction and occupation histories. The massive post circles had also been extensively rebuilt and then replaced, consistent with their use over a long period in recurring ritualized social activities. Thus, Poverty Point's construction history appears to have been a series of pulses, rapid constructions followed by longer intervals of occupation. How and why an early, presumably non-hierarchical society that relied on hunting, gathering, and fishing constructed massive earthworks is not clear, but our new evidence for a complex, variably paced construction history suggests that the Poverty Point phenomenon may have been more of a dynamic process than a singular event.

Acknowledgements

Our thanks to Dr Lewis Somers, Dr Bruce Bevan, Mr Thurman Allen, the Poverty Point World Heritage Site staff, volunteers, and the many students who participated in geophysical survey, lab analyses, and archaeological investigations.

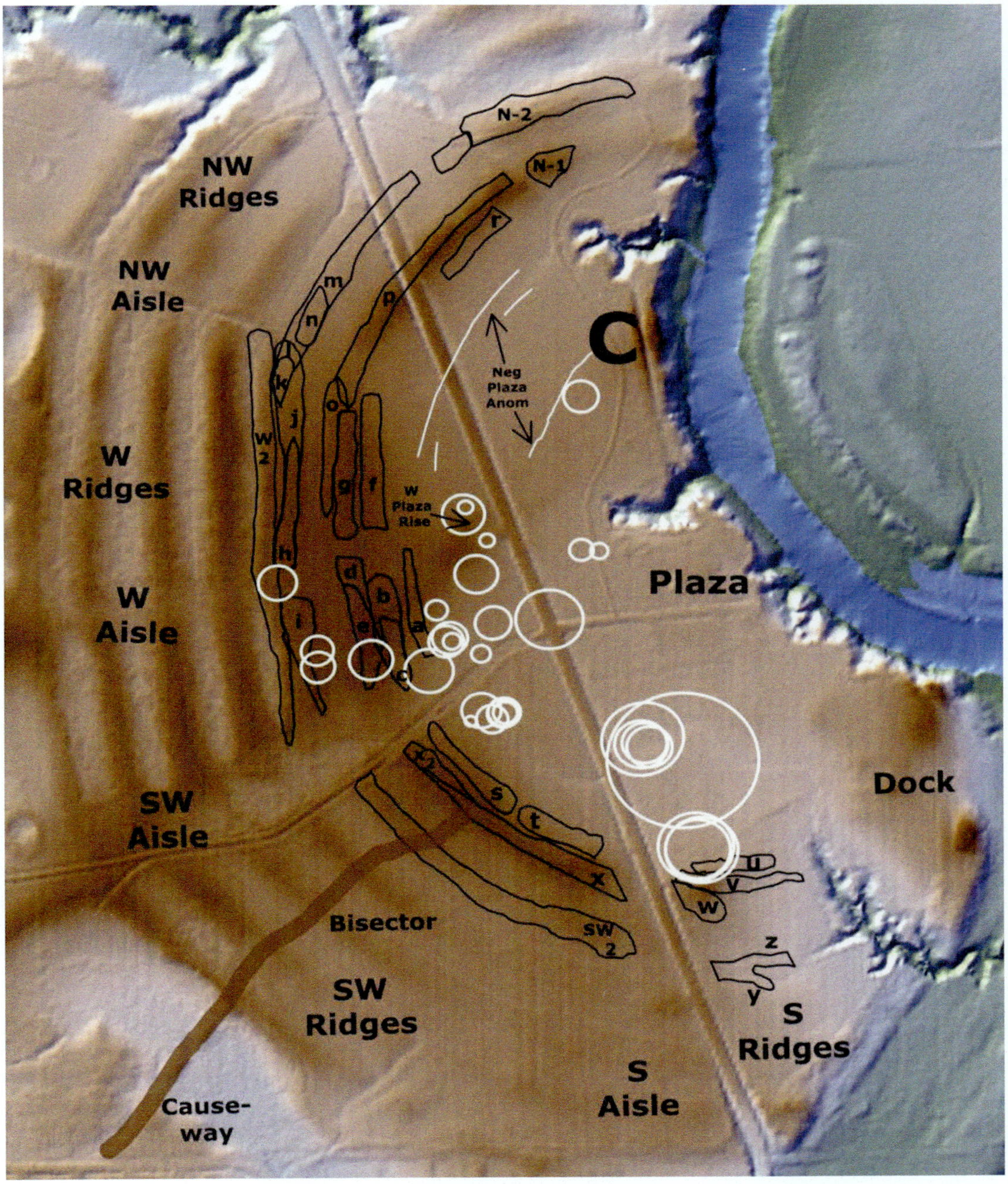

Fig. 3. Location of ridge components and post circles based on interpretation of the magnetic data. Traces of the Bisector ridge are seen in the magnetic data and early aerial photographs.

References

Britt, T., Hargrave, M.L. and Simms, J. 2002. *Geophysical archaeological survey at the Poverty Point State Historic Site (16WC5), West Carroll Parish, Louisiana*. ERDC SR-02-13.

Dalan, R.A., Beard, J., Blaha, A. and Cota, A. 2010. *Magnetic susceptibility studies at the Poverty Point State Historic Site*. Report to the Louisiana Division of Archaeology.

Dalan, R.A. and Sharp, J. 2012. *Magnetic susceptibility studies in excavation blocks 1-4*. Report to the Poverty Point Station Archaeology Program.

DOI (US Department of the Interior). 2013. *Nomination to the World Heritage List by the United States of America. Monumental Earthworks of Poverty Point*. Prepared by D. Greenlee.

Gibson, J.L. 2000. *The ancient mounds of Poverty Point place of rings*. University of Florida.

Greenlee, D.M. 2009. *2009 Annual report of the Station Archaeology Program at the Poverty Point State Historic Site*. Report to the Louisiana Division of Archaeology.

Greenlee, D.M. 2011. *2011 Annual report of the Station Archaeology Program at the Poverty Point State Historic Site*. Report to the Louisiana Division of Archaeology.

Hargrave, M.L., Britt, T. and Reynolds, M.K. 2007. Magnetic evidence of ridge construction and use at Poverty Point. *American Antiquity.* 72. 757-769.

Hargrave, M.L. and Clay, R.B. 2016. *Geophysical and LiDAR investigations at the Poverty Point World Heritage Site (16WC5)*. Report to the Louisiana Division of Archaeology.

Ortmann, A.L. 2010. Placing the Poverty Point mounds in their temporal context. *American Antiquity.* 75. 657-678.

Ortmann, A.L. and Kidder, T.R. 2013. Building Mound A at Poverty Point, Louisiana: monumental public architecture, ritual practice, and implications for hunter-gatherer complexity. *Geoarchaeology.* 28. 66–86.

The Origins of the West Plaza Rise at the Poverty Point World Heritage Site, Louisiana, USA

Rinita Dalan[1], Diana Greenlee[2], Michael L. Hargrave[3], R. Berle Clay[4], Thurman Allen[5], George Holley[1]
[1]Minnesota State University Moorhead, Moorhead, United States, [2]School of Sciences, University of Louisiana-Monroe, Monroe, United States, [3]CEERD-CN-C, Champaign, United States, [4]Retired, Lexington, United States, [5]NRCS/Morehouse SWCD, Bastrop, United States

Background

The Poverty Point World Heritage Site (see also Hargrave *et al.*, this volume), located in north-eastern Louisiana, is a 163ha complex of earthen monuments constructed by hunter-fisher-gatherers 3,700-3,100 years ago (Fig. 1). The landscape they created includes five earthen mounds (a sixth was added later), six enormous, concentric, semi-elliptical earthen ridges, and a nearly flat, 17.4ha interior plaza. Our study focuses on a broad topographic high within the plaza known as the West Plaza Rise (WPR), investigating its origins, whether natural or cultural, prehistoric or historic, and examining its significance and relationship to other aspects of the site.

The WPR is distinguished by topography; it is subtly-rounded, *c.* 40m x 50m, and elevated approximately 1m above an otherwise relatively flat plaza. Its strategic location, at the end of an aisle extending east from the largest mound (A), and along a ridge that may connect Mounds E and C, further accentuates this landform. The presence of monumental post circles in its vicinity, visible in an earlier gradiometer survey (Hargrave and Clay 2016), together with the density and variety of postholes documented in excavations within and just south of this landform in the 1970s (Haag 1990), indicate that the WPR may have been a feature of substantial prehistoric activity. Historic contributions to its form must also be considered as an early 1900s farmstead was located there. The WPR was not mapped as a discrete feature by the gradiometer survey although historic iron debris are clustered in this area.

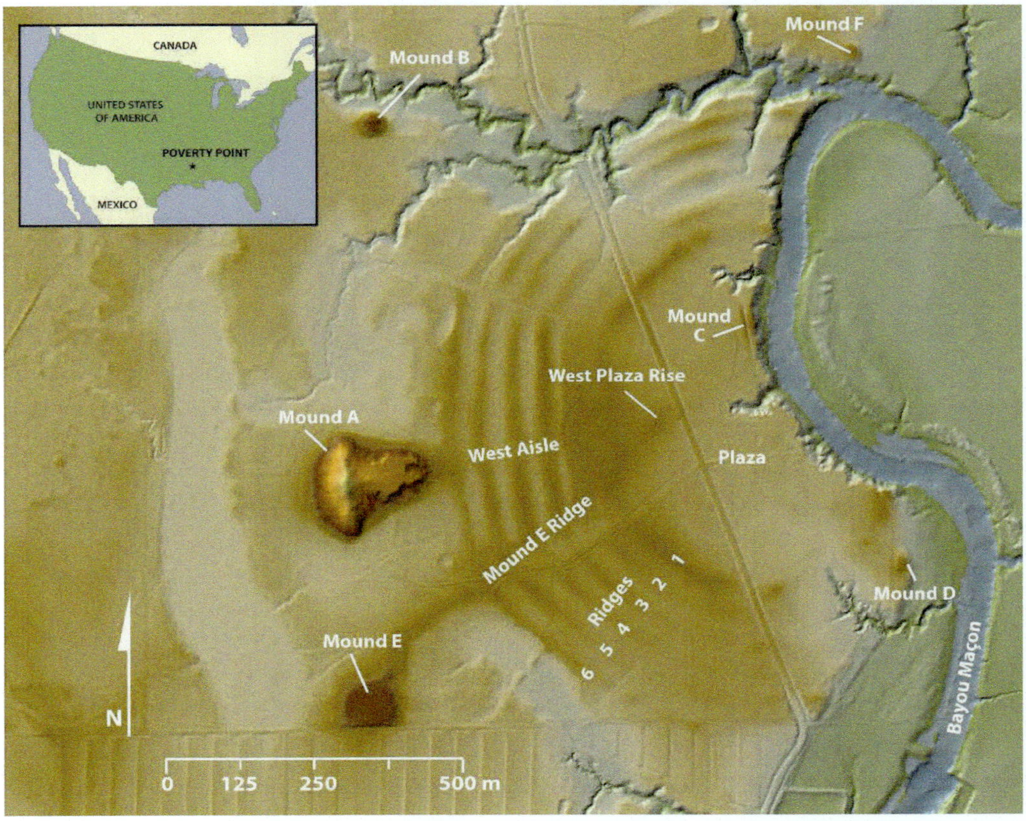

Fig. 1. LiDAR topographic model of Poverty Point WHS, showing the West Plaza Rise and other landscape features.

Methods

Since 2006, an approach combining coring, magnetic susceptibility studies, and soil physical and chemical analyses has been developed and applied at the site. This approach has been effective in tracking soil formation and horizonation and in distinguishing between cultural fills and undisturbed soils, information useful for understanding both feature formation and post-depositional processes.

An east-west and a north-south transect of cores and downhole susceptibility tests were completed across the WPR in November of 2018 (Fig. 2). The 15 2.5-cm diameter cores were extracted to depths of 92cm-175cm bgs by hand using a JMC push-tube corer and archived for later soil description and sampling. Downhole susceptibility measurements were conducted within the cored holes using a Bartington MS2 meter and MS2H down-hole susceptibility sensor connected to a PC tablet operating the Multisus FieldPro (v1.01) database program. Volume susceptibilities were recorded at 2cm depth increments, starting at 10cm bgs. A standard protocol was used for drift correction, zeroing the sensor in the air and then checking temperature-induced drift at intervals throughout the test and again at the end of the test.

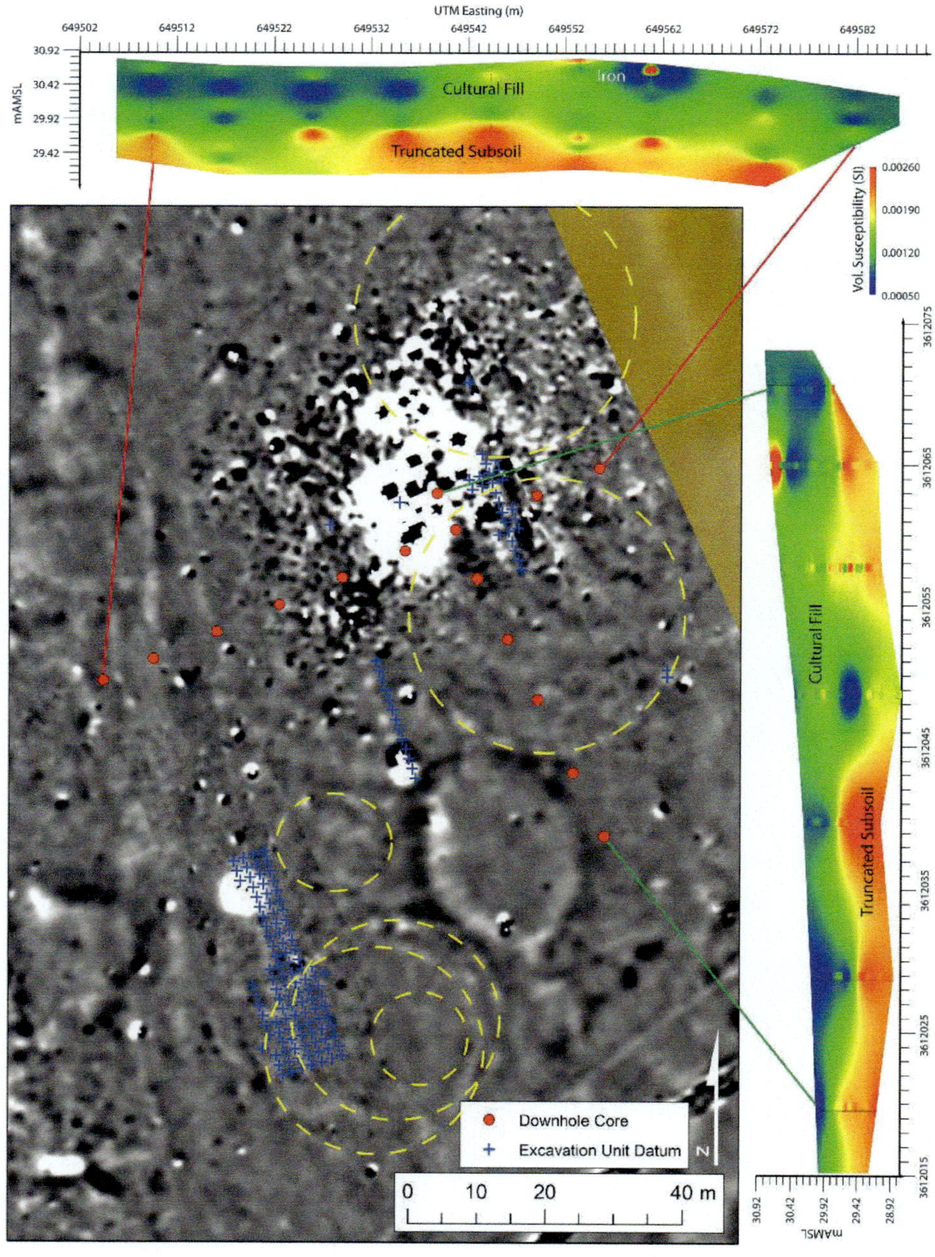

Fig. 2. Magnetic gradient basemap (± 4nT) of the WPR area, with locations of downhole transects, previous excavations, and circular magnetic anomalies. Magnetic susceptibility curtains at top and right correspond to west-east and north-south transects, respectively.

Results

Cross sections show susceptibility and lithology (Fig. 2 and Fig. 3). Upper, low susceptibility soils in all cores were identified as prehistoric cultural fill, with thickness ranging from 52cm to 150cm. In the majority of cores the fill was placed on a truncated, relatively high susceptibility natural argillic horizon; occasionally surface horizons of this soil were partially preserved. Thicker fill in the central portion of the landform produces the WPR.

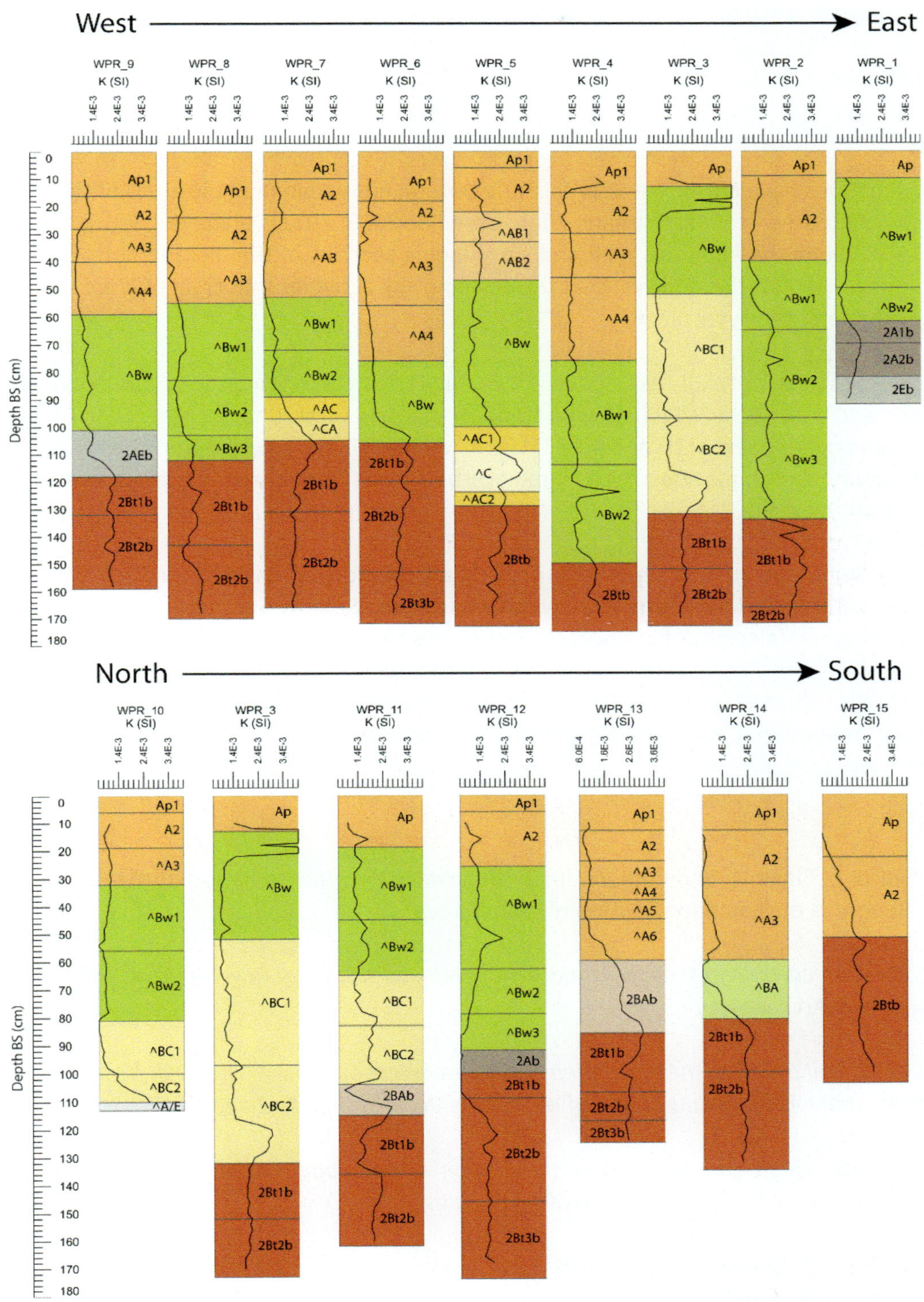

Fig. 3. Lithology and susceptibility plots of west-east and north-south core transects. Horizon designations based on USDA-NRCS (2012).

The west-east transect indicates that the WPR was built on a slope; rather than mimicking the topography of the truncated surface, soils were mounded higher in the area of the sloping ground to create the WPR. In contrast, the north-south transect indicates a truncated ground surface that was relatively level, with more fills placed on the north end. The WPR is more distinct north-south than east-west. On the south end of the WPR, a deposit of eroded materials suggests that at some point in the past this asymmetry was even more extreme.

The cross sections do not define a distinct edge for the WPR. Similar fill soils and the truncated surface continue for the length of each transect. Two cores with deep, anonymously low susceptibility measurements may relate to linear anomalies in the gradiometer data (Fig. 2) or to the collapse of core holes due to saturated conditions.

Conclusions

Previous investigations have documented prehistoric cultural fills within multiple areas of the plaza (Gibson 1984, Woodiel 1990, Greene 1992, Ortmann 2005, Greenlee 2009; 2011, Dalan *et al.* 2010). The absence of discontinuities between the fill in the WPR and surrounding sections of the plaza in both the gradiometer and downhole susceptibility data suggest that the earthmoving activities that created it are related to those that produced the relatively level plaza.

Although post circles were not explored as part of this project, excavations indicate a continuum of post circle construction on and around the WPR following stripping of the natural surface (Haag 1990, Greenlee 2009). Closely-spaced and overlapping posts and post circles in the plaza in the gradiometer data are also suggestive of continued activity, and this is supported by radiocarbon dates from several post pits that together span (2σ) 1630-1120 cal. years BC.

Several steps are being taken to test hypotheses about the formation and use of the WPR. We have submitted a grant proposal with the TEMAR (Terrestrial, Marine and Aerial Remote Sensing for Archaeology) research group at the Norwegian University of Science and Technology for a multichannel GPR survey. In the meantime we are subsampling cores to investigate soil development within the fill and at the fill/natural surface boundary. A parallel effort is underway to analyse field notes and collections from Haag's excavations for unpublished information that may shed light on the relationship of cultural activities to the creation of the plaza and the WPR.

References

Dalan, R.A., Beard, J., Blaha, A. and Cota, A. 2010. *Magnetic susceptibility studies at the Poverty Point State Historic Site*. Report to the Louisiana Division of Archaeology.

Gibson, J.L. 1984. *The earthen face of civilization: Mapping and testing at Poverty Point, 1983*. Report to the Louisiana Division of Archaeology.

Greene, G.S. 1992. *Soil coring in the plaza area and a three-year summary: The 1990 field season of the soil coring program*. Unpublished manuscript on file, Poverty Point Station Archaeology Program.

Greenlee, D.M. 2009. *2009 annual report of the Station Archaeology Program at the Poverty Point State Historic Site*. Report to the Louisiana Division of Archaeology.

Greenlee, D.M. 2011. *2011 annual report of the Station Archaeology Program at Poverty Point State Historic Site*. Report to the US National Park Service.

Haag, W.G. 1990. Excavations at the Poverty Point Site: 1972-1975. *Louisiana Archaeology.* 13. 1-36.

Hargrave, M.L. and R.B. Clay. 2016. *Geophysical and LiDAR investigations at the Poverty Point World Heritage Site (16WC5)*. Report to the Louisiana Division of Archaeology.

Ortmann, A.L. 2005. *Project 2/01: Results of the 2001 and 2002 field seasons at Poverty Point*. Report to the Louisiana Division of Archaeology.

USDA-NRCS. 2012. *Field book for describing and sampling soils*. Lincoln.

Woodiel, D. 1990. Investigations at the visitor center, Poverty Point State Commemorative Area, 1978. *Louisiana Archaeology*. 13. 41-71.

GPR Investigations in Earthlodges of the Northern Plains, USA

Kenneth L. Kvamme
University of Arkansas, Fayetteville, United States

Nearly two decades of GPR investigations have been carried out intermittently in large village sites once occupied by the Mandan, Hidatsa, and Arikara and their ancestors in the plains of the Dakotas (Kvamme 2003; 2007, Wiewel and Kvamme 2016). This work has been collaborative, generally with the PaleoCultural Research Group, a non-profit organisation headquartered in Colorado, and the State Historical Society of North Dakota, but often with other groups depending on the project and location, such as Rinita Dalan and Minnesota State University at Moorhead. The sites in question date between approximately AD 1250-1861. Most of our projects focus on broad-area magnetic gradiometry and electrical resistivity surveys while GPR is generally targeted to specific "lodges" of interest (a term meaning "house"). In the Northern Plains lodges were initially rectangular but eventually evolved into circular forms. Both types exhibit a maximum dimension of 14m-18m. While the former had earthen sides and post-supported gabled roofs likely covered with buffalo hides, the latter were hemispherical and composed of a timber frame that, except for an entryway and central smoke hole, was completely covered with a quarter-metre of earth. The circular forms are aptly referred to as "earthlodges". Both had linear entryways up to 3m long.

GPR Data Acquisition Problems

In many villages of the Northern Plains good-quality GPR results have been difficult to achieve. Significant site damage from ubiquitous plains rodents is generally present, a circumstance that introduces numerous anomalies and makes recognition of more patterned ones stemming from the archaeological record difficult. More problematic is climate. The Northern Plains is semi-arid with an average annual rainfall of only 40cm, and droughts are frequent. In these relatively dry conditions strong subsurface reflections and particularly contrasts between anomalous areas and undisturbed background in the mollisols and subsurface loess tend to be low. Yet, excellent reflection strengths and subsurface contrasts can be achieved when surveys are undertaken after a particularly wet spring where the greater moisture enhances reflection contrasts (compare Fig. 1a, and Fig. 1b).

Methods

Until the last five years, surveys have been conducted using a common half-metre transect separation, with 40 traces m^{-1}. Our interest generally focused on time or depth-slice maps to produce plan views of lodges for interpretational purposes—for signage in the state parks which hold many of these sites or to guide excavations. In recent years, and in view of the excellent high resolution results obtained by closely-spaced GPR arrays pulled by carts (e.g. Trinks *et al.* 2010), we have upped our game by moving to quarter-metre and even one-sixth-metre transect separations with a single antenna and clearer results have been achieved. Moreover, our focus has changed to a greater consideration of the GPR profiles themselves owing to the added insights they offer (Conyers *et al.* 2019, see also Conyers, this volume), especially concerning significant lodge elements such as identifying their floors, gaining evidence of central hearth complexity, recognising sub-floor storage pits, and sorting out rodent-caused anomalies (Fig. 1e).

Findings

GPR findings have been highly variable. When clear, they have been used to document rectangular or circular lodge forms, more accurately determine their sizes, investigate highly variable entryway shapes, locate important features such as hearths and occasional ancillary constructions such as storage pits (Fig. 1c, d, f). In a number of instances overlapping lodges have been discovered (Fig. 1d). A significant aspect of our GPR work has been the estimation of depth to significant features in advance of excavations, achieved through velocity estimates by hyperbola-fitting or through pipe tests using open excavation or cut-bank walls.

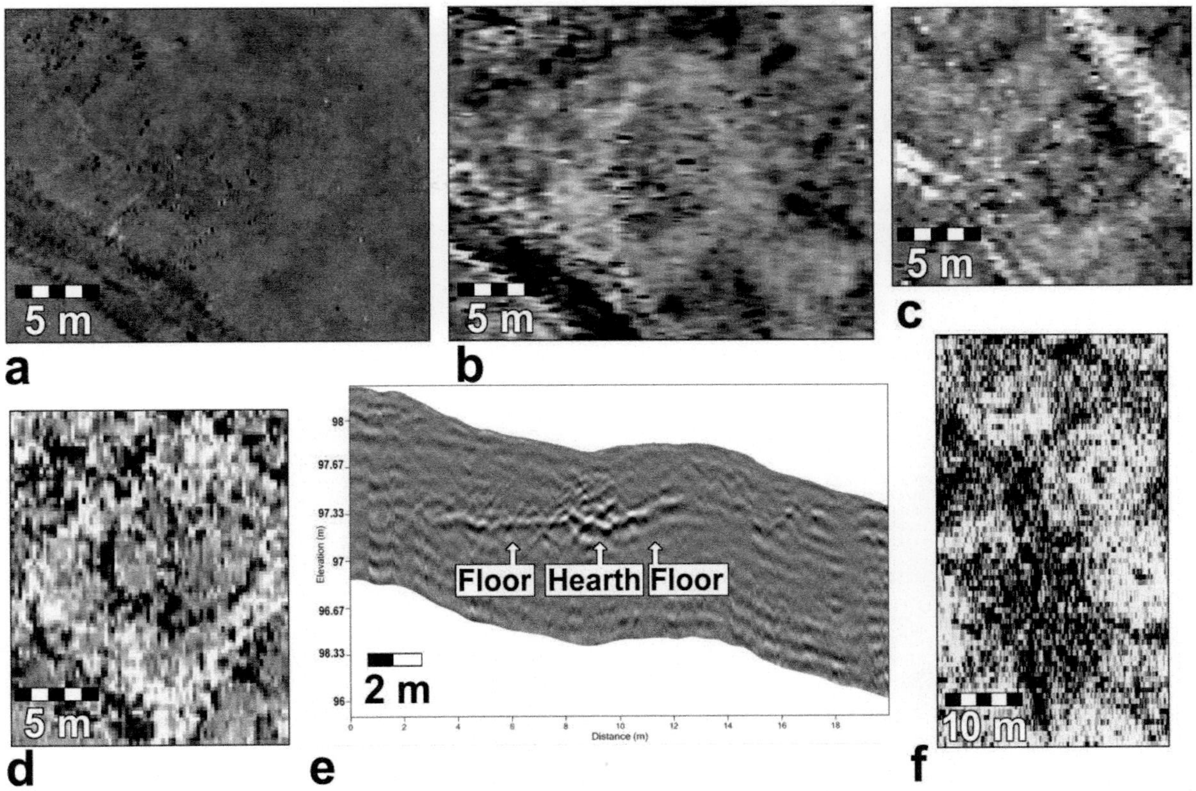

Fig. 1. Time-slice and GPR profile indications of lodges: a) shallow time-slice in a normal-moisture year, b) same area as (a) with identical settings but shortly after a very moist spring (note keyhole entryway, lower right), c) rectangular lodge extending under footpath (white line, upper right), d) small circular lodge atop larger rectangular one, e) terrain-corrected profile through lodge centre, f) segment through village showing multiple lodges and well-indicated central hearths.

Most of these villages occur on level ground although former lodge locations are frequently associated with subtle surface depressions. Yet, we have not been greatly concerned with terrain-correction of GPR profiles, especially since their floors lie only a few cm beneath the surface (typically 30cm-60cm). Recently, this perspective has begun to change. In the last few years we have had access to systematic high-resolution drone photography through collaborations with Arlo McKee of the University of Texas at Dallas. Through photogrammetry this work has generated Digital Elevation Models (DEM) with spatial resolutions as high as 2cm. The availability of such data recommended terrain-correction of GPR profiles. The real impetus, however, has been our research in 2017-18 at the ancestral Hidatsa village of Molander in North Dakota where a strong need for terrain-correction is obvious because lodges are buried by sloping colluvium in one part of the site (Fig. 1e).

R Language Software

We have begun a process of developing custom GPR processing software through the R language and RStudio. This software permits such basic operations as distance normalization, stacking, clipping at ground zero, background removal, time-to-depth conversion, display in a variety of forms including trace plots, terrain correction and extraction of time slices. Because terrain correction demands surface elevation data corresponding to each GPR transect, the software also includes DEM display, resampling to appropriate resolutions, and data extraction for alignment with GPR transects (Fig. 2).

Acknowledgements

Many thanks are offered to Mark Mitchell of the PaleoCultural Research Group and to Fern Swenson of the State Historical Society of North Dakota for their support through the years and to students from various universities in Arkansas, Colorado, Minnesota and Oklahoma who have provided assistance.

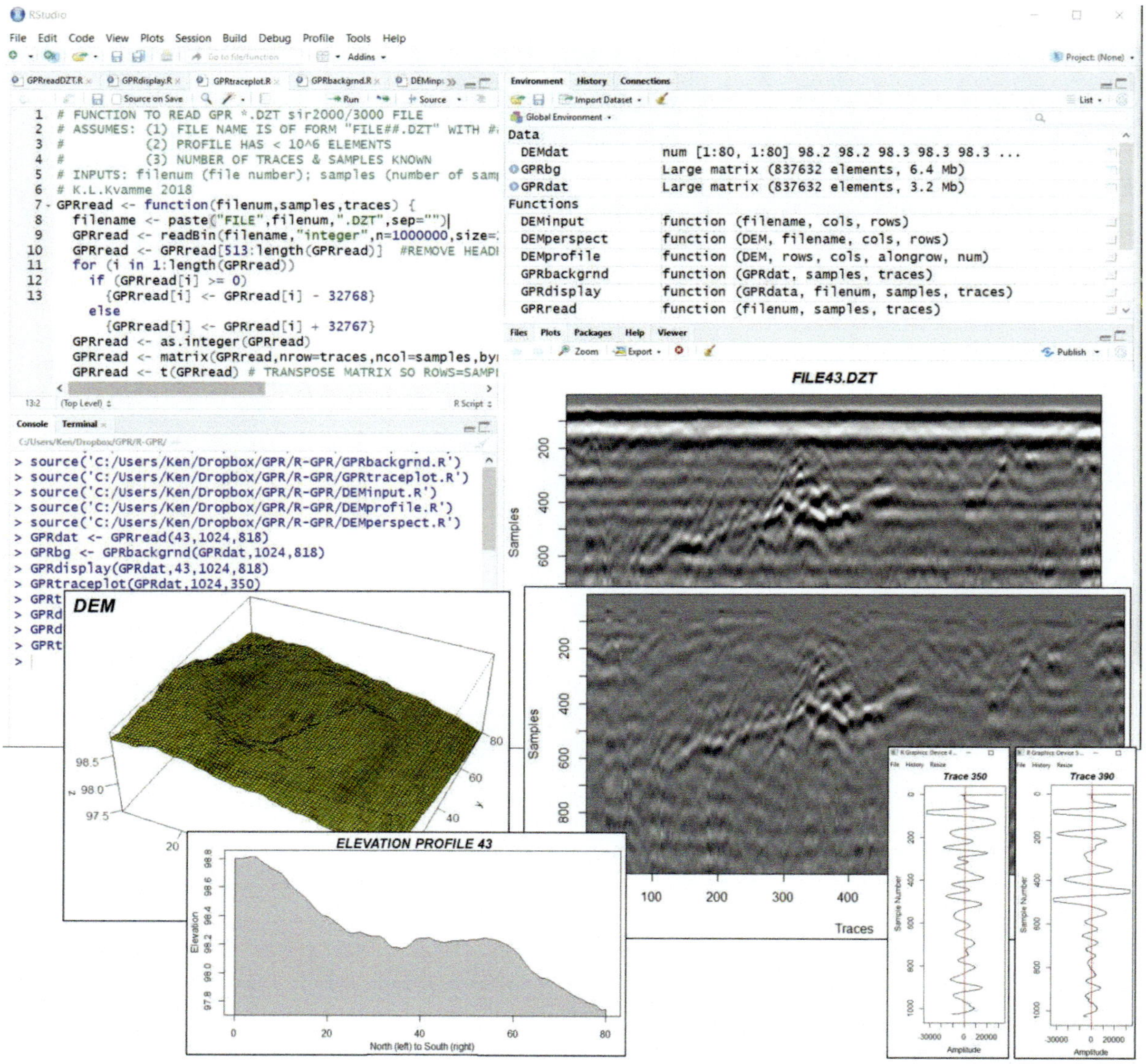

Fig. 2. Screen shot from RStudio showing various windows with R language code, command lines, defined functions, and output graphics depicting GPR and DEM findings.

References

Conyers, L.B., Sutto, M.J. and St. Pierre, E. 2019. Dissecting and interpreting a three-dimensional ground-penetrating radar dataset: An example from Northern Australia. *Sensors.* 19. 1239.

Kvamme, K.L. 2003. Multidimensional prospecting in North American Great Plains village sites. *Archaeological Prospection.* 10. 131-142.

Kvamme, K.L. 2007. Geophysical mappings and findings in Northern Plains village sites. In S.A. Ahler and M. Kay (eds.), *Plains Village Archaeology: Bison Hunting Farmers in the Central and Northern Plains.* 210-221. Salt Lake City.

Trinks, I., Johansson, B., Gustafsson, J., Emilsson, J., Friborg, J., Gustafsson, C., Nissen, J. and Hinterleitner, A. 2010. Efficient, large-scale archaeological prospection using a true three-dimensional ground penetrating radar array system. *Archaeological Prospection.* 17. 175-186.

Wiewel, A.S. and Kvamme, K.L. 2016. The nineteenth-century Mandan/Arikara village at Fort Clark: Results from a multi-instrument remote sensing investigation. *Plains Anthropologist.* 61(240). 449-468.

Part Six – Technical Aspects of Data Acquistion, Analysis, Processing and Visualisation

Drone radar: A new survey approach for Archaeological Prospection?

Roland Linck[1], Alen Kaltak[2]
[1]*Bavarian State Department of Monuments and Sites (BLfD), Munich, Germany,* [2]*Drone it GmbH, Munich, Germany*

Introduction

Over the last few decades Ground-Penetrating radar (GPR) has become one of the most common geophysical survey methods in archaeological prospection, alongside the magnetometry and resistivity methods. The capabilities of industrial survey drones have also advanced considerably over the last few years, and the question arises whether it would not also be possible to mount a radar antenna underneath a commercial drone instead of a simple camera. First attempts have already produced promising results. However, these instances concentrated on applications requiring extreme penetration depth rather than high resolution. Examples are the bathymetry of fresh water, geological surveys of soil layers and the detection of large objects like airplanes in the Greenland ice shield (UgCS.com, 2019). Here, we present for the first time a drone survey examining the applicability of this method to detailed archaeological prospection, which requires only a small penetration depth, but a high resolution.

Equipment Employed

For our test survey, we applied one of the most common drones in industrial applications, the DJI M600 (Fig. 1). This UAV can carry a payload of up to 6kg below the hexacopter flight system, which is sufficient for our radar antenna which weighed 4kg. The M600 offers a flight time of up to 15 minutes even with its full maximum take-off weight of 15kg. Together with a sophisticated flight control system, highly-capable safety equipment, and the ability to carry many different types of sensors, it represents the ideal platform for such test applications.

Fig. 1. Drone radar equipment consisting of DJI M600 hexacopter and Drone it GPR antenna (in black cylinder below drone) during survey in *Cambodunum*.

We used a specially adapted radar antenna, provided by Drone it GmbH, with a centre frequency of 80MHz (Fig. 1). This monostatic antenna can be easily attached to the payload compartment of the UAV and does not need any ground coupling like conventional GPR antennas. In addition, it requires little power - an essential feature to be used underneath a drone. Hence, this antenna is an excellent fit for such aerial applications. Another advantage is that this system does not need a special permit from the German licensing authorities (i.e. 'Bundesnetzagentur').

Due to the drone adaption, the same antenna could not be used for a comparative ground survey. Therefore, we checked the quality of the results by comparing them within the test area with a high-resolution radar survey executed in 2011 by the Bavarian State Department of Monuments and Sites (Linck and Kühne 2012, Linck 2013). These data were acquired with a GSSI SIR-3000 and 400MHz antenna in a sample density of 2cm x 25cm.

Test Site and Survey Parameters

Our test site is located in the Roman *forum* of *Cambodunum*, which contains a multitude of Roman buildings that can be clearly identified and mapped by standard ground-based GPR. As the area consists of grassland and was never built upon in later historical periods, it provides us with an outstanding opportunity to map the nearly complete official quarter of a Roman town in Bavaria by using geophysical techniques.

The drone radar survey was executed by an automatic flight plan with profiles of 1m separation. Due to the sophisticated safety concept of the DJI M600, it was possible to fly the complete survey at an altitude of 2m-3m AGL. At any time, an altimeter controlled the constant flight altitude above ground and corrected for the terrain relief. To improve the signal-to-noise ratio and to cope with the relatively low resolution provided by the available antenna, we acquired the data in a cross grid design. The georeferencing of the radar pulses is done directly by an RTK-corrected GPS system mounted to the drone. Because of the low flight altitude, this accuracy is sufficiently high to allow for a detailed and accurate survey to be made, and there was no need for ground markers to improve the data location.

The ground-based GPR survey was executed in summer 2011, and the drone radar test was done mid-March 2019. To ensure that the soil moisture content would not have any influence on the data quality, the present value was recorded by Time-Domain-Reflectometry and verified with official weather data provided by the 'Deutscher Wetterdienst'. The comparison of the two days shows that the soil moisture was nearly the same and no effect on the visibility of the archaeological remains due to signal attenuation caused by water in the soil should occur.

Results

The ground GPR survey reveals very clearly the Roman walls at a depth of 45cm-150cm below the modern surface. The test area for the drone radar is located in the southwest of the monumental official buildings of the *forum*. There the results show no dense archaeological construction by a regular Roman *insula*. Instead, some smaller single buildings and roads can be detected (Fig. 2) (Linck and Kühne 2012, Linck 2013).

As previously mentioned, the only available antenna for this test survey had a frequency of 80MHz. Hence, it is not possible to generate the usual high-resolution depth slices. The data can only be displayed by picking the origin of the hyperbolae of the reflections in the survey profiles. Nevertheless, the drone radar profiles show quite impressively a multitude of anomalies in the relevant depth range of 1m-2m below the surface (Fig. 3a). Georeferencing them reveals that many correspond very well to those detected by ground-based GPR, as each flight profile crossing perpendicularly to a Roman wall or street creates a distinct anomaly (Fig. 3b). Altogether, the apexes of 33 hyperbolae can be allocated to an archaeological feature mapped previously with GPR. Hence, a success rate of *c.* 65% is achieved. Some other hyperbolae are probably caused by single stones buried in the subsurface and are not of archaeological origin.

Conclusion

Despite the relatively low frequency of the available drone radar antenna, many relevant archaeological features can be clearly detected. Therefore, this method has been shown to provide an alternative for the archaeological prospection of areas with difficult terrain or those not accessible to conventional ground surveys. With drone radar, a quick survey of wide areas may become possible, especially with increasing flight time due to future battery technology or new propulsion systems, and the development of antennas enabling greater flight altitudes. A big disadvantage of drone radar compared with a ground-based one will always be the reduced resolution of faint structures in the subsurface, as, due to the greater distance between antenna and object, the signal cone will be wider and hence small objects cannot be resolved.

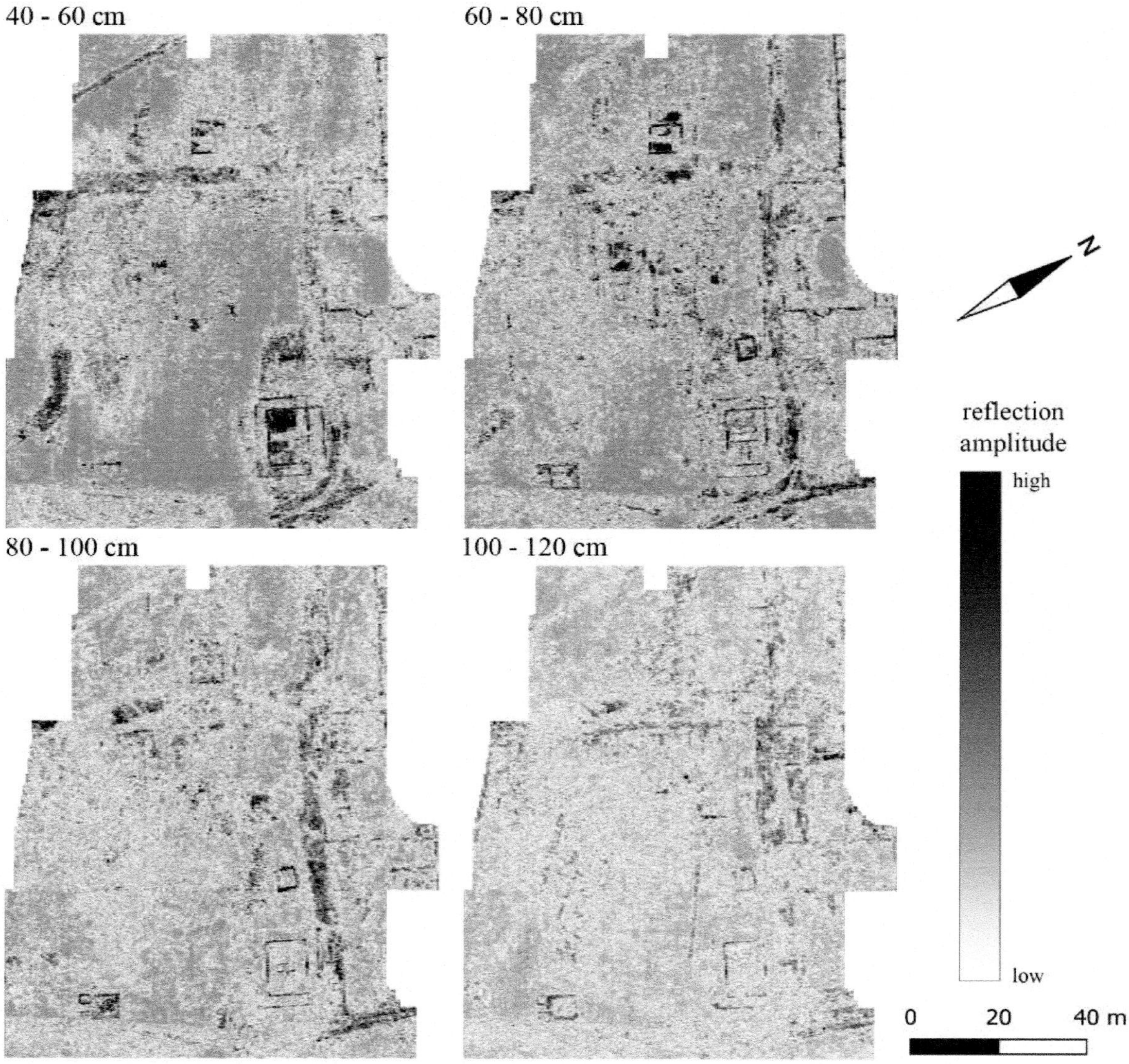

Fig. 2. Selection of depth slices of GPR survey covering the whole Roman *forum* of *Cambodunum*. GSSI SIR-3000 with 400MHz antenna, sample density: 2cm x 25cm, grid size: 94m x 117m.

One next step of our cooperation project will be a further test survey with a modified Drone it Cobra CBD antenna that has a frequency range of 200MHz-800MHz and hence is even more suitable for the detection of faint and shallow archaeological remains. This approach may also allow for the creation of depth slices to map buried features in detail.

Acknowledgements

The main author thanks Drone it GmbH (Munich) for the cooperation and Novatest S.r.l. from Italy for their technical support in drone radar equipment and data processing.

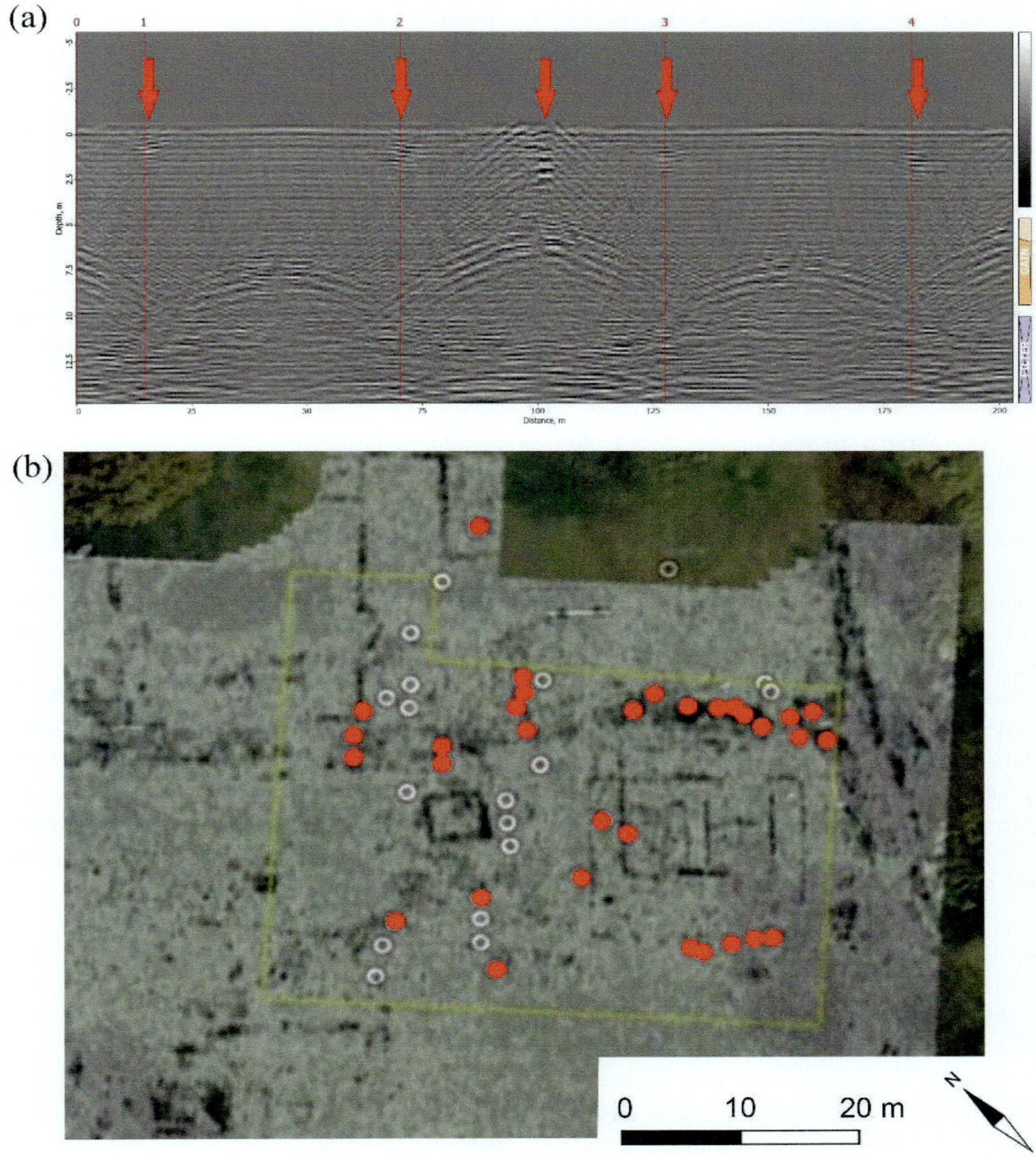

Fig. 3. (a) Sample profile of drone radar showing several anomalies at around 1m depth due to buried archaeological remains; relevant anomalies marked with red arrows. (b) Mapping of clear hyperbola apexes onto the 60cm-80cm depth slice of ground-based GPR survey. Red dots mark those hyperbolas fitting quite well to the location of known Roman walls or roads.

References

Linck, R. and Kühne, L. 2012. Römerstadt unterm Bodenradar: Prospektion für den "Archäologischen Park Cambodunum". *Das Archäologische Jahr in Bayern 2011*. 84-87.

Linck, R. 2013. *Methodische Untersuchungen zur Weiterentwicklung der Boden- und Satellitenradarprospektion in der Archäologie*. 109-114. Aachen.

UgCS.com. 2019. *New Approach to use Ground Penetrating Radar (GPR) integrated with a drone (UAV)*. Available at: https://www.ugcs.com/files/GPR/GPR_integrated_system_article-FIN.pdf (Accessed 29 March 2019).

The limits of a blob: geophysically informed automatic extraction of magnetometer anomalies

Armin Schmidt

Geodatawiz, Remagen, Germany

The rapidly increasing size of magnetometer surveys requires some form of automation for the analysis of measured anomalies (Linford and Linford 2017). This process can be separated into three stages: (a) the delineation of anomalies (b) an estimation of the shape of the causative features and (c) an archaeological interpretation. While the latter interpretative step requires human expertise, the estimation of respective feature shapes and in particular the delineation of individual anomalies can be greatly simplified by modern algorithms and software technologies.

Since images created from magnetometer data often look similar to 'ground plans' there have been attempts to treat them with common remote-sensing and GIS tools. However, most of these tools do not take the geophysical nature of measured data into account and therefore tend to produce misleading results. Satellite data, for example those depicting agricultural crops, often have to be classified based on the *similarity* of all data produced by one feature (e.g. a field with a single crop type) and the same is true for image segmentation approaches that separate data into areas that are in some way *homogeneous*. However, geophysical data show a characteristic *change* across a buried feature (e.g. as a bipolar magnetometer anomaly). In fact the buried feature manifests itself in the data through such a variation, not through its homogeneity and processes that search for the latter are hence less useful. GIS packages, on the other hand, allow investigations of geophysical raster data for example by creating contour diagrams. However, the characteristic contour level of each anomaly is dependent on its overall amplitude and it is impossible to select a single contour level to represent the whole dataset. Geophysical data analysis, therefore, uses specific raster processing (e.g. edge detection (Stampolidis and Tsokas 2012), reduction to the pole, analytical signal) to delineate and characterise anomalies. Unfortunately, vectorised mapping of the results remains difficult (Linford and Linford 2017).

The approach taken in this study uses advanced processing of vector data (Schmidt and Tsetskhladze 2013) to create polygons that outline the Full Width at Half Maximum (FWHM) of each anomaly in a magnetometer dataset (Aspinall *et al.* 2008). To this end, measured raster data are first converted to vectorised contours at 0.1nT intervals. These are then analysed to separate the individual anomalies and determine the amplitude of each anomaly. The latter is then used to select the contour polygon that represents half the value of the anomaly's maximum amplitude. The topological processing that is required for these steps has become possible with modern vector engines and PostGIS was used for this study. Fig. 1 shows the results for a fluxgate gradiometer survey (FM18) over an iron-age enclosure at Guiting Power, UK (Fig. 1). By calculating FWHM polygons it is possible to represent the data in the form of individual polygonal anomalies.

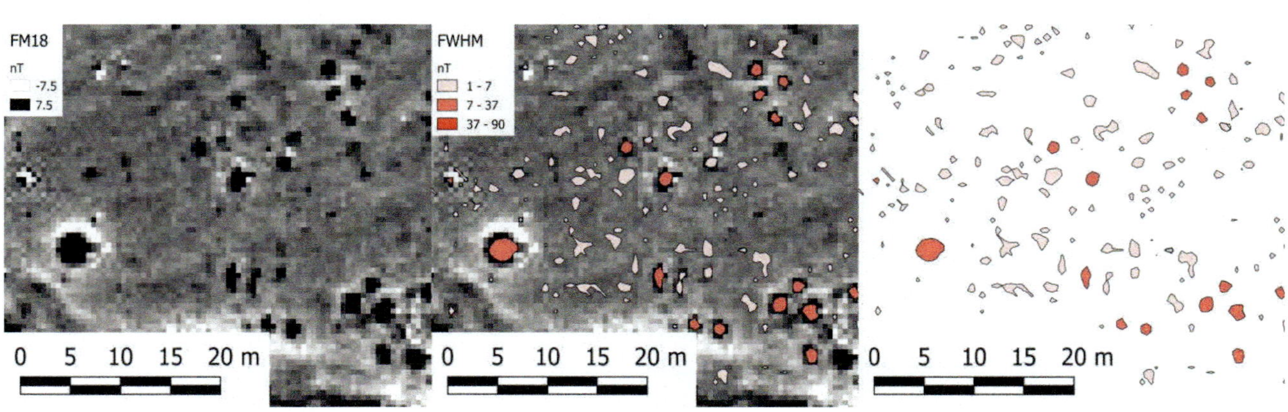

Fig. 1. (left) Fluxgate gradiometer data (FM18) from Guiting Power, (centre) overlaid with amplitude classified FWHM polygons and (right) the polygons on their own.

Modelling studies have shown that the FWHM is a reasonable estimator for a feature's lateral outline (Schmidt and Marshall 1997). The reason for this is the close proximity of the FWHM to the location of the maximum gradient over a magnetometer's anomaly. This, in turn, is often used as predictor for the underlying feature's edges (Blakely and Simpson 1986). Further analysis of the measured data at the location of the extracted polygons allows some estimates about the feature's characteristics, for example its approximate depth. Pre-processing of the magnetometer data with a pseudogravity transformation enhances the results further and the method is tested with synthetic data calculated for typical features.

While the processing presented here allows the creation of vector representations that can estimate underlying features, several challenges remain. For example, where strong singular anomalies exist within a broader anomaly (e.g. a horseshoe over a ditch) only the FWHM of the former will be extracted due to its high amplitude. In addition, with increasing depth of a feature its magnetometer response becomes smoother and edges more rounded. This effect is difficult to eliminate with the presented vector processing. Nevertheless, the extracted vector data are a geophysically correct representation of measurement results and a useful input for subsequent interpretation steps.

Acknowledgements

The data were collected and made available by Alistair Marshall.

References

Aspinall, A., Gaffney, C.F. and Schmidt, A. 2008. *Magnetometry for Archaeologists Geophysical methods for archaeology*. Lanham. AltaMira Press.

Blakely, R.J. and Simpson, R.W. 1986. Approximating edges of source bodies from magnetic or gravity anomalies. *Geophysics.* 51(7). 1494-1498.

Linford, N. and Linford, P. 2017. The application of semi-automated vector identification to large scale archaeological datasets considering anomaly morphology. In Jennings, B., Gaffney, C., Sparrow, T. and Gaffney, S. (eds.) *AP2017. 12th International Conference of Archaeological Prospection*. Archaeopress. Oxford. 138-139.

Schmidt, A. and Marshall, A. 1997. Impact of resolution on the interpretation of archaeological prospection data. In Sinclair, A., Slater, E. and Gowlett, J. (eds.) *Archaeological Sciences 1995*: 343-348. Oxford. Oxbow Books.

Schmidt, A. and Tsetskhladze, G. 2013. Raster was Yesterday: Using Vector Engines to Process Geophysical Data. *Archaeological Prospection.* 20(1). 59-65.

Stampolidis, A. and Tsokas, G.N. 2012. Use of Edge Delineating Methods in Interpreting Magnetic Archaeological Prospection Data. *Archaeological Prospection.* 19(2). 123-140.

Quantification of Daub Masses based on Magnetic Prospection Data

Natalie Pickartz[1,2], Robert Hofmann[1,2], Stefan Dreibrodt[1,2], Knut Rassmann[3], Liudmyla Shatilo[1,2], René Ohlrau[1], Dennis Wilken[1,2], Wolfgang Rabbel[1,2]

[1]*Kiel University, Germany,* [2]*CRC 1266 - Scales of Transformation, Kiel, Germany,* [3]*Römisch-Germanische Kommission, Deutsches Archäologisches Institut, Frankfurt, Germany*

Introduction and Archaeological Background

Many case studies of archaeological sites include magnetic surveys aiming to get a complete image of the area of interest. Magnetic gradiometry is well-established for this purpose since a variety of archaeological features can be mapped in different subsurface settings and data acquisition is fast, especially with motorized survey systems. Yet, frequently the interpretation of magnetic gradiometry data stops at the identification and classification of archaeological features. For large sites, spanning over several hundred hectares and thousands of archaeological features, a quantitative interpretation of site-covering prospection data is of great importance because only small excavations are admissible. We are presenting a new scheme for a quantitative interpretation of magnetic prospection data (Pickartz *et al.* 2019). With this scheme we first calculate the magnetization distribution constraining the inversion with *a priori* information of key target excavations. Secondly, we estimate the masses of the magnetic sources. Finally, we apply the approach to the Chalcolithic site Maidanetske (Ukraine).

The site Maidanetske (Fig. 1) is located in the Dnieper Upland and is one example of the so called 'Megasites' of the Cucuteni-Tripolye culture. Radiocarbon dates indicate settlement activities between BC 3950 and BC 3650 (Müller *et al.* 2016, Ohlrau 2018). With an area of about 200ha, Maidanetske is one of the largest megasites of the Tomashivka regional group and the advanced stage of Tripolye-development (Tripolye C1). The site comprises about 3000 - mostly burned - buildings which can be identified in recent and early magnetic prospection data (Rassmann *et al.* 2014, Дудкин 1978). Excavations showed that the remains of these buildings consist of a compact layer of daub which lies in a distinct depth range (e.g. Müller *et al.* 2017).

Methods

Via inversion computations, a subsurface setting, i.e. the spatial distribution of a physical parameter, is determined, which calculates the measured data to a certain accuracy. In general, the inversion of magnetic data suffers from ambiguity which means that different subsurface settings can reproduce the measured data with the same accuracy. To reduce the ambiguity assumptions about the spatial distribution, the parameter range can be made.

Using forward calculations of documented daub in the excavation, we showed that it is the main source of the magnetic anomalies. This enables us to constrain the depth range of the magnetic source layer to the depth range of the daub layer which is known from excavations and drillings. For the inversion the area is discretized in regular cells. For each cell the magnetization, which is in accordance with the measured data, is then determined. We assume that the total magnetization points into the direction of the earth's magnetic field at the time of the magnetic surveys. Smoothness constraints to each northern and eastern neighbouring cells are introduced. In the first step, the inversion is computed in a least-squares sense. All cells with positive magnetization are used as an initial guess in a subspace trust region interior reflective (STIR) algorithm (Branch *et al.* 1999) where the magnetization is constrained to be positive.

A comparison between calculated magnetization distribution (Fig. 2B) and documented daub distribution (Fig. 2C) shows that areas of increased magnetization correlate with the occurrence of daub and therefore confirm our computations.

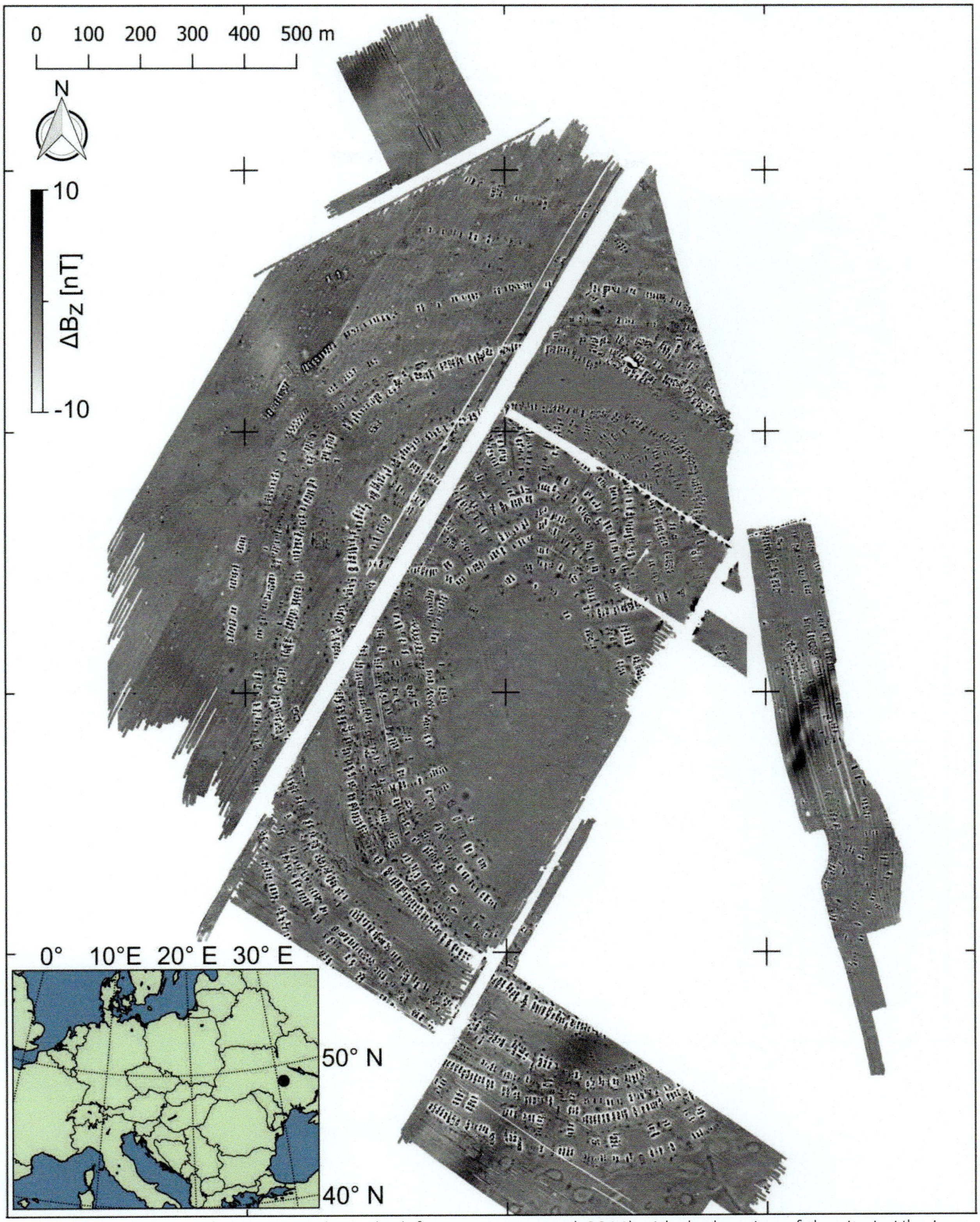

Fig. 1. Magnetic map of the site Maidanetske (after Rassmann *et al.* 2014) with the location of the site in Ukraine.

In the next step we relate the documented masses of daub and pottery within each excavation square (Fig. 2E) to the mean calculated magnetization for each square. Under the assumption of a linear relation, an orthogonal distance regression (Boggs and Rogers 1990) is performed (Fig. 2F). We then generalize the linear relation of three excavated buildings to finally use the generalized relation to determine the masses of unexcavated buildings. To quantify unexcavated buildings, we first compute the magnetization distribution and then use the mass-magnetization relation to estimate their total masses.

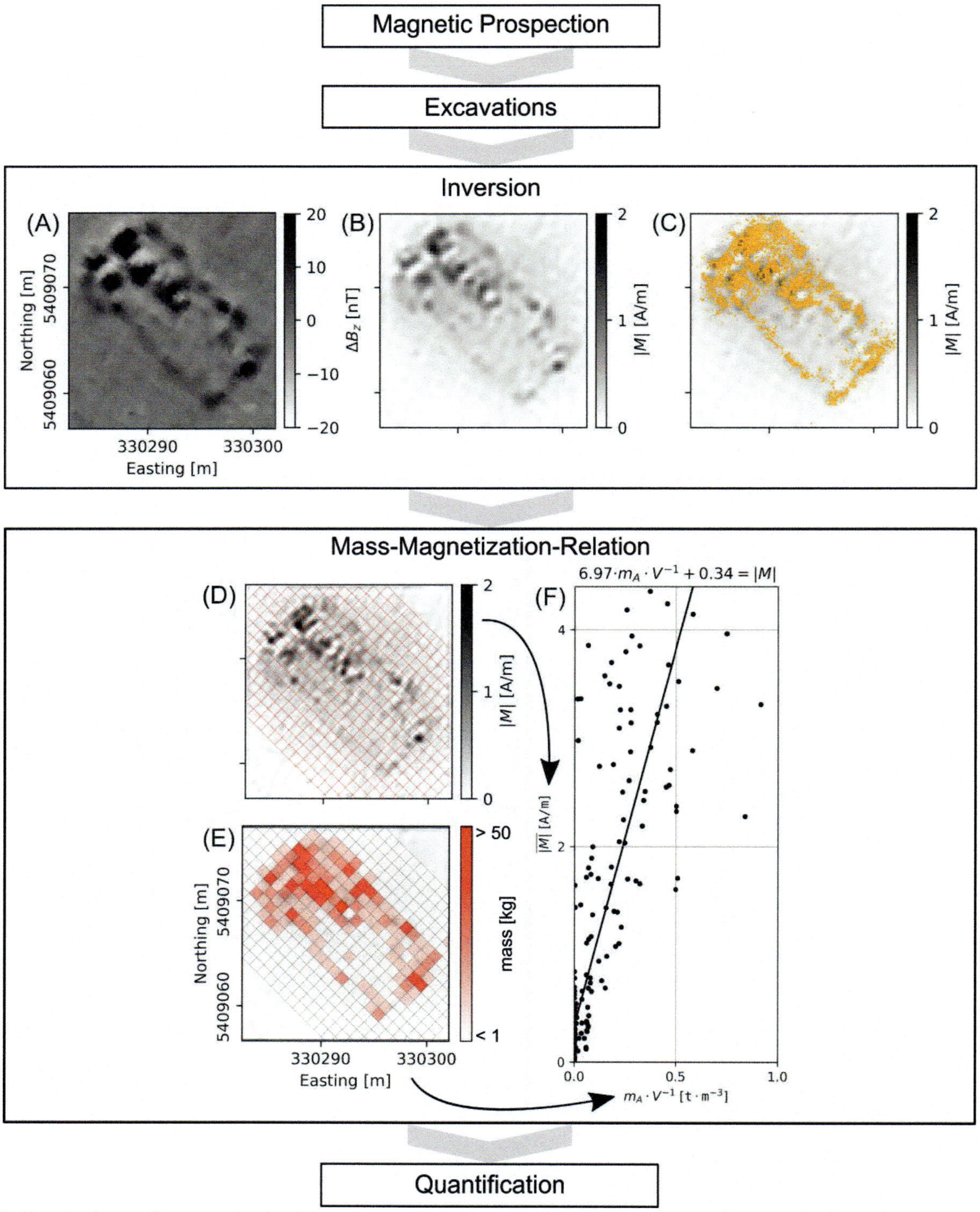

Fig. 2. Novel scheme for quantitative interpretation of magnetic prospection data with example data of the excavated megastructure. (A) Magnetic map, (B) calculated magnetization, (C) magnetization distribution overlaid with documented daub in the excavation (orange). Note that areas of increased magnetization overlap with the documented daub. (D) Magnetization distribution with excavation squares, (E) documented mass of daub and pottery per excavation square, (F) orthogonal distance regression of mass and mean magnetization per excavation square.

Results

For the site of Maidanetske, we have applied the interpretation scheme to 45 buildings. They have been chosen because either direct information on the depth range was known from excavations, test trenches and drillings or the information was available in the vicinity. Furthermore, we covered with these buildings a selection of the classifications 'burned', 'unburned/eroded' and 'megastructure' based on the interpretation of the magnetic prospection data.

For the chosen set of houses, the magnetization raises up to 5 A/m. We compared the magnetization distribution (Fig. 3A) with the standardized floor plans (Fig. 3B) of the houses belonging to the Tomashivka regional group presented by Chernovol (2012). This enables us to determine the orientation of the houses.

The application of the mass-magnetization relation yielded total masses from a few tens of kilograms for only slightly or partly burned houses and up to 7t for large burned dwellings and other building structures. Interpreting these with respect to the floor area of the houses, it tentatively indicates two different subsets of houses (Fig. 3C). Possible explanations might be differences in the construction, the inventory, the burning conditions or the preservation.

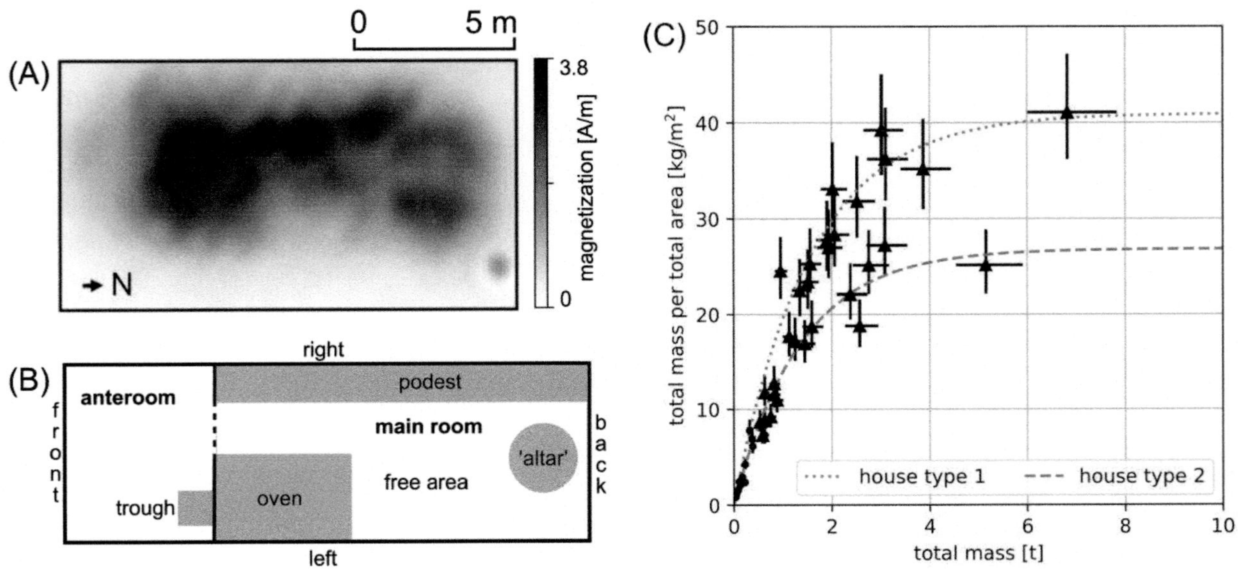

Fig. 3. Example results of the interpretation scheme. (A) Magnetization distribution for one unexcavated building in comparison to the schematic floor plan (B) (after Chernovol 2012). Note that areas of increased magnetization can match to the location of the immovable inventory. (C) Estimated masses of unexcavated buildings with a tentative interpretation of two different types of houses.

Conclusions

This is a new approach for a quantitative interpretation of magnetic prospection data. With this approach, we calculated the magnetization distribution of 45 unexcavated buildings and estimated the masses of their remains. This enabled us to deduce the orientation of the houses and give a preliminary interpretation of two different house types. With this approach, we demonstrated how excavation results can be extrapolated to a whole site by magnetically guided upscaling. Therefore, we can add new data to the analysis of complete sites that might develop a thorough understanding of settlement dynamics without invasive excavation techniques.

Acknowledgements

We are grateful to the German Research Foundation - Project Number 2901391021 – SFB 1266 for funding the presented research.

References

Boggs, P. and Rogers, J. 1990. Orthogonal Distance Regression. In Brown, P. and Fuller, W. (eds.) *Statistical analysis of measurement error models and applications: proceedings of the AMS-IMS-SIAM joint summer research conference held June 10-16, 1989*, 186–194.

Branch, M.A., Coleman, T.F. and Li, Y. 1999. A Subspace, Interior, and Conjugate Gradient Method for Large-Scale Bound-Constrained Minimization Problems. *SIAM Journal on Scientific Computing.* 21(1). 1–23.

Chernovol, D. 2012. Houses of the Tomashovskaya Local Group. In Menotti, F. and Korvin-Piotrovskiy, A. (eds.), *The Tripolye Culture Giant-Settlements in Ukraine. Formation, Development and Decline*, 183–209. Oxford and Oakville.

Müller, J., Hofmann, R., Brandtstätter, L., Ohlrau, R. and Videiko, M.Y. 2016. Chronology and Demography: How Many People Lived in a Mega-Site? In J. Müller, K. Rassmann and M.Y. Videiko M (eds.), *Trypillia Mega-Sites and European Prehistory 4100-3400 BCE*, 133–170. London/New York.

Müller, J., Hofmann, R., Kirleis, W. *et al.* 2017. *Maidanetske 2013: New excavations at a Trypilia mega-site.* Bonn.

Ohlrau, R. 2018. *Maidanets'ke. Development and decline of a Trypillian mega-site in Central Ukraine.* Unpublished PhD thesis, Kiel University.

Pickartz, N., Hofmann, R., Dreibrodt, S. *et al.* 2019. Deciphering Archaeological Contexts from the Magnetic Map: Determination of Daub Distribution and Masses of Chalcolithic House Remains. *The Holocene.* https://doi.org/10.1177/0959683619857238

Rassmann, K., Ohlrau, R., Hofmann, R. *et al.* 2014. High precision Tripolye settlement plans, demographic estimations and settlement organization. *Journal of Neolithic Archaeology.* 16. 96–134.

Дудкин, В.П. 1978. Геофизическая разведка крупных трипольских поселений. In Генинг, В.Ф. (ed.) *Использование методов естественных наук в археологии*, 35–45. Киев.

Applying Magnetic Depth Estimation Techniques to Archaeo-geophysics

Jeremy G. Menzer
University of Arkansas, Environmental Dynamics PhD Program, Fayetteville, United States

Magnetometry is possibly the most widely used archaeo-geophysical technique in the world. However, a major drawback is the lack of depth information to anomalous source bodies. In fact, many novice archaeo-geophysical users are under the impression that magnetometry does not or cannot provide depth information. Yet, depth estimation is an active area of research and is commonly applied in more traditional geologic studies. This study works towards a comprehensive review and real-world testing of various depth estimation techniques and their applicability to archaeology.

Recent trends in archaeo-geophysics have been to increase spatial data resolution and survey speeds with large motorized arrays and cart systems. This is an important area of research, however, this has occurred in conjunction with a shift in the user group of archaeo-geophysicists from early pioneers having more traditional physics training to archaeologists specializing in geophysics. Due to this, fewer users are focused on further developing the method and more are simply using the technology to map archaeological features. This research aims to rediscover and modernize depth estimation techniques for archaeological applications in the hope to increase the quantitative methodologies available to archaeo-geophysicists.

Previous evaluation of depth estimation techniques on modelled data provided promising results, therefore, this study evaluates real-world data collected at a controlled archaeological test site in Illinois and multiple pre-historic Native American sites in Arkansas and Tennessee, USA (Isaacson *et al.* 1999). The study focuses on applying simple depth estimation techniques including half-width rules and multi-height measurements (Weymouth 1976; 2000, Aspinall *et al.* 2009). However, more complex methods including Euler deconvolution and centre of magnetism are also considered (Helbig 1963, Reid *et al.* 1990, Desvignes *et al.* 1999). These techniques are performed on 2.5 dimensional magnetic data and verified through down-hole magnetic susceptibility measurements (Dalan *et al.* 2011). Some other depth estimation techniques include: complex attribute analysis, werner deconvolution, equivalent dipole, downward continuation of equivalent stratum, and certainly even more techniques, but these are not included in this study for brevity (Tsokas and Hansen 1995; 2000, Desvignes *et al.* 1999).

All depth estimation techniques are performed on deferentially corrected total-field data that has been reduced to the pole. Thus far, half-width rules over-estimate the depth to most features by more than 75cm, however, this is possibly due to the method showing the greater width of the features. Multi-height techniques provide better results for more point source like anomalies with possible erroneous estimations resulting from data collection errors between the two height surveys. Euler deconvolution estimates range from +/- 2cm-15cm from the peak of magnetism provided by the down-hole magnetic susceptibility.

A subset of magnetic data from the Tennessee site, Runion, shows a sub-selection of three archaeological features F, G, and I (Fig. 1a). An example of Multi-height and Euler deconvolution depth estimates using a structural index (SI) of both two and three provide a comparison between estimation techniques and down-hole magnetic susceptibility results for features F, G, and I (Fig. 1b, 1c, 1d). Feature F's depth is well categorized by both Multi-height and Euler methods when using a SI of three. This appears appropriate given visual interpretation of the magnetic feature, as it is slightly dipolar and quite small, compared to the other features examined, suggesting the magnetic feature is point-source like (e.g. spherical). Feature G's depth is over estimated when using an SI of three and underestimated with an SI of two, however, both techniques have similar results. Upon visual inspection this magnetic anomaly is larger and more broad than Feature F. The error in depth estimation is likely due to a more amorphous source body shape, but a valuable estimation, being off by 12cm, is still produced. Feature I contains two magnetic peaks, which are independently well categorized by Euler methods. Multi-height techniques fail to properly estimate the depth, likely due to measurement error and differences between multiple surveys.

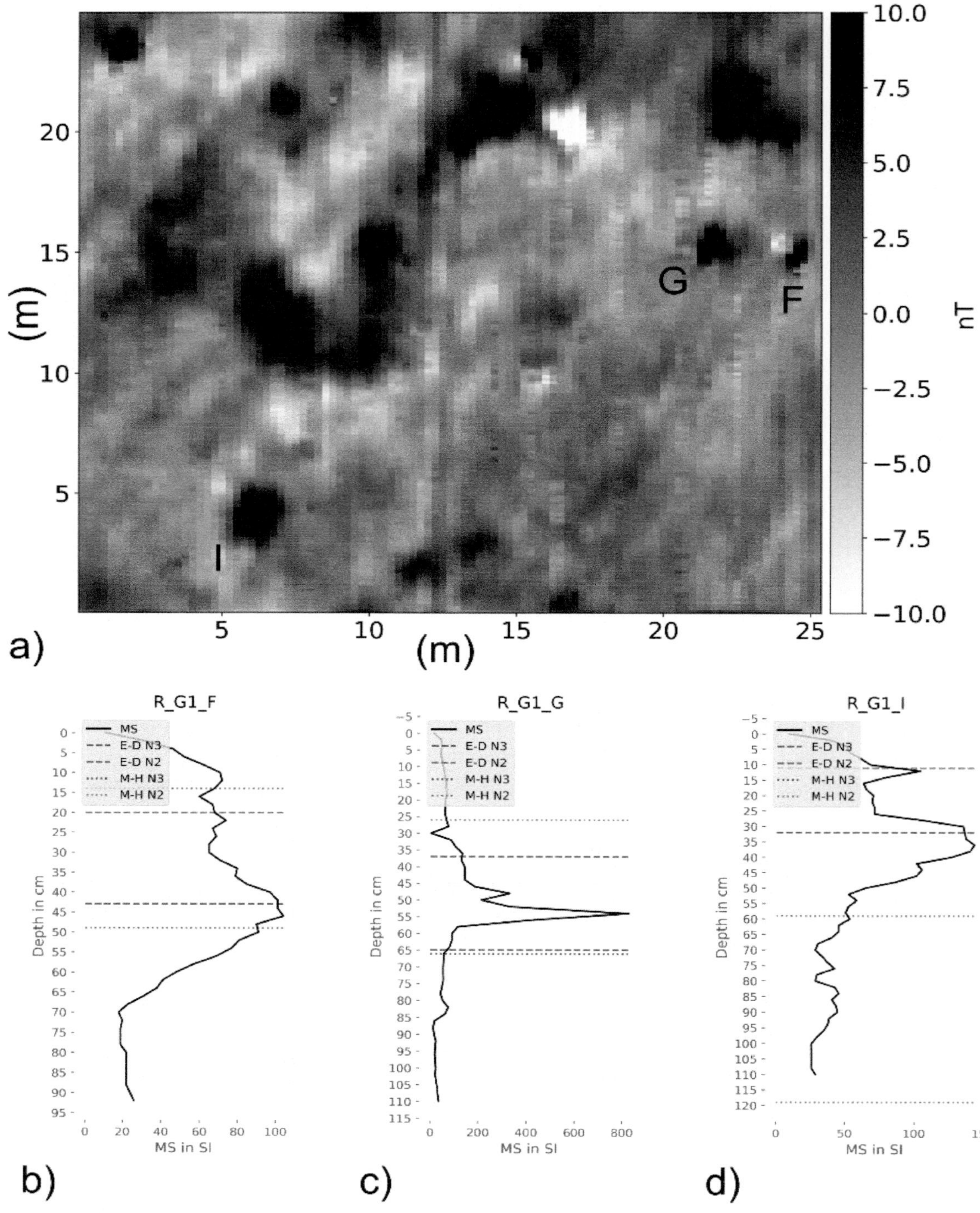

Fig. 1. (a) Shows a subset of the lower sensor survey at Runion, Tennessee with three features indicated G, F, and I. (b), (c), and (d) Black line shows the down-hole magnetic susceptibility compared to Multi-height (short dash) and Euler deconvolution (long dash) depth estimates using a structural index of two (green) and three (red), for features F, G, and I respectively.

Although further examination is required, early results show promise that depth information from magnetic surveys is attainable in an archaeological context. With further refining of the techniques, this methodology is sure to provide much needed and applicable depth information for academic research and cultural and heritage management practices alike. This advances magnetometry to a true three-dimensional method expanding the archaeo-geophysicist's toolbox.

References

Aspinall, A., Gaffney, C.F. and Schmidt, A. 2008. *Magnetometry for Archaeologists, Geophysical Methods for Archaeology*. AltaMira Press.

Dalan, R.A., Bevan, B.W., Goodman, D., Lynch, D., De Vore, S., Adamek, S., Martin, T., Holley, G. and Michlovic, M. 2011. The Measurement and Analysis of Depth in Archaeological Geophysics: Tests at the Biesterfeldt Site, USA. *Archaeological Prospection.* 18. 245–265.

Desvignes, G., Tabbagh, A. and Benech, C. 1999. The determination of the depth of magnetic anomaly sources. *Archaeological Prospection.* 6. 85–105.

Helbig, K. 1963. Some integrals of magnetic anomalies and their relation to the parameters of the disturbing body. *Zeitschrift fur Geophysik.* 29. 83–96.

Isaacson, J., Hollinger, R.E., Gundrum, D. and Baird, J. 1999. A Controlled Archaeological Test Site Facility in Illinois: Training and Research in Archaeogeophysics. *Journal of Field Archaeology.* 26. 227–236.

Reid, A.B., Allsop, J.M., Granser, H., Millett, A.J. and Somerton, I.W. 1990. Magnetic interpretation in three dimensions using Euler deconvolution. *Geophysics.* 55. 80–91.

Tsokas, G.N. and Hansen, R.O. 1995. A Comparison of Inversve Filtering and Multiple Source Werner Deconvolution for Model Archaeological Problems. *Archaeometry.* 37. 185–193.

Tsokas, G.N. and Hansen, R.O. 2000. On the Use of Complex Attributes and the Inferred Source Parameter Estimates in the Exploration of Archaeological Sites. *Archaeologcial Prospection.* 7. 17–30.

Weymouth, J.W. 1976. *A Magnetic Survey of The Walth Bay Site (39WW203).* Midwest Archaeological Center.

Weymouth, J.W. 2000. *On the Relationship Between the Profile of a Magnetic Anomaly and the Depth of the Source*. Unpublished Report.

Modelling the layer between topsoil and subsoil using magnetic prospection data

Alois Hinterleitner[1,2], Ralf Totschnig[1], Klaus Löcker[1,2], Hannes Schiel[1,2], Bendeguz Tobias[3]
[1]Central Institute for Meteorology and Geodynamics, Vienna, Austria, [2]Ludwig Bolzmann Institute for Archaeological Prospection and Virtual Archaeology (LBI ArchPro), Vienna, Austria, [3]Institut für Archäologien, Fachbereich Mittelalter- und Neuzeitarchäologie, Innsbruck, Austria

Introduction

Based on the detection and classification of magnetic anomalies identified by Hinterleithner *et al.* (2015), we now go one step further and try to model these detected structures. This method has classified two types of magnetic anomalies: near-surface iron objects and subsoil pits and ditch structures. The modelling is therefore also carried out in two steps with two different methods, one for near-surface iron objects and one for subsoil pit-like structures.

First, the classified iron objects are modelled as simple dipoles. This not only determines the exact location, but also the depth of the object. In addition, an approximate size of the magnetic object can be derived from the field vector obtained by the modelling.

In order to model the subsoil pit-like structures, a susceptibility distribution below the arable layer is calculated which best represents the measured magnetic anomalies. Although it is not possible to calculate the depth and volume of the (archaeological) structures, the exact position and size of the structures on the surface below the arable layer is possible. The resulting susceptibility distribution of the layer between topsoil and subsoil can be interpreted as a visualization of the first interface of an archaeological excavation.

Modelling of near-surface dipolar anomalies

The method for detection and classification of magnetic anomalies (Hinterleitner *et al.* 2015) delivers a list of near-surface dipole-like structures with an approximate position and a known maximum anomaly value. Based on these parameters, a field vector with a certain position, direction and field strength is assumed and used for starting the optimization process. In an iterative process, this field vector is randomly changed in position, direction and field strength until the magnetic anomaly caused by this field vector corresponds best to the measured data in the surrounding area. The approximation algorithm "leaped annealing" (Eder-Hinterleitner *et al.* 1996) is used to minimize the difference between the reconstructed and the measured magnetic anomaly.

Several parameters of the field vector of each reconstructed near-surface iron object are written to a shapefile and can therefore be displayed and further processed in any geographical information system. The horizontal position, the depth, the field strength, the inclination and the declination describe the magnetic field vector. The minimum and maximum values of the reconstructed anomaly, the absolute difference to the measured anomaly and the relative difference to the measured maximum value describe the reconstructed magnetic anomaly and the quality of the reconstruction. The reconstruction algorithm is fast enough to reconstruct several dipoles per second.

The magnetic anomalies of the reconstructed near-surface dipole-like structures are output as magnetic georeferenced raster data and subtracted from the magnetic data. The georeferenced raster data without surface-near dipole anomalies are now used for the second step, the modelling of the subsoil pit-like structures.

Modelling of the susceptibility distribution in the layer between topsoil and subsoil

A simple rasterized model is chosen to model the susceptibility distribution in the layer between topsoil and subsoil (Scollar 1990: 428-431). A grid with a side length of 10cm is assumed, which lies 50cm below the ground surface. Each grid point represents a magnetic dipole with a field vector whose direction corresponds to the earth's magnetic field vector. The strength of this field vector depends on the magnetic susceptibility of this raster point.

First all susceptibilities are set to zero. Now the optimization algorithm "leaped annealing" (Eder-Hinterleitner *et al.* 1996) is applied to determine the susceptibility of each raster point. In an iterative process, the susceptibility of each raster point is changed randomly until the magnetic anomalies thus calculated best match the anomalies measured. Since raster points with very high susceptibility are repeatedly formed next to many raster points with very low susceptibility, a selective average filter is applied several times to the susceptibility-raster during the optimization process so that these susceptibilities are distributed more evenly. This is necessary because otherwise no restrictive assumptions about the distribution of susceptibilities are assumed.

The optimization of this susceptibility distribution is more time-consuming than the reconstruction of the near-surface dipoles. Although only one parameter is varied here, one million grid points have to be optimized for an area of 1ha, which also influence each other if they are close enough to each other. In contrast, usually only a few hundred near-surface dipole structures per hectare have to be reconstructed in the first step.

Example at Podersdorf

The application of these methods is demonstrated at a part (area 68m x 64m) of the magnetic prospection of Podersdorf, Austria, (Fig. 1a) using a motorized Foerster Fluxgate system with a magnetic gradient of 0.65cm and 8 parallel sensors mounted at a distance of 25cm developed by LBI ArchPro and ZAMG Archeo Prospections. The very small magnetic anomalies show various archaeological structures: ditches, graves, pits and postholes. Since the site has been known for a long time and was frequently visited by hobby archaeologists with metal detectors, only a few near-surface iron parts are visible (Fig. 1b). These are modelled very differently here. Some anomalies disappear almost completely, others leave large anomalies in the difference image (Fig. 2a).

We interpret these results to mean that near-surface iron parts that can be very well reconstructed are very similar to a dipole and are therefore rather spherical as an object. Magnetic anomalies that cannot be well modelled by a dipole are not spherical but elongated, e.g. a long screw or a piece of wire. In order to model these objects better, magnetic models other than a simple dipole must be used in future developments.

The difference data (Fig. 2a) is the starting point for modelling the layer between topsoil and subsoil. The reconstructed susceptibility distribution (Fig. 3b) clearly shows all possible pits, ditches and graves. Even the very small anomalies of the ditches lead to recognizable structures. The graves are very well reconstructed and even the smallest post holes are modelled. The magnetic anomaly image of the reconstructed (archaeological) structures (Fig. 2b) is free from large dipole-like disturbances and from all influences of the arable layer. The archaeological structures are clearer and easier to recognize and therefore much easier to understand and to interpret archaeologically.

The difference image of the measured data and all modelled anomalies (Fig. 3a) shows everything that was not modelled. In addition to the poorly modelled strong anomalies, these are above all the influences of the arable layer. No anomalies of possible archaeological structures are visible in this image, which in turn shows that the modelling of the layer between topsoil and subsoil includes all essential archaeological structures.

The calculation of the total 4.8ha area models 4.8 million susceptibility values and takes about 4 minutes on a modern PC workstation by calculating 157 million changes. The mean deviation between measuring points and modelled values is 0.13nT.

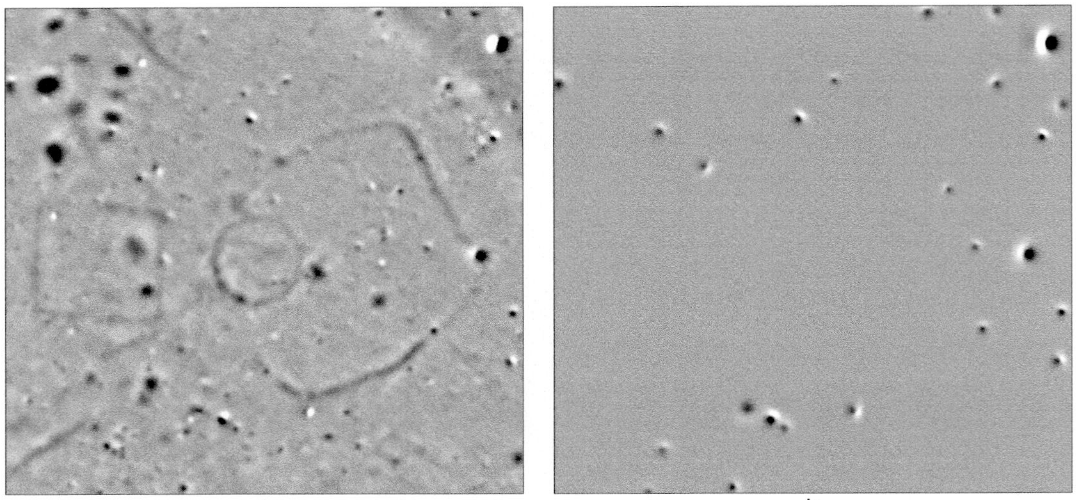

a b

Fig. 1. a) Part of the magnetic prospection of Podersdorf. b) Magnetic anomalies of the modelled near-surface dipoles. Visualization [white ... black]: [-2 ... +3nT].

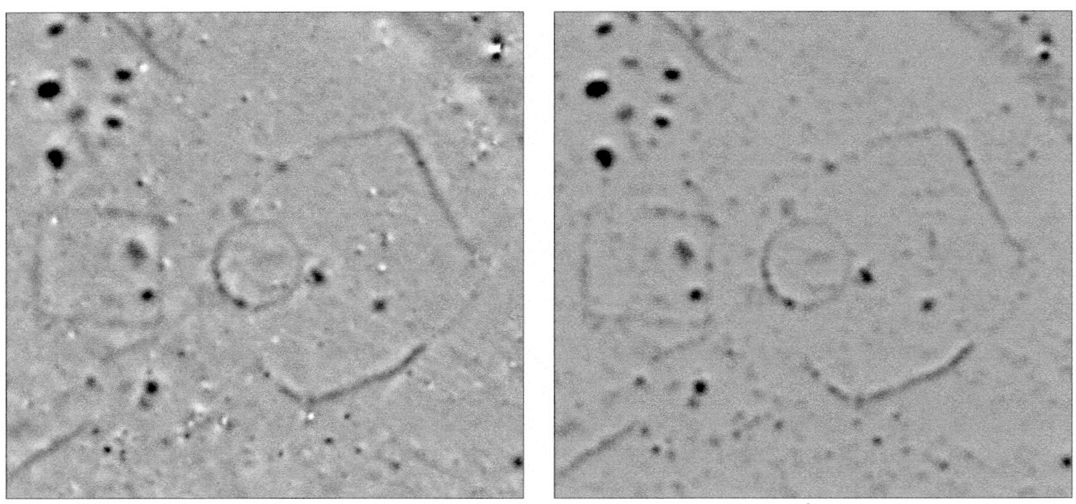

a b

Fig. 2. a) Magnetic anomalies minus the modelled near-surface iron parts. b) Magnetic anomalies of the modelled susceptibility distribution. Visualization [white ... black]: [-2 ... +3nT].

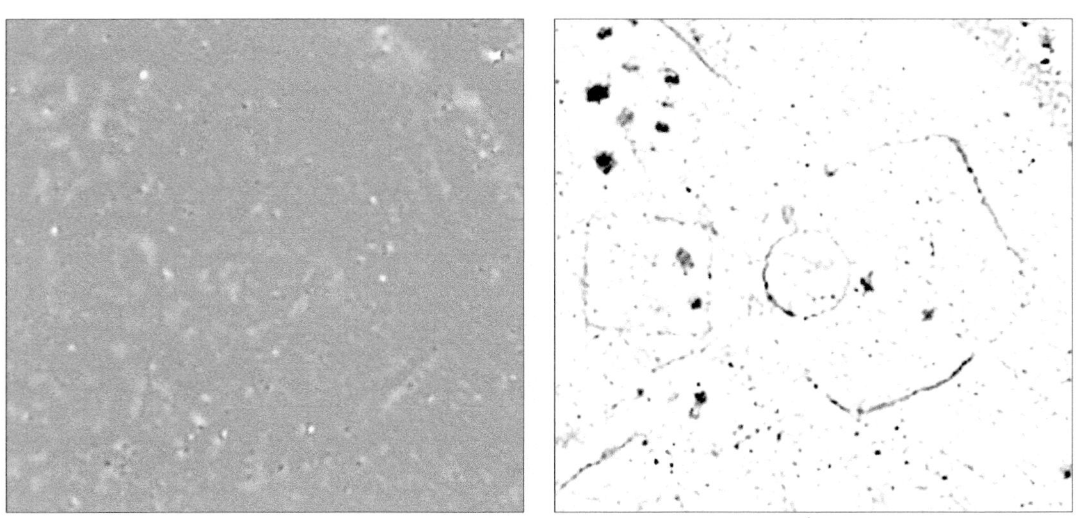

a b

Fig. 3. a) Magnetic data without the modelled magnetic anomalies. Visualization [white ... black]: -2 ... +3nT]. b) Relative susceptibility distribution. Visualization [white ... black]: [0 ... 3].

Conclusion

The modelling of near-surface iron parts and the layer between topsoil and subsoil supports the archaeological interpretation considerably. The reconstructed dipole anomalies provide accurate depth information and a rough estimate of the size of the object. The modelled archaeological structures are more accurate in extent and position than by drawing from anomaly visualizations. Weak structures are easier to recognize and above all their (spatial) connection becomes more recognizable. The modelling of the anomalies also provides a division of the magnetic measurement data according to their origin into three areas: strongly magnetic objects, influence of the arable layer and possible archaeological structures.

References

Eder-Hinterleitner, A., Neubauer, W., Melichar, P. 1996. Reconstruction of archaeological structures using magnetic prospection. In Kamermans, H. and Fennema, K. (eds.) *Interfacing the Past. Computer Applications and Quantitative Methods in Archaeology CAA95*. University of Leiden 1996. 131-137.

Hinterleitner, A., Löcker, K., Neubauer, W., Trinks, I., Sandici, V., Wallner, M., Pregesbauer, M. and Kastowsky-Priglinger, K. 2015. Automatic detection, outlining and classification of magnetic anomalies in large-scale archaeological magnetic prospection data. In A. Rzeszotarska-Nowakiewicz (ed.) Archaeological Prospection. 11th International Conference on Archaeological Prospection. Warsaw, Poland, 15-19.09.2015, The Institute of Archaeology and Ethnology, Polish Academy of Sciences. *Archaeologia Polona*. 53. 296-299.

Scollar, I. 1990. Prediction of magnetic anomalies. In *Archaeological Prospection and Remote Sensing*. Cambridge University Press, Topics in Remote Sensing. 423–450. Cambridge.

Pathways to prediction: quantifying the impact of soil moisture variations on electric and electromagnetic contrast

Philippe De Smedt, Daan Hanssens
Department of Environment, Ghent University, Ghent, Belgium

While seasonal influences on the electrical discrimination of archaeological features have been investigated frequently, such studies rarely provide a broadly applicable analytical framework. This leaves extrapolating findings from archaeological monitoring projects often up to generalized assumptions such as indicative estimates of optimal survey moments after periods of rainfall (Schmidt *et al.* 2017), or on the impact of moisture variations on soils with coarse or fine-grained sediments (Boddice 2015). Although findings of such studies contribute to the survey decision-making process, they do not offer a reliable quantitative assessment beyond the monitored sites (as, for instance, indicated by Fry (2014)), nor do these offer site- or instrument specific certainties.

The research we present, aims to bridge this gap through both an experimental and a theoretical approach. Starting from monitoring seasonal variations of a specific test-site, a synthetic environment was constructed to assess the impact of changes in moisture balance on the contrast observed with different types of electromagnetic and earth resistance instrumentation.

We present the results from a wetland site in the Netherlands (Wervershoof), where known ploughed out Bronze Age burial mounds were investigated both in wet and dry conditions (Van Der Heiden *et al.* 2018). One of these burial mounds was targeted to test the influence of moisture balance on the electrical discrimination potential of similar features in the area. By collecting information on relevant soil properties, and conducting frequency-domain electromagnetic (FDEM) survey at different moisture conditions, we aimed to develop a procedure to predict the contrast of archaeological features in the (quadrature phase and in-phase) response recorded by FDEM instruments.

Starting from a modified version of Archie's law for unconsolidated sediments (Generalized Archie's Law (Shah and Singh 2005)), we evaluated the volumetric water content, electrical conductivity of the pore solution (σ_w), volumetric moisture content (θ) and clay content (CL) from features rendering variable contrast under wet and dry conditions. Two reference profiles were established to evaluate the potential of a synthetic procedure to predict electrical contrast under varying moisture conditions.
The synthetic procedure builds on a Monte-Carlo analysis to evaluate the influence of varying θ, CL and σ_w on the electrical conductivity of soils (σ). Fig. 1 shows results of such an analysis, whereby the bulk electrical conductivity was determined for 15,000 synthetic soils with varying clay and moisture content and different soil water conductivity. This analysis clearly shows the governing influence of moisture on the bulk soil electrical conductivity.

Based on such relationships, the contrast between two soil profiles with *n* layers with a given θ, CL, and σ_w can be evaluated, and can be expanded by integrating layer-specific moisture permeabilities. FDEM responses were modelled using a 1D forward modelling algorithm (Hanssens *et al.* 2019b).

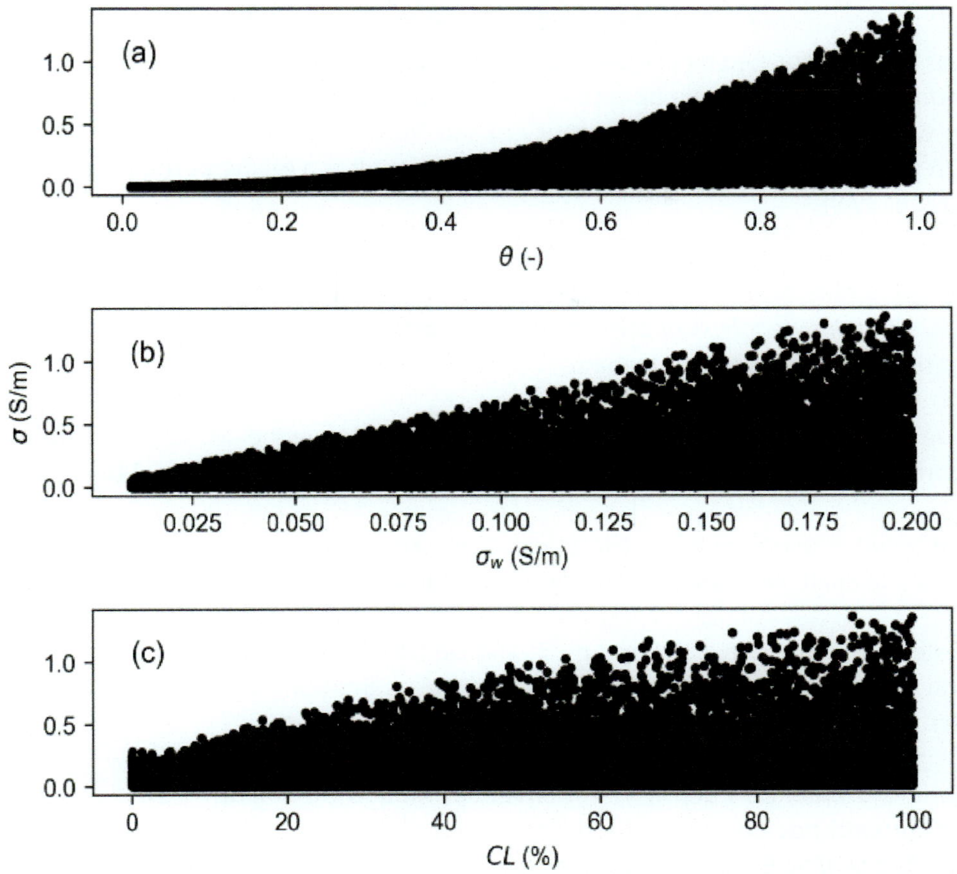

Fig. 1. Monte-Carlo (random uniform) sampling (15,000 samples) of the Generalized Archie's Law (Eq. 1). The soil's bulk electrical conductivity σ (S/m) is expressed in function of (a) the volumetric moisture content θ (-), (b) the electrical conductivity of the pore solution σ_w (S/m) and (c) the clay content CL (%).

Through FDEM measurements and soil samples obtained at a Wervershoof, we were able to evaluate the efficiency of this procedure to provide a reliable assessment of varying electrical contrast under changing moisture conditions. While the procedure allowed for accurately reproducing the contrasts observed in the FDEM surveys conducted under both dry and wet conditions at the site of Wervershoof, a discrepancy exists in the obtained absolute values. At Wervershoof, this resulted in an overestimation of the absolute modelled FDEM responses compared to the observed *in situ* variation. This inconsistency is related to a stacking of multiple errors: the influence of relative calibration of FDEM measurements (Delefortrie *et al.* 2014); effects of anisotropy on the measured responses as opposed to 1D modelled responses (Tølbøll and Christensen 2007); the inherent simplification of assumptions (linearity) in the Generalized Archie's Law (Shah and Singh 2005); theoretical simplifications in the 1D forward model (Ward and Hohmann 1987, Hanssens *et al.* 2019a; 2019b); and errors on measured soil parameters. For the latter, the influence of changing soil moisture conductivities through the soil profile cannot be underestimated.

Although such inherent limitations are difficult to surpass, our study shows the potential of modelling approaches as a fast, quantitative estimator of the detection potential of specific electrical contrasts under varying moisture conditions. By integrating a Monte-Carlo analysis, the entire range of possible variations of relevant parameters can be sampled, allowing to recreate any real-world scenario. Based on this broad application potential, we have integrated this analytical procedure into an open-source tool that enables users to evaluate the influence of moisture variations when a limited number of site-specific properties are known. By integrating the appropriate forward modelling code, this software forms a first straightforward and user-friendly tool to predict the optimal combination of survey time and instrumentation for electrical prospection. Alongside assisting in developing survey strategies, modelling procedures such as these can provide a measure of uncertainty by allowing us to investigate which types of soil variations remain likely undetected under the survey moisture conditions.

Acknowledgements

This research was partly supported by the ISAP Fund, granted for the project "Predicting the effect of soil and moisture variations on the interpretive potential of FDEM survey". Surveys at Wervershoof were conducted in collaboration with the Cultural Heritage Agency of the Netherlands (supervised by Mr Menno Van der Heiden, and with kind permission of the landowner). P. De Smedt is a fellow of the Research Fund Flanders (grant number 1288117N).

References

Boddice, D. 2015. *Changing geophysical contrast between archaeological features and surrounding soil*. PhD doctoral dissertation, University of Birmingham.

Delefortrie, S., De Smedt, P., Saey, T., Van De Vijver, E. and Van Meirvenne, M. 2014. An efficient calibration procedure for correction of drift in EMI survey data. *Journal of Applied Geophysics*. 110. 115-125.

Fry, R. 2014. *Time-lapse geophysical investigations over known archaeological features using electrical resistivity imaging and earth resistance*. PhD, University of Bradford.

Hanssens, D., Delefortrie, S., Bobe, C., Hermans, T. and De Smedt, P. 2019a. Improving the reliability of soil EC-mapping: Robust apparent electrical conductivity (rECa) estimation in ground-based frequency domain electromagnetics. *Geoderma*. 337. 1155-1163.

Hanssens, D., Delefortrie, S., De Pue, J., Van Meirvenne, M. and De Smedt, P. 2019b. Frequency-Domain Electromagnetic Forward and Sensitivity Modeling: Practical Aspects of Modeling a Magnetic Dipole in a Multilayered Half-Space. *IEEE Geoscience and Remote Sensing Magazine*. 7(1). 74-85.

Schmidt, A., Fry, R., Parkyn, A., Bonsall, J. and Gaffney, C. 2017. When the time is right: the impact of weather variations on the contrast in earth resistance data. In Jennings, B., Gaffney, C., Sparrow, T. and Gaffney, S. (eds.) *AP2017. 12th International Conference of Archaeological Prospection*. Archaeopress. Oxford.

Shah, P.H. and Singh, D.N. 2005. Generalized Archie's Law for Estimation of Soil Electrical Conductivity. *Journal of ASTM International*. 2. 1-20.

Tølbøll, R.J. and Christensen, N.B. 2007. Sensitivity functions of frequency-domain magnetic dipole-dipole systems. *Geophysics*. 72. F45-F56.

Van Der Heiden, M., Feiken, H., Mauro, G.V., De Smedt, P., Verschoof-Van Der Vaart, W., Brinkkemper, O., Van Os, B., Laarman, F., Theunissen, E.M., Van Doesburg, J. and Brounen, F. (eds.) 2018. *Een grafheuvellandschap nader bekeken. Geofysisch en validerend boor- en proefputtenonderzoek van het rijksmonument Wervershoof-Eendenkooi (gemeente Medemblik)*, Amersfoort. Rijksdienst voor het Cultureel Erfgoed.

Ward, S.H. and Hohmann, G.W. 1987. *Electromagnetic Theory for Geophysical Applications*, Society of Exploration Geophysicists.

A 3D imaging procedure for subsurface magnetic susceptibility: application to the basaltic foundations of a Gallo-Roman villa in Auvergne, France

Julien Guillemoteau[1], Francois-Xavier Simon[2], Bertrand Dousteyssier[3], Marion Dacko[3], Guillaume Hulin[2], Jens Tronicke[1]

[1]University of Potsdam, Potsdam-Golm, Germany, [2]INRAP, France, [3]University of Clermont-Ferrand, Clermont-Ferrand, France

Background

Rigid-boom frequency domain electromagnetic induction (EMI) loop-loop sensors are commonly used for sensing the electrical conductivity and the magnetic permeability (or susceptibility) of the subsurface. The loop-loop (also referred to as Slingram) geometry consists of two coils, which respectively act as a transmitting and a receiving magnetic dipole. The transmitter generates an oscillating primary electromagnetic field interacting with the subsurface and the receiver measures the resulting secondary magnetic field. The obtained datum is by convention provided in the frequency domain as a complex ratio between the secondary magnetic field and the known primary magnetic field at the location of the receiver. The real part of this complex number thus reflects the in-phase response, while the imaginary part corresponds to the quadrature or out-of-phase response (both with regards to the phase of the primary excitation).

Portable EMI sensors operate at low and moderate induction numbers (<1). In these cases, the out-of-phase response is principally sensitive to the subsurface electrical conductivity and the in-phase response is typically used for sensing the magnetic permeability of the subsurface, although it contains a rather small "induction" fraction that is dependent on the electrical conductivity (Tite and Mullins 1970; Tabbagh 1986). Among efficient interpretation approaches, allowing the inversion of entire datasets, several joint 1D multi-layer inversions of (in-phase/out-of-phase) data are often carried out. These processing schemes use either a simultaneous (Farquharson *et al.* 2003) or a sequential (Guillemoteau *et al.* 2016) inversion framework. However, they are, by definition, limited for imaging local 3D targets as often expected in archaeo-geophysical prospections.

A rapid 3D linear inversion/deconvolution procedure was presented by Thiesson *et al.* (2017) for the case of a well- and even-determined inverse problem where, in practice, the number of fixed-depth layers is limited by the number of non-redundant data at each sounding location. This parametric restriction on the vertical direction limits the proper modelling of subsurface structures. To overcome this problem uisng a dense grid in the vertical direction, we propose to apply the 3D full-grid multi-channel deconvolution (MCD) procedure presented in Guillemoteau *et al.* (2017) to in-phase data. This method can handle a 3D voxel-based linear under-determined inversion subject to spatial smoothness constraints for a dense model grid (100 million parameters) and for extended and/or spatially densely sampled datasets as collected by portable EMI sensors (hundred thousand data points). The aim of this study is to evaluate the 3D MCD imaging procedures for interpreting EMI in-phase data in a context of archaeological exploration.

Methods

We remove the induction fraction from the recorded total in-phase response by modelling it with the subsurface conductivity information present in the out-of-phase data. The remaining magnetization fraction of the in-phase response, sensitive to the surrounding magnetic permeability, follows the magneto-static equations similar to the magnetic method but with an active magnetic source. The use of a source actually offers the possibility to collect data with several loop-loop configurations, which illuminate the subsurface with different sensitivity patterns and depth/volume extents. A good consequence is that the EMI in-phase data are not affected by the classical depth ambiguity, as it is encountered for the passive geomagnetic method. Moreover, due to the use of a time-varying source, the in-phase response is not sensitive to the static remnant magnetization, which often represents a source of complexity for the inversion of passive geomagnetic data. Klose *et al.* (2018) have evaluated and demonstrated the applicability of a 3D linear forward modelling approach and have applied it to reconstruct field in-phase data collected with a portable multi-configuration EMI sensor across a controlled permeable target.

Here, we apply the 3D MCD approach of Guillemoteau *et al.* (2017) - initially developed to interpret out-of-phase data - to in-phase data, by using the 3D sensitivity functions as derived in Klose *et al.* (2018). In Fig. 1, we show a synthetic imaging example for an anomalous four-configuration in-phase dataset computed for two perpendicular (PERP) geometries with intercoil distances of 1.1m and 2.1m, and two horizontal co-planar geometries (HCP), with intercoil distances of 1m and 2 m.

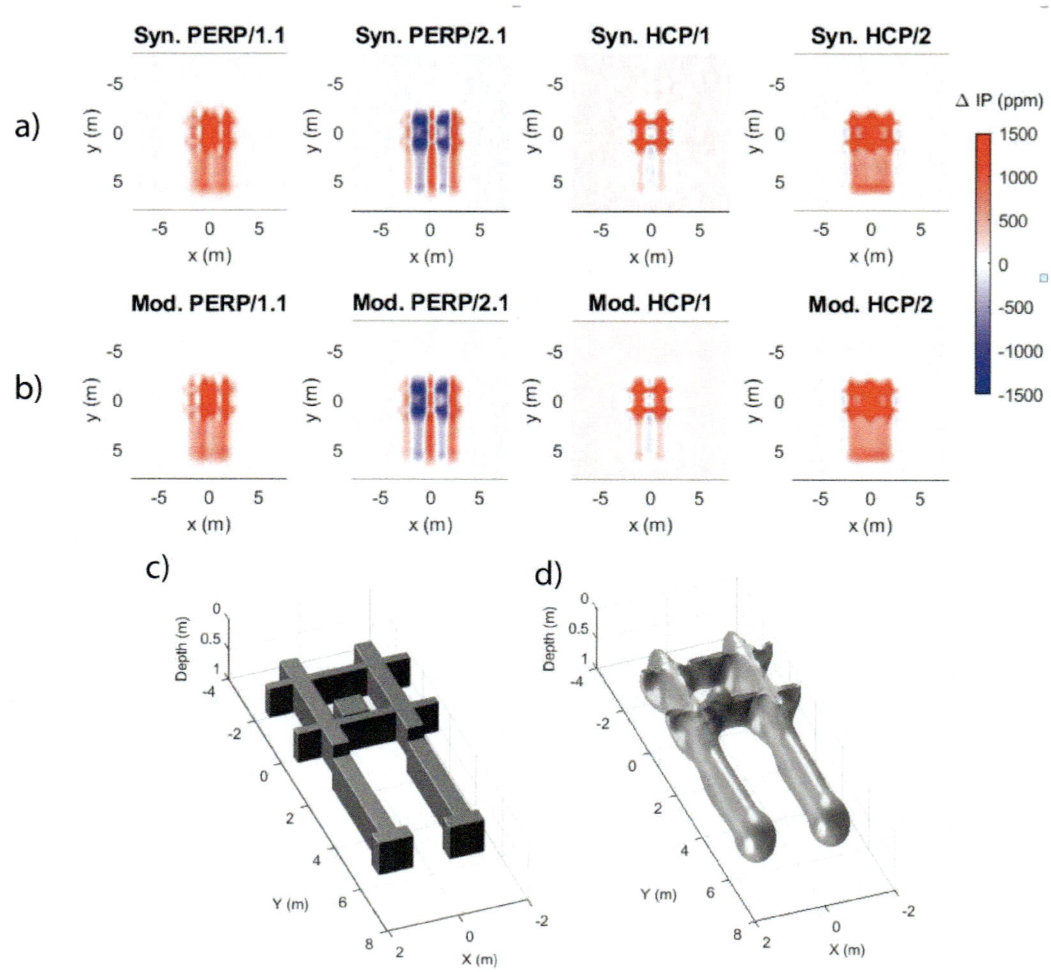

Fig. 1. Synthetic test with noisy EMI four-configuration in-phase data modelled at a height of 0.2m above ground. The data are interpreted in terms of an anomalous in-phase field. a) Synthetic dataset with an added 50 ppm random noise. b) Modelled data after 3D MCD imaging. c) True structure, which served to generate the synthetic dataset, with a relative magnetic permeability contrast of the background. d) Reconstructed subsurface model shown with an iso-surface representation.

Results

We applied our 3D MCD method to field data collected with a DUALEM21s sensor maintained at a height of 0.2m above ground, at the archaeological site of Lieu Dieu, located in La Sauvetat, Auvergne (France). This site holds the foundations of a Gallo-Roman Villa, which was mainly built with highly magnetic basalt rocks by the former inhabitants due to the proximity of the lava dome chain "*Chaîne des Puys*".

For the data acquisition, we followed the procedure described in Guillemoteau and Tronicke (2015) to efficiently record densely sampled EMI datasets (interline of 0.5m and in-line sampling of 0.1m) with a mobile acquisition setup, which is continuously positioned at a cm-precision by an auto-tracking total-station. This approach allows us to properly sample the 2D and 3D anomalies associated to each loop-loop configuration. The resulting pre-processed anomalous in-phase maps are shown in Fig. 2a for a fraction of the Lieu Dieu's villa.

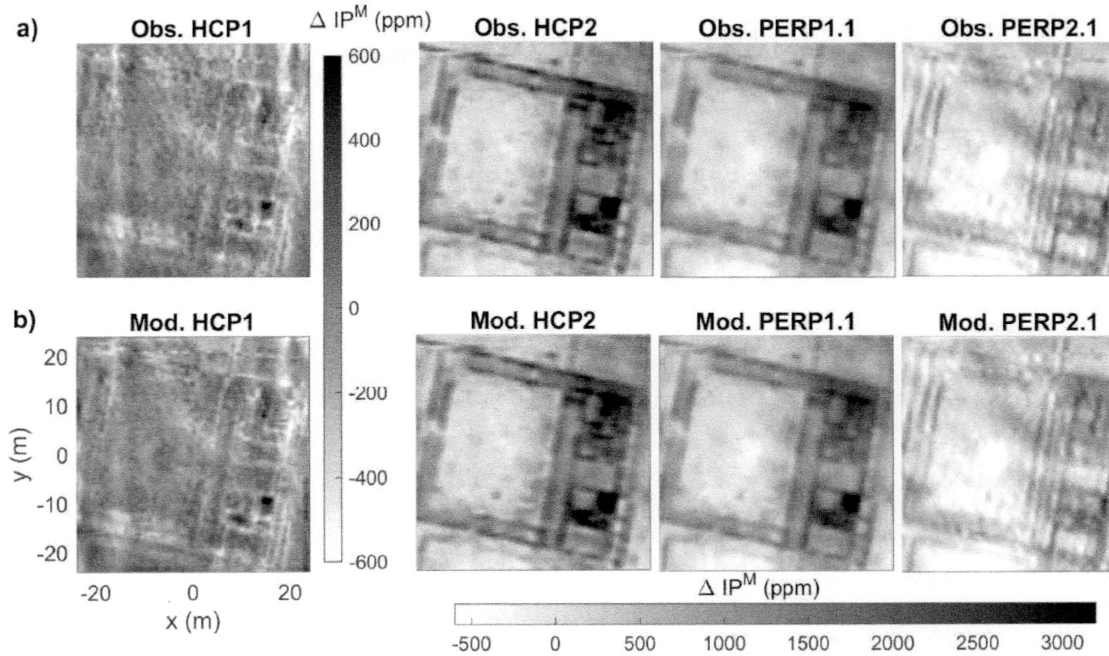

Fig. 2. Comparison between (a) the observed field data and (b) the modelled data computed after the 3D MCD imaging approach. The global RMS error including the four channels is 59 ppm.

The result of the 3D MCD method could provide rather detailed depth and lateral information about the different components of the villa. In Fig. 3, we show one slice of the resulting 3D voxel-based model of magnetic permeability at a depth of 0.4m.

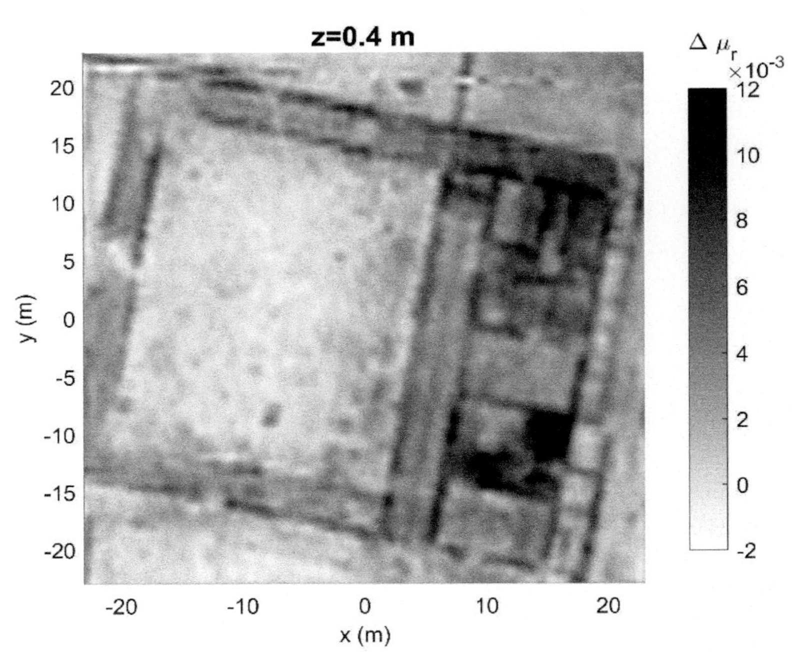

Fig. 3. Horizontal slice, at z=0.4m below ground surface, of the 3D model of magnetic permeability contrast resulting from the 3D MCD imaging approach.

Conclusion

The synthetic test has shown that our imaging algorithm is robust and versatile, for a complex 3D target with an extent that is smaller than the lateral footprint of the method. Our procedure finds direct applications on detailed imaging of the subsurface in permeable bedrock environments as in volcanic areas. The real field case study is an example of these applications, which illustrates how this method allows us to characterize the depth and shape of basaltic foundations buried in the first 2m of the subsurface with a sub-meter resolution.

Thanks to the on-going and expected future progresses in terms of the in-phase signal/noise ratio collected by EMI instruments, our method also appears to be a promising approach to image environments with low permeability contrasts, as found in non-volcanic areas.

References

Farquharson, C.G., Oldenburg, D.W. and Routh, P.S. 2003. Simultaneous 1D inversion of loop–loop electromagnetic data for magnetic susceptibility and electrical conductivity. *Geophysics*. 68.6. 1857-1869.

Guillemoteau, J. and Tronicke, J. 2015. Non-standard electromagnetic induction sensor configurations: Evaluating sensitivities and applicability. *Journal of Applied Geophysics*. 118. 15-23.

Guillemoteau, J., Simon, F-X., Lück, E. and Tronicke, J. 2016. 1D sequential inversion of portable multi-configuration electromagnetic induction data. *Near Surface Geophysics*. 14. 423-432.

Guillemoteau, J., Christensen, N.B., Jacobsen, B.H. and Tronicke, J. 2017. Fast 3D multichannel deconvolution of electromagnetic induction loop-loop apparent conductivity datasets acquired at low induction numbers. *Geophysics*. 82. E357-E369.

Klose, T., Guillemoteau, J., Simon, F-X. and Tronicke, J. 2018. Toward subsurface magnetic permeability imaging with electromagnetic induction sensors: Sensitivity computation and reconstruction of measured data. *Geophysics*. 83. E335-E345.

Tabbagh, A. 1986. Applications and advantages of the Slingram electromagnetic method for archaeological prospecting. *Geophysics*. 51. 576-584.

Thiesson, J., Tabbagh, A., Simon, F-X. and Dabas, M. 2017. 3D linear inversion of magnetic susceptibility data acquired by frequency domain EMI. *Journal of Applied Geophysics*. 136. 165-177.

Tite, M.S. and Mullins, C. 1970. Electromagnetic prospecting on archaeological sites using a soil conductivity meter. *Archaeometry*. 12. 97-104.

WuMapPy an open-source software for geophysical prospection data processing: 2019 milestone

Quentin Vitale[1,2], Lionel Darras[2], Christophe Benech[2], Julien Thiesson[3]
[1]*Éveha International, Lyon, France,* [2]*UMR-5133 Archeorient, Lyon, France,* [3]*UMR-7619 METIS, Paris, France*

Introduction

In the last few years, Python has shown a growing importance in the geophysical community, with an increasing number of available packages and plugins for general geophysical modelling and inversion (PyGIMLI — Cockett *et al.* 2015, SimpEG — Rücker *et al.* 2017) or more specifically dedicated to archaeological applications (ArchaeoPy — Pope-Carter *et al.* 2014, AGT plugin for QGIS — Hulin *et al.* 2017). In 2015, Marty *et al.* introduced WuMapPy, an open-source software implemented in the Python language for visualization and processing of near-surface geophysical data. It is the outcome of successive software developments initially achieved by the *Centre de Recherche Géophysique de Garchy* (that successively became the UMR 7619-Sisyphe and now the UMR 7619-METIS). WuMapPy is the successor of Wumap (implemented in Fortran, developed by Jeanne Tabbagh), whose development stopped in the late 2000s. WuMapPy's development started in 2015 by UMR 7619-METIS and UMR 5133-Archéorient. Since 2017, the main development has been taken over by Geo-Heritage, a partnership between the research laboratory UMR 5133-Archéorient and Éveha international, an archaeological research and investigations company.

Features and usage

WuMapPy benefits from the advantages of the Python language. It is entirely open-source, with an easy-to-read source code. Each processing step is documented with references to the corresponding scientific literature (Fig. 1). One can see how a specific processing step is implemented, modify it, or adapt it if needed. It is freely distributed on the "pypi.python.org" website and is cross-platform (Windows, Linux, MacOS, etc.). It can be upgraded with the variety of scientific packages freely available by the large "Python community" on the "pypi.python.org" website.

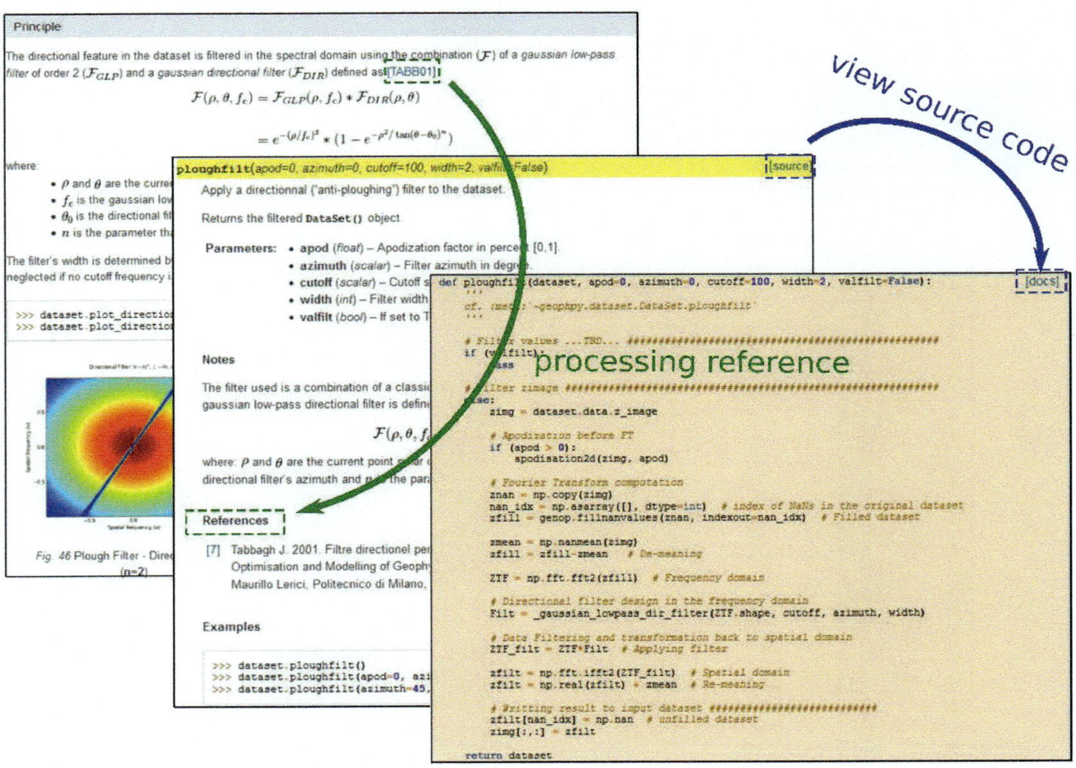

Fig. 1. Extract from GeophPy's documentation.

While still in active development, WuMapPy currently integrates a large number of processing and display features partly inherited from Wumap. Typically, you can:

- Import data files from ASCII Delimiter-Separated Values (.csv, .dat, .xyz, etc.) or binary format (NetCFD and Surfer grid).
- Grid the data using different interpolation algorithms (linear, cubic, etc.) or process the un-gridded data directly.
- Display the data with different plot types (scatter-plot, 2D-surface, contour-plot, etc.), colour scales (linear or logarithmic) and colour maps.
- Enhance the data using general processing such as data thresholding, data destaggering, outlier removal, median filtering, data destriping, directional filtering (anti-ploughing filter), FFT spectrum computation, regional trend elimination, contrast enhancement via Wallis filtering.
- Use specific filters for the magnetic method such as reduction to the pole, 3-D analytical signal computation, magnetic source depth estimation using Euler deconvolution, upward and downward continuation, conversion between different sensor configurations (total field, total field gradient and fluxgate).
- Georeference the data using GPS control points (for instance local grid corners with known GPS positions).
- Export the processed dataset in .kml, .png+tfw, .grd, .asc or other raster formats so it can be imported into GoogleEarth or other GIS software (ArcGis, QIS) without requiring third party application.

To ease maintenance and address different users' needs, the software is split into two sub-projects: the GeophPy module, where all the processing is actually done, and the WuMapPy module that gathers all the GUI features.

On one hand, users can use the available commands in GeophPy to create their own routines, scripts or notebooks to process and display the data.

On the other hand, non-programmers can work with WuMapPy's user-friendly GUI to process their data (Fig. 2). The GUI is multilingual and, even though only French and English are available at the moment, the use of a simple dictionary file eases future translations.

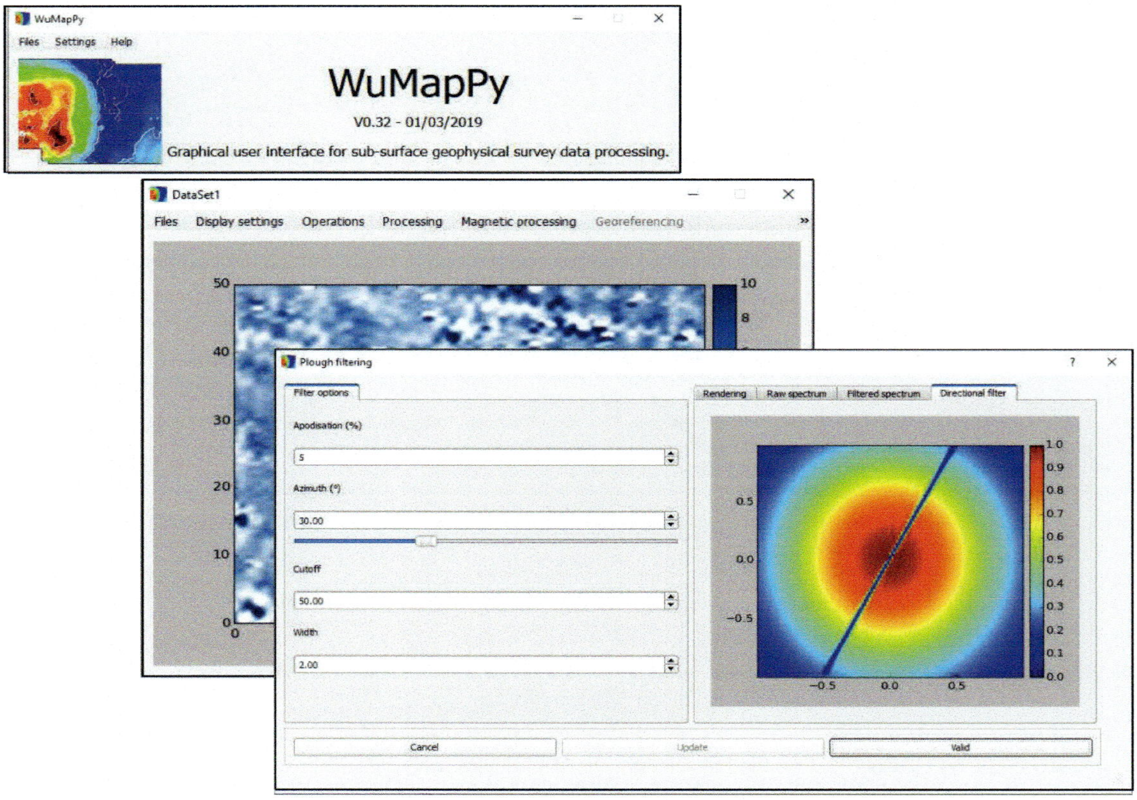

Fig. 2. Example of the WuMapPy graphical interface.

Future work and perspectives

Although one can process various types of data (magnetic, EM, resistivity) via the available general processing, WuMapPy is mainly devoted to process magnetic data for now, as only filters specific to the magnetic method have been implemented so far. To enrich the software scope, other method-specific filters such as the transformation of raw EM in-phase and quadrature data into physical properties (electrical conductivity, magnetic susceptibility, etc.) or 1-D resistivity sounding inversion will be added in the near future.

One of WuMapPy's objectives is to provide an accessible processing software that unveils the black box by revealing the scientific content; it is hence suited for students in their academic work.

WuMapPy is an open-source and multi-platform software project to display, process and georeference surface and sub-surface geophysical survey data in a user-friendly manner. It can be used by anyone wishing to post-process geophysical datasets, or just perform basic processing and display the geophysical image in a mapping application.

Acknowledgements

WuMapPy is a collaborative project whose development greatly benefits from all the feedback from the scientific community. The authors thank all the individuals that help through their advice, reviews, enhancement ideas and beta-testing.

Special thanks to Claire Bouligand from the University of Grenoble and her students, who recently tested the software during a geophysical tutorial class.

References

Cockett, R., Kang, S., Heagy, L.J., Pidlisecky, A. and Oldenburg, D.W. 2015. SimPEG: An open source framework for simulation and gradient based parameter estimation in geophysical applications. *Computers & Geosciences*. 85. 142–154.

Hulin, G., Simon, F.-X. and Hatami, N. 2017. AGT – Archaeological Geophysics Toolbox for QGIS. *ISAPNEWS*. 51. 7-8.

Marty, P., Darras, L., Tabbagh, J., Benech, C., Simon, F.-X. and Thiesson, J. 2015. WuMapPy – an open-source software for geophysical prospection data processing. *Archaeologia Polona*. 53. 563–566.

Pope-Carter, F., Harris, J.C., Sparrow, T. and Gaffney, C. 2014. ArchaeoPY: Constructing and Utilising Open Source Software for Archaeological Geophysics. *Recent Work in Archaeological Geophysics & Forensic Geoscience*. Future Horizons, Geological Society of London, Burlington House, Piccadilly, London.

Rücker, C., Günther, T. and Wagner, F.M. 2017. pyGIMLi: An open-source library for modelling and inversion in geophysics. *Computers & Geosciences*. 109. 106–123.

Automated segmentation of archaeo-geophysical images by convolutional neural networks

Melda Küçükdemirci[1,2], Apostolos Sarris[1,3]

[1]Laboratory of Geophysical - Satellite Remote Sensing & Archaeo-environment (GeoSat ReSeArch Lab), Foundation for Research & Technology (F.O.R.T.H.), Crete, Greece, [2]Istanbul University- Cerrahpasa Engineering Faculty Geophysical Engineering Department, Istanbul, Turkey, [3]The Archaeological Research Unit, Department of History and Archaeology, University of Cyprus, Nicosia, Cyprus

Background

Artificial Intelligence (AI), is most simply described as the ability of a computer program or a machine to think and learn as humans do, has gone from a science-fiction dream to a part of our daily life and scientific researches. For instance; extracting information out of images, carrying out segmentation aiming towards feature detection, classification and making the whole process more automated, have become very crucial for many scientific disciplines, which deal with image processing, such as medicine, dental analysis, microbiology and earth sciences. However, this stage of automation is computational and time intensive, especially when dealing with a large size or a number of images. Nowadays, in order to enhance the efficiency and accuracy of the automatic analysis of features in the images, deep artificial neural networks called Convolutional Neural Networks (CNNs) take the primary role. Although various approaches have been used in the past for the automatic analysis of aerial and satellite remote sensing data related to archaeological feature analysis (Toumazet *et al.* 2017, Trier *et al.* 2016; 2018, Traviglia *et al.* 2016), there has been less extensive experimentation of these techniques for geophysical prospection data (Leckebusch *et al.* 2008; Pregesbauer *et al.* 2014; Bescoby *et al.* 2016; Green *et al.* 2017; Linford and Linford 2017). Taking account of the large spatial extent of areas covered by archaeo-geophysical surveys - especially those using multi-sensor arrays driven by motorized carts - and the increasing quantity of the collected data, automated / semi-automated analysis seems to be beneficial for the preliminary interpretation. Starting from this point of view, in this paper, a specific Convolutional Neural Network, which has been generated via Python programming language and the Deep Learning Library of Keras, trained from scratch for automatic segmentation of anomalies in the archaeo-geophysical maps/images, is discussed.

Methodological Approach

Image segmentation is an interpretive process of image processin used to partition an image into meaningful parts, which have similar features or properties. Convolutional Neural Networks attempt to comprehend the relationship between the input and the output images and store the learned experience in their weights. Thus, once they are trained based on labelled / annotated specific images, they can recognize, classify, segment, cluster and make predictions for images they have never seen before. In this work, the basic principles of U-net architecture, developed for biomedical analysis (Ronneberger *et al.* 2015), is adapted to our network. Although there are a lot of network perspectives for image analysis, U-shape architecture and its corresponding training has an advantage when dealing with a limited amount of data (Ronneberger *et al.* 2015).

Briefly, the cycle of our network is designed based on 1) Convolutional layers; 2) Activation layers; 3) Pooling Layers and 4) Up sampling steps, then the system employs neural networks for proceeding to the prediction stage. To understand how the network is performing, the cost function is checked at each epoch, aiming towards the minimization of the prediction deviation or error. After the calculation of the error, it is back propagated through the network in the opposite direction. Upon the adjustment of weights and kernels, the whole process starts again to optimize the network. A block diagram of the network which describes the basic flowchart is given in Fig. 1.

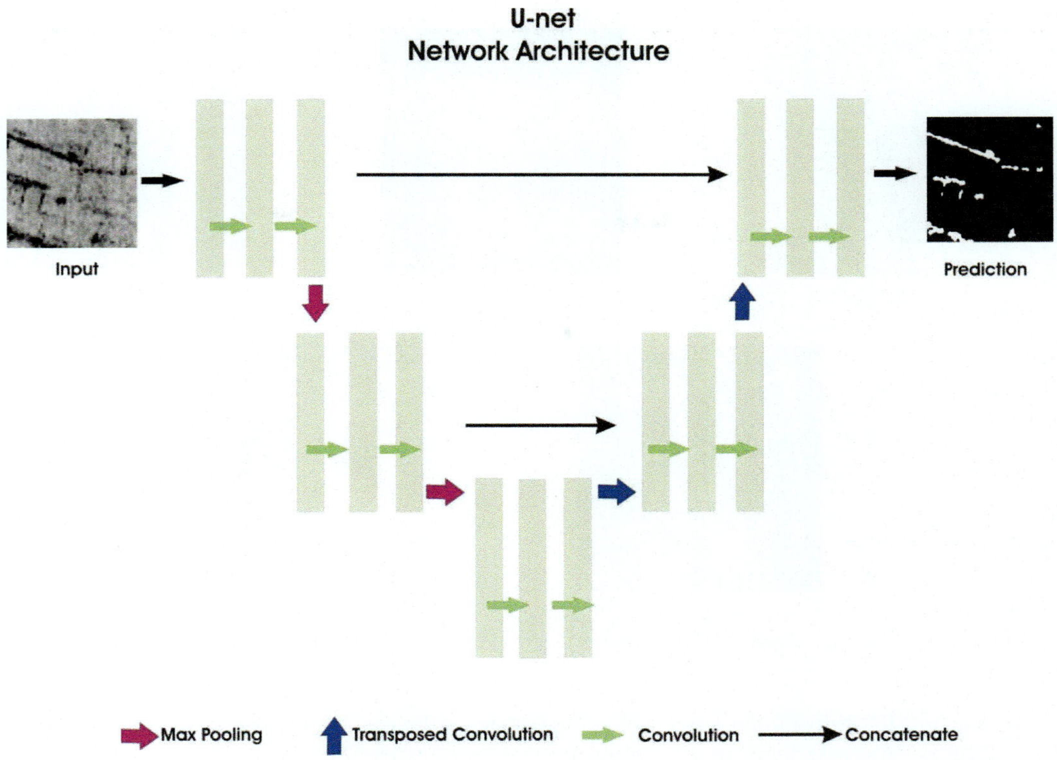

Fig. 1. Block diagram of U-net Architecture

Results

In this study, for carrying out the training network, 4000 annotated images for a training set and 800 images for a test set were generated. To avoid overfitting, a data augmentation was applied to the dataset with random shifting, rotation and shearing ranges. Choosing the optimal hyper-parameter of a network is a vital phase to obtain better prediction results. The parameters of a CNN model are optimized to minimize the loss and we expect the prediction of a CNN model to match with the ground truth with high accuracy. Thus, after several experiments on network parameters, 1 batch size, 100 epochs, and 1e-4 learning rate were adapted to maximize the performance of the training model. With these network parameters, we achieved reasonable results with 92% of Dice-coefficient. As an example, the small scale image results for the prediction with different hyper parameters, are presented in Fig. 2.

Conclusion

The convolutional neural network is the most widely used deep learning approach in feature learning for large-scale image classification and recognition. Considering the accuracy of the results, the performance of this network can be admissible and it can be quite helpful when adapted into the challenging task dealing with the processing of images resulting from archaeological prospection surveys. This is also especially useful for automated anomaly detection or segmentation when dealing with extensive high-resolution maps. On the other hand, the advantage of U-net architecture relies on the fact that it can be used for training even with a relatively small amount of datasets, especially because of the lock of labelled training sample issue for archaeological prospection. The particular experiments proved the potential for the application of the CNN approach in the classification of archaeo-geophysical data. This work includes the first attempt of generating large labelled pre-trained archaeo-geophysical datasets and developing a segmentation tool. But considering the future foreseeable technological developments, there is still much more to be done by fusing a number of different datasets originating from different sensors and using different training and learning approaches.

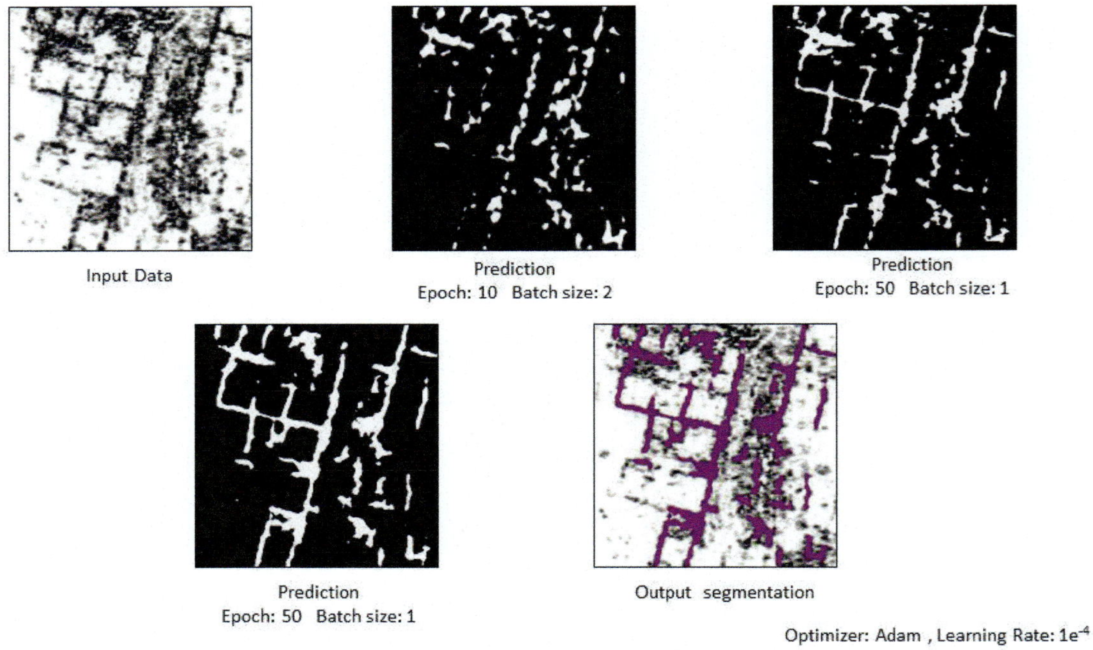

Fig. 2. The results of prediction process on a sample image created from a GPR (Sensors & Software, Nogin Plus, 250MHz antenna) depth slice from the archaeological site of Demetrias, close to Volos, Greece.

Acknowledgements

Part of this work was financially supported by the Stavros Niarchos Foundation within the framework of the project ARCHERS ("Advancing Young Researchers' Human Capital in Cutting Edge Technologies in the Preservation of Cultural Heritage and the Tackling of Societal Challenges").

References

Bescoby, D., Cawley, G. and Chroston, P. 2003. Interpretation of geophysical surveys of archaeological sites using artificial neural networks. *Proceedings of the International Joint Conference on Neural Networks*, v2. 1132–1137.

Green, A., Cheetham, P. and Darvill, T. 2017. Automation, automation, automation: a novel approach to improving the pre-excavation detection of inhumations, In Jennings, B., Gaffney, C., Sparrow, T. and Gaffney, S. (eds.) *AP2017. 12th International Conference of Archaeological Prospection*. Archaeopress. Oxford. 89-92.

Leckebush, J., Weibel, A. and Bühler, F. 2008. Semi-automatic feature extraction from GPR data for archaeology. *Near Surface Geophysics*. 6(2). 75-84.

Linford, N. and Linford, P. 2017. The application of semi-automated vector identification to large scale archaeological datasets considering anomaly morphology, In Jennings, B., Gaffney, C., Sparrow, T. and Gaffney, S. (eds.) *AP2017. 12th International Conference of Archaeological Prospection*. Archaeopress. Oxford. 138–139.

Opitz, R. and Herrmann, J. 2018. Recent Trends and Long-standing Problems in Archaeological Remote Sensing. *Journal of Computer Applications in Archaeology*. 1(1). 19–41.

Pregesbauer, M., Trinks, I. and Neubauer, W. 2014. An object oriented approach to automatic classification of archaeological features in magnetic prospection data, *Near Surface Geophysics*. 12(5). 651-656.

Ronneberger, O., Fischer, P. and Brox, T. 2015. U-Net: Convolutional Networks for Biomedical Image Segmentation. In Navab, N., Hornegger, J., Wells, W. and Frangi, A. (eds.) *Medical Image Computing and Computer-Assisted Intervention – MICCAI 2015. MICCAI 2015. Lecture Notes in Computer Science*, vol 9351. Springer, Cham.

Toumazet, J-P., Vautier, F., Roussel, E. and Dousteyssier, B. 2017. Automatic detection of complex archaeological grazing structures using airborne laser scanning data. *Journal of Archaeological Science*. 12. 569–579.

Traviglia, A., Cowley, D. and Lambers, K. 2016. Finding common ground: Human and computer vision in archaeological prospection. AARGnews-The newsletter of the Aerial Archaeology Research Group. 53. 11–24.

Trier, Ø.D., Salberg, A-B., Holger Pilø., L, Tonning, C., Marius Johansen, H. and Aarsten, D. 2016. Semi-automatic mapping of cultural heritage from airborne laser scanning using deep learning. *EGU General Assembly Conference Abstracts*. 18. 5716.

Trier, Ø.D., Cowley, D.C. and Waldeland, A.U. 2018. Using deep neural networks on airborne laser scanning data: Results from a case study of semi-automatic mapping of archaeological topography on Arran, Scotland. *Archaeological Prospection*. 26. 1–11.

Rise of the Machines: Improving the identification of possible graves in GPR data with interactive survey guidance and machine learning

Ashely Green, Paul Cheetham
Bournemouth University, Poole, United Kingdom

This research explores the use of computational methods and machine learning as tools for the interpretation of archaeo-geophysical data. Archaeo-geophysical survey methods and data interpretation have developed since the 1960s, yet one pioneering technique, ground-penetrating radar (GPR), is still unable to meet the needs of the archaeological commercial and research communities with respect to burials. GPR data is one of the most difficult archaeo-geophysical datasets to interpret, especially for non-specialists. This difficulty is often attributed to the number of GPR reflections within any given volume of deposit, which often leads to an overwhelming number of anomalies present within a dataset. Detecting archaeologically simple graves and wooden coffin interments is also problematic. As the surviving physical evidence for an inhumation is the grave cut, grave fill, and skeletal remains, the geophysical contrast between the grave and surrounding materials is ephemeral thus rendering it difficult, and in some cases impossible, to detect geophysically. While GPR has proven one of the most successful geophysical techniques for detecting graves and has a high rate of detection, some graves are still overlooked in the interpretation stage of an investigation.

Guidelines for conducting archaeological geophysical surveys have been established at international and national levels (Historic England 2008, Bonsall 2014, Bonsall *et al.* 2014, Schmidt *et al.* 2015). The European Archaeological Council established a set of guidelines for geophysical prospection in archaeology in 2015 (Schmidt *et al.*). These guidelines outline the most suitable techniques for the most common targets, controllable factors, and uncontrollable factors encountered during surveys in Europe. Historic England established guidelines for the minimum resolution and guidance for choosing suitable techniques for archaeo-geophysical surveys with respect to England (David *et al.* 2008). In 2014, Transport Infrastructure Ireland (TII, formerly the National Roads Authority) produced guidelines for geophysical surveys of road corridors in Ireland (Bonsall *et al.* 2014).

Using these existing guidelines, this project developed an interactive tool for determining appropriate parameters for a given survey, which serves as a simple solution for the immediate improvement of data acquired in the field and subsequently the interpretability of such data. The tool provides an output which details the appropriate survey parameters (e.g. traverse interval, sampling interval, and survey size) that would allow for the maximum potential detection of the designated target(s), which includes graves. The interactive tool uses established survey guidelines for the UK, Ireland, and Europe which account for a range of input variables including controllable factors (e.g. traverse interval, sampling interval), uncontrollable factors (e.g. local geology, weather conditions), and potential targets. A limited version of the tool is available as a web app and, on request, a full version is available as a geodatabase. The outputs were determined through a decision table which uses the freely available bedrock geology, superficial geology, soil, and land cover data to determine which technique(s) would be most suitable for a site and any known targets. The online version is limited by file size and, therefore, only provides the user with the outputs. The full desktop version provides the user with the full dataset used to determine the outputs. An example of the web app user interface is shown in Fig. 1.

Equally, this research also provides an automatic binary classification tool, *Reilig*, which identifies grave-like responses in GPR data. *Reilig* uses convolutional neural networks to analyse GPR B-scan data. Existing models were retrained primarily on Irish medieval grave types, with a small selection of data from similar grave types in southern Britain. Illustrations of the primary grave types included in the training data are shown in Fig. 2.

Training data were primarily derived from survey data held by Bournemouth University and the limited available geophysics data with corresponding test trench evaluations identified through a comparison of the available records from the 2005–2017 geophysical survey licence database and archaeological excavation reports held by the Irish National Monuments Service. Additional training data were simulated using GPRISM and gprMax (Warren *et al.* 2016) using known characteristics of graves and local soils and geologies. The simulated geophysical characteristics of graves are based on the relative dielectric permittivity, conductivity, and attenuation constant for materials expected to comprise grave fills, a selection of which are detailed in Table 1. Morphological characteristics of the simulated graves were based on excavation reports and records held by the National Museum of Ireland (Cahill and Sikora 2011). As with the real data, the simulated data also includes simple and complex negative examples which represent hyperbolic responses which would not result from graves. Simple simulations include the ideal representation of one reflector, while the complex simulations include multiple reflectors/objects and simulate 'real-world' scenarios. Only B-scans, or radargrams, were used in the initial stages of the project as processing can vary on time slice data; however, interpretations of the B-scans can be extrapolated to interpret C-scans.

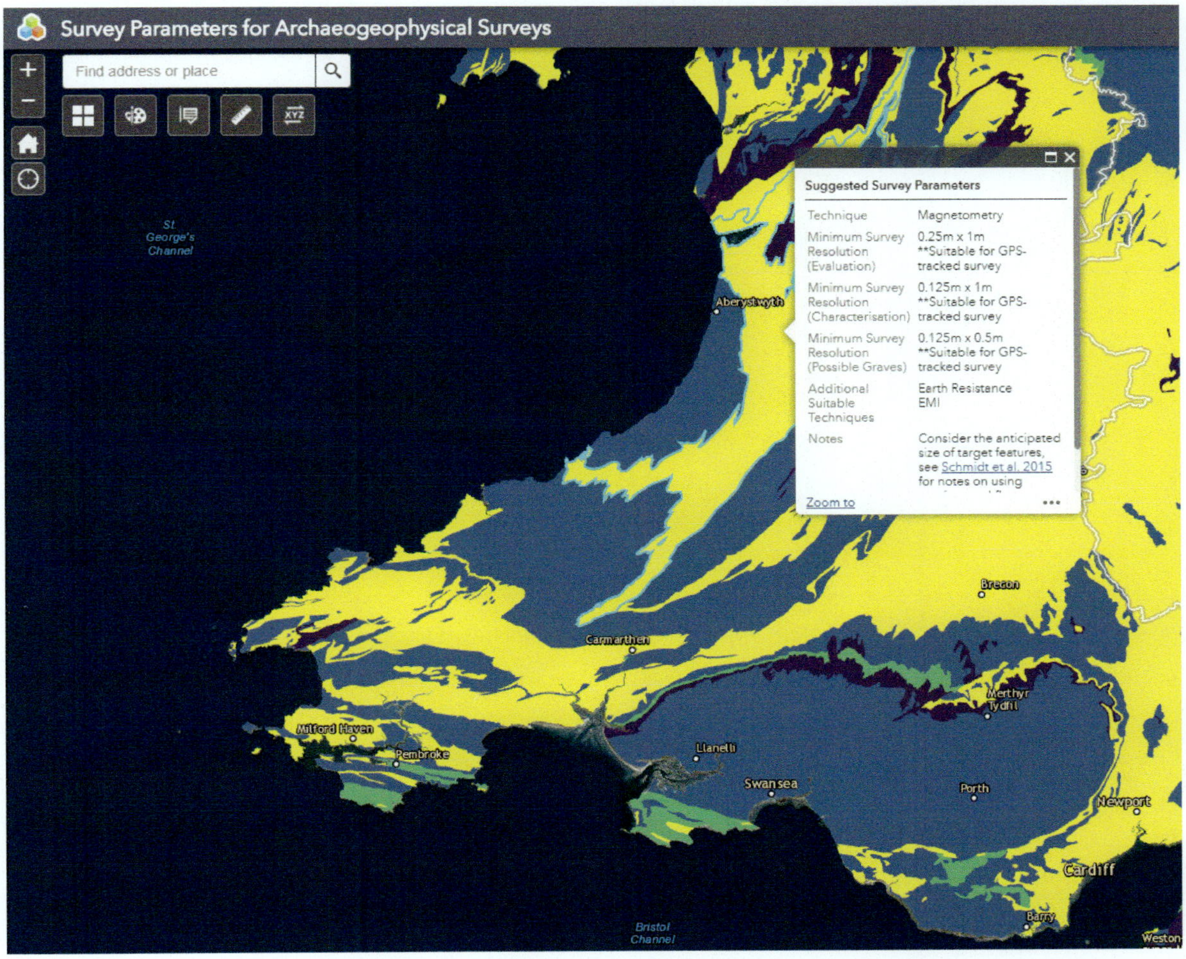

Fig. 1. Example of the survey parameters tool web app user interface.

Images, similar to those shown in Fig. 3, were randomly cropped to 150 x 150 pixels and labelled based on the presence or absence of a grave in the image. Images were then used to retrain existing TensorFlow Inception models for convolutional neural networks (Goldsborough 2016) and ResNet models for residual networks (He *et al.* 2016). Greyscale plots of radargrams are inputted into the network and responses are highlighted and given a probability score of their likelihood of being a grave. Outputs from the network are either a classified image with a probability score, or the original input returned with a bounding box surrounding the detected grave-like response or a note that no detection was made in the instance there is no grave-like response in the image. Both outputs can be saved as image files of quality suitable for inclusion in technical reports.

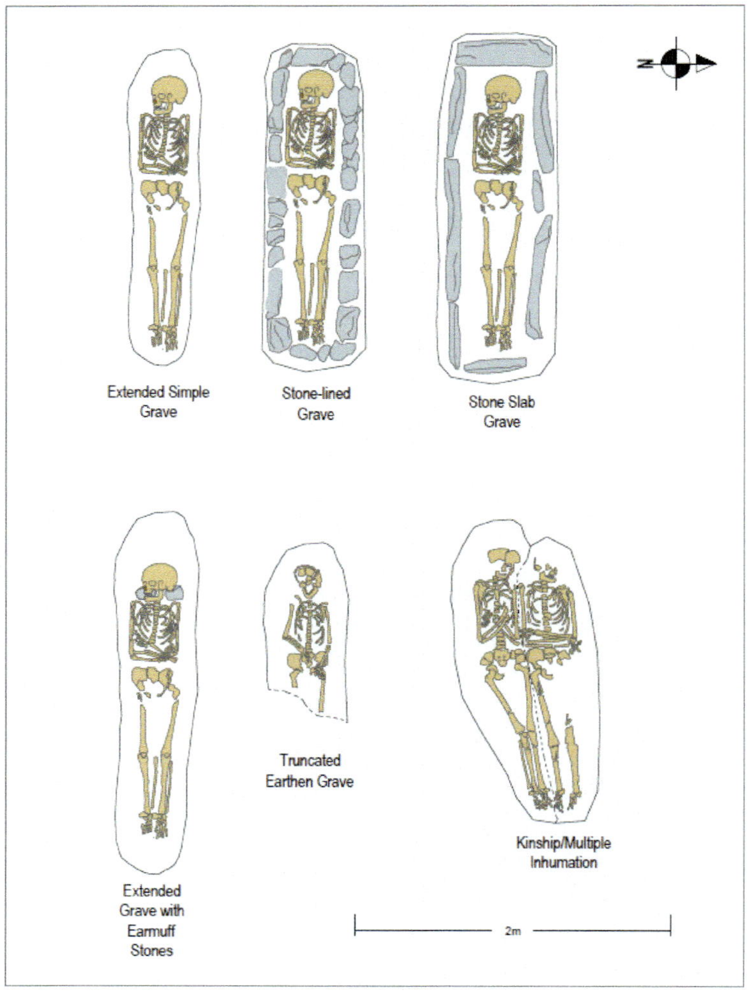

Fig. 2. Illustrations of the primary grave types included in the training dataset.

Material	ε_r	Conductivity (S/m)	Attenuation Constant (dB/m)
Air	1	0	0
Fresh water	80	10^{-6}-10^{-2}	0.01
Dry sand	3-6		
Wet sand	10-30		
Dry sandstone	2-3		
Wet sandstone	5-10		
Limestone	4-8		
Dry limestone	7		
Wet limestone	8		
Shale	5-15		
Wet shale	6-9		
Silts	5-30		
Clays	5-40		
Dry clay	2-6	10^{-3}-10^{-1}	10-50
Wet clay	15-40	10^{-1}-10^{-0}	20-100
Dry sandy soil	4-6	10^{-7}-10^{-3}	0.01-1
Wet sandy soil	15-30	10^{-3}-10^{-2}	0.5-5
Dry loamy soil	4-6		
Wet loamy soil	10-20		
Dry clayey soil	4-6		
Wet clayey soil	10-15		
Granite	4-6		
Dry granite	5		
Wet granite	7		
Dry salt	4-7		

Table 1: A selection of the relative dielectric permittivity, conductivity values, and attenuation constants used in the simulated GPR training data.

TensorFlow models demonstrated high training accuracy and low training loss but low testing accuracy and high training loss, indicating overfitting of the model which could not be adequately reduced through altering the training dataset or changing the optimizer. However, the ResNet models were more accurate and did not overfit. ResNet152 was the most accurate model for classification (maximum accuracy: 93.75%, F_1 score: 0.936, Matthew's Correlation Coefficient: 0.877). When using a training dataset labelled with bounding boxes and custom head added to the model, the maximum accuracy of the ResNet152 model increased to 96.5%. Heatmaps delineating areas of interest within the test images the ResNet152 model was most and least confident in, are shown in Fig. 3.

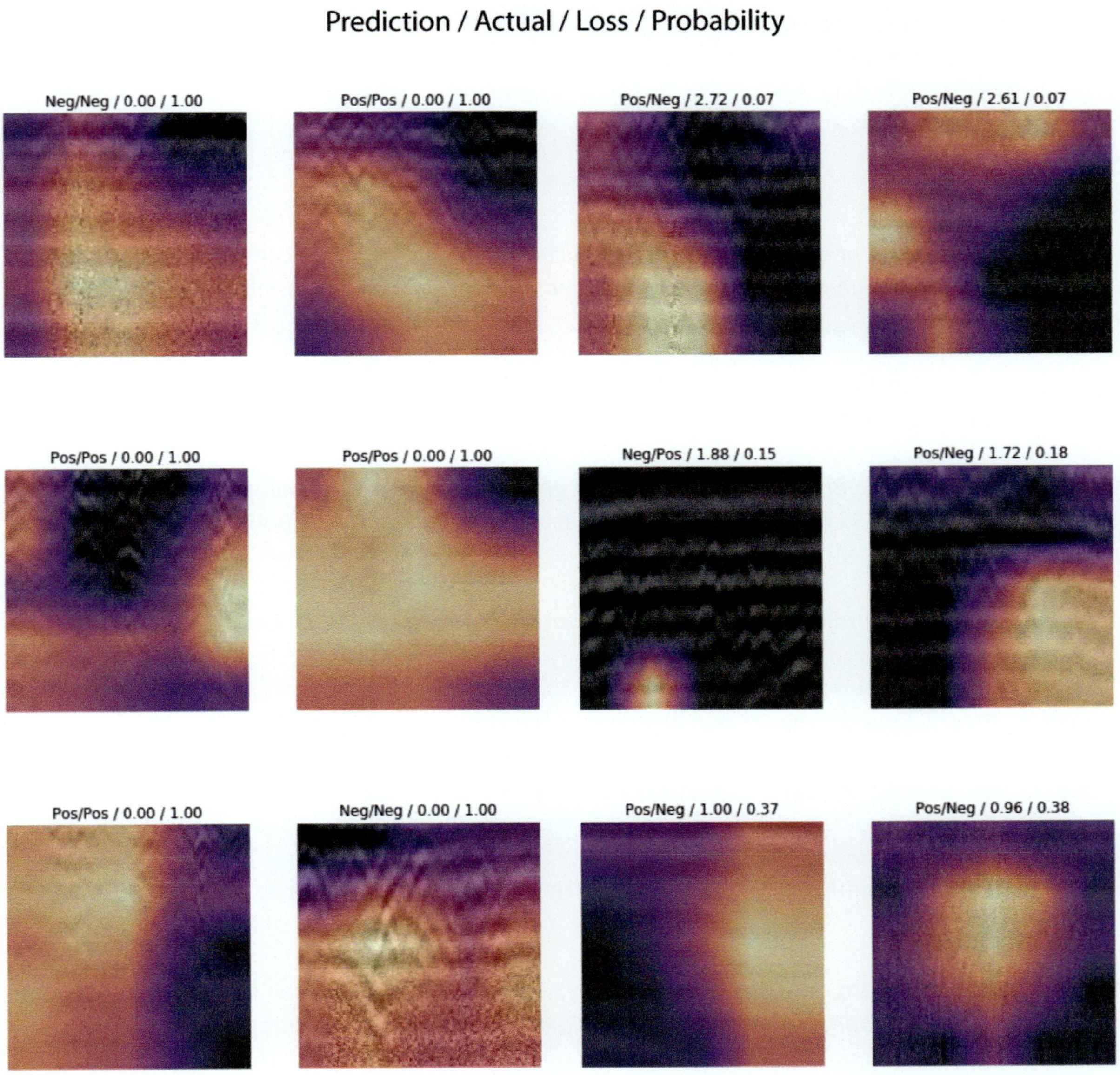

Fig. 3. Heatmaps indicating the areas of interest identified using the retrained ResNet152 model and their predictions and probability score.

The final model was tested on recently collected data from the survey of five sites in Ireland which was facilitated by the directors of the Irish Heritage School (trading as Irish Archaeology Field School) and Blackfriary Archaeology Field School. The five case study sites are located in the Republic of Ireland and constrained to three counties – Offaly, Meath and Wexford. While the superficial drift geologies of each site are a similar glacial till with silty clay to clay loam soils, the bedrock and superficial geologies differ. Each site also has recorded medieval (*c.* AD 500–1600) activity, though this varies from excavation evidence to historical evidence between sites. Inference was performed on a batch of greyscale radargrams from each site with successful classification of and detection on the images. There was a high rate of false positive detections; however, it can be argued that in commercial archaeology false positives are more acceptable than false negatives.

This research aimed to increase the rate and confidence of which graves can be identified prior to excavation, improve the accuracy and speed of commercial and research GPR survey outputs and reporting, and subsequently improve the potential to identify and recover inhumations prior to their destruction by machine excavation, ploughing, and other modern activities. While the automatic classification and detection on images still requires human confirmation, the high accuracy of the models indicate that a machine learning approach can be successful in improving the rate of pre-excavation detection of graves. As large-scale landscape geophysical surveys are becoming more integrated into the commercial framework, it is becoming increasingly important to produce reliable methods to identify and interpret small features within large datasets. Yet, further training data is necessary for the successful detection of additional grave types and the use of time slices or animations and inputs.

References

Bonsall, J. 2014. *A reappraisal of archaeological geophysical surveys on Irish road corridors 2001-2010, with particular reference to the influence of geological, seasonal and archaeological variables*. Unpublished PhD thesis, University of Bradford.

Bonsall, J., Gaffney, C. and Armit, I. 2014. *Preparing for the future: A reappraisal of archaeo-geophysical surveying on Irish National Road Schemes 2001-2010*. Dublin.

Cahill, M. and Sikora, M., 2011. *Breaking ground, finding graves: Reports on the excavation of burials by the National Museum of Ireland*, 1927-2006. Wordwell Ltd., in association with the National Museum of Ireland. Bray.

David, A., Linford, N. and Linford, P. 2008. *Geophysical Survey in Archaeological Field Evaluation.* Swindon, English Heritage.

Goldsborough, P. 2016. A Tour of TensorFlow: Proseminar Data Mining. *arXiv: 1610.01178.*

He, K., Zhang, X., Ren, S. and Sun, J. 2016. Deep Residual Learning for Image Recognition. In *Proceedings of the IEEE Computer Society Conference on Computer Vision and Pattern Recognition*. Washington, DC: IEEE Computer Society. Available from: https://www.cv-foundation.org/openaccess/content_cvpr_2016/papers/He_Deep_Residual_Learning_CVPR_2016_paper.pdf

Schmidt, A., Linford, P., Linford, N., David, A., Gaffney, C., Sarris, A. and Fassbinder, J.W.E. 2015. *EAC Guidelines for the Use of Geophysics in Archaeology: Questions to Ask and Points to Consider*. Belgium, Europae Archaeologia Consilium.

Warren, C., Giannopoulos, A. and Giannakis, I. 2016. gprMax: Open source software to simulate electromagnetic wave propagation for Ground Penetrating Radar. *Computer Physics Communications.* 209. 163-170.

Automated detection and analysis of diffraction hyperbolas in ground-penetrating radar data

Lieven Verdonck
Ghent University, Ghent, Belgium

Although raw ground-penetrating radar (GPR) data give high-resolution information on the 3-D geometry and location of buried features, travel times need to be converted to depth in order to make more precise assumptions. The wave velocity needed for this can strongly vary laterally and with depth. It can be determined by directly measuring the depth to the reflectors in an excavation, or estimating the dielectric permittivity in the lab or in the field, e.g. by means of time domain reflectometry. Alternatively, a common midpoint survey can give velocity information. All these methods require considerable effort. Therefore methods based on the analysis of diffraction hyperbolas (hyperbola fitting, migration velocity analysis) remain important, although also these are time-consuming if the aim is to construct a detailed 3-D velocity model.

Therefore it was explored if the detection and analysis of diffraction hyperbolas could happen in a more automated way. Several papers have addressed computer-aided methods for the segmentation and classification of archaeological features from GPR data (e.g. Leckebusch *et al.* 2008, Schmidt *et al.* 2013, Verdonck 2016, Green *et al.* 2017, Linford *et al.* 2017, Verdonck *et al.* 2019). For the automated detection and analysis of diffraction hyperbolas, different methods exist (Mertens *et al.* 2016). One strategy is based on machine learning (e.g. Maas and Schmalzl 2013), and usually consists of two steps. First, regions of interest (ROIs) are extracted from the GPR profile using a neural network. Subsequently, a hyperbola fitting algorithm such as the Hough transform is applied. In line with this, a hyperbola detection method that enables the creation of 3-D GPR velocity models for archaeological prospection is presented here.

The first step is based on a Faster Regions with Convolutional Neural Networks (R-CNN) object detector. As opposed to traditional CNNs, which can be used for the classification of complete images, the R-CNN approach allows localizing objects within an image, by generating region proposals (regions in the image which may contain an object), which the CNN then classifies. Faster R-CNNs are more efficient implementations of R-CNNs (Ren *et al.* 2017). To train the detector, ~300 sample images of hyperbolae, with a fixed size, were selected from GPR data collected at two Roman towns in Lazio, Italy (Falerii Novi and Interamna Lirenas). A rectangle defining the hyperbolaes was manually marked in the images (Fig. 1a, 1b). These were split into a training set and a test set.

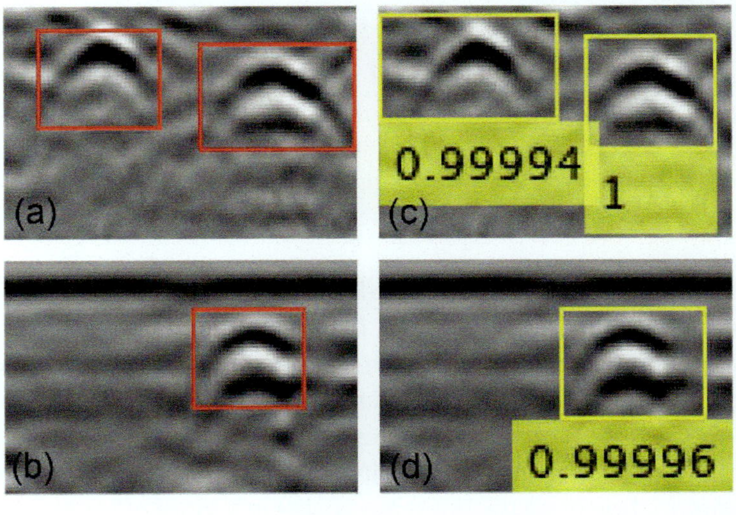

Fig. 1. Sample images used to test the Faster R-CNN detector after it had been trained. (a-b) Rectangles defining hyperbolae, marked manually. (c-d) Hyperbolae extracted by the Faster R-CNN detector, with the confidence score indicated.

The second step in the detection procedure aims to extract the coordinates (horizontal position and two-way travel time, TWT) of the hyperbola apex, as well as the velocity of the medium at that location. In the ROIs selected by the Faster R-CNN algorithm, edges were detected using the Canny edge detector. To the resulting edge points, a theoretical hyperbola was fitted, to verify if the edge point corresponds to a hyperbola apex. This occurred using a number of criteria implementing tolerance levels that account for imperfectly shaped hyperbolas (Mertens *et al.* 2016). These criteria relate, for example, to the symmetry of the hyperbola or the presence of gaps.

To evaluate the Faster R-CNN detector, it was run on the test set (Fig. 1c, 1d), yielding an average precision of 0.8891 (Zhang and Zhang 2009). To apply the detector to complete GPR profiles, these were divided in regions with the same size as the training data. Multiple detections of the same hyperbola were removed. Only the detections with the highest confidence score (>= 0.9999) were selected.

When comparing the capability of the proposed procedure to detect hyperbolae and to extract their characteristics (location, TWT and wave velocity) with manual migration velocity analysis, there are still important differences, as is illustrated in Fig. 2 and Table 1. In this GPR profile, the automated detection algorithm extracts two hyperbolae which were not used during the manual analysis since they interfere with other hyperbolae (Nos. 1 and 2 in Fig. 2). Conversely, it did not detect two other hyperbolae (Nos. 6 and 8) which were used during manual analysis. However, the general trend of a gradually decreasing velocity is present in both the manual and automated detection results.

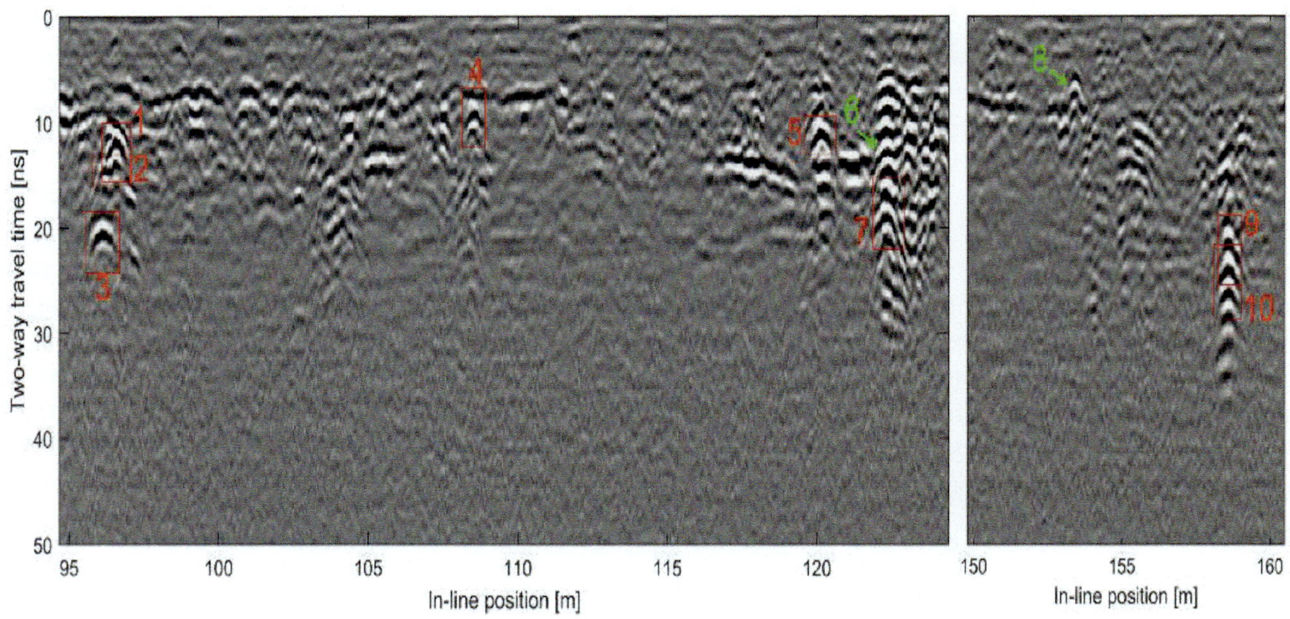

Fig. 2. Two parts of a GPR profile from Falerii Novi. The automatically detected hyperbolas are indicated in red, the ones used for the manual velocity analysis but not detected automatically are indicated in green. The numbers correspond to the ones in Table 1, and are explained in the text.

Several factors can cause the velocity obtained by diffraction hyperbola analysis to deviate from the true wave velocity. An example is the radius of the object producing the hyperbola (Ristić *et al.* 2009). For this reason, hyperbola No. 3 in Fig. 2 may result in a too high velocity both in the manual and automatic detection. Other problems are the estimation of time zero (Koyan *et al.* 2018), and lateral velocity variations causing pull-up and push-down effects (Leckebusch 2007). These problems are not solved by automated detection. However, when compared to manual migration velocity analysis, the presented procedure may in the near future produce velocity information of nearly the same quality in a more efficient way.

Hyperbola No. (see Fig. 2)	Two-way travel time (ns)	Velocity derived from manual analysis (m/ns)	Velocity derived from automated procedure (m/ns)
1	11.3	Hyperbola not analysed	0.0750
2	13.3	Hyperbola not analysed	0.0675
3	20.1	0.1000	0.1075
4	9.5	0.1125	0.1075
5	10.5	0.1100	0.1050
6	11.8	0.1025	Hyperbola not detected
7	18.5	0.0900	0.0925
8	7.2	0.1100	Hyperbola not detected
9	23.5	0.0700	0.0700
10	24.5	0.0700	0.0675

Table 1. Two-way travel time and velocity of the hyperbolas extracted manually and automatically, indicated in Fig. 2.

Acknowledgements

This work was made possible by a postdoctoral fellowship of the Research Foundation – Flanders (FWO). The GPR prospections were undertaken by Ghent University and the University of Cambridge (Faculty of Classics), in collaboration with the British School at Rome, the Soprintendenza Archeologia, Belle Arti e Paesaggio (SABAP) per le Province di Frosinone, Latina e Rieti, and the SABAP per l'Area Metropolitana di Roma, la Provincia di Viterbo e l'Etruria Meridionale, with generous support from the Arts and Humanities Research Council (grant Ref. AH/M006522/1—Beneath the Surface of Roman Republican Cities) and the Comune di Pignataro Interamna. The author would like to thank Martin Millett, Alessandro Launaro and Frank Vermeulen.

References

Green, A., Cheetham, P. and Darvill, T. 2017. Automation, automation, automation: a novel approach to improving the pre-excavation detection of inhumations. In Jennings, B., Gaffney, C., Sparrow, T. and Gaffney, S. (eds.) *AP2017. 12th International Conference of Archaeological Prospection*. Archaeopress. Oxford. 90–91.

Koyan, P., Tronicke, J., Allroggen, N., Kathage, A. and Willmes, M. 2018. Estimating moisture changes in concrete using GPR velocity analysis: potential and limitations. In *2018 17th International Conference on Ground Penetrating Radar (GPR), Rapperswil*, 18–21 June 2018, 664–669.

Leckebusch, J. 2007. Pull-up/pull-down corrections for ground-penetrating radar data. *Archaeological Prospection*. 14. 142–145.

Leckebusch, J., Weibel, A. and Bühler, F. 2008. Semi-automatic feature extraction from GPR data for archaeology. *Near Surface Geophysics*. 6. 75–84.

Linford, N. and Linford, P. 2017. The application of semi-automated vector identification to large scale archaeological datasets considering anomaly morphology. In Jennings, B., Gaffney, C., Sparrow, T. and Gaffney, S. (eds.) *AP2017. 12th International Conference of Archaeological Prospection*. Archaeopress. Oxford. 138–139.

Maas, C. and Schmalzl, J. 2013. Using pattern recognition to automatically localize reflection hyperbolas in data from ground penetrating radar. *Computers & Geosciences*. 58. 116–125.

Mertens, L., Persico, R., Matera, L. and Lambot, S. 2016. Automated detection of reflection hyperbolas in complex GPR images with no a priori knowledge on the medium. IEEE Transactions on Geoscience and *Remote Sensing*. 54. 580–596.

Ren, S., He, K., Girshick, R. and Sun, J. 2017. Faster R-CNN: Towards Real-Time Object Detection with Region Proposal Networks. *IEEE Transactions on Pattern Analysis and Machine Intelligence*. 39. 1137–1149.

Ristić, A., Petrovacki, D. and Govedarica, M. 2009. A new method to simultaneously estimate the radius of a cylindrical object and the wave propagation velocity from GPR data. *Computers & Geosciences*. 35. 1620–1630.

Schmidt, A. and Tsetskhladze, G. 2013. Raster was yesterday: using vector engines to process geophysical data. *Archaeological Prospection*. 20. 59–65.

Verdonck, L. 2016. Detection of buried roman wall remains in ground-penetrating radar data using template matching. *Archaeological Prospection*. 23. 257–272.

Verdonck, L., De Smedt, P. and Verhegge. J. 2019. Making sense of anomalies: practices and challenges in the archaeological interpretation of geophysical data. In R. Persico, S. Piro and N. Linford (eds.), *Innovation in Near-Surface Geophysics. Instrumentation, application and data processing methods*, 151–194. Amsterdam.
Zhang, E. and Zhang, Y. 2009. Average Precision. In L. Liu and M.T. Özsu (eds.), Encyclopedia of Database Systems. Boston.

Novel volume visualisation of GPR data inspired by medical applications

Wolfgang Neubauer[1,2], Alexander Bornik[3], Mario Wallner[1], Geert Verhoeven[1]
[1]*LBI ArchPro, Wien, Austria,* [2]*VIAS-University of Vienna, Wien, Austria,* [3]*LBI Clinical Forensic Imaging, Graz, Austria*

Introduction

The analysis and interpretation of high-resolution GPR datasets (Trinks *et al.* 2018) is a time-consuming and complex process and requires not only three-dimensional imagination but also a broad understanding of the archaeological remains (Poscetti *et al.* 2015).

The interpretation of 3D high-resolution GPR datasets (min. 1 kvoxel/m^3) is particularly challenging, because of the amount and level of detail of the respective data. The primary task of an archaeological interpretative mapping approach is to identify archaeologically relevant structures to graphically define and classify them by adding respective attributes using a GIS environment (Neubauer *et al.* 2002). The conventional way to analyse animated stacks of 2D GPR depth-slices is similar to the way radiologists read CT or MRI datasets. However, unlike in radiology where physicians try to identify deviations from known healthy patient anatomy, the 3D nature of an archaeological site is unknown due to its unique stratification imaged in great detail by the GPR arrays.

We present a novel integrated visualisation approach inspired by medical applications, which supports conjoint visualisation of scenes composed of heterogeneous data including GPR volumes and 3D models of interpretations and reconstructions (Bornik *et al.* 2018). The approach was used in the first instance to check for the accuracy of the conventional interpretative mapping in contrast to the enhanced 3D visualisation.

High-resolution GPR datasets and data preprocessing

The high-resolution GPR datasets used for this example were recorded at the LBI ArchPro test site at the forum of the Roman town of Carnuntum (Neubauer 2014) by a motorised 400MHz MIRA from Malå Geoscience AB with a spatial resolution of 8cm x 4cm. The recorded GPR waves were band-pass filtered and migrated. The absolute amplitude of the wave was calculated in the time domain corresponding to the depth range of the 3D data cube, here set to 5cm. The resulting dataset counts 4096 voxels x 4096 voxels x 52 voxels at a resolution of 10cm x 10cm x 4cm.

Direct 3D volume visualisation shows all the details in the data, but fails to isolate meaningful large-scale structures. Contiguous visualisations of, e.g. building foundations, require a higher degree of dataset homogeneity, which can be facilitated using state-of-the-art denoising algorithms, which remove high-frequency noise while edges of large structures are preserved. Dataset filtering is performed on the GPU alongside visualisation. Filter parameters can be adjusted at any time to meet the visualisation goal (Urschler *et al.* 2014).

Data visualisation

To facilitate conventional GIS-based archaeological interpretative mapping, the GPR datasets are visualised as accumulated greyscale images for specific depth ranges (Fig. 1) or as animated sequences of the 5cm depth slices. GIS tools as developed by the LBI ArchPro are frequently used to integrate and visualise GPR datasets since they facilitate the integrated visualisation, analysis and mapping of georeferenced datasets. However, 2.5D visualisations are standard in most GIS-software and support for full 3D datasets is limited.

Software used in radiology offers direct volume rendering based on a global mapping from dataset values to colour and transparency (transfer functions), but neither offers support for non-volumetric data nor functionality for local control of the visualisation style. Other tools, like *Voxler* or *VG Studio,* are more flexible. However, their functionality to locally control the visualisation outcome and to integrate non-volumetric datasets like 3D interpretation or reconstruction models is limited.

Fig. 1. Greyscale image representing the depth range between 1.0m and 2.0m of the high-resolution GPR dataset from the forum of the Roman town Carnuntum.

The proposed visualisation is based on a hybrid volume and surface data rendering algorithm (Fig. 2) inspired by developments for medical applications (Kainz *et al.* 2009). 3D models like interpretation or virtual reconstructions are rendered to a Hashing A-buffer, which provides a fast and memory efficient sorted representation of visible object entry and exit points traversed by rays from the viewpoint through each pixel on the screen (Fig. 3). These lists are traversed, maintaining a set of active objects. Colour/transparency values from the current list entry are seamlessly combined with the results from volume ray casting all active GPR objects by blending in front-to-back order. Sampling multiple filtered and unfiltered versions of the dataset using different transfer functions combined with domain control using polyhedral objects allow to visually and flexibly combine multiple features.

Fig. 2. View using the proposed visualisation techniques showing the complete GPR dataset including geometric and volumetric interpretation representations.

Fig. 3. Integrated visualisation of the GPR dataset, polyhedral interpretation models, a reconstructed building and textual annotations.

Conclusions and future work

The first experiments have clearly shown the potential of the proposed visualisation approach inspired by medical applications (Bornik *et al.* 2018). So far we were impressed by its fidelity over the state-of-the-art 3D visualisations applied for archaeological GPR datasets and by the possibilities to evaluate interpretations elaborated on a slice-by-slice basis comparing them to the 3D data evidence in the respective regions. Our first investigations revealed numerous interpretation inaccuracies of the results of earlier interpretative mapping approaches, which can be related to the limitations of 2D views of 3D data. Future work will, therefore, address 3D visualisation support in the data analysis and interpretative mapping phase.

Acknowledgements

The geophysical survey was conducted by the LBI ArchPro in cooperation with the ZAMG as part of the research project *ArchPro Carnuntum* funded by the Amt der Niederösterreichischen Landesregierung. The reconstructed 3D surface model was provided by Michael Klein, *7reasons*.

References

Bornik, A., Urschler, M., Schmalstieg, D., *et al.* 2018. Integrated computer-aided forensic case analysis, presentation, and documentation based on multimodal 3D data. *Forensic Science International.* 287. 12–24.

Kainz, B., Grabner, M., Bornik, A., Hauswiesner, S., Muehl, J. and Schmalstieg, D. 2009. Ray casting of multiple volumetric datasets with polyhedral boundaries on manycore GPUs. *ACM Trans, Graph.* 28(5). 152:1–152:9.

Neubauer, W., Eder-Hinterleitner, A., Seren, S. and Melichar, P. 2002. Georadar in the Roman civil town Carnuntum, Austria: an approach for archaeological interpretation of GPR data. *Archaeological Prospection.* 9(3). 135–156.

Neubauer, W. 2014. Die Entdeckung des Forums der Zivilstadt. In *Carnuntum. Wiedergeborene Stadt der Kaiser*, Humer F. (ed.). von Zabern.

Poscetti, V., Zotti, G. and Neubauer, W. 2015. Improving the GIS-based 3D mapping of archeological features in GPR data. In *Archaeological Prospection. 11th Intl. Conf.* (2015), Rzeszotarska-Nowakiewicz A., (ed.), Polish Academy of Sciences. 603–607.

Trinks, I., Hinterleitner, A., Neubauer, W., *et al.* 2018. Large-area high-resolution ground-penetrating radar measurements for archaeological prospection. *Archaeological Prospection*. 25. 171– 195.

Urschler, M., Leitinger, G. and Pock, T. 2014. Interactive 2d/3d image denoising and segmentation tool for medical applications. In *Proceedings MICCAI Workshop Interactive Medical Image Computation (IMIC)*.

Strategies for the optimization of 3-D electrical resistivity tomography data using the Jacobian matrix

Kleanthis Simyrdanis, Nikos Papadopoulos
Institute for Mediterranean Studies, Rethymno, Greece

Introduction

In the last twenty years Electrical Resistivity Tomography (ERT) exhibited increased attention boosted by the improvement of the hardware's capabilities, the development of multichannel resistivity systems and the compilation of automated inversion and reconstruction algorithms. A complete 3-D ERT survey involves the arrangement of a number of electrodes on the nodes of a regular or irregular grid. In such a case the linear independent apparent resistivity measurements increase exponentially with the respective number of electrodes (Noel and Xu 1991). Thus, even today the routine practice relies on the collection of tomographic data using standard electrode arrays (Dahlin and Zhou 2004) and the integration of dense parallel or orthogonal 2-D lines to characterize the 3-D resistivity structure (Papadopoulos *et al.* 2011).

A step forward to exploit the full capability spectrum of the modern acquisition instruments was firstly introduced by Stummer *et al.* (2004) followed by Wilkinson *et al.* (2006), who presented experimental optimization methods to select electrode arrays that maximize the resolving capabilities of the inversion image, based on the appraisals of the model resolution matrix. These methodologies are directly related to the calculation of the generalized inverse, which can be extremely time consuming in large 2-D and 3-D problems. Athanasiou *et al.* (2009) proposed an alternative method to select the optimal measurements solely on the examination of the sensitivity matrix entries. The extension of the optimization procedure to generate a set of optimum measurements for 3-D survey grids poses new challenges in order to balance between the time constraints and the need to increase the resolution analysis of the 3-D subsurface inversion images.

Methodology

The present study is focused on generating an optimum 3-D dataset composed of multiple parallel lines, based on the Jacobian matrix. The calculation of the Jacobian matrix is part of the solution of the forward resistivity problem, thus the proposed optimization technique is faster than the current methods that use the definition of the resolution matrix. The initial comprehensive dataset for each array, from which the optimum dataset is selected, is composed of the most commonly used initial measurements (dipole-dipole, pole-dipole, pole-tripole and gradient), respectively. Furthermore, an extra 'Full' array is created after merging all the previous protocols which is used as an input to the algorithm to produce the optimised protocol.

With the Jacobian matrix method (Fig. 1a), the sensitivity value of each measurement is calculated for each parameter, where the most sensitive is considered the one with the higher absolute value. Each model parameter is characterized as "weak" or "strong" considering the absolute cumulative sensitivity values corresponding to the specific parameter. By 'parameter' it is considered a block in space with predefined size and boundaries characterised by a specific property (e.g. resistivity, sensitivity value). Hence, all the model parameters are sorted from the weakest to the strongest, based on the respective cumulative sensitivity value, so that the strongest measurement ('MAX') can be accordingly selected for a specific weak parameter. In case a 'MAX' measurement is already chosen for another parameter, the next 'MAX' measurement will be chosen for that parameter. The final optimum protocol contains as many measurements as the number of the parameters exhibiting only the "strongest" measurements for each parameter.

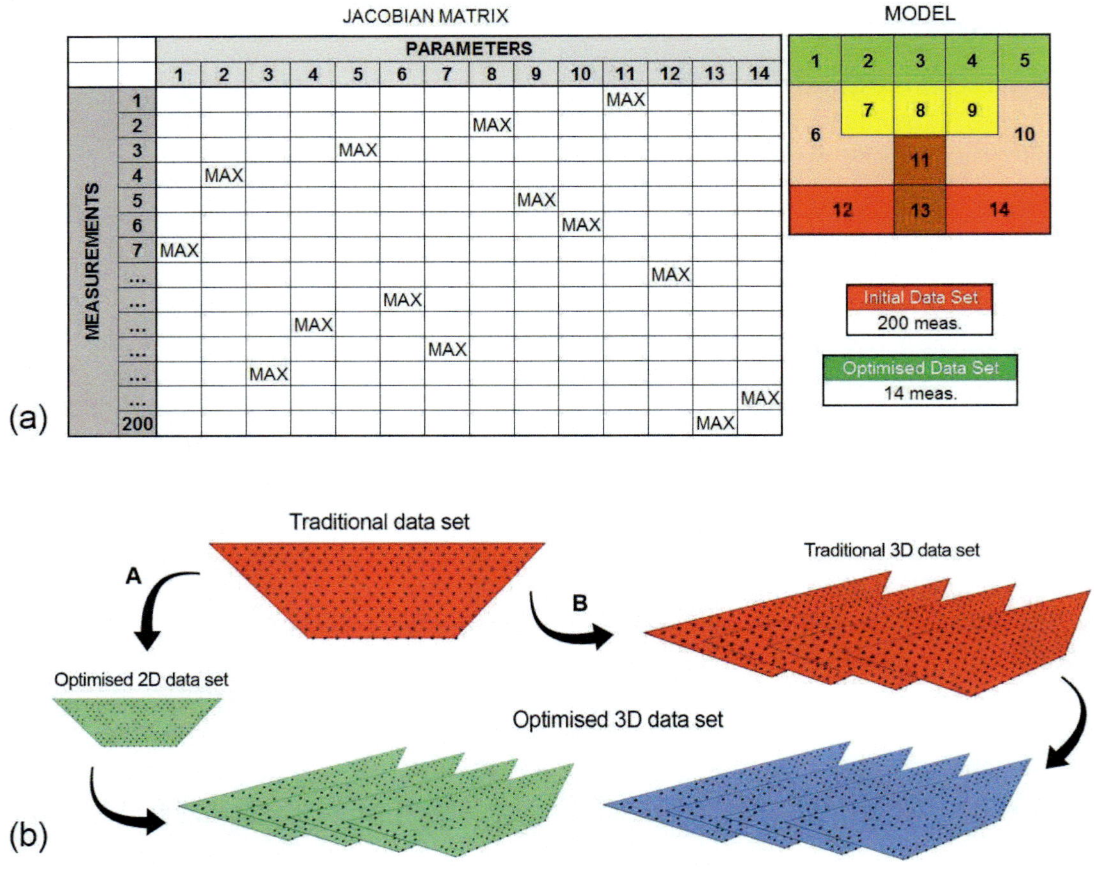

Fig. 1. (a) Technique for compiling the 3-D optimum dataset, (b) using 2-D (A) or 3-D (B) algorithm.

Two different approaches were investigated to compile the optimum 3-D datasets. Firstly, a 2-D algorithm incorporating the technique of the Jacobian method was used to extract the optimum protocol from one individual 2-D line and then the 3-D optimum protocol was created by combining multiple optimized 2-D lines (Fig. 1b - A). Secondly, the Jacobian method was materialized in a 3-D algorithm, where the original 3-D data made by multiple lines was used to extract the optimized data points in a complete 3-D context (Fig. 1b - B).

Synthetic Data

The methodological approach was tested in a synthetic model composed by two targets, one resistive 'A' (10,000 Ohm.m) and one conductive 'B' (10-100 Ohm.m), buried at different depths in a two layered background (100 and 1000 Ohm.m) model. A survey layout with eleven parallel 2-D lines and twenty one electrodes (1m apart) in each line was simulated. The initial comprehensive dataset was defined by the traditional data of Dipole-Dipole, Pole-Dipole, Pole-Tripole, Gradient and the Full array. The aforementioned scenarios are tested with synthetic data a) between the traditional and the optimum dataset and b) between two different optimum datasets that are generated either from a 2-D or a 3-D software (although both are used for 3D optimization analysis).

Results

Fig. 2 shows the comparison between the traditional array (23,670 measurements), the optimum 2-D protocol (3390 measurements) and the optimum 3-D protocol (3315 measurements). The exact position of the target position is indicated with black rectangular boxes. The calculated RMS error is less than 4%. It should be mentioned that although 14% of the traditional array is used the targets can be located on both optimum array datasets. Furthermore, at the optimum 3-D data the target 'A' is reconstructed with resistivity values closer to the true model's values.

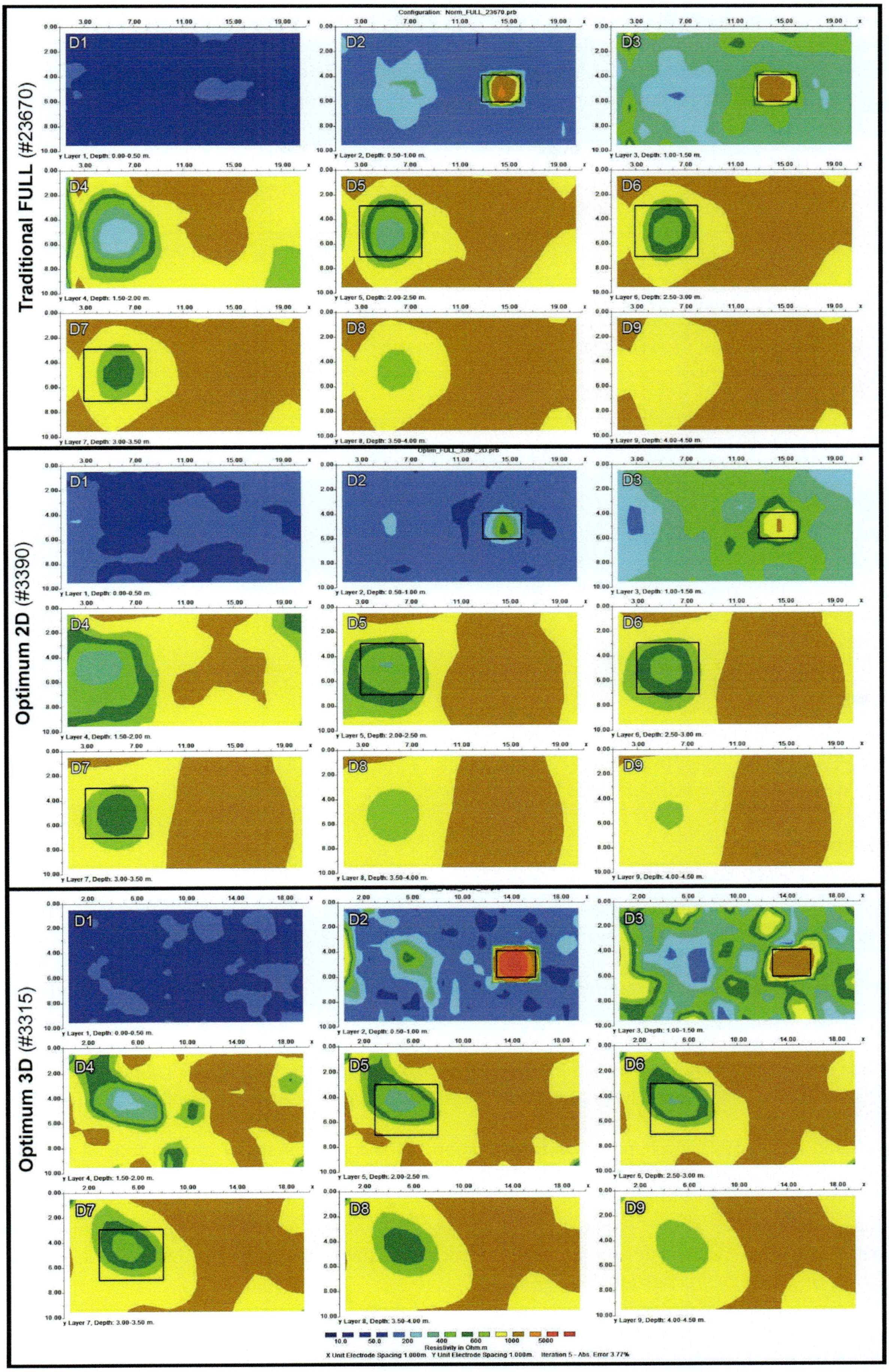

Fig. 2. Inversion results with Traditional dataset, Optimum 2-D and Optimum 3-D.

Additionally, for further verification of the results, the model misfit (Table 1, Loke *et al.* 2015) is calculated where the resistivity values between the true and the calculated model are compared only for depth layers where the targets exist. The model misfit values for the optimum 3-D technique is lower at all depths (except D6) showing a superiority in comparison with the traditional and the optimum 2-D technique.

LEVELS	FULL (Traditional)	FULL (Optim2D)	FULL (Optim3D)
D2 (0.5-1.0m)	0.3303	0.3379	0.2462
D3 (1.0-1.5m)	0.5085	0.4762	0.3899
D5 (2.0-2.5m)	0.2525	0.2656	0.2496
D6 (2.5-3.0m)	0.3827	0.3931	0.3841
D7 (3.0-3.5m)	0.2814	0.278	0.2762

Table 1. Model Misfit values for specific depths between the true and the calculated resistivity values.

Conclusions

The results show that the optimization technique based on the sensitivity matrix is able to preserve comparable quality and resolving capability with the traditional arrays, where the optimum protocol is using only part of the initial dataset. The Jacobian matrix technique seems to provide a robust alternative to extract the resistivity measurements that carry the maximum resolving capability to illuminate the subsurface properties.

Acknowledgements

This work was performed in the framework of the "ARCHERS: Advancing Young Researchers' Human Capital in Cutting Edge Technologies in the Preservation of Cultural Heritage and the Tackling of Societal Challenges" project Exclusively funded by the Stavros Niarchos Foundation.

References

Athanasiou, E.N., Tsourlos, P.I., Papazachos, C.B. and Tsokas, G.N. 2009. Optimizing Electrical Resistivity Array Configurations by Using a Method Based on the Sensitivity Matrix. *Near Surface 2009 - 15th EAGE European Meeting of Environmental and Engineering Geophysics*, 7 September 2009.

Dahlin, T. and Zhou, B. 2004. A numerical comparison of 2D resistivity imaging with 10 electrode arrays. *Geophysical Prospecting*. 52(5). 379-398.

Loke, M.H., Wilkinson, P.B., Chambers, J.E., Uhlemann, S.S. and Sorensen, J.P.R. 2015. Optimized arrays for 2-D resistivity survey lines with a large number of electrodes. *Journal of Applied Geophysics*. 112. 136-146.

Noel, M. and Xu, B. 1991. Archaeological investigation by electrical resistivity tomography: a preliminary study. *Geophysical Journal International*. 107. 95–102.

Papadopoulos, N.G., Tsourlos, P., Papazachos, C., Tsokas, G.N., Sarris, A. and Kim, J.H. 2011. An Algorithm for the Fast 3-D Resistivity Inversion of Surface Electrical Resistivity Data: Application on Imaging Buried Antiquities, *Geophysical Prospecting.* 59. 557-575.

Stummer, P., Maurer, H. and Green, A.G. 2004. Experimental design: Electrical resistivity datasets that provide optimum subsurface information. *Geophysics*. 69(1). 120-139.

Wilkinson, P.B., Meldrum, P.I., Chambers, J.E., Kuras, O. and Ogilvy R.D. 2006. Improved strategies for the automatic selection of optimized sets of ERT measurement configurations. *Geophysics Journal International*. 167. 1119-1126.

Processing Strategies for 3-D Marine Dynamic Electrical Resistivity Tomography Data

Dimitrios Oikonomou[1,4], Nikos Papadopoulos[1], Kleanthis Simyrdanis[1], Gianluca Cantoro[1], Julien Beck[2], Meng Heng Loke[3]

[1]Institute for Mediterranean Studies, Rethymno, Greece, [2]University of Geneva, Geneva, Switzerland, [3]Geotomo Software, Penang, Malaysia, [4]Aristotle University of Thessaloniki, Thessaloniki, Greece

Introduction

The Electrical Resistivity Tomography (ERT) geophysical method has been extensively employed for mapping on-shore buried antiquities (e.g. Papadopoulos *et al.* 2007), although its usage in reconstructing the cultural relics in littoral and ultra shallow marine environments is rather limited (Simyrdanis *et al.* 2019). A marine ERT survey involves: a) the static mode where the electrodes remain fixed and stable during data acquisition and b) the dynamic mode where the electrodes are constantly moving along profiles inside the area of interest.

The main advantage of the static mode is the accurate positioning of the electrodes along a dense network of parallel lines, though at the expense of increased efforts for data collection. On the other hand, dynamic mode can cover a relatively larger marine area in a limited time along a mixture of semi-parallel, diagonal and crossing transects, due to the inability of the boat carrying the electrode streamer to navigate along perfectly parallel lines. In both cases, the sea water layer is a crucial parameter that controls the final resolving capabilities of the resistivity images (Simyrdanis *et al.* 2016, Papadopoulos and Simyrdanis 2017).

Synthetic data approach

The evalution of the efficiency and the potential use of dynamic ERT composed of randomly distributed lines was firstly checked through 3-D numerical and inversion resistivity modelling. A synthetic model describing an archaeological U-shape wall made of two linear parallel sectors of 10m length and a connecting arc of similar length with 0.5m width and 1m height, buried below the sea bottom with 3m water depth was created (Fig. 1 top). The resistivity values were set to 0.17Ohm.m for the seawater, 0.5Ohm.m for the homogenous sea bottom and 5Ohm.m for the wall foundations. The direction of the lines were chosen based on a uniformly random distribution in order to simulate the characteristics of an actual dynamic survey mode (Fig. 1 bottom-left).

Fig. 2a shows the depth slices that have been extracted from the 3-D resistivity inversion model in the case of the random distribution of the lines. Despite the relatively small dimensions of the submerged target the inverted slices clearly outline the shape and the burial depth of the original structure. Fig. 2b indicates the respective model in the case where the randomly distributed electrodes were moved on the nodes of a regular grid (Fig. 1 bottom-center), thus minimizing the total number of the point sources and its subsequent processing time. It is clear that both inversion models show comparable accuracy thus indicating that ERT results are rather insensitive to the movement of the electrodes.

Real Data

During 2018 field season, a moving dipole-dipole array along predefined survey lines (dynamic mode) was used to map the sub-bottom electrical properties at Lambayanna, a submerged Early Bronze Age settlement in the Bay of Kiladha, Greece (Beck *et al.* 2017, Surdez *et al.* 2018) (Fig. 3 top). The cable, composed of 13 graphite electrodes with 1m separation, was sunk in the sea and the main instrumentation was kept protected on a boat. The use of the multichannel resistivity device, that is able to log ten potential readings from different layers with a single current injection, facilitated the field application of the dynamic mode. The one end of the underwater cable and the GPS were connected to the resistivity instrument, which in turn was synchronized to a toughpad running a special managing software (SYSMAR). The boat was navigated along predefined transects keeping a relatively small speed and an effort was made to keep a constant distance between the individual lines. This goal was not always achievable since even small water-currents were slightly drifting the boat thus deviating from the ideal predefined path. Despite the *in situ* difficulties, the area was covered with substantial resolution, reaching the needs and the objectives of the survey.

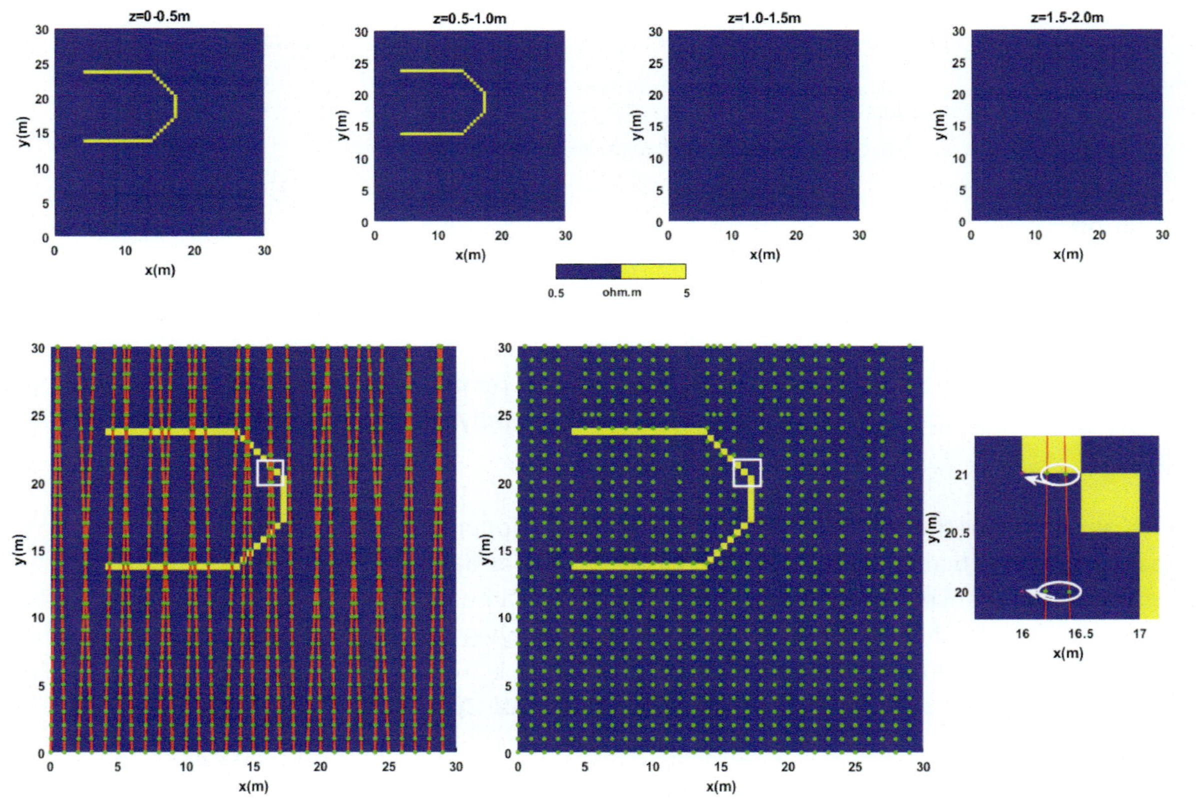

Fig. 1. Model that describes a U-shape wall buried below the sea bottom (top). Dynamic ERT lines (red lines) and electrodes (green dots) in their correct position (bottom-left). Electrodes moved in regular grid of 1m (bottom-centre) and schematic representation of this movement (bottom-right).

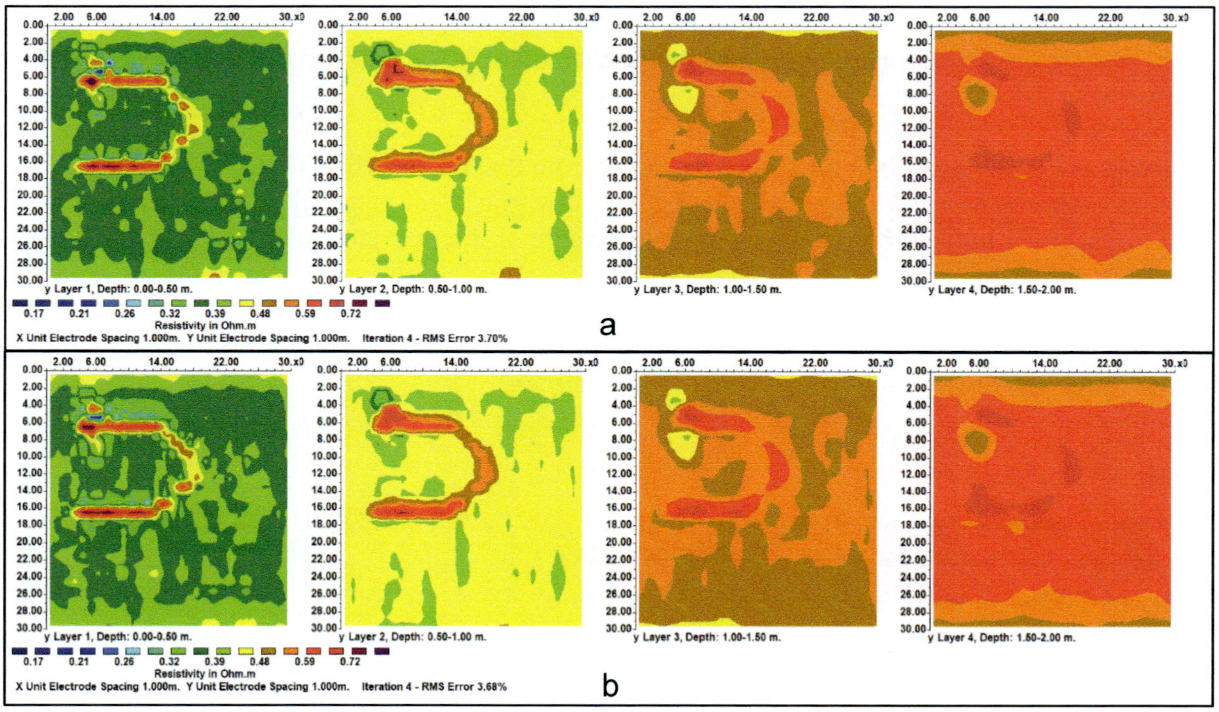

Fig. 2. a) Inverted section with electrodes at their actual position. b) Inverted section with electrodes moved to fit in a regular grid of 1m.

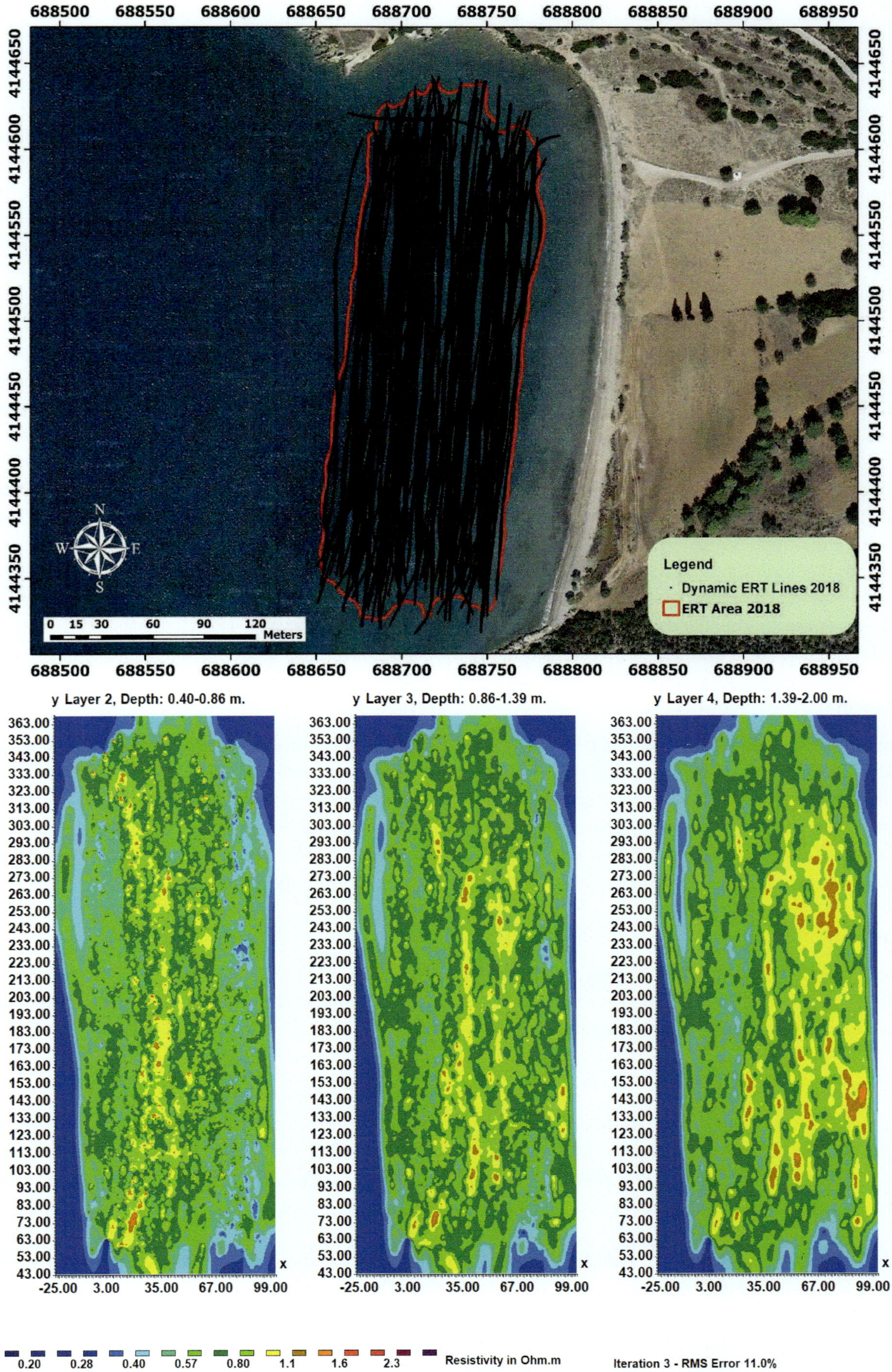

Fig. 3. Area covered with 3-D electrical resistivity tomography (red polygon) and the individual dynamic ERT lines during 2018 field season in Lambayanna (top). Horizontal depth slices extracted from the 3-D robust resistivity inversion model from Lambayanna using the dynamic dipole-dipole measurements (bottom).

The submerged dynamic ERT lines from Lambayanna posed specific challenges for the data processing. In a first stage, an algorithm that simulated the movement of the underwater cable was developed in order to reconstruct the UTM global coordinates of the electrodes from all the lines with a certain degree of accuracy. The operation of the algorithm was based on three hypotheses. Firstly, during the survey the electrode that was closest to the boat (the 13th electrode in our case) was attached on the boat with a rope of known length, so the distance between the specific electrode and the GNSS antenna was known. This meant that the 13th electrode was inside the sphere with radius equal to that distance and centre the position of the antenna. The second hypothesis was that the movement of any electrode couldn't be longer than the movement of its previous electrode or the GNSS movement for the 13th electrode plus a relatively small distance due to gravity. Finally, each electrode would make the smallest movement to reach a position that satisfies the two previous conditions.

Based on the above assumptions, a data file with 216,653 different electrode positions and more than 166,520 data points describing the 3-D variation of the apparent resistivity below the sea bottom at Lambayanna was created. By applying the same technique as in the synthetic data and moving the electrodes to fit in a regular grid of 0.5m, the survey ended up with 106,580 different electrode positions, which are 50% less than the original dataset.

The underwater resistivity imaging outlined a few linear resistive segments as well as compact resistive regions. The linear resistive segments appear with a strike direction similar to that of the profiles which could lead to the hypothesis of potential artefacts introduced by the inversion procedure. However, the fact that these anomalies don't extend in the whole section and that the expected orientation of the archaeological deposit to that strike makes them more likely to be related to walls or roads. The compact resistive regions are either related to collapsed remains or gathered piles of rocky material. The ERT depth slices up to 1m below the sea floor also defined the western boundary of the submerged settlement (Fig. 3 bottom).

Conclusions

The possibility to apply a dynamic electrical resistivity tomography method to shallow-water archaeological contexts was very important from the methodological point of view. Although the processing of such data presented some challenges and required the creation of novel algorithms and customization of commonly used software, the final results support such efforts and provide useful information. Ultimately, the results of this work can be regarded as a step toward the development of an effective method that could be applied to similar archaeological surveys in coastal or shallow-water environments.

Acknowledgements

This work was performed in the framework of the "ARCHERS: Advancing Young Researchers' Human Capital in Cutting Edge Technologies in the Preservation of Cultural Heritage and the Tackling of Societal Challenges" project Exclusively funded by the Stavros Niarchos Foundation.

References

Beck, J., Sakellariou, D. and Koutsoumba, D. 2017. Submerged Neolithic Landscapes off Franchthi Cave: The Measurements from the *Terra Submersa* Expedition and their Implications. *Communities, Landscapes, and Interaction in Neolithic Greece. Proceedings of the International Conference, Rethymno 29-30 May, 2015*, Edited by A. Sarris, E. Kalogiropoulou, T. Kalayci and L. Karimali. 261-268.

Papadopoulos, N.G, Tsourlos, P., Tsokas, G.N. and Sarris, A. 2007. Efficient ERT Measuring and Inversion Strategies for 3D Imaging of Buried Antiquities. *Near Surface Geophysics.* 5(6). 349-362.

Papadopoulos, N. and Simyrdanis, K. 2017. Dynamic 3D Electrical Resistivity Tomography for shallow off-shore archaeological prospection. In Jennings, B., Gaffney, C., Sparrow, T. and Gaffney, S. (eds.) *AP2017. 12th International Conference of Archaeological Prospection*. Archaeopress. Oxford. 178-180.

Simyrdanis, K., Papadopoulos, N. and Cantoro, G. 2016. Shallow off-shore archaeological prospection with 3-D electrical resistivity tomography: The case of Olous (modern Elounda), Greece. *Remote Sensing*. 8. 897.

Simyrdanis K., Bailey M., Moffat I., Roberts A., van Duivenvoorde W., Savvidis A., Cantoro G., Bennett K. and Kowlessar J. 2019. Resolving Dimensions: A Comparison Between ERT Imaging and 3D Modelling of the Barge Crowie, South Australia, 3D Recording and Interpretation for Maritime Archaeology, *Coastal Research Library*. 31. 175-186. Springer.

Surdez, M., Beck, J., Sakellariou, D., Vogel, H., Birchler Emery, P., Koutsoumba, D. and Anselmetti, F. 2018. Flooding a landscape: impact of Holocene transgression on coastal sedimentology and underwater archaeology in Kiladha Bay (Greece). *Swiss Journal of Geosciences*. 111. 3. 573-588.

Improving the lithological significance of shear wave tomograms through coring and pressure correction

Dennis Wilken[1], Wolfgang Rabbel[1], Tina Wunderlich[1], Martina Karle[2]
[1]*Institute of Geosciences, Kiel University, Kiel, Germany,* [2]*Lower Saxony Institute for Historical Coastal Research, Wilhelmshaven, Germany*

Introduction

Geophysical prospection methods are increasingly used for investigating palaeolandscapes. Particularly in wetlands that usually exhibit a high soil moisture content and salinity, the range of available methods is limited, because depth resolving methods suffer from bad penetration in wetlands due to the increased conductivity. We thus investigated the possibilities of seismic shear-wave traveltime tomography to detect sedimentological boundaries in a fluvial context. There are various possibilities of seismic wave analysis that allow a spatial resolution on a metre scale. The most common seismic method is reflection seismics. Unfortunately, shallow reflections coming from the first few metres of the subsoil can mostly be considered to be in the near field of the seismic source. In order to resolve shallow layer boundaries, surface wave seismics (e.g. Wunderlich *et al.* 2017) or refraction seismic methods can be applied. There are a lot successful applications of refraction seismics in landscape reconstruction at archaeological sites such as Rabbel *et al.* (2004). The seismic velocity generally shows sensitivity to temperature (which can be neglected in the shallow domain), pressure, and the porosity of the subsoil (Schön 1983). In unconsolidated sediments the different grain contact conditions between non-cohesive sands and gravels and cohesive clay or till need to be considered as well. The porosity in turn depends on grainsize, clay and silt content. Generally, an increasing porosity leads to a reduced seismic velocity. With increasing pressure, the seismic velocity increases.

We present a method of suppressing the pressure induced depth trend in the seismic tomograms by using lithology descriptions of a single core on the seismic profile. The method aims at improving the dynamics of the tomographic images to emphasize the changes in shear-wave velocity due to changes in grain size distribution.

The feasibility of this method is tested on two seismic profiles crossing the medieval river harbour site of Bardy/Świelubie, Poland (Fig. 1), a site of strong riverbed dynamics. The site was investigated to reconstruct the old riverbed situation to locate possible landing sites close to an early medieval fortification.

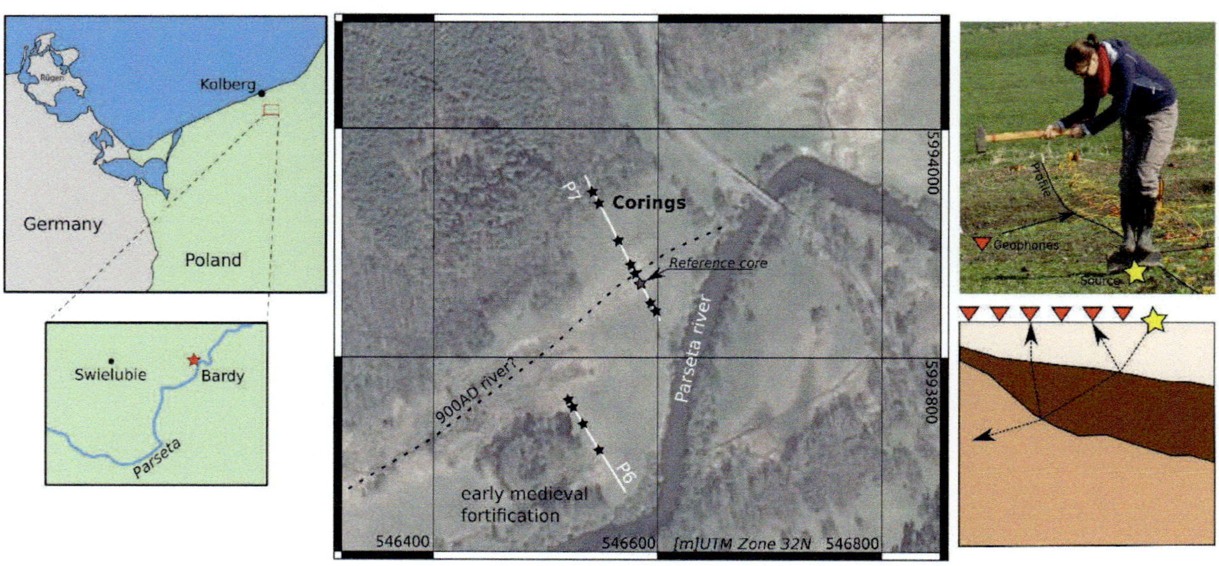

Fig. 1. Map of the investigation area. White lines indicate the seismic lines, black stars indicate the coring locations and the dark grey star the location of the reference core. Right: sketch and picture of seismic field setup.

Methodology

The seismic profiles were measured using horizontal geophones with a distance of 2m in SH wave mode orientation. Shots were performed every 2m. For all shots, the first arrival traveltimes were used for tomographic analysis. Tomography is based on the simultaneous iterative reconstruction technique (SIRT) (Dines and Lytle 1979).

The proposed method to suppress the pressure effect in the resulting tomograms includes the following steps (Fig. 2):

1. Extracting a velocity depth curve $V_s(z)$ from the tomogram at the position of a reference core. Define center depths, z_i, of the lithological units in the core.
2. We assume that inside each lithological unit, the change in seismic velocity only depends on the pressure effect, thus the sediment of one unit is assumed to be homogeneous in all other parameters affecting seismic velocity, especially grain size. This is only a suitable assumption for non-cohesive materials and becomes less valid in mixed materials. In such a case, the observed slope of seismic velocity inside a unit corresponds to the gradient of the pressure induced "velocity with depth" curve, $V_c(z)$. We thus performed a linear regression fit inside each unit to obtain an average gradient value $\frac{dV_c}{dz(z_i)}$.
3. A $V_c(z)$ model based on Hamilton (1976), $V_c(z) = a \cdot z^b$, was differentiated and fitted to the values $\frac{dV_c}{dz(z_i)}$.
4. Integration of the obtained gradient model results in a $V_c(z)$ estimate.
5. Subtraction of $V_c(z)$ from the original $V_s(z)$ curve along the whole profile

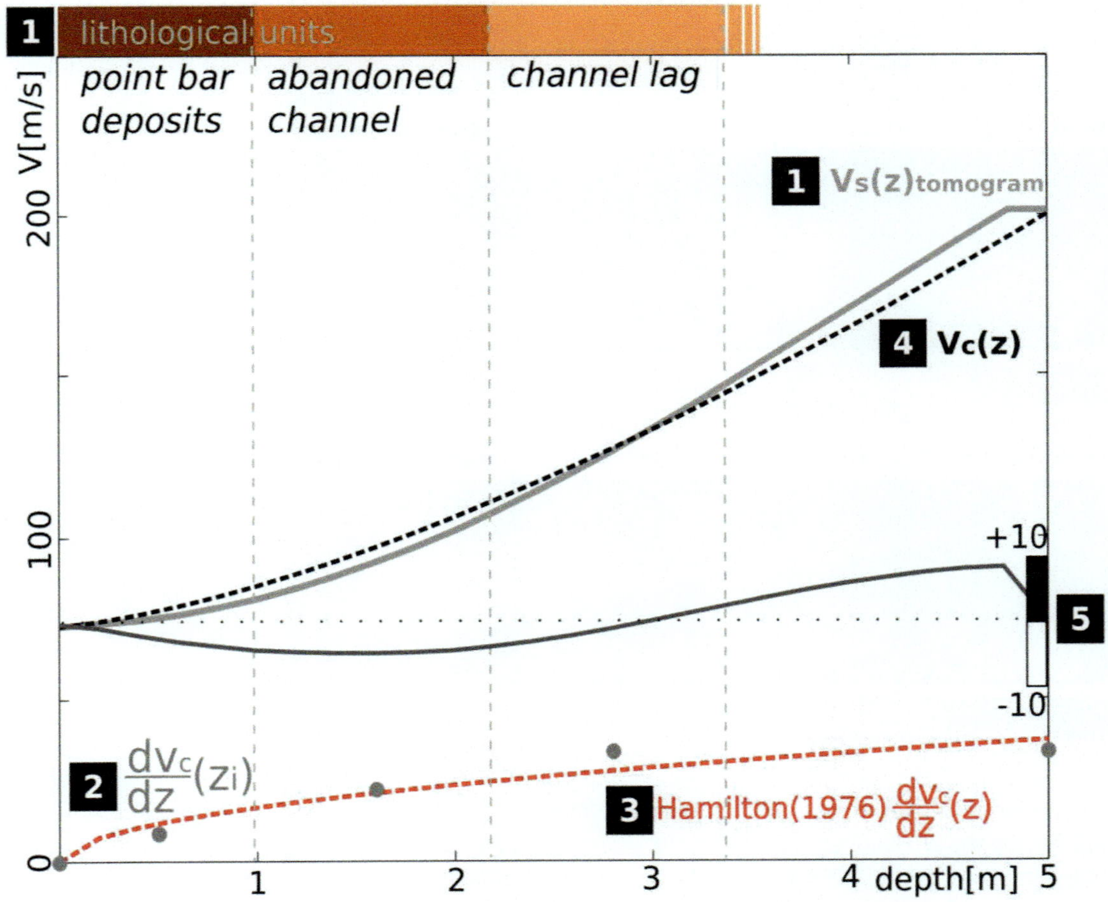

Fig. 2. Steps involved in the pressure correction methodology, using the velocity depth curve example located at the position of the reference core (numbers refer to the steps detailed in the text).

Results

Fig. 3 shows a comparison of the original seismic tomograms and corrected seismic tomograms together with the changes in lithological units in all cores. Black lines and labels show a basic geological interpretation based on the core information. The results show that especially the abandoned channel in profile P7 and the terrace sands become visible. Profile P6 shows a better emphasized channel fill, and sharpens the transition between floodplain deposits and abandoned floodplain sediments. The cultural layer cannot be identified.

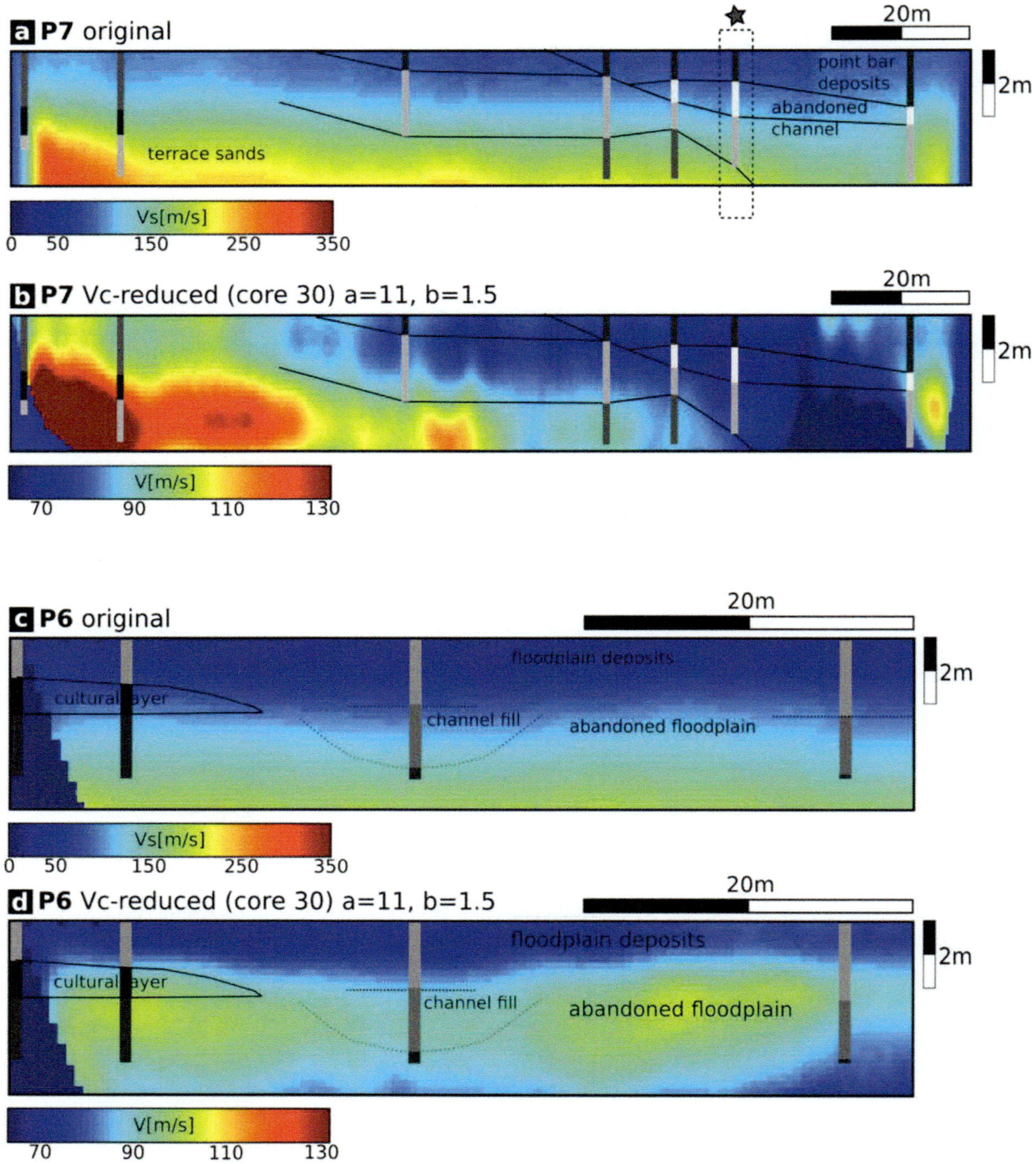

Fig. 3. a) shear wave velocity tomogram of profile P7 and lithological coring description. Dashed box indicates the core used for determining the pressure effect curve. b) Vp-substracted tomogram of profile P7. c) northern part of the shear wave velocity tomogram of profile P6 and lithological coring description. d) Vp-substracted tomogram of the northern part of profile P6.

Discussion

When discussing the feasibility of seismic methods for palaeolandscape investigations, one has to keep in mind that the seismic spatial resolution is limited. If we assume an average velocity of 200m/s, a center frequency of 80Hz and a travelpath of the wave corresponding to a half circle of 4m radius, we can calculate the width of the first Fresnelzone to 2.5m (Kvasnicka and Vlastislav 1996). This resolution estimate represents the deepest parts of the tomograms. For a depth of 1m we get 1.4m, respectively. This limits the feasibility of land based seismic measurements, which can clearly be seen in the presented examples (e.g. the cultural layer in profile P6). To enhance the seismic resolution the application of seismic full waveform inversion should be considered (e.g. Köhn *et al.* 2018). For traveltime tomographic data, these results show that the proposed method significantly enhances the seismic image quality and emphasizes sedimentological facies boundaries, which are usually overprinted by the pressure/depth effect. Nevertheless, the exponent derived for the Hamilton curve is much larger than expected and indicates a cohesive fraction in the sediments, and an influence of density and cementation.

Acknowledgements

We are grateful to the German Research Foundation (DFG) grants RA 496/26-SPP1630 and 2901391021 (SFB 1266) for funding the presented research.

References

Dines, K.A. and Lytle, R.J. 1979. Computerized geophysical tomography. *Proceedings of the IEEE*. 67(7). 1065-1073.

Hamilton, E.L. 1976. Shear wave velocity vs. depth in marine sediments: a review. *Geophysics*. 41(5). 985-996.

Köhn, D., Wilken, D., De Nil, D., Wunderlich, T., Rabbel, W., Werther, L., Schmidt, J., Zielhofer, C. and Linzen, S. 2018. Comparison of time-domain SH waveform inversion strategies based on sequential low and bandpass filtered data for improved resolution in near-surface prospecting. *Journal of Applied Geophysics*. 160. 69-83.

Kvasnicka, M. and Vlastislav, C. 1996. Analytical expressions for Fresnel volumes and interface Fresnel zones of seismic body waves. Part 2: transmitted and converted waves. Head waves. *Studia geophysica et geodaetica*. 40. 381-397.

Rabbel, W., Stuempel, H. and Woelz, S. 2004. Archeological prospection with magnetic and shear-wave surveys at the ancient city of Miletos (western Turkey). *The Leading Edge*. 23(7). 690-703.

Schön, J. 1983. *Petrophysik*. Ferdinand Enke Verlag. Stuttgart.

Wunderlich. T., Wilken, D., Erkul, E., Rabbel, W., Vött, A., Fischer, P., Hadler, H. and Heinzelmann, M. 2017. The river harbour of Ostia Antica - stratigraphy, extent and harbour infrastructure from combined geophysical measurements and drillings. *Quaternary International*. 473. 55-65.

Aerial thermal imaging from UAV in archaeology, a case study: the abandoned medieval town of Montecorvino (Foggia, Italy)

Luca d'Altilia
University of Foggia - Department of Humanities, Foggia, Italy

Developments in thermographic technology and in the production of more affordable and reliable Unmanned Aerial Vehicles (UAVs) over the last decade have led to a new era for thermal imaging in archaeology. New generation thermal cameras and drones are no longer out of reach for many archaeological projects and can be effectively used for revealing subsurface remains or better understand previous archaeological investigations.

Montecorvino is a deserted medieval settlement, located near Volturino (Foggia), in south-eastern Italy. Since 2006, archaeological investigations have been conducted on this site by the University of Foggia, Italy (Department of Humanities), under the supervision of Prof. Pasquale Favia and Prof. Roberta Giuliani (Favia *et al.* 2012). In August 2018 a series of infrared (thermal) pictures was acquired on this site by means of a FLIR Vue Pro R drone-mounted thermal camera[1], flown at different times of the day (mostly dawn or dusk). The principle behind this methodology is that archaeological remains under the surface may radiate or absorb heat differently than the nearby ground and this process can be recorded by the thermal camera as a difference in temperature. To obtain good results there should be sufficient contrast in the thermal properties of archaeological features and the soil matrix, also the archaeological materials should be close enough to the surface to be affected by heat flux; and the image should be acquired at a time when such differences are pronounced (Casana *et al.* 2017).

Image acquisition has been focused on two main areas of the medieval settlement: east of the motte and bailey castle (including the ditch area) and around the Cathedral. Thermal pictures taken in the first area at dusk show some possible archaeological features north-east of the motte (higher temperature, "positive", mark), where the city walls could join the castle proper and east of the ditch (both "positive" and lower temperature, "negative", marks), where the common people lived (Fig. 1, Fig. 2). Pictures taken around the Cathedral show some features ("positive" marks) that could be related to the settlement's main road (west of the church) and to some structures (east of the church), possibly related to the medieval city's main gate.

These data will be useful, starting from the 2019 excavation season, in combination with high resolution aerial photogrammetry, for orienting future excavation strategies at Montecorvino and better understanding the development of this strategically important medieval town in south-eastern Italy. Some excavation trenches will be located in the aerial thermal survey areas, for a verification of the preliminary results. Meanwhile, new aerial thermal survey areas will be located on the archaeological site.

References

Casana, J., Wiewel, A., Cool, A., Hill, A., Fisher, K. and Laugier, E. 2017. Archaeological Aerial Thermography in Theory and Practice. *Advances in Archaeological Practice*, 5.4. 310-327.

Favia, P., d'Altilia, L. and Laccone, R. 2018. *Accumulazione delle risorse e difesa degli spazi signorili nella città di Montecorvino: l'apporto integrato dei dati da indagine aerofotogrammetrica e scavo archeologico*. VIII Congresso nazionale di archeologia medievale. (Matera, 12-15 settembre 2018). 2.

Favia, P., Giuliani, R. and De Venuto, G. 2012. La ricerca archeologica sul sito di Montecorvino. Le campagne di scavo 2009-2010, In A. Gravina (a c. di), *Atti 32° Conv. Naz. Preistoria– Protostoria- Storia della Daunia* (San Severo 12-13 novembre 2011), San Severo 2012. 331-354.

[1] Resolution: 640 x 512 Pixels, 13mm lens, 7.5-13.5μm Spectral Response

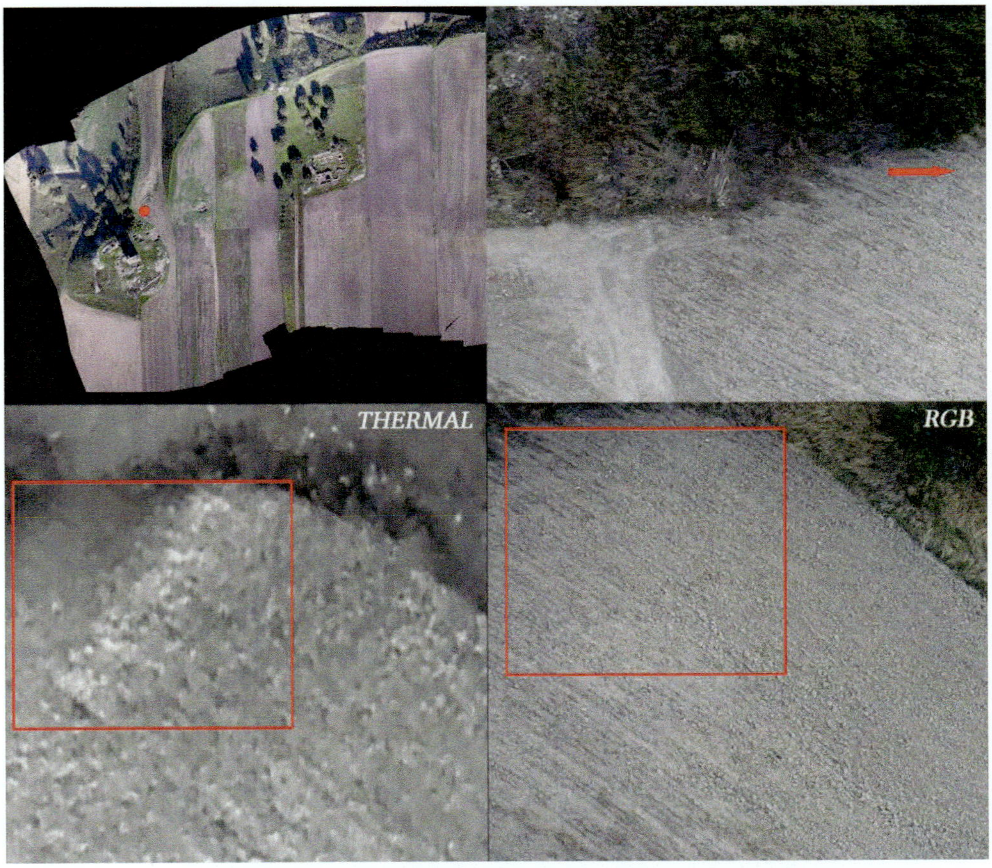

Fig. 1 (pictures taken at 8:30 PM).

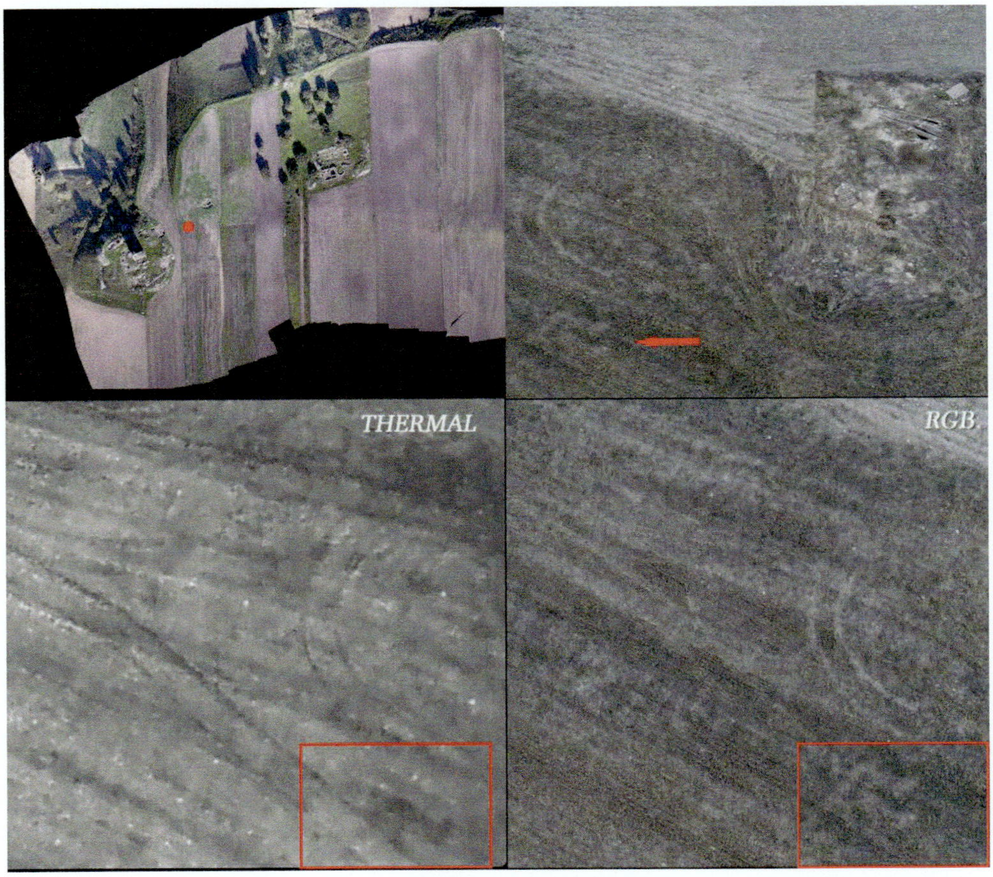

Fig. 2 (pictures taken at 8:45 PM).

Multi-spectral, multi-temporal survey in the American Midwest

Helen McCreary[1, 3], Jarrod Burks[2], Alex Corkum[2], Chris Gaffney[1], Andy Wilson[1]
[1]*University of Bradford, Bradford, United Kingdom,* [2]*Ohio Valley Archaeology, Inc,* [3]*Environmental Research Group, LLC*

Recent work has shown that commercial drone-mounted multispectral sensors developed for precision agriculture can be effectively adopted by archaeologists as prospection tools (Moriarty *et al.* 2018). The principal author's PhD research focuses on using just such a sensor (the Parrot Sequoia+) to survey multiple sites in southern Ohio over the course of a calendar year to track changes in spectral responses over known and unknown buried archaeological features. These layers of multispectral data will be integrated with legacy geophysical data and excavations to ground-truth interpretations.

The research will examine several sites in Chillicothe, Ohio, that have already seen extensive geophysical survey (Burks and Cook 2011). The figures below show Anderson Works, a large Hopewell earthwork on active agricultural land (Lynott *et al.* 2009). Survey was carried out on 5 May 2019 using a DJI Inspire 1 and the Parrot Sequoia+, which simultaneously collects green, red, red-edge and near-infrared bands. The field was bare earth and freshly ploughed, and the shape of the earthwork can be seen in the red, red-edge, and near-infrared bands (though only the red is shown in Fig. 1). Further surveys will track the earthwork as crops grow on top of it, documenting the efficacy of this prospection method over time.

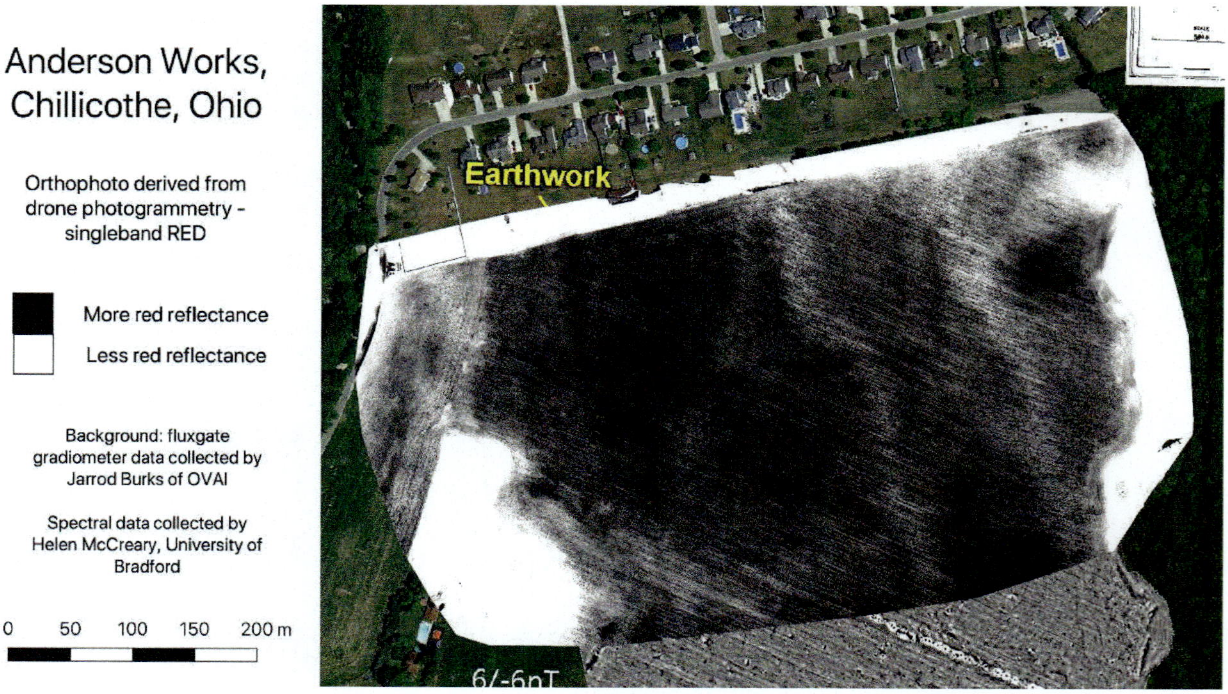

Fig. 1. Red-band orthophoto of a bare-earth Anderson Works, data collected 5 May 2019.

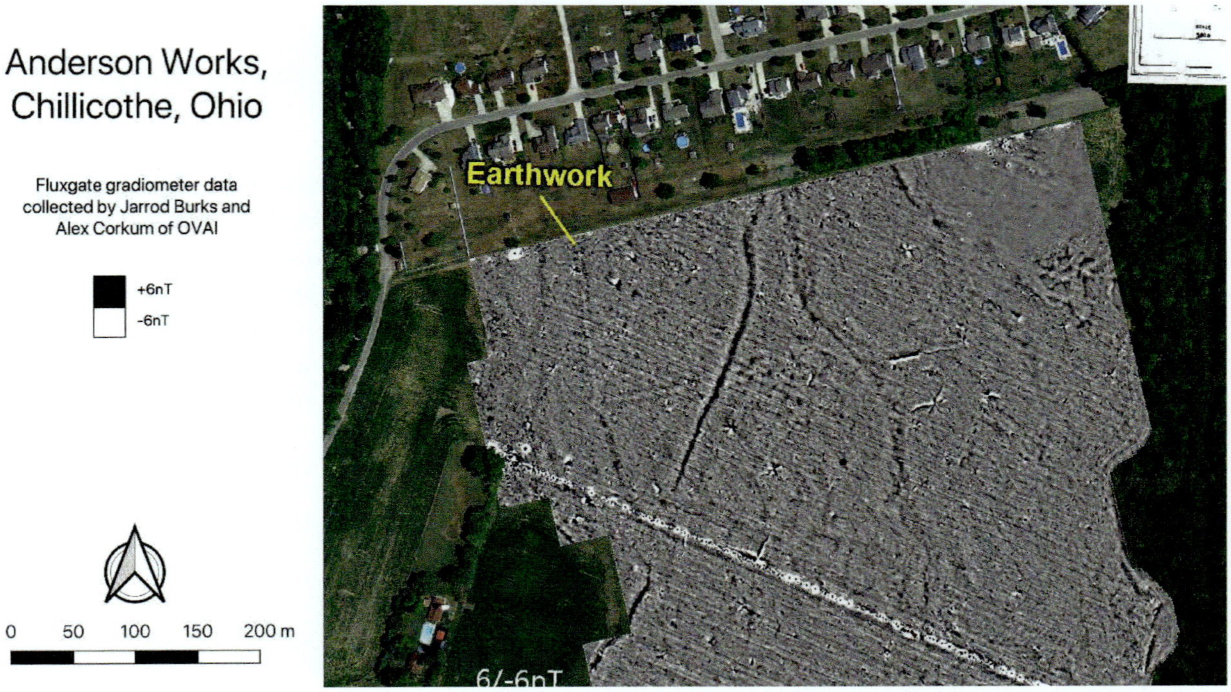

Fig. 2. Fluxgate gradiometer data of Anderson Works, collected by Jarrod Burks and Alex Corkum of Ohio Valley Archaeology, Inc.

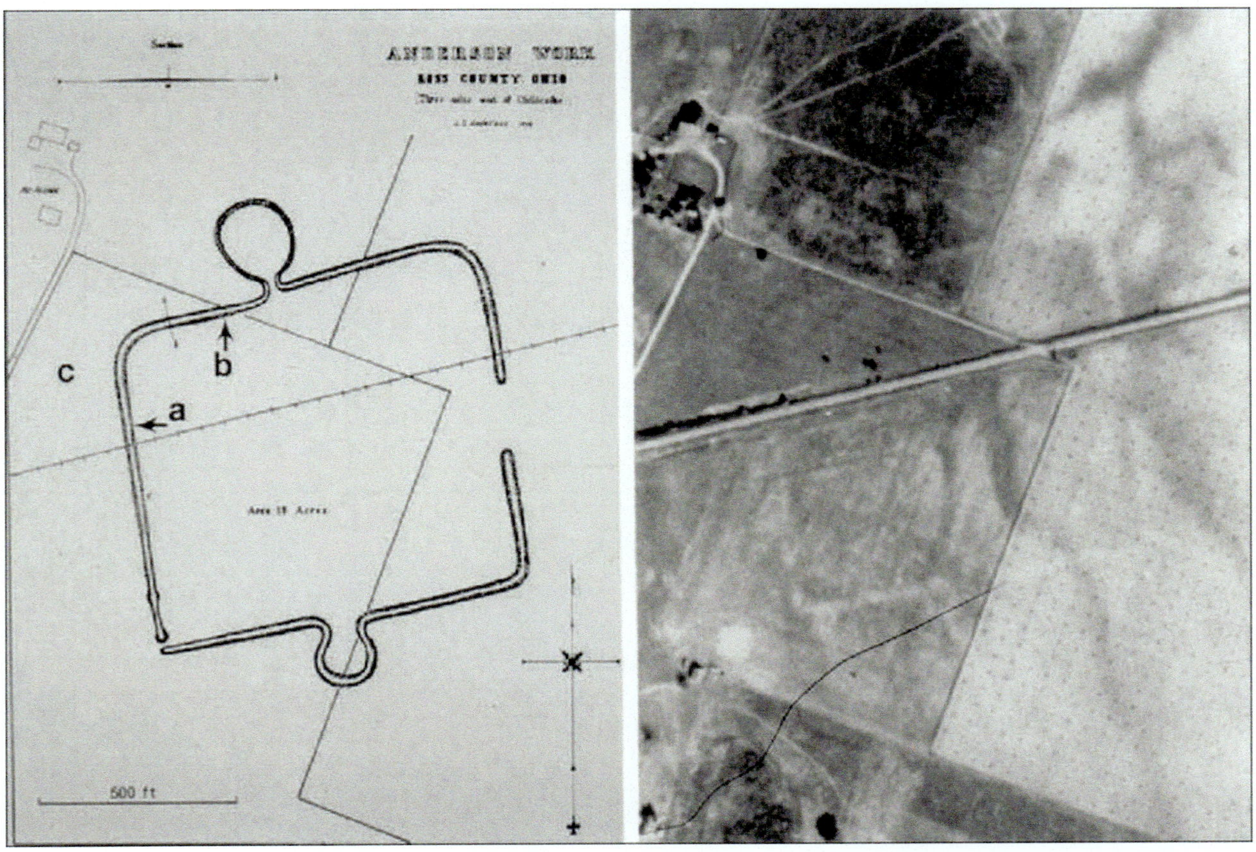

Fig. 3. Jerrel Anderson's 1979 map (left) and the 1938 U.S.D.A. aerial photograph of the Anderson Earthwork (right) (Lynott *et al.* 2009).

References

Burks, J. and Cook, R.A. 2011. Beyond Squier and Davis: Rediscovering Ohio's Earthworks Using Geophysical Remote Sensing. *American Antiquity.* 76(4). 667- 689.

Lynott, M.J., Pederson-Weinberger, J., Greber, N.B., Shane, O.C., Ruby, B.J., Pickard, W.H., Weinberger, J.W., Dancey, W.S., Burks, J., Walter Gagliano, D., Weymouth, J., Bevan, B., Dalan, R., Mandel, R.D. and Spielmann, K.A. 2009. *Footprints- In the Footprints of Squier and Davis: Archeological Fieldwork in Ross County, Ohio*. Midwest Archaeological Center Special Report Lincoln, Nebraska: National Park Service.

Moriarty, C., Cowley, D.C., Wade, T. and Nichol, C.J. 2019. Deploying multispectral remote sensing for multi-temporal analysis of archaeological crop stress at Ravenshall, Fife, Scotland. *Archaeological Prospection.* 26(1). 33-46.

Evaluating the capability of a SUAS mounted multispectral sensor for the mapping of archaeological resources in an alluvial landscape

Nicholas Crabb[1], Matthew Brolly[2], Niall Burnside[2], Andy Howard[3], Robin Jackson[4], Chris Carey[2]
[1]*EPSRC Centre for Doctoral Training in Science & Engineering in Arts, Heritage & Archaeology, University College London, United Kingdom,* [2]*University of Brighton, United Kingdom,* [3]*Landscape Research and Management, Stanmore, United Kingdom,* [4]*Worcestershire County Council, United Kingdom*

Introduction

In the UK, the use of airborne remote sensing techniques has primarily focused on the identification of cropmarks from aerial photographs and large areas have now been recorded as part of Historic England's National Mapping Programme (Bewley 2003). Despite notable exceptions (Baker 2003), there are smaller numbers of examples where this has been successfully used to detail complex geomorphological areas, such as alluvial environments. This is primarily because large parts of these landscapes are covered with a thick layer(s) of fine-grained alluvium that prevents the effective visualisation of any remains that may be deeply buried. However, such settings are attractive locations for archaeological activity and when discovered, they can be exceptionally well preserved. Moreover, the valley floor contains an assemblage of geomorphological landforms such as palaeochannels, terraces and gravel islands, which record the evolution of the river system (Brown 1997). These geomorphological features often contain important ecofactual and archaeological remains, and understanding their location, morphology, and sedimentary sequences is important for understanding archaeological potential. Thus, whilst the geoarchaeological investigation of alluvial landscapes is well established (e.g. Needham and Macklin 1992, Howard *et al.* 2003), the application of appropriate remote sensing technologies to determine archaeological potential within complex depositional environments requires more research (Challis and Howard 2006).

The use of LiDAR has been highly effective at mapping geomorphological features that are expressed as extant topographic variation (Carey *et al.* 2006, Challis *et al.* 2009, Stein *et al.* 2017). However, as alluvial deposition can blanket important geomorphological features and subsequent ploughing can also smooth out topography, the identification of geomorphological components can be problematic. The use of complementary information from geoarchaeological coring/test-pitting can go a long way towards reducing this, but normally requires the use of costly intrusive ground investigations. Geophysical survey methods and deposit modelling from pre-existing geotechnical datasets can provide a non-intrusive means of identifying geomorphological features that are not expressed topographically, but there has been limited consideration of how other techniques such as multispectral sensing can be deployed to assist in this regard.

Panchromatic imagery is sensitive to a broad spectral range covering the visible part of the electromagnetic spectrum, whereas multispectral sensors co-collect imagery from discrete (narrow) wavelength ranges over different parts of the spectrum (Beck 2011: 88). This can be advantageous as crop stress and vigour variations that may relate to subsurface archaeological/geomorphological features, are sometimes better expressed in non-visible wavelengths (e.g. Powlesland *et al.* 1997). Though archaeological applications of satellite and airborne multispectral sensors are by no means new, there has been a relatively limited uptake of this technology for the study of alluvial valleys. This is largely due to the cost of deploying systems that can provide suitable spatial resolution for the definition of individual features. However, with the development of lightweight multispectral sensors that can be mounted on Small Unmanned Aerial Systems (SUAS), imagery can now be provided at very high spatial resolution and relatively low cost. Whilst the spectral resolution of these sensors is low, being limited to portions of the visible and near-infrared parts of the spectrum, they have potential to assist in the analysis of surface landform assemblages. Furthermore, some studies have shown that spectral content appears to be of less importance for the detection of features than the spatial resolution of the imagery (De Laet *et al.* 2015) and recent research has also shown enormous potential for archaeological applications (Colomina and Molina 2014, Themistocleous *et al.* 2015, Agudo *et al.* 2018, Moriarty *et al.* 2019). Consequently, there is a need to further demonstrate the capability of this technology, particularly with reference to how best to exploit its potential within complex geomorphological areas, and this paper presents initial results of such a study.

To evaluate the capability of the SUAS mounted multispectral sensor, a study area was established in the Lower Lugg Valley in Herefordshire, UK. This was selected as it contains a diverse array of well documented archaeological remains, as well as a complex geomorphological history. To establish the nature of the known archaeological resource, a full search of the county Historic Environment Record and desk-based assessment was undertaken. LiDAR data from the Environment Agency was acquired and a series of Digital Terrain Models (DTMs) were produced and interpreted, highlighting palaeochannels and other alluvial landforms. By comparing the visibility of these landforms with those visible in the imagery from the SUAS mounted sensor, its capability could be assessed, and the relative successes and failures of the technology considered. Although it is not possible achieve the same area coverage as the LiDAR dataset, key areas that contained topographically extant features, and those that contained no visible features were targeted.

The multispectral sensor used for this study was a Parrot Sequoia. This has been primarily developed for use in agriculture and provides RGB photography and data from four bands of the electromagnetic spectrum (red, green, NIR and red-edge). This was mounted on a fixed-wing senseFly eBee, which is a fully autonomous SUAS fitted with GPS. Image processing was then undertaken using photogrammetric software (Pix4Dmapper) that enables data collected as a series of overlapping images to be matched using tie-points to create a single dataset. It also facilitates the generation of a point-cloud, mesh and orthomosaic and subsequently the production of raster datasets of each spectral band. Preliminary results have shown that the high spatial resolution of the SUAS mounted sensors enables the clear visualisation of small-scale individual archaeological features (Fig. 1). In addition, it also established that alluvial landforms such as palaeochannels could also be identified, although these can sometimes be hard to define, emphasising the importance of topography when understanding their morphology.

As use of a SUAS platform also enables the production of elevation models through Structure from Motion (SfM) photogrammetry, a comparison with the LiDAR constrained colour DTMs was also undertaken. In some cases, this enabled an improved definition of the features, but the reduced area coverage meant that this was only possible on a small scale. However, targeted use of additional high-resolution remote sensing techniques such as these SUAS mounted multispectral sensors, can be advantageous for the mapping of geoarchaeological resources.

The alluvial valleys of the Rivers Lugg and Wye form the focus of this research, investigating the efficacy of using multispectral data to model geomorphological components of these complex valley systems. The modelled interpretations were further tested through ground-based sediment sampling; reconstructing the sediment sequences of the valley system and examining their relationship to near surface and sub-surface sediment variability. The spectral responses of different valley components were analysed using computational image analysis and a detailed assessment of the potential for this to map surface sediment conditions can then be more accurately provided. Ultimately this will enable a more complete analysis of the capability of contemporary remote sensing techniques and it is hoped that this will provide an increased understanding of subsurface sediment architectures, allowing predictions to be made of their archaeological potential.

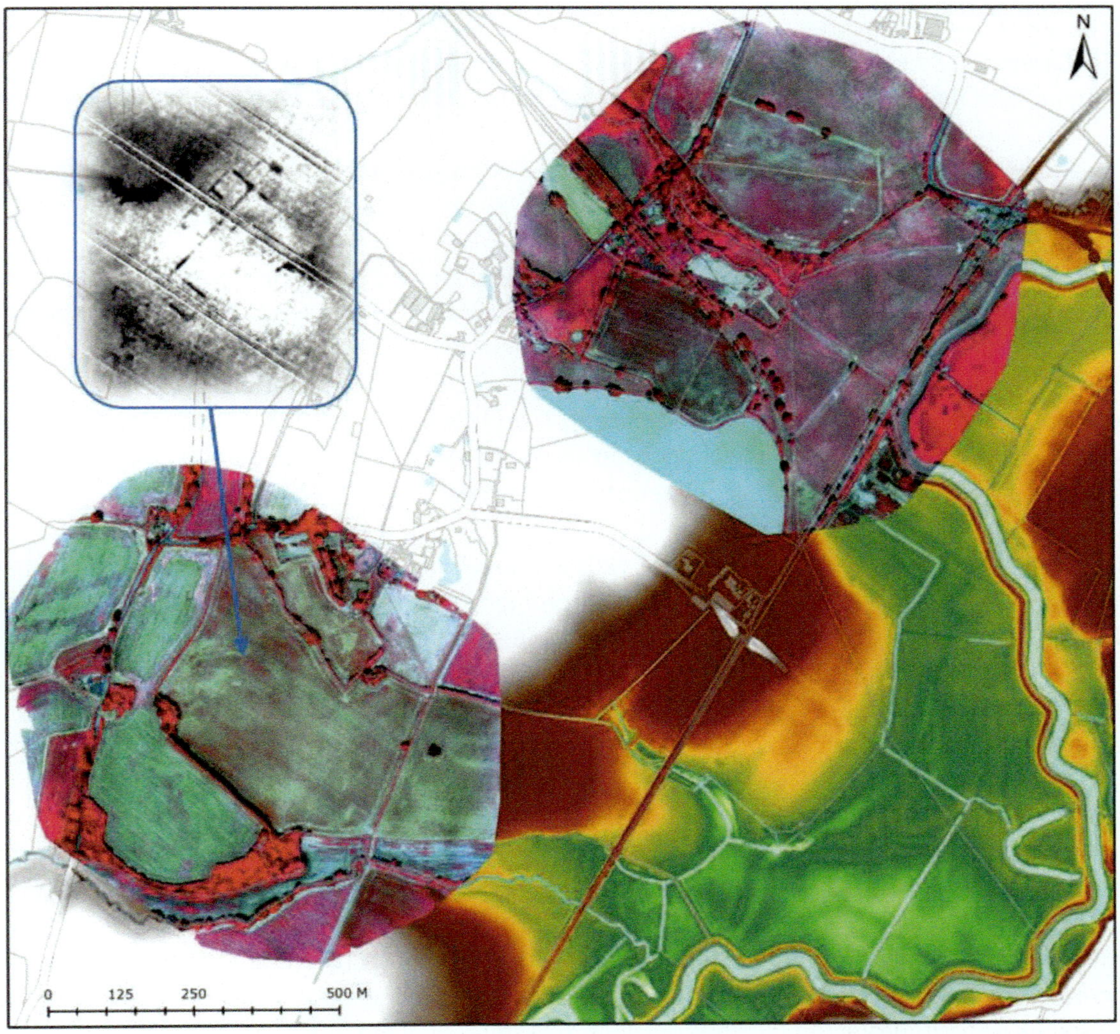

Fig. 1. LiDAR DTM constrained to 2m-9m (aOD), with false colour composite imagery overlain (R=NIR, G=Red, B=Green) and a detailed view of a Romano-British Villa (inset; greyscale NDVI).

References

Agudo, P., Pajas, J.A., Pérez-Cabello, F., Vicente Redó, J. and Ezquerra Lebr, B. 2018. The Potential of Drones and Sensors to Enhance Detection of Archaeological Cropmarks: A Comparative Study Between Multi-Spectral and Thermal Imagery. *Drones*. 2(3). 29.

Baker, S. 2003. *The Trent Valley: Palaeochannel Mapping from Aerial Photographs, Trent Valley Geoarchaeology Research Report*. Trent and Peak Archaeological Unit.

Beck, A. 2011. Archaeological applications of multi/hyper-spectral data - challenges and potential. In Cowley, D. (ed.), *Remote sensing for archaeological heritage management: Proceedings of the 11th EAC Heritage Management Symposium*, Reykjavík, Iceland, 25-27 March 2010 (EAC occasional paper; no. 5). Brussel: Europae Archaeologiae Consilium. 87–97.

Bewley, R. H. 2003. Aerial survey for archaeology. *Photogrammetric Record*. 18(104). 273–292.

Brown, A. G. 1997. *Alluvial Geoarchaeology: Floodplain archaeology and environmental change*. Cambridge Manuals in Archaeology.

Carey, C.J., Brown, T.G., Challis, K.C., Howard, A.J. and Cooper, L. 2006. Predictive Modelling of Multiperiod Geoarchaeological Resources at a River Confluence: a Case Study from the Trent-Soar, UK. *Archaeological Prospection*. 13. 241–250.

Challis, K., Carey, C., Kincey, M. and Howard, A.J. 2009. Airborne Lidar Intensity and Geoarchaeological prospection in River Valley floors. *Archaeological Prospection*. 18. 1–13.

Challis, K. and Howard, A.J. 2006. A Review of Trends within Archaeological Remote sensing in Alluvial Environments. *Archaeological Prospection*. 13. 231–240.

Colomina, I. and Molina, P. 2014. Unmanned aerial systems for photogrammetry and remote sensing: A review. *ISPRS Journal of Photogrammetry and Remote Sensing*. 92. 79–97. International Society for Photogrammetry and Remote Sensing, Inc. (ISPRS).

De Laet, V., van Loon, G., Van der Perre, A., Deliever, I. and Willems, H. 2015. Integrated remote sensing investigations of ancient quarries and road systems in the Greater Dayr al-Barshā Region, Middle Egypt: A study of logistics. *Journal of Archaeological Science*. 55. 286–300.

Howard, A.J., Macklin, M.G. and Passmore, D.G. 2003. *Alluvial archaeology in Europe: Proceedings of the Alluvial Archaeology of North-West Europe and Mediteranian*, 18-19 December 2000, Leeds, UK / edited by Andy J. Howard, Mark G. Macklin, David G. Passmore. Lisse; Abingdon: A.A. Balkema.

Moriarty, C., Cowley, D.C., Wade, T. and Nichol, C.J. 2019. Deploying multispectral remote sensing for multi-temporal analysis of archaeological crop stress at Ravenshall, Fife, Scotland. *Archaeological Prospection.* 26. 33–46.

Needham, S. and Macklin, M.G. 1992. *Alluvial archaeology in Britain.* Proceedings of a conference sponsored by the RMC Group plc, 3-5 January 1991, British Museum. Oxford: Oxbow (Oxbow monograph; 27).

Powlesland, D., Lyall, J. and Donohoe, D. 1997. Enhancing the record through remote sensing: the application and integration of multi-sensor, non-invasive remote sensing techniques for the enhancement of the Sites and Monuments Record. Heslerton Parish Project, N. Yorkshire, England. *Internet Archaeology*. 2.

Stein, S., Malone, S., Knight, D., Howard, A.J. and Carey, C. 2017. New Approaches to Mapping and Managing Palaeochannel Resources in the Light of Future Environmental Change: A Case Study from the Trent Valley, UK. *Historic Environment: Policy and Practice*. 8.2. 113–124.

Themistocleous, K., Agapiou, A., Cuca, B. and Hadjimitsis, D.G. 2015. Unmanned aerial systems and spectroscopy for remote sensing applications in archaeology. *International Archives of the Photogrammetry, Remote Sensing and Spatial Information Sciences*. ISPRS Archives. 40 (7W3). 1419–1423.

Beneath the Stains of Time: The physiochemical prospection of multiperiod sites in southern Britain - a geophysics approach to geochemistry?

Paul Cheetham
Bournemouth University, Poole, United Kingdom

Despite the undoubted success of geophysical survey prospection techniques in locating some types of archaeological evidence, how well these are serving as tools for informing us where to deploy more intrusive, destructive, and expensive interventions, whether in a commercial or research context, is rather more debatable. Many archaeological sites in the arable zone in southern Britain have been severely truncated and so lack the occupation surfaces and upstanding deposits required to understand many of the activities that do not result in cut features. The latter, while damaged, often do partially survive the ravages of the plough and so turn up clearly on geophysical surveys, but such differential survival will inevitably bias our views of the archaeological character and potential of a site.

This paper considers whether the use of topsoil geochemical data is effective on these types of sites by informing improved interpretations and so excavation strategies. Importantly, this is a geophysicist's view of geochemical data, in which observed distribution and elemental morphology transcend any analysis of actual geochemical processes, the complexities of which (see Oonk *et al.* 2009) have, to an extent, blighted the use of geochemical data in archaeological prospection.

Topsoil geochemical surveys using portable X-ray fluorescence (pXRF) data are compared with the excavated features and their geochemical signatures to help characterise the underlying nature of the archaeology or fill in gaps in knowledge from the geochemical stains left in the topsoil from otherwise physically destroyed surfaces, structures and deposits. It will assess if such an approach can significantly contribute to the understanding of archaeological sites and landscapes, and therefore should be considered being deployed more routinely alongside geophysical surveys to provide a more nuanced interpretation of the archaeological potential of sites. Fig. 1 illustrates the potential considering more the 'texture' of a suite of elements over the absolute values of individual elements obtained using a Niton pXRF.

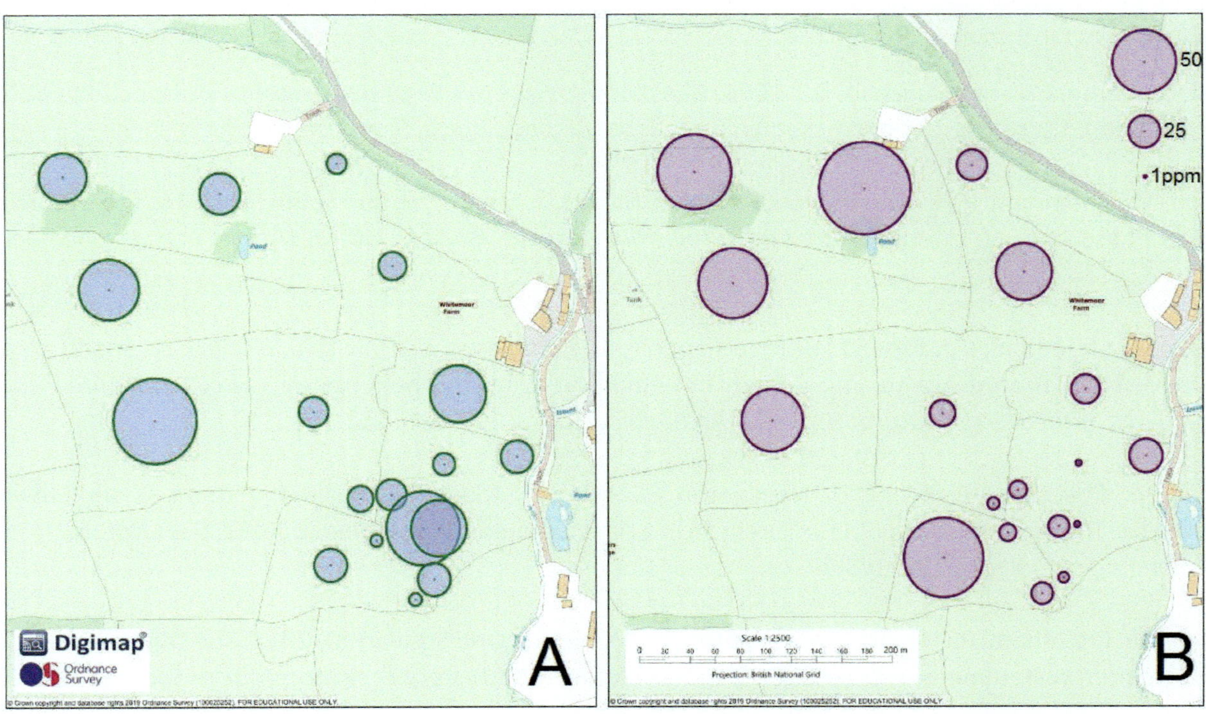

Fig. 1. (A) An example of topsoil geochemical variation (TGV) (the variability within a suite of 18 elements, excluding phosphorous) and (B) against total phosphorus. The site in the SE exhibits lower more consistent levels of P, against a greater range of variability within a suite of 18 elements. All data obtained using a pXRF on soil samples from Whitemoor Farm (Somerset) collected by BU student Bethany Lowman.

A case study from a large multiperiod site in south Dorset (Russell *et al.* 2014; 2015; 2016; 2017; 2019) illustrates elemental geochemical survey effectively integrated into site assessment and interpretation, and extrapolation, assessing its effectiveness at a scale and resolution rarely undertaken as part of the site evaluation process. These sites characteristically produce little surface evidence and the site shown in Fig. 2 was only located on the strength of a single Roman Republican coin found by metal detectorists. That such large and complex sites can go undetected without geophysical survey is one issue - what such sites represent functionally, the geophysical survey alone can only partially address. In such cases topsoil geochemical survey can practically contribute, however does this need to be approached with a more empirical experiential ethos?

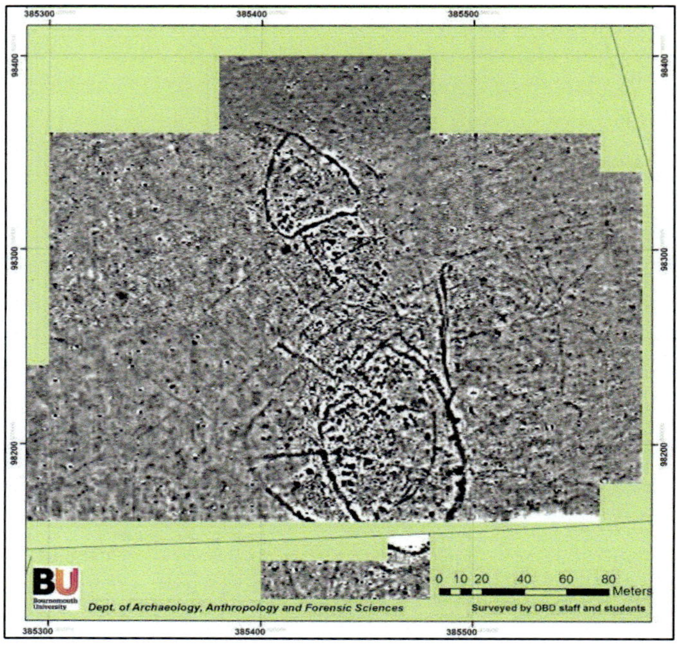

Fig. 2. Example magnetometry results from a site that is within the multiperiod landscape at Winterborne Kingston, Dorset, UK. Bartington 601-2 fluxgate gradiometry at 0.125m x 1m plotted -5 to + 5nT (white to black).

References

Oonk, S., Slomp, C.P. and Huisman, D.J. 2009. Geochemistry as an Aid in Archaeological Prospection and Site Interpretation: Current Issues and Research Directions. *Archaeological Prospection*. 16(1). 35–51.

Russell, M., Cheetham, P., Evans, D., Hambleton, E., Hewitt, I., Manley, H. and Smith, M. 2014. The Durotriges Project, Phase one: an interim statement. *Proceedings of the Dorset Natural History and Archaeological Society*. 135. 217-221.

Russell, M., Cheetham, P., Evans, D., Gerdau-Radonic, K., Hambleton, E., Hewitt, I., Manley, H., Speith, N. and Smith, M. 2015. Bronze Age to Late Roman settlement at Winterborne Kingston - Second Interim Report. *Proceedings of the Dorset Natural History & Archaeological Society*. 136. 157-161.

Russell, M., Cheetham, P., Evans, D., Hambleton, E., Hewitt, I., Manley, H., Gale, J., Pitman, D. and Stewart, D. 2016. The Durotriges Project, Phase Three: an interim statement. *Proceedings of the Dorset Natural History & Archaeological Society*. 137. 173-177.

Russell, M., Cheetham, P., Hewitt, I., Hambleton, E., Manley, H. and Stewart, D. 2017. The Durotriges Project 2016: an interim statement. *Proceedings of the Dorset Natural History & Archaeological Society*. 138. 105-111.

Russell, M., Cheetham, P., Barrass, K., Evans, D., Hambleton, E., Manley, H., Pitman, D. and Stewart, D. 2019. The Durotriges Project 2017: an interim statement. *Proceedings of the Dorset Natural History & Archaeological Society*. 139. 127-133.

Three-dimensional modelling of petroglyphs of South Siberia

Vladislav Kazakov[1,2], Vasily Kovalev[1], Kair Zhumadilov[1], Lyudmila Lbova[1], Aleksandr Simukhin[1]
[1]*Novosibirsk State University, Novosibirsk, Russian Federation,* [2]*Novosibirsk State University of Economics and Management, Novosibirsk, Russian Federation*

Traditional methods of documenting petroglyphs have several disadvantages. For example, redrawing onto a transparent film gives an inaccurate and subjective reproduction of the drawings, and when photographing, rock drawings are usually of insufficient resolution and geometrically distorted. When photographing, a favorable side light is needed to illuminate the petroglyph. This is difficult to achieve, and if it is present, the colour space of the object is distorted.

Over the past few years, attempts have been made to document petroglyphs in digital form, including 3D modelling (Anderson *et al.* 2018, Williams and Twohig 2015). These methods allow the production of a digital image with better quality and accuracy. Up to now, methods of digital documentation of petroglyphs, including 3D, have been tested on single objects. A large-scale project which involves working on thousands of objects has not yet been attempted.

At the same time, using digital 3D modelling for documentation would make it possible to increase the accuracy of documentation of rock art monuments, i.e. with high resolution, without geometrical distortions - including its third dimension, and without the influence of a subjective factor. Also, work with the rock art in the field could be non-invasive and less time-consuming, and the results of the laboratory work would become reproducible on the basis of primary photographic materials. With a three-dimensional model, it is possible to create a side-illumination of a drawing with programmed methods, measure the geometric parameters of an object with software tools, including the depth of the mark, etc. On the basis of the 3D model, it is possible to automatically construct a two-dimensional projection of a plane with drawings, devoid of geometric distortions – an orthophoto. Thus, orthophotography as a two-dimensional form of fixation for rock art is preferable rather than traditional photofixation.

In recent studies Monna *et al.* (2018) proposed the next step in the direction of automation of rock art documenting: the authors of the paper proposed a technique for creating a black and white redrawing of stone carvings in a software graphical editor on the basis of a preliminary digital selection of the image using topographic and graphical algorithms. The result of such a drawing is comparable in quality with the drawings made manually using traditional methods, but unlike them, it is geometrically precise, not invasive, and the biggest part of the work is undertaken in the laboratory.

Over a few years, the authors of this paper have tested various approaches and techniques for the documentation of rock art in Southern Siberia (Cheremisin *et al.* 2015; 2016; 2017). For the digitization of petroglyphs, both photogrammetry methods and scanning on structured light technology were used. The report represents the work of 2018 which was undertaken by the team of authors as a part of the artemiris.org project. As a result of these works more than 100 models of petroglyphs of various rock art sites of Southern Siberia were obtained.

To represent three-dimensional models, a specialized online information system was developed with a 3D viewer based on the open source three.js library. Using such a self-written software product, it was possible to make a flexible adjustment of the functionality and design of the information system, including the implementation of an individual role-based data access model, tools for analysing archaeological images in 3D form, etc. At the core of the data storage model is a modern library - and one of the most lightweight 3D formats – Draco from Google Open Source.

Acknowledgements

This work was supported by the Russian Science Foundation (project No. 18-78-10079).

References

Anderson, H., Galvin, E. and de Torres Rodriguez, J. 2018. Museological Approaches to the Management of Digital Research and Engagement: The African Rock Art Image Project. *African Archaeological Review.* 35. 321–337.

Cheremisin, D., Kazakov, V., Kovalev, V. and Zharikova, M. 2015. Experience of three-dimensional scanning of rock paintings using the technology of structured illumination. *Problems of archeology, ethnography, anthropology of Siberia and adjacent territories*. Novosibirsk: IAET SB RAS XXI: 446-450.

Cheremisin, D., Kazakov, V., Kovalev, V. and Zhumadilov, K. 2016. 3D modeling of Altai petroglyphs using three-dimensional scanning and photogrammetry. *Problems of archeology, ethnography, and anthropology of Siberia and adjacent territories*. Novosibirsk: IAET SB RAS XXII: 466–469.

Cheremisin, D., Kazakov, V., Kovalev, V. and Zhumadilov K. 2017. Three-dimensional scanning of Altai rock paintings in the engraving technique. *Problems of archeology, ethnography, anthropology of Siberia and adjacent territories*. Novosibirsk: IAET SB RAS XXIII: 449–453.

Monna F., Esin Yu., Magail J., Granjon L., Navarro N., Wilczek J., Saligny L., Couette S., Dumontet A. and Chateau, C. 2018. Documenting carved stones by 3D modelling – Example of Mongolian deer stones. *Journal of Cultural Heritage.* 34. 116–128.

Williams, K. and Twohig, E. 2015. From sketchbook to structure from motion: Recording prehistoric carvings in Ireland. *Digital Applications in Archaeology and Cultural Heritage.* 2(2–3). 120–131.

Part Seven – Improving our Legacy: Reviewing the Key Outcomes of Archaeological Prospection Data

Introducing the 'Soil science & Archaeo-Geophysics Alliance' (SAGA): a new interdisciplinary network in archaeo-geophysics

Carmen Cuenca-Garcia[1], Kayt Armstrong[2], Apostolos Sarris[3], Philippe De Smedt[4], Clare Wilson[5], Elina Aidona[6], Anne Roseveare[7], Martin Roseveare[7], Petra Schneidhofer[8], Jörg Faßbinder[9], Ian Moffat[10], Marion Scheiblecker[11], Abir Jrad[12], Martijn van Leusen[13], Kelsey Lowe[14], and SAGA's Management Committee[15]

[1]Norwegian University of Science and Technology (NTNU), NTNU University Museum, Department of Archaeology and Cultural History, Erling Skakkes Gate 47b, Trondheim, Norway, [2]Magnitude Surveys Ltd, Bradford, United Kingdom, [3]Institute for Mediterranean Studies, Foundation for Research and Technology, Rethymno, Greece, [4]Department of Environment, Faculty of Bioscience Engineering, Ghent University, Ghent, Belgium, [5]Biological and Environmental Sciences, University of Stirling, Stirling, United Kingdom, [6]Department of Geophysics, School of Geology, Aristotle University of Thessaloniki, Thessaloniki, Greece, [7]TigerGeo, Hereford, United Kingdom, [8]Vestfold fylkeskommune, Vestfold, Norway, [9]Bayerisches Landesamt für Denkmalpflege, München, Germany, [10]Flinders University, Adelaide, Australia, [11]Department of Earth and Environmental Sciences, LMU, München, Germany, [12]Gabès University, Gabès, Tunisia, [13]Department of Archaeology, University of Groningen, Groningen, Netherlands, [14]School of Social Science, The University of Queensland, St Lucia, Australia, [15]SAGA's Management Committee.

Summary

The 'Soil science & Archaeo-Geophysics Alliance: going beyond prospection' (SAGA) is a new interdisciplinary network of scientists. During a four year period (October 2018-October 2022), SAGA will develop, promote and facilitate research activities bringing together archaeo-geophysics and soil science with the overall goal of maximising interpretation of proxy data for archaeological purposes. The network and related activities are funded by the European Cooperation in Science and Technology (COST) and the grant is administrated by the Norwegian University of Science and Technology (NTNU). SAGA is coordinated by a Management Committee (MC) currently composed of 85 experts from 34 countries and integrates experts working in industry, management and academia. Participation in the network and its activities, is open to institutions and individuals with strong interests in contributing towards SAGA's objectives.

SAGA: going beyond prospection

Archaeological sites can be discovered and recorded in a high-resolution and non-invasive manner using geophysical methods. These measure the spatial variation of a range of physical properties of the soil, which may be representative proxies of the subsurface archaeology. Less-invasive and cost-effective field procedures have become a top-priority in efforts to mitigate the destructive effects on our cultural heritage (CH) from intensified land use, climate change and conflict.

In the last decade, the introduction of multi-sensor and motorised geophysical instrumentation has revolutionised the discipline of archaeo-geophysics by allowing extremely fast and high-resolution surveys. This critical technological development has enabled a new dimension in landscape prospection.

At a time when many organisations are investing in advanced geophysical equipment, a major issue is that our ability to fully interpret the information available from geophysical datasets is still very limited. This limitation arises from an incomplete understanding of the relationships between soil properties, soil dynamics, archaeological features and related geophysical signatures. Overcoming this limitation is a prerequisite for maximising the cost-effectiveness of geophysical methods, realising the expected benefits of technological investment and allowing a broader utility of geophysical methods in the CH sector.

Bridging this gap requires significant efforts in research coordination in order to develop fine-tuned and multi-disciplinary teams testing novel field and analytical methods, plus scholarly discussion to collate the outcomes of projects on this topic.

SAGA's interdisciplinary network efforts aim to identify, explore and disseminate novel methods for data collection, integration and analysis. The overall goal is to move archaeo-geophysics beyond basic prospection and to become a significant tool for answering nuanced questions about archaeology and their host landscapes. SAGA's objectives include the following:

- Set up a network of infrastructure and expertise to facilitate research collaborations.
- Facilitate cross-disciplinary discussion to establish the state-of-the-art in understanding how soil properties, dynamics and processes affect geophysical signatures. This will be done by synthesising existing knowledge in archaeological geophysics and incorporating the outcomes achieved in other geophysical applications.
- Identify and promote integrated strategies for data collection, visualisation and modelling and promote them beyond academia. A close dialogue with stakeholders (from practitioners to policy makers) will be established to make sure these strategies reach the 'real word' in CH management and that their potential is also known by the general public.
- Provide training to students and other interested stakeholders by involving Early Career Investigators (ECI) and high profile experts employed by the industry and public organisations through training schools and Short-Term Scientific Missions (STSM) to ensure the longevity of the Action outcomes.

The journey: from an Odyssey to SAGA

The idea behind SAGA emerged during the first workshop on 'Interactions between Soil Science and Geophysics in Archaeological Prospection (ISSGAP)' held in Rethymno by the GeoSatResearch Lab on 17-18 June 2015 (Armstrong *et al.* 2015). The aim of this meeting was to bring together researchers working on the integration of soil analysis and archaeo-geophysics to discuss the outcomes of past and on-going projects. One of the key outputs from this meeting was the creation of a self-sustained network of researchers with interest in these type of interactions and the COST Action program was identified as the ideal source of funding.

The follow up of this was a second ISSGAP workshop held again in Rethymno (5-7 June 2017) to define SAGA's concept and draft a proposal. The proposal was submitted to the COST Association in September 2017 with the support of an initial network of 32 proposers from 25 countries and selected by the COST Association in April 2018 (the proposal and further details can be found in Cuenca-Garcia *et al.* 2018). Formation of the MC of SAGA then began through the nomination of national representatives of COST countries.

SAGA was officially established on 26 October 2018 during the 'First MC Meeting', held in Brussels at the COST Association headquarters. During this meeting, the NTNU was elected as the grant holder institution and it is responsible for managing the grant.

The first scientific activity of SAGA was the 'First Joint Working Group (WG) and MC Meeting' held in Rethymno from 4-6 March 2019 (Fig. 1). During this 3-day meeting, detailed action plans were developed by each WG. This event also included a brainstorming session and a second MC meeting to decide on future activities to be developed by SAGA during the next grant period.

The current MC of SAGA has 85 members (including substitute members) from 34 countries and it is composed of experts working in industry, academia and CH management (Fig. 2).

Fig. 1. SAGA Management Committee members during the 'First Joint Working Group and Management Committee Meeting' held in Rethymno (4-5-6 March 2017).

SAGA's Working Groups

WG1 focuses on structuring the existing research on fundamental soil parameters involved in the detection of archaeological features using geophysical techniques. Tasks include the discussion and synthesis of findings from past studies into how land use practices, high variability in soil properties, soil post-depositional and other taphonomic processes have affected geophysical signatures. Based on these results, the WG will produce recommendations for further research.

WG2 explores, discusses and evaluates combined approaches using geophysical, archaeological and soil/geological sampling methods. WG2 will assemble a list of test-sites, instrumentation and labs to facilitate field experimentation and provide a pool of equipment to share between SAGA members. The main outcome is to provide standard guidelines for field solutions for data collection.

WG3 will identify and evaluate innovative solutions for multivariate proxy data analysis. WG3 seeks ways of integrating data from soil analytical techniques, environmental parameterisation and sensor signals registered by various instrumentation. Beyond co-display, this WG is working to find solutions for 'real' data integration and considering the potential of forward modelling in archaeo-geophysics.

By integrating the outcomes from WG1-3, the objective of WG4 is to demonstrate the benefits of incorporating soil science and geophysics in all stages of archaeological field practice to curators, field archaeologists and students.

SAGA's way forward

SAGA will organise a series of scientific and management meetings and will open calls for funding via STSM and conference grants for ITC (COST Inclusiveness Target Countries). Unlike other EU-funded projects, COST Actions are open networks so it is possible to get involved in SAGA in different ways. Information about how to participate, the network, its members and activities is available on SAGA's website: www.saga-cost.eu.

Fig. 2. Overall management structure of SAGA.

Acknowledgements

This paper is based upon work from COST Action SAGA (CA 17131), supported by COST (European Cooperation in Science and Technology).

References

Armstrong, K., Cuenca-García, C. and Moffat, I. 2015. Interactions between Soil Science and Geophysics in Archaeological Prospection (ISSGAP). *ISAP News.* 43. 7

Cuenca-García, C., Armstrong, K., Aidona, E., De Smedt, P., Roseveare, A., Roseveare, M., Schneidhofer, P., Wilson, C., Faßbinder, J., Moffat, I., Sarris, A., Scheiblecker, M., Jrad, A., van Leusen, M. and Lowe, K. 2018. The Soil science & Archaeo-Geophysics Alliance (SAGA): going beyond prospection. *Research Ideas and Outcomes.* 4. e31648.

Laying the geophysical groundwork: *in situ* measurements as a framework for strategizing archaeological prospection

Philippe De Smedt[1], Ynse Declercq[1], Daan Hanssens[1], Wouter B. Verschoof-van der Vaart[2], Tom Hamburg[3], Walter Laan[3], Pieter Lalloo[4], Frédéric Cruz[4], Karsten Lambers[2]

[1]Department of Environment, Ghent University, Ghent, Belgium, [2]Faculty of Archaeology, Leiden University, Leiden, the Netherlands, [3]Archeologisch Onderzoek Leiden (Archol), Leiden, the Netherlands, [4]GATE Archaeology, Ghent, Belgium

Within and beyond development-led archaeological frameworks, strategizing archaeological prospection approaches that incorporate geophysical methods remains a daunting task for many archaeologists and managing agencies. While national (David *et al.* 2008) and international guidelines (Schmidt *et al.* 2015) provide useful support, decisions on a site-specific level often remain difficult to make, particularly in regions with complex and heterogeneous geologies. Although data such as soil and geological maps, and general information about land-use and the archaeology of studied areas help more robust decision-making, the leap to assessing geophysical discrimination potential remains large, particularly when only so-called 'negative' features are targeted (e.g. ditches or pits).

To provide a more robust framework, we conceptualized a practical approach that starts from the current implementation of Malta-archaeology. As part of a collaboration between the universities of Ghent (BE) and Leiden (NL) and two commercial archaeological units: Archol (NL) and GATE Archaeology (BE), an *in situ* geophysical measurement programme has been set up. During trial trenching and test pitting campaigns, measurements of geophysical properties are conducted on excavated profiles. Hereby, the electrical conductivity, dielectric permittivity, and magnetic susceptibility are recorded on characteristic ('negative') archaeological features in these profiles, in addition to recording these properties on governing natural profiles within the studied area.

Our aim is to expand this approach to regional and national scales, with reference frameworks for understanding geophysical soil properties of governing geologies and frequently occurring archaeological features.

References

David, A., Linford, N. and Linford, P. 2008. *Geophysical Survey in Archaeological Field Evaluation*. Swindon, English Heritage.

Schmidt, A., Linford, P., Linford, N., David, A., Gaffney, C., Sarris, A. and Fassbinder, J.W.E. 2015. *EAC Guidelines for the Use of Geophysics in Archaeology: Questions to Ask and Points to Consider*. Belgium, Europae Archaeologia Consilium.

Review strategies for archaeological prospection, incorporating excavation and research

Cian Hogan
Earthsound Geophysics Ltd., Claremorris, Ireland

Introduction

Archaeological geophysics is used universally to investigate and record real and potential archaeological features beneath the soil. The resulting data informs commercial and research excavations, site-specific investigations and broader landscape projects. But how accurate are these outcomes? Can data be used beyond these outputs? Could geophysical and excavation results be combined to produce other datasets relevant to archaeological practices? A workflow for reviewing prospection interpretations will take a theoretical look at data acquired from archaeological geophysical surveys, and metadata from archaeological excavations, as a means of predicting the length of time required for an intrusive archaeological excavation.

Reviewing geophysical surveys

Despite the use of archaeological geophysics ahead of many excavation projects, there is little to no feedback between these two phases in Ireland – this situation may differ from region to region in other countries (Jones and Sharpe 2006, Neubauer 2004). Feedback is a necessary step for improving the provision of archaeological geophysical services. However, there are numerous considerations in undertaking a review that generates a helpful outcome.

Previous systems have proposed a review strategy running between the geophysical prospectors and the archaeological excavators (Neubauer 2004). This facilitates feedback on a site-by-site basis. By bringing in desk-based research that compiles information from numerous sites, feedback on overall patterns can be incorporated. Geophysical data can be related back to site typologies based on excavated confirmations (Fig. 1).

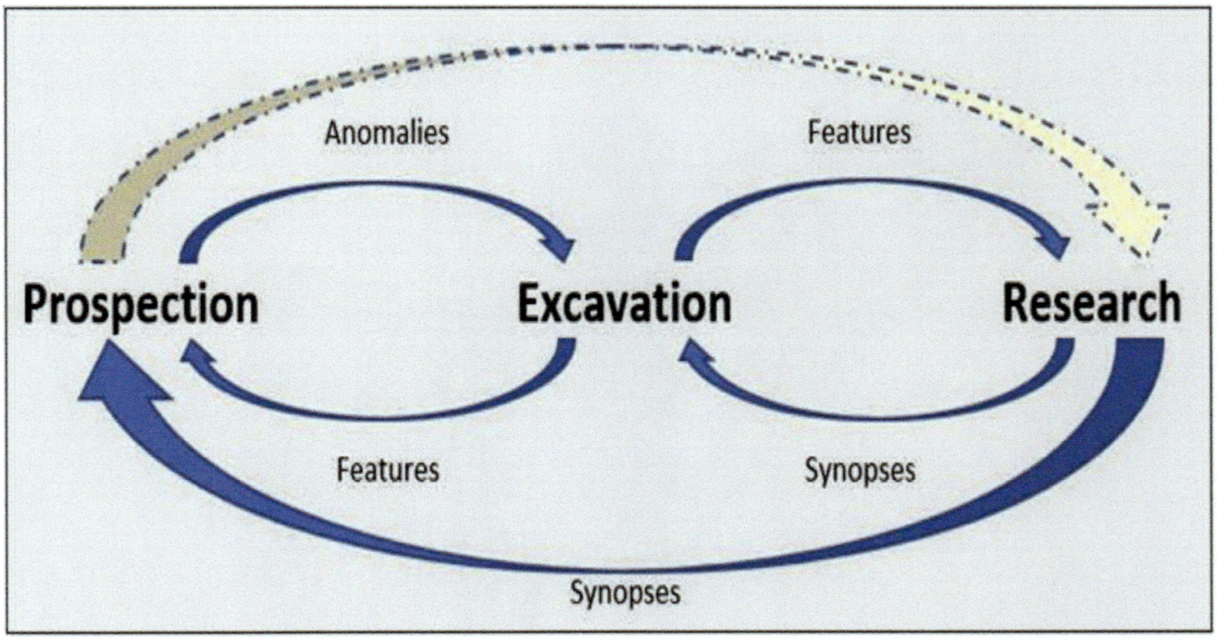

Fig. 1. Feedback loop incorporating geophysical prospection, archaeological excavation and archaeological research.

Developing an effective methodology for this process would be the first step in using geophysics to a greater degree in archaeological practices. It would allow for the accuracy of interpretations to be tested, and for a review of indeterminate anomalies to take place.

Data in = data out – the quality of any queries sought from GIS and databases will be guaranteed by the quality of the form that is constructed and the information recorded by it. A review and update of the standards system would have to be implemented in order to achieve the desired model. Under current standards, the necessary fields are not recorded, or at least not in an easily accessible format.

Discussions have taken place before on how to measure the achievement of a geophysical survey (Bonsall 2014: 137-141). Without a proper review strategy in place, it is impossible to qualify the outcome. Success cannot solely be based on the interpretation of archaeological, definite or potential, anomalies, but in conjunction with the expectations of clients and the parameters of a survey. Other factors to be taken into account include: techniques used, methodology implemented, and the quality of processing and interpretation.

Case Study – Ranelagh, Co. Roscommon

In 2015 magnetometer and electromagnetic induction surveys, at 0.5m traverses, were carried out along a road corridor (Bonsall and Gimson 2015). The survey identified a triple-ditched enclosure with internal features at one location. This was resurveyed with extensions outside the corridor to identify its full extent (Fig. 2) (Hogan and Gimson 2015). The surveys identified adjoining enclosures, along with other archaeologically-derived anomalies. The anomalies within the corridor were subject to excavation, which confirmed the enclosure, but also uncovered several hundred burials. There was no indication of the burials in the geophysical data, despite duplicating the original survey (Bonsall 2017). The resolution and methodology were inappropriate for identifying burials, a feature type that was not accounted for at the procurement stage. Recent research on multi-vallated enclosures has presented them not solely as habitation sites but with an additional ceremonial role, with use for burial a part of this. Therefore, if such a site presents itself through anomalies, further prospection targeting burials can be recommended.

The feedback informing this case study is purely informal. The knowledge of the burials came back to the surveyors by hearsay. It is by chance that a presentation (Gleeson 2016) covering such enclosures was attended by employees and could therefore be incorporated.

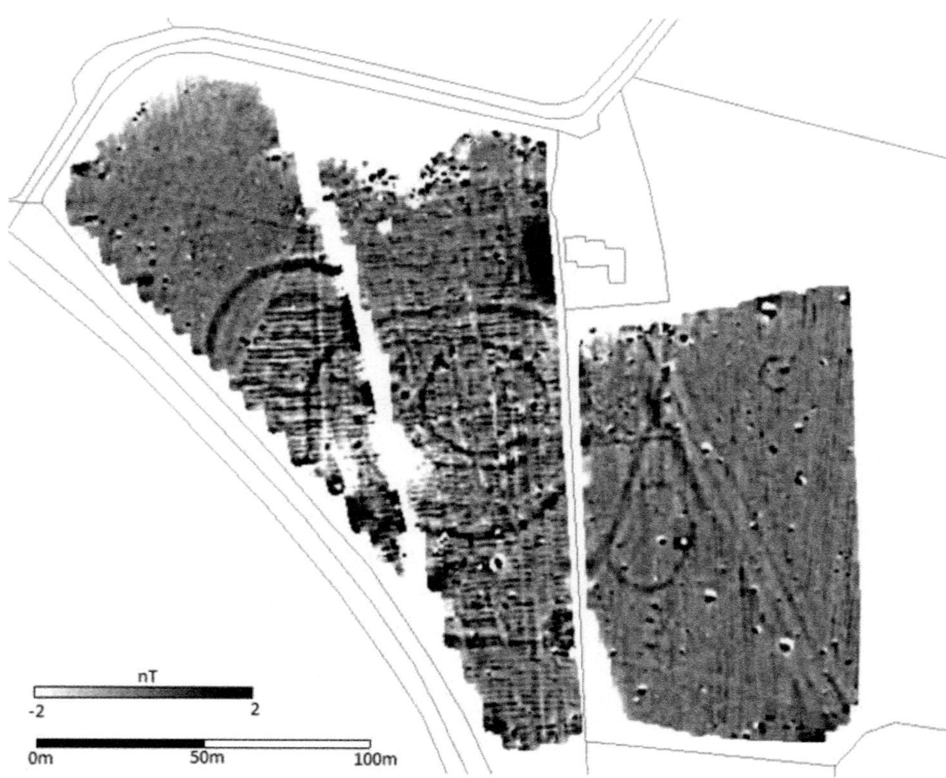

Fig. 2. Magnetometer data of the enclosure at Ranelagh along the N61 Coolteige Road Project, County Roscommon.

Indicating (future) work

At its most simple level, a geophysical survey commissioned in the development-led sector, is an indicator of further, anticipated, archaeological investigation. Where surveys are requested ahead of a development, they are an indication that archaeological excavation could occur to some extent. This is a basic meter, but could certainly be used to infer some levels of archaeological excavation that will follow, or not as the case may be. In some instances, the results of the geophysical survey cause developers to cancel a construction project, negating any potential employment.

In many cases, an estimation of excavation time will be based on geophysical results. For example, at Ranelagh, an excavation of the enclosure elements were predicted to take three months. Due to the presence of burials (unforeseen by the geophysical surveys), the excavation ran to twelve months.

The resolution of geophysical surveys determines the features identified. A resolution of 1m traverse intervals used for standard magnetometer surveys will not identify many small-scale features (Bonsall *et al.* 2014: 14); in Ireland, recommended higher resolution surveys provide more potential information, though may still leave some low-/no-contrast features undetected/uninterpreted (Bonsall 2014: 464). This means that archaeological sites, as presented in geophysical datasets - including those taking advantage of multi-method datasets - should be viewed as a partial picture and that when excavating a greater level of complexity should be expected.

In order to use geophysical surveys to provide reliable indications of excavation timeframes, input is needed from several stakeholders – curators, geophysicists, excavation companies, academic researchers and relevant third-party bodies. The necessary information, in a suitable format, needs to be collated and worked upon collaboratively.

Concluding Remarks

This paper has presented a deliberately vague proposal aimed at stimulating discussion around the potential for using data, that is already being generated and stored, to greater effect. The use of digital methods for recording archaeological excavations are becoming more frequent. The capabilities of databases and GIS programs are constantly improving. The functions discussed above have been tried to some extent before but need further development.

Acknowledgements

I would like to thank Heather Gimson and Ursula Garner, Earthsound Geophysics Ltd.; Matt Seaver, National Museum of Ireland; Dr Rob Sands, University College Dublin; and Transport Infrastructure Ireland.

References

Bonsall, J. 2014. *A reappraisal of archaeological geophysical surveys on Irish road corridors 2001-2010, with particular reference to the influence of geological, seasonal and archaeological variables*. Unpublished PhD thesis, University of Bradford.

Bonsall, J. 2017. 'Living in a post-workhorse world: observations learnt from rapidly acquired electromagnetic induction surveys in Ireland (...when magnetometry just won't do...)'. In Jennings, B., Gaffney, C., Sparrow, T. and Gaffney, S. (eds.) *AP2017. 12th International Conference of Archaeological Prospection*. Archaeopress. Oxford. 28-30.

Bonsall, J., Gaffney, C. and Armit, I. 2014. A Decade of Ground Truthing: Reappraising Magnetometer Prospection Surveys on Linear Corridors in light of Excavation Evidence 2001-2010. In Kamermans, H., Gojda, M. and Posluschny, A.G. (eds.). *A Sense of the Past: Studies in current archaeological applications of remote sensing and non-invasive prospection methods*. BAR International Series 2588. Archaeopress, Oxford. 3-17.

Bonsall, J. and Gimson, H. 2015. *N61 Coolteige (Phase 1) Road Project, County Roscommon. Earthsound Geophysics Ltd.*: EAG 253, unpublished report.

Gleeson, P. 2016. Chronologies and Historical Archaeologies of Early Medieval Ireland. In *IAI 2016 Conference Brochure: A Matter of Time*.

Hogan, C. and Gimson, H. 2015. *Multi-vallate Enclosure, Ranelagh and Barnhill, County Roscommon. Earthsound Geophysics Ltd.*: EAG 274, unpublished report.

Jones, R.E. and Sharpe, L. 2006. *Going over old ground. Perspectives on archaeological geophysical and geochemical survey in Scotland*. British Archaeological Report 416, Oxford.

Neubauer, W. 2004. GIS in Archaeology – The Interface between Prospection and Excavation. *Archaeological Prospection.* 11. 159-166.

Methodological framework to automatically compare large-scale magnetometry measurements with excavation datasets

Gábor Mesterházy

Castle Headquarters Integrated Regional Development Centre, Budapest, Hungary

Over the past years, large-scale geophysical measurements became more and more common as a survey method for different scientific and development-led archaeological projects. The collected large datasets opened new opportunities in archaeological prospection and in archaeology as well. Meanwhile, we have to keep in mind that the limitations and restrictions of the applied methods have also became larger due to the increased extent of investigation.

In the past five years we have integrated magnetometer surveys into the Hungarian cultural resource management as a standard tool, covering more than 30km^2 with magnetometry on different planned motorways, industrial parks and various investment projects. Usually large-scale preventive excavations had also occurred on these projects, creating the possibility to compare these datasets for large areas. As these methods are a vital part of Hungarian development-led archaeological mitigation, it is expected to have a constantly growing database to work with.

The aim of this paper is to introduce a method, which measures and automatically analyses the physical and cultural properties of archaeological phenomena by comparing magnetometry, survey interpretation and excavation features. Each of these elements, one by one, have different limitations and biases. Magnetometer surveys are affected by measurement settings (probe distance, sensitivity, spatial accuracy, feature detectability, etc.) and other factors such as soil properties. Survey interpretation is mostly based on the interpreter's experience and knowledge, meanwhile the skill of the field archaeologist has a huge impact on the excavated features (Seren *et al.* 2013). By analysing and quantifying these uncertainties and biases on large-scale real datasets the general aim during the development of this toolset was to achieve a better understanding of measurement, interpretation and excavation datasets.

The data processing is aided by an ArcGIS toolbox, which offers shorter scripts and models to prepare and standardize the input data (preliminary stage), meanwhile the rest of the calculations (first to fourth steps) are carried out as an automatic black-box solution (Fig. 1).

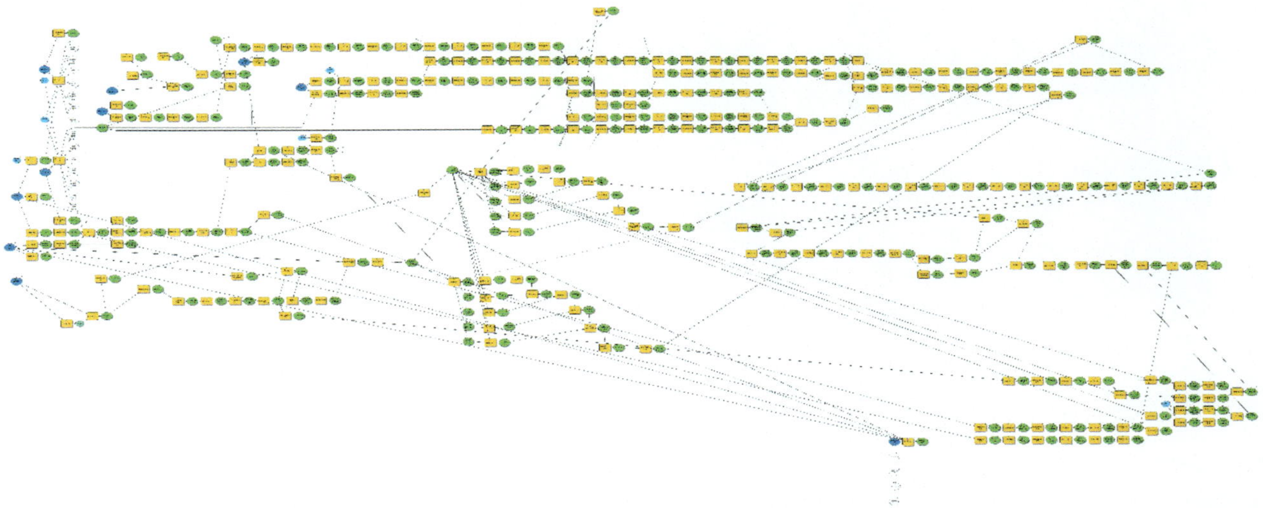

Fig. 1. Overview of the feature-based workflow in ArcGIS Toolbox.

The statistical analysis was conducted in a pre-written Excel spreadsheet, which automatically computed general and detailed statistics about the dataset.

The preliminary stage of the proposed workflow is the structural data exchange from excavation map (AutoCAD) to shapefile data format in a semi-automatic way, where the excavation, feature borders and different heights (original surface elevation, excavation surface, arhcaeological features) are extracted and separated in a standardised form.

During the first stage the different heights are used to interpolate three surfaces in order to calculate height related properties of the interpretation and excavated feature polygons (top-soil depth, feature depth, cubic content, etc.). Other variables (size, overlap percent) are also derived from the raw dataset to collect the feature's analysable physical properties.

The second stage connects the measurement values with the interpretation and excavatedn feature polygons. Based on the filtered and regularly gridded (25cm x 25cm) measurement file, minimum, maximum, average and standard deviation values are subtracted inside the interpretation and excavated feature polygons and four user-defined buffer zones around them.

The third step overlaps the geophysical interpretation and the excavated feature polygons to measure the accuracy and success ratio of the interpretation, and the first and second step are completely repeated on the overlapping polygon parts (Fig. 2.). All calculated and derived data are stored in one output shapefile (180 columns).

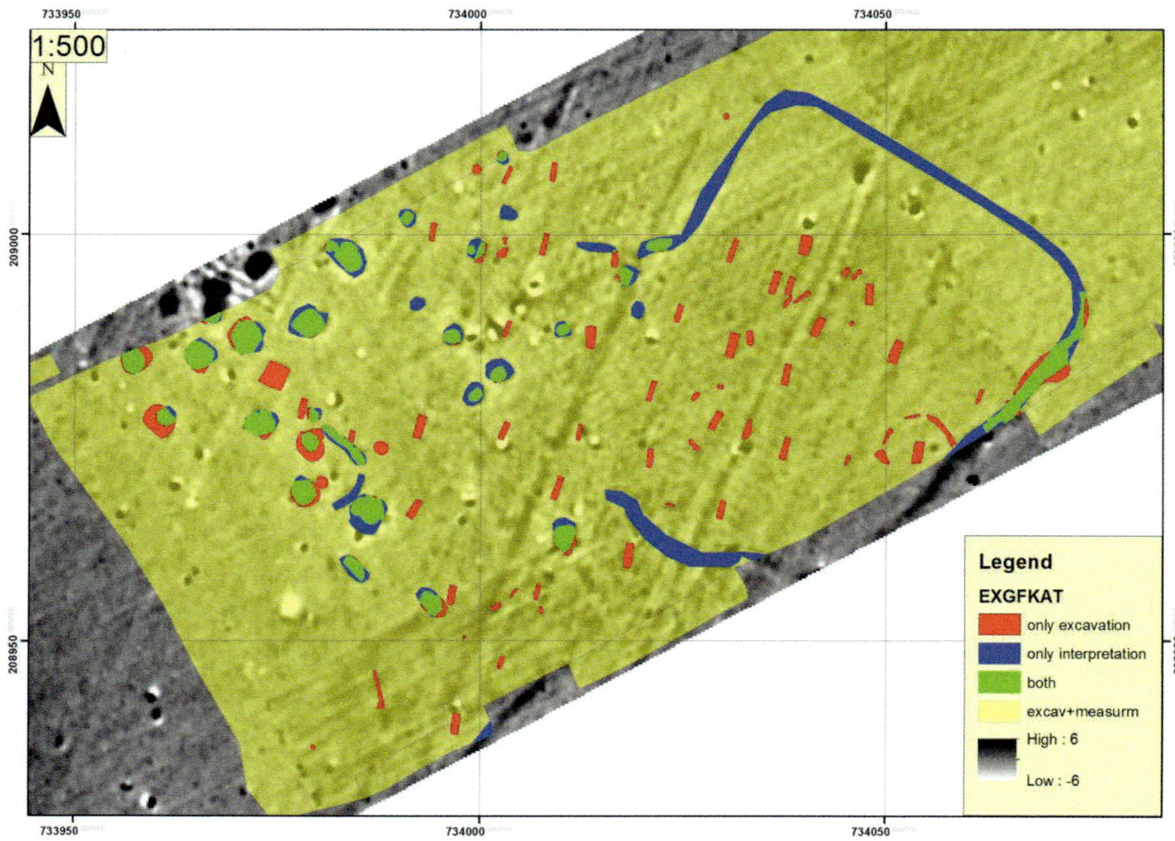

Fig. 2. Results of overlapping interpretation and excavation feature polygons.

The fourth step is an implemented sub-model, aiming to define the interpretation's site level accuracy by comparing the colocated area of interpretation and excavation features. A buffer zone, which takes into consideration the feature's intensity, is defined between 7m and 20m around the interpretation and excavation features. By overlapping these, comparison between overlapping and non-overlapping areas can be made based on general attributes (Fig. 3.). The separate output shapefile contains 36 columns.

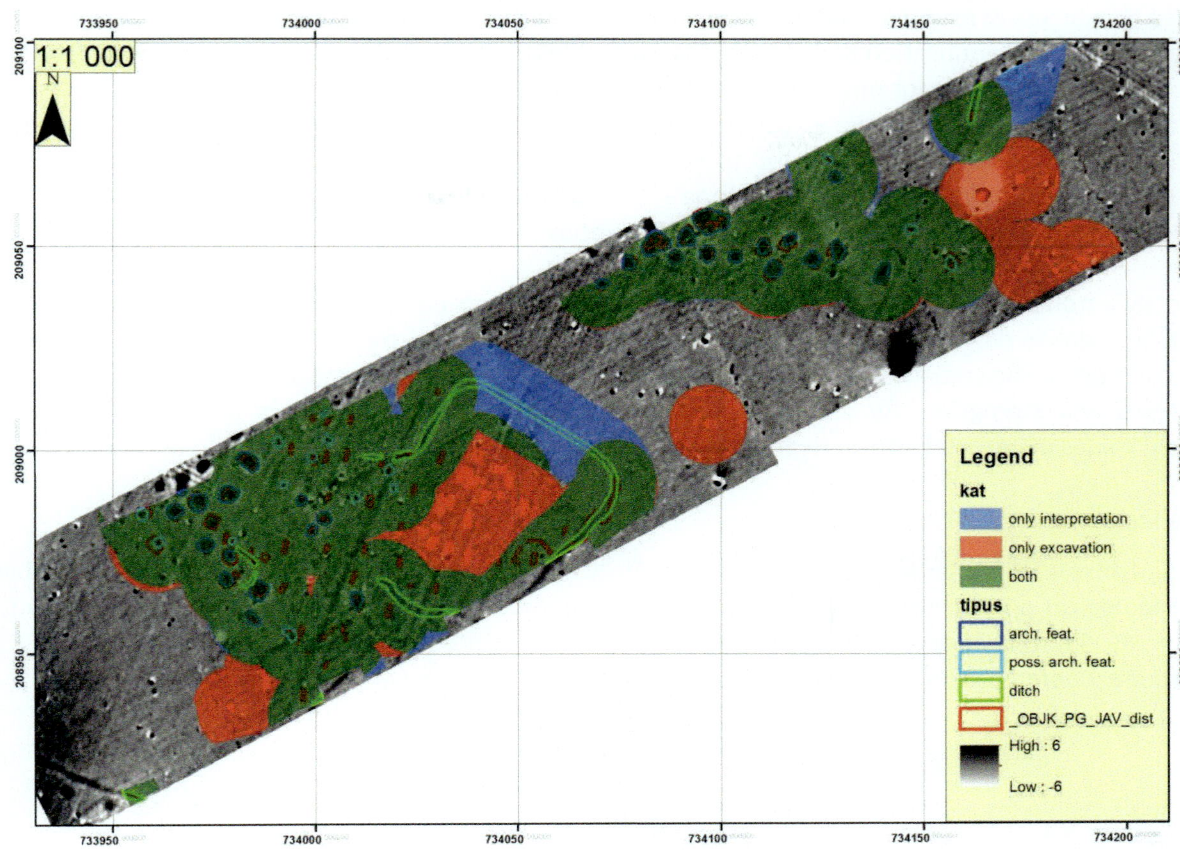

Fig. 3. Results of the area-based comparison

Up to this point the input geophysical and archaeological datasets were transformed into an archaeologically and geophysically understandable, measurable and quantifiable database. The size and the complex nature of these output files requires statistical analysis to recognize basic and underlying patterns. By simply copying the databases (.dbf) into a prewritten Excel spreadsheet, all the calculations and diagrams are automatically generated. To preserve the flexibility of the analysis threshold, values are easily adjustable and spreadsheets are automatically recalculated.

The feature-based statistics exam in detail the three datasets (measurement, interpretation, excavation). Firstly, feature depth, topsoil depth, cubic content and feature size is analysed, separately handling period types and feature types (all, identified, not identified). Secondly, a pairwise comparison is made between the above mentioned four parameters, where the whole dataset and a selected period can be studied. Thirdly causes of misidentifications can be examined in six categories (high topsoil depth, overlain by recent feature, small feature size, small feature depth, no contrast, misinterpretation). Fourthly, the spatial relationship of interpretation and excavation features are compared by size, centroid distance and overlap percentage. Fifthly, average and standard deviation of measurement values are plotted in various combinations.

The methodology was applied over a 19km long section of the M4 motorway (Abony-Fegyvernek) where 26ha were measured with an Overhauser GSM-19 gradiometer (~25m x 50cm sampling interval). After the trial trenching and large-scale excavations, 7.1ha overlapped with the geophysical surveys on four sites, where 702 archaeological features were found from the Neolithic to the Middle Ages. The overall positive feedback rate was 40% (278 features). The identification of postholes (6%) and graves (6%) was almost completely unsuccessful, pits presented a mediocre detectability (43%), meanwhile sunken buildings and kilns (75%) were mostly recognized. Shallow feature depth (under 0.5m) resulted in poor detectability (32% - 554 features), meanwhile features larger than the sampling interval were mostly identified (71% - 142 features). Interestingly, positive feedback rates improved with thicker topsoil conditions (0.5m – 30%; 0.75m – 37%; 1m – 56%). Similarly, linear trends were observable with different feature sizes (<0.25 m^2 - 7%; 0.25-2 m^2 - 16%; 2-10 m^2 - 35%; 10+ m^2 - 56%).

The large proportion of negative evidence rate (60% - 426 features) made it crucial to analyse the underlying patterns. The failed identification had in most cases multiple causes (1 cause - 27%; 2 causes – 60%; 3+ causes - 13%). The primary reason for misidentification was a small feature depth (<0.33m – 40%) and no contrast (<2nT/m difference of average values inside the feature and its 2m buffer zone – 33%). Small feature sizes (<0.375 m^2 – 7%) were mostly connected to these also. Misinterpretation (>2nT/m difference of average values inside the feature and its 2m buffer zone – 12%) and high topsoil depth (<1m - 5%) were defined as less frequent causes. For 18 features (3%), none of the above mentioned criteria was fulfilled.

The mostly automatized workflow aims to be a modifiable, user-friendly framework to process large-scale datasets in order to define limitations of the instruments and human errors in different environmental and survey conditions. The statistical analysis based on the smaller dataset has highlighted and quantified certain weaknesses and strengths. Our long-term aim by processing similar datasets is to define factors and threshold values of detectability on regional-scale.

References

Seren, S., Trinks, I., Hinterleitner, A. and Neubauer, W. 2013. The anomaly that wasn't there – on the visibility of Archaeological prospection anomalies and causative structures in the subsurface. In W. Neubauer, I. Trinks, R. B. Salisbury and C. Einwögerer (eds.), *Archaeological Prospection – Proceedings of the 10th International Conference on Archaeological Prospection*, 252-254. Vienna.

Archaeological guidelines for geophysical survey in the urban environment

Guglielmo Strapazzon[1], Dimitrios Oikonomou[1], Evan Alevizos[1], Kleanthis Simyrdanis[1], Melda Küçükdemirci[2], Apostolos Sarris[1]
[1]*Laboratory of Geophysical – Satellite Remote Sensing & Archaeo-environment, Foundation of Research & Technology, Rethymno, Hellas,* [2]*Istanbul University Geophysical Engineering Department, Istanbul, Turkey*

From a methodological point of view, the contemporary urban environment is one of the most complex and challenging contexts for archaeological research. When an urban site is inhabited for a prolonged period of time, deeply complex multiphase archaeological deposits are generated. The study of the corresponding architectural residues, both buried or standing, provide a unique opportunity for the comprehension of the evolution of urban sites.

Nevertheless, the subsurface deposits are in continuous transformation as natural and anthropogenic post-depositional processes keep on modifying them, usually in a destructive way. Among these processes the one that threatens archaeological deposits the most, and especially the more superficial layers of them, is the ongoing and extensive urbanisation.

The study of such archaeological evidence in an urban environment (urban archaeology) has been developed significantly over the past 40 years, focusing mainly on the integration of building-archaeology and archaeological excavation. Specifically, building-archaeology focuses on the reconstruction of the history of existing buildings, using direct observations of the building themselves, while stratigraphic excavation is used to extract historical information from deeply stratified archaeological buried deposits of different occupational phases.

However, this approach has many limitations, as archaeological excavation in such a complex urban environment is extremely time consuming, expensive and usually limited in size. The use of non-invasive approaches, and in particular of high-resolution geophysical techniques, offers a unique opportunity to push further the development of the conventional methodologies of urban archaeology.

Over the past twenty years, geophysical surveys in urban environments have continued to increase all over Europe. The experience acquired has allowed archaeo-geophysicists to establish that a combination of GPR and ERT are usually the most effective in these contexts (Schmidt *et al.* 2015), that contain high environmental noise levels and accessibility problems.

Still, in order to achieve an archaeological interpretation of the acquired geophysical data, specific methodological approaches must be developed. This paper proposes specific guidelines for archaeo-geophysical surveys in urban environments based on the methodology developed from several historical urban centres of the Mediterranean Basin. Two cases of studies will be used to demonstrate this: the historical city centre of Padua in Italy (by the University of Padua) (Strapazzon *et al.* 2013) and the city centres of Rethymno and Heraklion in Crete, Greece (by the GeoSat ReSeArch Lab of FORTH) (Papadopoulos *et al.* 2009).

Those sites, both located in the Mediterranean Basin, present extremely different geomorphological and historical backgrounds. Padua is located in the alluvial plain of the Po valley where from the 10th century BC a sequence of settlements have formed up to 7m of anthropogenic deposits on a river bend (Mozzi *et al.* 2017), while Rethymno and Heraklion developed on the northern shore of Crete, mainly from the early middle ages, on a limestone shoreline.

The same workflow, designed to be adapted to specific characteristics of each area to be surveyed, has been used for each cases study. Specifically, all available archaeological (i.e. archaeological excavation, aerial pictures, historical architecture and 3-D models of the area of interest), geological (geological and geomorphological maps) and documentary (historical maps and written sources) information have been gathered and georeferenced into a GIS environment. Such a fusion of information within GIS has been used to construct an indispensable tool both at the preliminary stage and the final stage.

In the preliminary stage this tool provides information for an evaluation of the archaeological potential of the target area and it also allows for the formulation of archaeological questions. This phase was crucial for targeting areas not disturbed by contemporary activities and to define the survey and acquisition strategy of the geophysical measurements. In the final stage, the processed geophysical datasets are accurately georeferenced into the GIS to proceed with their interpretation. This step is important for archaeo-geophysical datasets acquired in such environments due to the complexity of deeply stratified urban anthropogenic deposits. In this perspective, geophysical datasets cannot be interpreted alone and need to be accurately put in a dialogue with all the information previously gathered into the GIS.

Part of this work was financially supported by the Stavros Niarchos Foundation within the framework of the project ARCHERS ("Advancing Young Researchers' Human Capital in Cutting Edge Technologies in the Preservation of Cultural Heritage and the Tackling of Societal Challenges").

References

Mozzi, P., Ferrarese, F., Zangrando, D., Gamba, M., Vigoni, A., Sainati, C., Fontana, A., Ninfo, A., Piovan, S., Rossato, S. and Veronese, F. 2017. Modeling of Archaeological and Geomorphic Surfaces in a Multi-Stratified Urban Site in Padua, Italy. *Geoarchaeology*. 33. 67-84.

Papadopoulos, N., Sarris, A., Yi, M. and Kim, J. 2009. Urban archaeological investigations using surface 3D Ground Penetrating Radar and Electrical Resistivity Tomography methods. *Exploration Geophysics*. 40. 56-68.

Schmidt, A., Linford, P., Linford, N., David, A., Gaffney, C., Sarris, A. and Fassbinder, J.W.E. 2015. *EAC Guidelines for the Use of Geophysics in Archaeology: Questions to Ask and Points to Consider*. Belgium, Europae Archaeologia Consilium.

Strapazzon, G., Deiana, R. and Brogiolo, G.P. 2013. Extensive geophysical investigation in urban environment: the case of the episcopal center of Padua. *Non-destructive approaches to complex archaeological sites in Europe: a round-up, Closing event radio-past Ghent, Belgium*, January 15–17, 2013.

Addressing archaeological research questions using geophysical surveys – a landscape case study

Jakob Kainz
University of Vienna, Vienna, Austria

Introduction

The paper forms part of a PhD at the University of Vienna using archaeological prospection methods for the Kreuttal case study, Austria (Ludwig Boltzmann Institute for Archaeological Prospection, LBI Arch Pro, http://lbi-archpro.org). The focus lies on the application of geophysical methods, especially magnetometry in archaeological research, for landscape studies. Often these methods are used to simply locate archaeology rather than to answer archaeological questions, so a framework addressing archaeological questions that geophysical data can address was adapted. Additionally, chi-square tests were used to examine relationships between the physical properties of anomalies and their distribution. These steps allowed characterising the archaeological landscape and hence a qualitative and quantitative assessment of the magnetometry data. This improved the investigation of the built environment for the continuity and discontinuity in the use of space and architecture for the case study area. Furthermore, this enabled an assessment of the representation of the archaeological record in the magnetometry data.

Background and work

The case study area lies *c.* 20km to the north of Vienna, Austria containing a multi-period landscape. The aim is to study the landscape using non-invasive methods, including remote sensing, geophysical prospection, virtual reality and GIS (http://lbi-archpro.org/cs/kreuttal/). A motorised magnetometry survey of *c.* 4km^2 was collected by the LBI Arch Pro using eight fluxgate sensors with a measuring raster of 25cm x *c.* 12.5cm (Fig.1). Vertical photographs were used to complement the magnetometry data while ALS data provided the topographic background. The datasets were collated within a GIS system and used for the initial mapping of features and structures. A research framework relating to geophysical data allowed the investigation of archaeological questions and chi-square tests identified relationships between the physical properties of features and structures and their distribution.

Small-area magnetometry and MS surveys were undertaken prior to and during excavations. Firstly, this helped improve the magnetic definition of features before and after topsoil removal. Secondly, it enabled the identification of new features and especially features not visible to the eye during the excavation and topsoil magnetometry surveys. This enabled an assessment of feature visibility, especially the impact of the topsoil and other deposits, highlighting the biases and deficiencies in the topsoil magnetometry data (Kainz 2016, Kainz and Cotter 2018). An example is a palisade with postholes that appeared only partially in the magnetometry data; the full extent was only visible magnetically after the removal of the topsoil, and the postholes were only identifiable in the MS data (Fig.2).

Geophysical surveys in archaeological research

The application of geophysical methods in archaeology is increasing and changing, continually so due to technological advancements. Their current role in archaeological research however is still affected by their past use (Neubauer 2001, Kvamme 2003, Gaffney 2008, Thompson *et al.* 2011, Verhoeven 2017). Hence excavation and field survey prevail in answering archaeological research questions. The lack of an archaeological research framework for geophysics exacerbates this and has multiple reasons. Firstly, geophysical methods are primarily seen as a locating tool. Secondly, the understanding of non-geophysicists has been limited by mainly presenting successful surveys rather than demonstrating the strengths and weaknesses of geophysical methods. Thirdly, the overt focus on vertical and stratigraphic succession over horizontal sequences in archaeology (Bradley 2002, Lucas 2002). This has limited geophysical based archaeological research, as most geophysical data provides more horizontal than vertical insights.

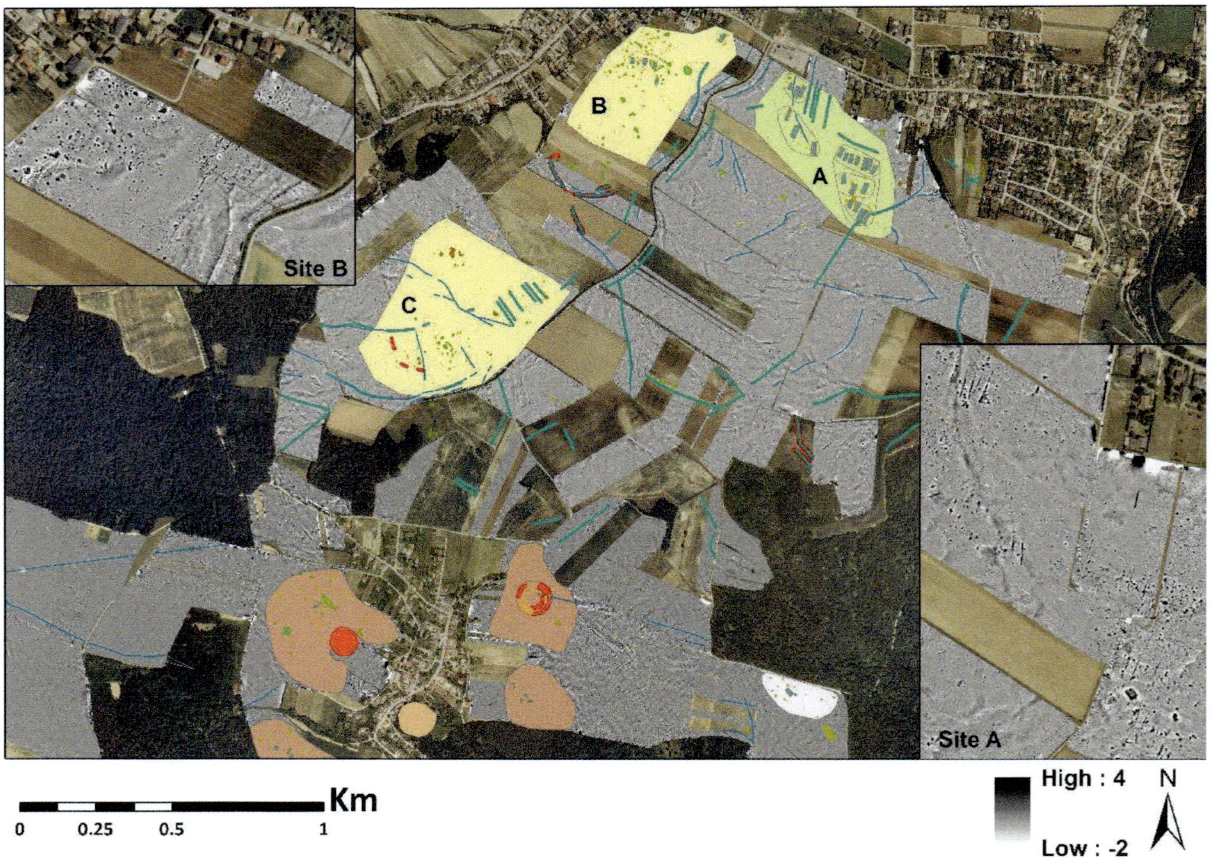

Fig. 1. Magnetometry data overlain with the aerial and magnetic interpretation and the three sites (A, B & C) used in the study.

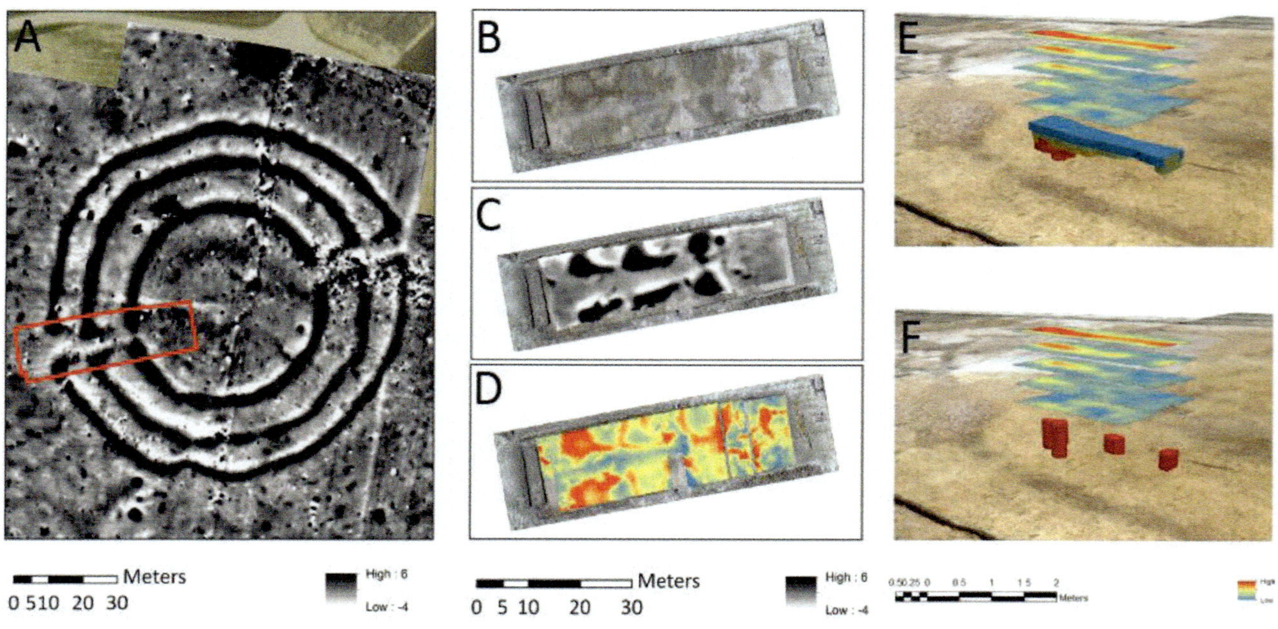

Fig. 2. Topsoil magnetometry survey (A), and surveys after the removal of the topsoil: aerial photo (B), magnetometry (C), magnetic susceptibility (D), and detailed magnetic susceptibility survey of the palisade showing the outline (E), and postholes (F).

Approaching geophysical data for archaeological research

Firstly, the anomalies identified within the magnetometry data were mapped in a GIS-system. Secondly, these were examined using a framework adapted from previous works by Thompson *et al.* (2011) and Kvamme (2003). This allowed archaeological questions applicable to geophysical questions to be addressed. The magnetometry data were then examined according to the following themes:

- Identifying construction variation in terms of the built environment
- Identifying continuity/discontinuity in the use of space
- Identifying natural and cultural modifications
- Identifying regularities in the use of space and architecture at the local and regional level

Thirdly, chi-square tests were used to identify relationships within the data by creating analytic units to identify non-random associations between attributes of artefacts and emic traits that are intended and consistently produced (VanPool and Leonard 2011). These attributes related to the physical properties of features and structures and their occurrence throughout the landscape. Structures (e.g. houses), and features (e.g. pits, ditches, postholes etc.) were analysed for their quantity, magnetic values, orientation, size in relationship to their location (Fig.3).

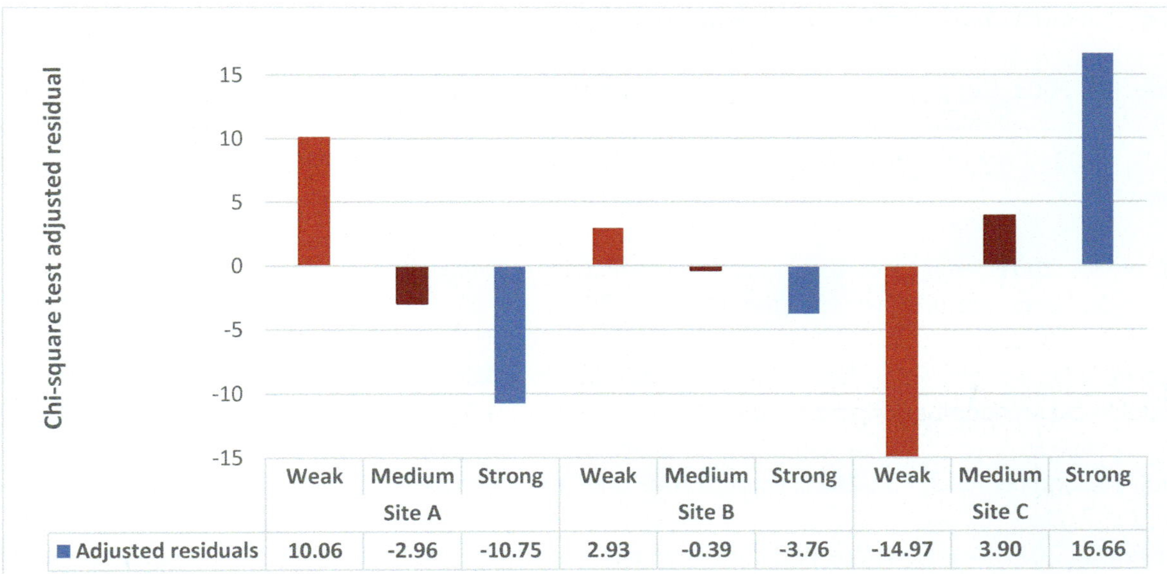

Fig. 3. The chi-square adjusted residuals for the frequencies of weak (<5nT), medium (5-10nT) and strong (>10nT) magnetic anomalies at sites A, B and C showing the different magnetic characteristics for these, with a trend for weak anomalies at site's A and B and medium and strong anomalies at site C.

Outcome

The mapping of the magnetometry data within a GIS-system enabled the application of the research framework. This identified similarities and differences for the built environment, continuity/discontinuity and regularities in the use of space and architecture. This underlined the changing behaviour of people in the past throughout the landscape and identified the changing use of space throughout time. The chi-square tests provided a statistical quantification of the magnetometry data, emphasising different trends of past human behaviour within the landscape. The geophysical surveys during the excavations provided a qualitative assessment of the magnetometry data addressing the (in)visibility of certain features. In conjunction with the chi-square results this raised the biases and deficiencies of the archaeological record within the magnetometry data. The applied steps allowed for characterising past human behaviour between and within sites, especially the changing settlement structures during the early- and mid-Neolithic and later sites. This enabled a better comparison of phases within settlements, especially during the early-Neolithic and in relation to mid-Neolithic settlements. Additionally, features within and between sites were investigated, allowing for a comparison between them. The analysis raised new questions for further research, including why sites show different characteristics in the magnetometry data and in variation in the magnetic properties, quantity, size, orientation of features and structures. This allowed insights to be gained for the preservation and distribution of archaeological remains throughout the landscape. Overall the research framework and chi-square tests better facilitated the investigation of archaeological questions using the magnetometry data, and the small-area of focused geophysics provided a qualitative and quantitative assessment of the magnetometry data.

References

Bradley, R. 2002. *The past in prehistoric societies*. Routledge.

Gaffney, C. 2008. Detecting trends in the prediction of the buried past: a review of geophysical techniques in archaeology. *Archaeometry*. 50(2). 313-336.

Kainz, J. 2016. An integrated archaeological prospection and excavation approach at a Middle Neolithic circular ditch enclosure in Austria. In Forte, M. and Campana, S. (eds.), *Digital Methods and Remote Sensing in Archaeology*. 371-403. Springer.

Kainz, J. and Cotter, J.P. 2018. A broad band magnetic susceptibility test study—The magnetic spectroscopy of a Neolithic ditch. *Journal of Archaeological Science*. 18. 139-150.

Kvamme, K.L. 2003. Geophysical surveys as landscape archaeology. *American Antiquity*. 68(3). 435-457.

Lucas, G. 2002. *Critical approaches to fieldwork: contemporary and historical archaeological practice*. Routledge.

Neubauer, W. 2001. Images of the invisible-prospection methods for the documentation of threatened archaeological sites. *Naturwissenschaften*. 88(1). 13-24.

Thompson, V.D., Arnold, P.J., Pluckhahn, T.J. and Vanderwarker, A.M. 2011. Situating remote sensing in anthropological archaeology. *Archaeological Prospection*. 18(3). 195-213.

VanPool, T.L. and Leonard, R.D. 2011. *Quantitative analysis in archaeology*. John Wiley & Sons.

Verhoeven, G. 2017. Are we there yet? A review and assessment of archaeological passive airborne optical imaging approaches in the light of landscape archaeology. *Geosciences*. 86. 134–166.

Up-skilling and Up-scaling: the realities of adapting to the challenges of the current environment in British commercial geophysics

Chrys Harris
Magnitude Surveys Ltd, Bradford, United Kingdom

Geophysics in the Britain is a well-established investigation technique within the planning framework for commercial developments. This has resulted in a dynamic, competitive market for geophysical survey covering a vast range of type and scale of projects (from 1ha to over 1000ha size). Despite the expansion of the archaeological sector in the UK (Landward Research 2019), the UK's Migration Advisory Committee, recently added the profession to the 'UK Shortage Occupation List' due to a skills gap in qualified archaeologists. A review of the UK's archaeological market in 2018 (Landward Research 2019) revealed an increase in respondents investing in training from the previous year.

As a small start-up geophysical contractor, Magnitude Surveys faced the challenge of up-skilling less experienced staff while expanding from 3 founders to 20+ staff over the course of four years, in order to survive in a competitive market. In the early years of the company's development, our research and development work focused on streamlining fieldwork and data processing efficiencies. This was accomplished by adopting a GNSS-based approach and live-streaming data live from the field to be processed in real-time through cloud-based processing. Data could then be monitored, and quality-control checked by more experienced staff in the office. The technical elements of our field-side workflow have been discussed before (Pope-Carter *et al.* 2017), however, the streamlining of fieldwork efficiency highlighted a workflow bottleneck at the reporting stage, where productivity decreases. This paper examines the data processing, interpretation, and analysis aspects of commercial geophysics in a UK context, and explore some of the methods we have experimented with to address the challenges of being able to report on sites of a variable scale, while up-skilling new members of staff. The key questions we sought to answer were:

1. How can we report faster? Large surveys produce large datasets and often are collected using rapid quad-based systems. As data are processed automatically, this results in data collection outpacing data analysis. To meet the rapid turnaround time for client's timetables, this requires multiple people to digitise, interpret, and write up sites while fieldwork is ongoing. Interpretations are digitised directly into PostGIS, which allows for multiple users to digitise the same site simultaneously without producing different shapefiles. Efficiency is further improved by pulling in all supplementary datasets (e.g. LiDAR, historic mapping, soils and geology) directly into GIS from a webmapping server.

2. How can we be more effective? Rapidly upskilling staff was the key factor to cope with the demands of a rapid expansion. While all new staff members had degree qualifications and/or field experience related to archaeological geophysics, none had specific training in theoretical or field aspects of geophysical survey. To compensate for this lack of experience, a collaborative working environment (a 'POD' system) was developed, to facilitate less experienced staff to work alongside more experienced staff on projects. New staff were given the opportunity to directly work on interpreting data, which would be checked and discussed at an 'interpretation meeting' — an open round-table discussion that staff members of all experience and positions took part in. The final preliminary output would then be checked by a more experienced staff member, who would go through the interpretation.

3. How can we be more efficient? An internal Wiki was created to track interpretations while digitisations were being produced to reduce handover time on large pieces of work. This means that everyone can see what is being worked on and the thought processes behind it. For larger projects, that extends across multiple weeks, this reduced the redundancies of having to revaluate the processes behind an earlier version of the work.

The benefits of this approach saw new staff members rapidly developing their skills and confidence in analysing data and reporting on it. Collaborative interpretations provide an inbuilt quality assurance checking system and means that different sets of skills and knowledge are applied to the data at all stages. This leads to an improved quality in the interpretation as a whole, as it can be understood holistically. Collaborative working and tracking reporting on Wiki aids in writing with a consistent voice, which is crucial for larger projects that extend across multiple weeks. Despite the advantages, there are pitfalls to this approach, one being the tendency for interpretations being produced too digital. It can be difficult to get staff to step back a little from the data in the GIS environment and to consider the results on different mediums. To encourage examining data from different mediums, the interpretation meetings are run with both digital projections and A3 printings of the results. Different coloured markers were provided to encourage staff to highlight and colour directly on the paper. We are currently planning on trialling tablets and pens for sketches as well.

References

Landward Research. 2019. *State of the Archaeological Market 2018*. London.

Pope-Carter, F., Harris, C., Attwood, G. and Eyre, T. 2017. Live-streaming data for the real-time monitoring of geophysical surveys. In Jennings, B., Gaffney, C., Sparrow, T. and Gaffney, S. (eds.) *AP2017. 12th International Conference of Archaeological Prospection*. Bradford: Archaeopress.